SINGLE AND MULTIPLE NUMBER SERIES

Addressing a significant gap in the study of number series, this book presents an in–depth theory of multiple number series and an exhaustive examination of one–dimensional series. It incorporates overlooked yet essential results alongside recent research advancements. Much of the text is based on the authors' original contributions, particularly in the development of relaxed monotonicity concepts, which have become fundamental tools in Fourier and functional analysis. Each chapter concludes with the historical context, aiding readers in understanding the theory's evolution.

The book is aimed at a wide audience, ranging from undergraduate students to experts in the field. It offers a modern perspective on the theory along with detailed introductory chapters that make complex concepts accessible to students. The audience will find the novel contributions enriching and inspiring.

Encyclopedia of Mathematics and Its Applications

This series is devoted to significant topics or themes that have wide application in mathematics or mathematical science and for which a detailed development of the abstract theory is less important than a thorough and concrete exploration of the implications and applications.

Books in the **Encyclopedia of Mathematics and Its Applications** cover their subjects comprehensively. Less important results may be summarized as exercises at the ends of chapters. For technicalities, readers can be referred to the bibliography, which is expected to be comprehensive. As a result, volumes are encyclopedic references or manageable guides to major subjects.

All the titles listed below can be obtained from good booksellers or from Cambridge University Press. For a complete series listing visit

www.cambridge.org/mathematics.

139 M. Fiedler *Matrices and Graphs in Geometry*
140 N. Vakil *Real Analysis through Modern Infinitesimals*
141 R. B. Paris *Hadamard Expansions and Hyperasymptotic Evaluation*
142 Y. Crama and P. L. Hammer *Boolean Functions*
143 A. Arapostathis, V. S. Borkar and M. K. Ghosh *Ergodic Control of Diffusion Processes*
144 N. Caspard, B. Leclerc and B. Monjardet *Finite Ordered Sets*
145 D. Z. Arov and H. Dym *Bitangential Direct and Inverse Problems for Systems of Integral and Differential Equations*
146 G. Dassios *Ellipsoidal Harmonics*
147 L. W. Beineke and R. J. Wilson (eds.) with O. R. Oellermann *Topics in Structural Graph Theory*
148 L. Berlyand, A. G. Kolpakov and A. Novikov *Introduction to the Network Approximation Method for Materials Modeling*
149 M. Baake and U. Grimm *Aperiodic Order I: A Mathematical Invitation*
150 J. Borwein *et al. Lattice Sums Then and Now*
151 R. Schneider *Convex Bodies: The Brunn–Minkowski Theory (Second Edition)*
152 G. Da Prato and J. Zabczyk *Stochastic Equations in Infinite Dimensions (Second Edition)*
153 D. Hofmann, G. J. Seal and W. Tholen (eds.) *Monoidal Topology*
154 M. Cabrera García and Á. Rodríguez Palacios *Non-Associative Normed Algebras I: The Vidav–Palmer and Gelfand–Naimark Theorems*
155 C. F. Dunkl and Y. Xu *Orthogonal Polynomials of Several Variables (Second Edition)*
156 L. W. Beineke and R. J. Wilson (eds.) with B. Toft *Topics in Chromatic Graph Theory*
157 T. Mora *Solving Polynomial Equation Systems III: Algebraic Solving*
158 T. Mora *Solving Polynomial Equation Systems IV: Buchberger Theory and Beyond*
159 V. Berthé and M. Rigo (eds.) *Combinatorics, Words and Symbolic Dynamics*
160 B. Rubin *Introduction to Radon Transforms: With Elements of Fractional Calculus and Harmonic Analysis*
161 M. Ghergu and S. D. Taliaferro *Isolated Singularities in Partial Differential Inequalities*
162 G. Molica Bisci, V. D. Radulescu and R. Servadei *Variational Methods for Nonlocal Fractional Problems*
163 S. Wagon *The Banach–Tarski Paradox (Second Edition)*
164 K. Broughan *Equivalents of the Riemann Hypothesis I: Arithmetic Equivalents*
165 K. Broughan *Equivalents of the Riemann Hypothesis II: Analytic Equivalents*
166 M. Baake and U. Grimm (eds.) *Aperiodic Order II: Crystallography and Almost Periodicity*
167 M. Cabrera García and Á. Rodríguez Palacios *Non-Associative Normed Algebras II: Representation Theory and the Zel'manov Approach*
168 A. Yu. Khrennikov, S. V. Kozyrev and W. A. Zúñiga-Galindo *Ultrametric Pseudodifferential Equations and Applications*
169 S. R. Finch *Mathematical Constants II*
170 J. Krajíček *Proof Complexity*
171 D. Bulacu, S. Caenepeel, F. Panaite and F. Van Oystaeyen *Quasi-Hopf Algebras*
172 P. McMullen *Geometric Regular Polytopes*
173 M. Aguiar and S. Mahajan *Bimonoids for Hyperplane Arrangements*
174 M. Barski and J. Zabczyk *Mathematics of the Bond Market: A Lévy Processes Approach*
175 T. R. Bielecki, J. Jakubowski and M. Niewęgłowski *Structured Dependence between Stochastic Processes*
176 A. A. Borovkov, V. V. Ulyanov and Mikhail Zhitlukhin *Asymptotic Analysis of Random Walks: Light-Tailed Distributions*
177 Y.-K. Chan *Foundations of Constructive Probability Theory*
178 L. W. Beineke, M. C. Golumbic and R. J. Wilson (eds.) *Topics in Algorithmic Graph Theory*
179 H.-L. Gau and P. Y. Wu *Numerical Ranges of Hilbert Space Operators*
180 P. A. Martin *Time-Domain Scattering*
181 M. D. de la Iglesia *Orthogonal Polynomials in the Spectral Analysis of Markov Processes*
182 A. E. Brouwer and H. Van Maldeghem *Strongly Regular Graphs*
183 D. Z. Arov and O. J. Staffans *Linear State/Signal Systems*
184 A. A. Borovkov *Compound Renewal Processes*
185 D. Bridges, H. Ishihara, M. Rathjen and H. Schwichtenberg (eds.) *Handbook of Constructive Mathematics*
186 M. Aguiar and S. Mahajan *Coxeter Bialgebras*
187 K. Broughan *Equivalents of the Riemann Hypothesis III: Further Steps towards Resolving the Riemann Hypothesis*
188 W. Lay *Higher Special Functions*

Encyclopedia of Mathematics and its Applications

Single and Multiple Number Series

ALBERTO DEBERNARDI PINOS
Universitat Autònoma de Barcelona, Spain

ELIJAH LIFLYAND
Bar–Ilan University, Israel

SERGEY TIKHONOV
ICREA, Spain

MARIA ZELTSER
Tallinn University, Estonia

CAMBRIDGE
UNIVERSITY PRESS

CAMBRIDGE
UNIVERSITY PRESS

Shaftesbury Road, Cambridge CB2 8EA, United Kingdom

One Liberty Plaza, 20th Floor, New York, NY 10006, USA

477 Williamstown Road, Port Melbourne, VIC 3207, Australia

314 – 321, 3rd Floor, Plot 3, Splendor Forum, Jasola District Centre, New Delhi – 110025, India

103 Penang Road, #05—-06/07, Visioncrest Commercial, Singapore 238467

Cambridge University Press is part of Cambridge University Press & Assessment, a department of the University of Cambridge.

We share the University's mission to contribute to society through the pursuit of education, learning and research at the highest international levels of excellence.

www.cambridge.org
Information on this title: www.cambridge.org/9781009562997
DOI: 10.1017/9781009562973

When citing this work, please include a reference to the DOI 10.1017/9781009562973

First published 2026

A catalogue record for this publication is available from the British Library

A Cataloging-in-Publication data record for this book is available from the Library of Congress

ISBN 978-1-009-56299-7 Hardback

To our families.

Contents

Introduction

The theory of number series is an essential and integral part of real analysis. Almost every analyst and numerous mathematicians working in other areas systematically use series. Needless to say that various crucial concepts in analysis, such as functional series, Fourier series, or series in other orthogonal systems, are based on the theory of number series. Undoubtedly, the study and use of double and multiple series are as important and interesting as that of usual one-dimensional series. Examples can be found in many areas of mathematics, especially in real and complex analysis (harmonic, numerical, functional), approximation theory, and more.

It is common that any calculus textbook contains certain basics of one-dimensional number series. There are several recognized sources devoted to this topic: [B-16], [B-54], Fichtenholz's [B-31] and [B-32] (translation of the corresponding chapters in his calculus course), and more recent textbooks [B-11] and [B-15], to mention some. Moreover, number series is an integral part of books on summability and power and trigonometric series. The collection is big: [B-1, B-4, B-7, B-12, B-14, B-18, B-19, B-22, B-23, B-25, B-29, B-34, B-35, B-40, B-41, B-44, B-45, B-59, B-62, B-64, B-65, B-66, B-70, B-72, B-76, B-83], and it is not safe to say that this list is complete. Some of these books that mainly deal with single number series also study the two-dimensional case; see, for instance, [B-16], [B-32], [B-33], or even a very concise book [B-49]. However, it seems that there is no book on multiple series providing a modern view on this topic and treating the theory in detail. As a result, in many books on various areas of analysis the authors have to dig for some basics and sometimes advances of the theory. The classical book of Körner [B-56] and the recent book by Gauthier [B-36] are good illustrations of this issue.

The subject of number series has a fascinating history. Without presenting a complete historical picture (the recent book [B-55] may be an appropriate read for this purpose; see also [B-30, B-48, B-2]), we outline some important names and landmarks. First, we touch on those in dimension one. It can be said that as far back as 200 years ago, Cauchy in his classical book on analysis [B-20] gave a special impact to this subject by indicating a strict border between the convergent and divergent number series. His predecessors, even so great as Newton, Leibniz, or Euler, "were too intoxicated by the mighty stream of learning springing from the newly-discovered sources to feel obliged to criticize the fundamentals" (a quotation from [B-54]). What should be mentioned is that development of the theory of number series marched in step with the development of the knowledge of fundamentals of real analysis, first of all a rigorous understanding of the nature of real numbers. Such mathematicians as Abel, Gauss, Weierstrass, Cantor, and Dedekind were involved in this study.

As in many other theories, one of the main questions in number series is establishing necessary and sufficient conditions for the accuracy of the principal propositions. Attempts to determine such conditions for convergence/divergence of a series were undertaken already at the dawn of the topic. In 1835, Kummer obtained a test for positive series, which in the modern language reads as follows [157]: a series

$$\sum_{k=1}^{\infty} a_k,$$

with positive terms a_k, converges if and only if there exist a positive sequence $\{p_k\}$ and $c > 0$ such that for all $k \geq k_0$, there holds

$$p_k - \frac{a_{k+1}}{a_k} p_{k+1} \geq c.$$

From a general point of view, Kummer's test provides the complete solution of the problem for positive series, as it fully characterizes their convergence. However, finding an appropriate sequence $\{p_k\}$ and constant c for a given $\{a_k\}$ may be extremely difficult, which makes this test not that practical. Another example of such theoretical but less practical criterion is due to Lobachevskii, who established it even earlier than Kummer (1834). After this, people were mostly concentrated on either sufficient or necessary conditions. Some general results were due to d'Alembert, Bertrand, Bugaev, Chebyshev, Raabe, and others.

Another perhaps unnoticed within the scope of number series yet interesting characterization of divergent series was given by Müntz in 1914 [198] and simplified by Szász in 1916 [249]. Even more surprising is the connection of this criterion with approximation theory. Their results state that, given a positive decreasing sequence $\{a_k\}$, the system $\{1\} \cup \left\{x^{\frac{1}{a_k}}\right\}_{k \in \mathbb{N}}$ is complete in $C[0, 1]$ if and only if $\sum a_k$ diverges.

Importantly, among many tests of convergence a big and substantial group of results consists in a study of the series $\sum_{k=1}^{\infty} a_k$ with *monotone* terms $\{a_k\}$. Such series are frequently called *monotone series*. As examples, we mention the very well-known tests by Abel and Cauchy (several named after them), de la Vallée Poussin, Dedekind, Dirichlet, Leibniz, Maclaurin, and Olivier. Less known are tests of du Bois-Reymond, Ermakov, Sapogov, and Schlömilch. For the sources, see, for example, [B-54], [B-16], [B-31], or [B-11].

It turns out that in various applications it is very much desirable to have such or similar assertions without so restrictive condition as monotonicity. A typical example is the characterization of the Besov norm in terms of best approximations, see in particular [B-26, Ch. 7, §9]. Some attempts to relax the monotonicity condition were made already in the first half of the twentieth century. In particular, replacing this condition by weak monotonicity for the Maclaurin–Cauchy test was the goal in the older papers [250] and [272], where quasi-monotone sequences were studied in the former, and functions of bounded variation in the latter. In the present book we systematically study the options of relaxing monotonicity. More precisely, it turns out that in many tests monotonicity can be significantly eased down by weak monotonicity (i.e., satisfying $a_k \le Ca_n$ for any $k \in [n, 2n]$) or general monotonicity conditions. The former concept is very general and includes many recognized classes of sequences. In particular, such classical tests as the condensation test and the Maclaurin–Cauchy test hold true under the assumption of weak monotonicity rather than usual monotonicity. A more delicate class of general monotone sequences deals with variation, unlike *WMS*, which only deals with magnitude. The idea is to assume some quantitative control over the variation of the sequence in terms of the magnitude of the sequence itself. A nonnegative null sequence $\{a_k\}$ is called *general monotone*, written $\{a_k\} \in GMS$, if it satisfies

$$\sum_{k=n}^{2n-1} |a_{k+1} - a_k| \le Ca_n,$$

for any n and some absolute constant C. General monotone sequences of this type were introduced by the third author in [254, 255]. Similar notions for functions were introduced in [176] and in [178].

Regarding the theory of multiple number series, many researchers realized that extending the theory to the case of several dimensions could be an important and beneficial step. It was apparently Pringsheim who first understood one of the basic difficulties of the theory of multiple number series as compared with number series in dimension one: a variety of the ways of ordering partial sums. Pringsheim's paper [221] is concerned with several types of summation, quite different in their properties. One of the main ways of ordering partial sums, the rectangular one, is frequently called Pringsheim's summation. In all fairness, it has to be added that this type of convergence can be found already in the paper [245] by Stolz, as early as in 1884. We would also like to mention the work by Jordan of 1881 [139] as an illustration of the respectable age of the topic. Further, basics of the theory of double number series analogous to those one-dimensional ones were given in [222] of 1900, and in the first edition of Bromwich's textbook [B-16] of 1908. In particular, it has been pointed out in [221] and [222] that for the series with nonnegative terms, all types of convergence are equivalent. In the textbook by Fichtenholz [B-32, Ch. 2] (the original Russian version goes back to 1939), one can find a more detailed presentation of certain elements of the theory of multiple series. For example, it is shown there that the space of all absolutely convergent multiple series is isomorphic to the space of one-dimensional absolutely convergent series. As mentioned above, there are more textbooks containing the same or somewhat different aspects of the study of multiple number series, in particular [B-49], [B-46], [B-21], and [B-37]. Let us also notice the works of Zhizhiashvili [284] and [B-85], in each of which an immense list of references is given. It is written in [B-85] that in most of the aforementioned and other works only double series were studied, which does not always give a complete picture in higher dimensions.

Before discussing the book's structure, we would like to stress that our work contains many classical and a variety of new and sometimes surprising results. We bring the reader's attention to the results in number series with important applications in functional analysis, in particular $L \log L$-type equivalences in Subsection 2.4.7 and Stechkin's lemma in Subsection 2.4.8. A substantial part of the discussion concerns the results in number series where monotonicity can be relaxed: first to weak monotonicity and then, in cases where weak monotonicity does not fit, to general monotonicity. As an outcome, many recognized tests are applicable to a broader set of sequences. Concerning multiple series, we believe that many results presented in the book are new. We, of course, discuss classical tests, which have usually been given only for double series.

Structure of the book. It is divided into three parts, each consisting of three chapters. Part I is devoted to single number series. In the first chapter we present basic facts for sequences and functions with generalized monotonicity properties. Chapter 2 unites a variety of tests for convergence/divergence of number series, always with monotonicity relaxed and replaced by a certain generalized monotonicity if possible. Chapter 3 grew from our intention to supplement classical results by their modern refinements. Many of the results in this chapter were obtained relatively recently.

Part II is devoted to double series. We intentionally separate the two-dimensional setting in Part II and the higher-dimensional analogues in Part III. For the readers' convenience, many significative multidimensional results in Part III are supplemented with their two-dimensional versions in Part II. The readers can easily see two-dimensional illustrations to the higher-dimensional theorems, whereas the latter sometimes have a more cumbersome form. We are aware that for many interested readers the results in dimension two will be enough for both their advanced knowledge and applications. With this part separated, such readers will be able to easily find the needed material rather than digging in the more sophisticated part. On the other hand, we mainly do not give the proofs in this part. Instead we refer to the corresponding multidimensional result and its proof in full generality in the following part.

Part II consists of three chapters. In Chapter 4 we discuss various types of convergence in detail, a fundamental difference with the one-dimensional case. In Chapter 5, we present main extensions of classical convergence/divergence tests. The book by Chelidze [B-21] had a serious impact on our presentation of Chapters 4 and 5. Chapter 6 presents our novelties concerning relaxed monotonicity in the double number series case. These will be displayed in full measure and proved in the next part for arbitrary dimension.

Part III also contains three chapters and is purely multidimensional; two-dimensional assertions may appear accidentally as counterexamples. In Chapter 7, basic notions and notation needed for dealing with multiple number series are introduced, as well as certain classical tests are given for arbitrary dimension. Even though these results are formally new, they are easily extended from the two-dimensional results as a continuation of its detailed treatment in the double case. In Chapter 8, we present multivariate versions of classical convergence/divergence results. What must be mentioned is that multivariate variations come into play. These represent one more important difference between one and several variables. In general, this chapter can be considered as an extended version of Chapter 5, with its specific notation and more detailed study, including proofs. In Chapter 9, we introduce the notions of weak monotonicity and general monotonicity for any dimension.

We can say that there are two different books under one cover: a modern study of single and multiple number series. Each part can be considered and read (almost) independently. We think that this form is more reader-friendly and enjoys some advantages. Part I is a comprehensive study of single number series up to date. In Part II, we announce several extensions of the one-dimensional results to two dimensions in a survey-like manner, leaving most of the proofs for the general multidimensional case. Indeed, Part III is to a large extent a novel study of the classical topic by applying modern and developing tools and, to the best of our knowledge, the first detailed presentation of the theory of multiple number series.

The historical information related to a specific notion or result can be found in a special section at the end of each chapter. Each of these special sections is titled Notes. Notes are split into enumerated items. We refer to such a section within the same chapter as just Notes. In the cases where we refer to a different chapter's Notes, we include the corresponding chapter number. Every reference to Notes is accompanied with the number of an item, like Notes-1 or Notes-23. For the readers' convenience, the whole book is supplemented with a Table of correspondences of the same theorems in different dimensions and a List of symbols. We believe that the reference list is rather comprehensive and reflects all the sides of the undertaken work: history, results, methods, and more. Separating books and papers into different lists also seems useful.

Acknowledgments. Kristina Oganesyan undertook the laborious work of reading a draft of the manuscript of this book. Actually she was an editor of this work in full value. Her help was invaluable, and the authors, assuming that all the responsibility for possible misprints or weak points rests with them, are indebted to Kristina for numerous corrections, suggestions and improvements.

The authors express their gratitude to A. Podkorytov for his meticulous review of the manuscript. They also appreciate the helpful remarks and positive feedback from A. Aasma, M. Dyachenko, M. Goldman, C. González-Riquelme, V. Kadets, D. Karp, E. Kosov, T. Leiger, A. Lukashov, M. Saucedo, and R. Trigub.

A disclaimer. Let us mention that the book does not assume to present a comprehensive treatment of all the areas in analysis where infinite series are involved (for example, we skip power series, functional series, or such a topic within the theory of number series as multiplication of series; these results can be easily found in other textbooks). We concentrated on the theory of number series itself.

To whom is this book addressed? We think that this book is addressed to everybody, from undergraduate students to specialists. For some, the whole text can be attractive, while others might be interested in certain pieces. Some may use Part I as a handbook with more complete collection of facts than other sources suggest, while some others will find open problems in the multivariate case interesting.

Part I

Single Series

The first part of this book is devoted to presenting a comprehensive theory of number series in dimension one, with significant improvements in several classical topics covered in most of the literature on number series. In particular, we do not only include the basic theory (Sections 2.1, 2.2, and 2.3), but we also present new extensions of classical convergence tests for series with monotone terms, with relaxed monotonicity (Section 2.4). To this end, Chapter 1 serves as a starting point both to motivate the given extensions and to establish the necessary background for the sequel.

A selection of typically less known (though interesting) results is also exhibited in Chapter 3.

1
Generalized monotonicity

We begin with a thorough study of modern monotonicity-type conditions. These are used for deriving generalizations of several convergence tests, as well as new ones, in Chapter 2.

There are many problems in mathematics where monotonicity assumptions for a sequence or a function play a crucial role. Such ones can easily be found in differential equations, approximation, probability, number series, harmonic analysis, and so on. Recently certain efforts have been made in extending the class of monotone sequences (and functions). A property (or, more precisely, a variety of them) that characterized those new classes can, in an integrated way, be called *generalized monotonicity*.

In the first section of this chapter we present a simple observation that leads us to this concept. Giving priority to sequences, we will also point out the corresponding analogues for functions, where possible and needed.

In this short chapter we will implicitly use some basic concepts and properties of number series, which will formally and correctly be introduced only in Chapter 2.

1.1 Getting started

Recall that a number sequence $\{a_k\}_{k=1}^{\infty} \subset \mathbb{C}$ is of bounded variation ($\{a_k\} \in BVS$) if

$$\sum_{k=1}^{\infty} |\Delta a_k| < \infty, \qquad \Delta a_k := a_k - a_{k+1}, \tag{1.1}$$

where $\{\Delta a_k\}$ is called the sequence of differences of $\{a_k\}$. Sequences of bounded variation are the main object in many topics, for instance, the classical theory of integrability of trigonometric series. We, however, mention that the class of sequences (the same is true for functions) of bounded variation is too broad to control local behavior and/or guarantee monotone-type properties. In such cases one has to reduce oneself to more delicate properties. For example, one can study quantitative characteristics of (1.1) by analyzing the rate of convergence to zero of the sums

$\sum_{k=n}^{2n-1} |\Delta a_k|$. While choosing a suitable majorant for these partial sums, let us keep in mind that monotone sequences must be a proper part of any such new class.

Before proceeding to details, recall that a real-valued sequence $\{a_n\}$ is monotone decreasing (respectively increasing) if $a_m \le a_n$ (respectively $a_m \ge a_n$) for $n \le m$ (unless explicitly stated, we will use the term "monotone" to mean monotone decreasing; such sequences are also referred to as just decreasing, or nonincreasing). The counterpart for a monotone decreasing (respectively increasing) real-valued function $f(x)$ defined on $\mathbb{R}_+ := [0, \infty)$ is $f(x) \le f(y)$ (respectively $f(x) \ge f(y)$) if $y \le x$. As for sequences, a "monotone" function will be understood as a monotone decreasing one; we will also call such functions decreasing, or nonincreasing. A slightly more general class is defined as follows. We say that a nonnegative sequence $\{a_k\}$ (respectively nonnegative function $f(x)$) is *almost monotone* if there exists a constant $C \ge 1$ such that

$$a_m \le C a_n \quad \text{for } n \le m \qquad (\text{respectively } f(x) \le C f(y) \quad \text{for } y \le x). \tag{1.2}$$

We have included the definition of almost monotonicity merely for instructive purposes, since we will not use such a property. As we will see, the concept of weak monotonicity (cf. Section 1.4), which is the central one in this book, is much more general and still inherits several useful properties from usual monotonicity.

We first remark that the classes of monotone sequences/functions always satisfy the relation

$$a_n \le 2 \sum_{k=\lfloor n/2 \rfloor}^{n} \frac{a_k}{k} \qquad \text{or} \qquad f(x) \le 2 \int_{x/2}^{x} f(t) \frac{dt}{t},$$

which we call the *monotonicity type* property (here and in what follows, $\lfloor c \rfloor$ denotes the integer part of $c > 0$). Secondly, in order to introduce the *counter-oscillation type* of monotone functions, we first introduce the concept of *variation*.

Definition 1.1 Let $f\colon [a,b] \subset \mathbb{R} \to \mathbb{C}$, and for $n \in \mathbb{N}$, let $\mathcal{P}([a,b])$ be the set of all partitions $\{x_k\}_{k=0}^n$ of $[a,b]$, that is, the set of all finite sequences satisfying $a = x_0 < x_1 < \cdots < x_n = b$. If the supremum

$$\sup_{\{x_k\}_{k=0}^n \subset \mathcal{P}([a,b])} \sum_{k=0}^{n-1} |f(x_{k+1}) - f(x_k)| \tag{1.3}$$

is finite, then it is called the *variation* of f on $[a,b]$, or the *(absolute) Riemann–Stieltjes integral of* f on the interval $[a,b]$, and it is denoted by

$$\int_a^b |df(t)|;$$

see Notes-2 for appropriate references on the Riemann–Stieltjes integral.

Definition 1.2 Let $f\colon [a,b] \subset \mathbb{R} \to \mathbb{C}$. We say that f is of bounded variation on $[a,b]$, written $f \in BV([a,b])$, if

$$\int_a^b |df(t)| < \infty.$$

If $J \subset \mathbb{R}$ is a noncompact interval, we say that f is of bounded variation on J, written $f \in BV(J)$, if

$$\sup_{[a,b]\subset J} \int_a^b |df(t)| < \infty.$$

On the other hand, we say that f is locally of bounded variation on J if it is of bounded variation in each compact subinterval of J.

If f is *absolutely continuous* on an interval $J \subset \mathbb{R}$, then

$$\int_J |df(t)| = \int_J |f'(t)|\,dt.$$

Recall that if a function is of bounded variation on an interval J, then it is differentiable almost everywhere in that interval and its derivative is Lebesgue integrable.

The counter-oscillation type property for sequences or functions reads as

$$\sum_{k=n}^{2n} |\Delta a_k| \le a_n \qquad \text{or} \qquad \int_x^{2x} |df(t)| \le f(x),$$

respectively.

Therefore, combining both ideas (the monotonicity type and counter-oscillation type properties), we observe that monotone sequences satisfy

$$\sum_{k=n}^{2n-1} |\Delta a_k| \le C\beta_n, \tag{1.4}$$

with both $\beta_n = |a_n|$ and $\beta_n = \sum_{k=\lfloor n/2\rfloor}^{n} |a_k|/k$, while monotone functions satisfy

$$\int_x^{2x} |df(t)| \le C\beta(x), \tag{1.5}$$

with both $\beta(x) = |f(x)|$ and $\beta(x) = \int_{x/2}^{x} |f(t)|\,\frac{dt}{t}$. Constants C in (1.4) and (1.5) are independent of n and x, respectively. Note that all absolute values above are redundant, but they will appear naturally when generalizing monotone sequences/functions through variation estimates in the real or even complex-valued cases, and thus it may be convenient to already incorporate them, even in the nonnegative case.

The above observations give rise to the following definitions. The first one is concerned with sequences.

Definition 1.3 Let $a = \{a_n\}_{n=1}^{\infty}$ and $\beta = \{\beta_n\}_{n=1}^{\infty}$ be two sequences of complex and nonnegative numbers, respectively. The couple (a, β) determines a *general monotone* sequence a with majorant β, written $(a, \beta) \in \mathcal{GMS}$, if there exists $C > 0$ such that (1.4) holds for all integers n.

Being guided by the concept of variation, we can now define the classes of general monotone functions. A function f defined in $(0, \infty)$ is said to be *admissible* if it is locally of bounded variation on $(0, \infty)$, and moreover $f(x) \to 0$ as $x \to \infty$.

Definition 1.4 Let $f\colon (0, \infty) \to \mathbb{C}$ be an admissible function, while $\beta\colon (0, \infty) \to [0, \infty)$. The couple (f, β) determines a *general monotone* function f with majorant β, written $(f, \beta) \in \mathcal{GM}$, if there is a constant $C > 0$ such that (1.5) holds for all $x \in (0, \infty)$.

We emphasize that the majorants β in Definitions 1.3 and 1.4 depend either on a or on f, respectively. Moreover, such cases are pivotal to us. The simplest cases are those where $\beta_n = |a_n|$ and $\beta(x) = |f(x)|$ for which we write, for brevity, $a \in GMS$ and $f \in GM$.

Remark 1.5 It is also worth noting that if $(a, \beta) \in \mathcal{GMS}$ and we define

$$\widetilde{\beta}(x) = \beta_1, \quad x \in (0, 1], \quad \widetilde{\beta}(x) = \beta_n, \quad x \in (n, n+1],\ n \geq 1,$$

$$f(x) = a_1, \quad x \in (0, 1], \quad f(x) = a_n, \quad x \in (n, n+1],\ n \geq 1,$$

then trivially $(f, \widetilde{\beta}) \in \mathcal{GM}$.

Before proceeding further, let us introduce certain basic terms. A sequence $\{a_k\}$ is called a *null sequence*, or a *sequence vanishing at infinity*, if

$$\lim_{k \to \infty} a_k = 0.$$

On the other hand, a function f defined in a neighborhood of a point $a \in \mathbb{R} \cup \{\pm\infty\}$ is said to be *vanishing at* a if

$$\lim_{x \to a} f(x) = 0.$$

To conclude this introductory section, we present some basic facts about the bounded variation property.

First, we note that any constant sequence is trivially *BVS*, and decreasing null sequences $\{a_k\}$ are also *BVS*, since in this case

$$\sum_{k=n}^{\infty} |\Delta a_k| = \sum_{k=n}^{\infty} \Delta a_k = a_n,$$

for every $n \in \mathbb{N}$.

It is also well known that *any sequence* $\{a_k\} \in BVS$ *can be written as a difference of two decreasing sequences*. Indeed, it is easily checked that when we write

$$a_k = a_k + \sum_{n=k}^{\infty} |\Delta a_n| - \sum_{n=k}^{\infty} |\Delta a_n|, \tag{1.6}$$

we have that both sequences $\{a_k + \sum_{n=k}^{\infty} |\Delta a_n|\}$ and $\{\sum_{n=k}^{\infty} |\Delta a_n|\}$ are nonincreasing. Of course, a similar relation is satisfied in the context of functions of bounded variation, where the sums $\sum |\Delta a_n|$ are replaced by the Stieltjes integrals $\int |df(t)|$.

Before proceeding further, we introduce additional notation.

Definition 1.6 Let f, g be nonnegative functions defined on a set X. If $f(x) \le Cg(x)$ (respectively $f(x) \ge Cg(x)$) for every $x \in X$ and some constant $C > 0$ independent of x, we denote $f \lesssim g$ (respectively $f \gtrsim g$). If $f \lesssim g$ and $f \gtrsim g$ simultaneously (with different constants, of course), we write $f \asymp g$.

For instance, with this notation in hand, conditions (1.4) and (1.5) may be written as

$$\sum_{k=n}^{2n-1} |\Delta a_k| \lesssim \beta_n \qquad \text{and} \qquad \int_x^{2x} |df(t)| \lesssim \beta(x),$$

respectively. On the other hand, if $y > 0$, we can write

$$\int_y^{2y} x^{\alpha}\, dx \asymp y^{\alpha+1}, \qquad \alpha \in \mathbb{R}.$$

Definition 1.7 Let $\{a_k\}, \{b_k\} \subset \mathbb{C}$ be arbitrary sequences. We write $a_n = o(b_n)$ if

$$\lim_{n\to\infty} \frac{a_n}{b_n} = 0,$$

and $a_n = O(b_n)$ if

$$a_n \le Cb_n, \qquad \text{for all } n \ge n_0.$$

For $f, g\colon \mathbb{R} \to \mathbb{C}$, we write $f(x) = o(g(x))$ as $x \to a \in \mathbb{R}$ if

$$\lim_{x\to a} \frac{f(x)}{g(x)} = 0.$$

The notation $f(x) = O(g(x))$ as $x \to a \in \mathbb{R}$ stands for the property that there exists $\delta > 0$ such that

$$\frac{|f(x)|}{|g(x)|} \le C, \qquad \text{for all } x \text{ satisfying } 0 < |x-a| < \delta.$$

Finally, in the case where $a = +\infty$, the definition of $f(x) = o(g(x))$ is the same if a is replaced by $+\infty$ in the limit. As for $f(x) = O(g(x))$, the corresponding condition is understood as the existence of $x_0 \in \mathbb{R}$ such that

$$\frac{|f(x)|}{|g(x)|} \le C, \qquad \text{for all } x \ge x_0.$$

For instance, under this notation, $a_n \to 0$ is equivalent to $a_n = o(1)$.

We are now in a position to proceed to a more detailed study of these and certain related generalizations of monotonicity. General monotone sequences will be considered in the following section while the corresponding classes of functions in Section 1.3.

1.2 Various generalizations of monotone sequences

Prior to our approach, let us first overview some earlier generalizations of the class of monotone sequences. Two main ways to extend this class are widely known.

The first one is to study quasi-monotone type sequences. The standard class of quasi-monotone sequences is defined as follows: for $\alpha > 0$, we set

$$QMS(\alpha) = \Big\{\{a_n\} : a_n \geq 0, \, \{n^{-\alpha} a_n\} \text{ is decreasing}\Big\}, \tag{1.7}$$

and

$$QMS = \bigcup_{\alpha>0} QMS(\alpha).$$

We observe that an equivalent definition of QMS is the following: Let $\{a_n\}$ be a nonnegative sequence. Then $\{a_n\} \in QMS$ if and only if there exists $\gamma > 0$ such that

$$\Delta a_n \geq -\frac{\gamma a_n}{n}. \tag{1.8}$$

Indeed, suppose that $\{a_n\} \in QMS(\alpha)$ for some $\alpha > 0$. We may assume that $\alpha \geq 1$. It follows from the definition of $QMS(\alpha)$ that

$$a_n - a_{n+1} \geq -a_n\left(\frac{(n+1)^\alpha}{n^\alpha} - 1\right) = -a_n \alpha \int_1^{1+\frac{1}{n}} t^{\alpha-1}\, dt. \tag{1.9}$$

If $\alpha = 1$, then (1.8) is obvious with $\gamma = 1$. If $\alpha > 1$, we have to estimate the integral from above. We have

$$\int_1^{1+\frac{1}{n}} t^{\alpha-1}\, dt \leq \left(1 + \frac{1}{n}\right)^{\alpha-1} \frac{1}{n} \leq \frac{2^{\alpha-1}}{n}.$$

Thus, (1.8) holds with $\gamma = \alpha 2^{\alpha-1}$.

On the other hand, assume that (1.8) holds with $\gamma > 0$. Then, the inequality

$$\frac{\gamma}{n} \leq \alpha \int_1^{1+\frac{1}{n}} t^{\alpha-1}\, dt$$

trivially holds with $\alpha = \gamma + 1$. Thus, (1.9) is valid.

Example 1.8 Condition (1.8) gives an easy way to construct a sequence $\{a_n\} \in QMS \backslash M$ (where M denotes the class of decreasing sequences). It is enough to take $\{b_n\} \in M$ and define

$$a_n = \begin{cases} b_n, & \text{if } n \notin I, \\ b_{n-1}\left(1 + \frac{\gamma}{n+1}\right), & \text{if } n \in I, \end{cases}$$

where $I \subsetneq \mathbb{N}$ and $\gamma > 0$.

A more general class than QMS is given by

$$QMS^* = \Bigg\{ \{a_n\} : a_n \geq 0 \text{ and there exists a positive increasing sequence } \{\lambda_n\} \\ \text{such that } \sup_n \frac{\lambda_{2n}}{\lambda_n} < \infty \text{ and } \left\{\frac{a_n}{\lambda_n}\right\} \text{ is decreasing} \Bigg\}.$$

Clearly, $QMS \subset QMS^*$ by choosing $\lambda_n = n^\alpha$ in the QMS^* condition.

The second way is to study the class of sequences "with rest of bounded variation", written $RBVS$, which is defined as

$$RBVS = \left\{ \{a_n\} : a_n \in \mathbb{C}, \quad \sum_{k=n}^{\infty} |\Delta a_k| \leq C|a_n| \right\}, \tag{1.10}$$

where the constant C does not depend on n. We now show that the classes QMS (or QMS^*) and $RBVS$ are incomparable (when restricted to nonnegative sequences; note that in the definition of $RBVS$ we allow sequences to be complex-valued in general). In order to prove $QMS \not\subset RBVS$ we consider $a_n = n$. It is clear that $\{a_n\} \in QMS$, since $n^{-2}a_n = n^{-1}$ is decreasing. On the other hand,

$$\sum_{k=n}^{\infty} |a_k - a_{k+1}| = \sum_{k=n}^{\infty} 1 = \infty \not\leq C|a_n| = Cn,$$

that is, $\{a_n\} \notin RBVS$. In order to prove that $RBVS \not\subset QMS$, let $0 < q < 1$ and

$$a_n = \begin{cases} 2q^m, & \text{if } n = 2^m,\ m \in \mathbb{N} \cup \{0\}, \\ q^m, & \text{if } 2^m < n < 2^{m+1},\ m \in \mathbb{N}. \end{cases}$$

For $2^m < n \leq 2^{m+1}$, we have

$$\sum_{k=n}^{\infty} |a_k - a_{k+1}| \leq 3 \sum_{\ell=m}^{\infty} q^\ell = \frac{3}{1-q} q^m \leq \frac{3}{1-q} a_n,$$

so $\{a_n\} \in RBVS$. On the other hand, $\{n^{-\alpha} a_n\}$ is not decreasing for any $\alpha > 0$, since for $m \in \mathbb{N}$, the inequality

$$2q^m (2^m)^{-\alpha} = a_{2^m} (2^m)^{-\alpha} \leq a_{2^m - 1} (2^m - 1)^{-\alpha} = q^m (2^m - 1)^{-\alpha}$$

is equivalent to

$$\left(1-\frac{1}{2^m}\right)^{\alpha} \leq \frac{1}{2},$$

which does not hold for m sufficiently large, and so we conclude $\{a_n\} \notin QMS$ (in fact, $\{a_n\} \notin QMS^*$).

The above classes of sequences represent important steps toward more advanced generalizations of monotonicity. Here we present the class of general monotone sequences (*GMS*), which contains all the previously mentioned classes and, moreover, has a simple description. The *GMS* class is defined as

$$GMS = \left\{\{a_n\} : a_n \in \mathbb{C}, \quad \sum_{k=n}^{2n-1} |\Delta a_k| \leq C|a_n|\right\}.$$

The following is a characterization of *GMS*.

Proposition 1.9 *A sequence $\{a_k\}$ belongs to the class GMS if and only if*

(i) *for any $n \leq m \leq 2n$,*

$$|a_m| \lesssim |a_n|, \tag{1.11}$$

(ii) *and for any $N > n$,*

$$\sum_{k=n}^{N} |\Delta a_k| \lesssim |a_n| + \sum_{k=n+1}^{N} \frac{|a_k|}{k}. \tag{1.12}$$

Proof If $\{a_k\} \in GMS$ and $n < m \leq 2n$,

$$C|a_n| \geq \sum_{k=n}^{2n-1} |\Delta a_k| \geq \left|\sum_{k=n}^{m-1}(a_k - a_{k+1})\right| = |a_n - a_m| \geq |a_m| - |a_n|,$$

which proves (1.11). In order to prove (1.12), let $\ell \in \mathbb{N} \cup \{0\}$ be such that $2^\ell n \leq N < 2^{\ell+1}n$. Using (1.11) and the *GMS* condition, we obtain

$$\begin{aligned}\sum_{k=n}^{N} |\Delta a_k| &\leq \sum_{m=0}^{\ell} \sum_{k=2^m n}^{2^{m+1}n-1} |\Delta a_k| \leq C \sum_{m=0}^{\ell} |a_{2^m n}| \\ &\leq C\left(|a_n| + \sum_{m=1}^{\ell} |a_{2^m n}| \sum_{k=2^{m-1}n+1}^{2^m n} \frac{1}{k}\right) \leq C\left(|a_n| + \sum_{k=n+1}^{N} \frac{|a_k|}{k}\right),\end{aligned}$$

where in the last inequality we used (1.11) (the constant C above may vary in each occurrence). In order to prove that (1.11) and (1.12) imply $\{a_k\} \in GMS$, we simply have

$$\sum_{k=n}^{2n-1} |\Delta a_k| \leq C\left(|a_n| + \sum_{k=n}^{2n-1} \frac{|a_k|}{k}\right) \leq C\left(|a_n| + C|a_n| \sum_{k=n}^{2n-1} \frac{1}{k}\right) \leq C|a_n|. \qquad \square$$

Remark 1.10 Although the class *GMS* (respectively $\mathcal{GMS}$) has been defined for complex-valued sequences, we would like to compare it with other classes (for obtaining embeddings between them), as, for instance, *QMS*, which consists only of nonnegative sequences. To this end, when studying these embeddings, we will restrict ourselves to *nonnegative GMS* (respectively $\mathcal{GMS}$), without mentioning this explicitly. On the other hand, comparing two different classes of the type $\mathcal{GMS}$, we will consider complex-valued sequences.

The first thing to observe is that $QMS \subsetneq GMS$ (we emphasize that here and in what follows we are restricting ourselves to nonnegative sequences). The fact that $GMS \backslash QMS \neq \emptyset$ follows from the facts that $RBVS \not\subset QMS$ (proved above) and $RBVS \subset GMS$ (which we will prove below). On the other hand, the inclusion $QMS \subset GMS$ is once again a consequence of the broader situation $QMS \subset QMS^* \subsetneq GMS$ (the last proper inclusion will be proved below).

Let us now study the interrelations between QMS^*, *RBVS*, and *GMS*. Under established definitions, the relation

$$QMS^* \cup RBVS \subsetneq GMS \tag{1.13}$$

holds. Let us first prove inclusions for both classes on the left-hand side, and later on their properness by means of corresponding counterexamples. Note that $RBVS \subset GMS$ is trivial. We now show that $QMS^* \subset GMS$. Let $\{a_k\} \in QMS^*$. Considering that $\{\frac{a_n}{\lambda_n}\}$ is decreasing and $\{\lambda_n\}$ is increasing, we obtain

$$\begin{aligned}|\Delta a_k| &= \left|\lambda_{k+1}\left(\frac{a_k}{\lambda_k} - \frac{a_{k+1}}{\lambda_{k+1}}\right) + (\lambda_k - \lambda_{k+1})\frac{a_k}{\lambda_k}\right| \\ &\le \lambda_{k+1}\Delta\left(\frac{a_k}{\lambda_k}\right) + (\lambda_{k+1} - \lambda_k)\frac{|a_k|}{\lambda_k}.\end{aligned}$$

Therefore,

$$\begin{aligned}\sum_{k=n}^{2n-1} |\Delta a_k| &\le \lambda_{2n} \sum_{k=n}^{2n-1} \Delta\left(\frac{a_k}{\lambda_k}\right) + \frac{|a_n|}{\lambda_n}\sum_{k=n}^{2n-1}(\lambda_{k+1} - \lambda_k) \\ &\le |a_n|\frac{\lambda_{2n}}{\lambda_n} + |a_n|\frac{\lambda_{2n} - \lambda_n}{\lambda_n} \le C|a_n|,\end{aligned}$$

where in the last inequality we used $\frac{\lambda_{2n}}{\lambda_n} \le C$.

Let us now give two examples. The first one shows explicitly the importance of the term $\sum_{k=n+1}^{N} \dfrac{|a_k|}{k}$ in the inequality (1.12) – contrary to (1.10) (in particular, such an example proves that $GMS \backslash RBVS \neq \emptyset$). In the second one, we show that the embedding (1.13) is proper.

Example 1.11 Let us define

$$c_n = \frac{1}{p_m} + \frac{n - 2^m}{q_m 2^m}, \qquad 2^m \le n \le 2^{m+1} - 1,$$

where p_m and q_m are monotone increasing sequences. Then, if $p_m \le Cq_m$ with C independent of m, we have $\{c_n\} \in QMS$ (and consequently $\{c_n\} \in GMS$). With the assumption $p_m \le Cq_m$ in hand, we observe that if

$$\sum_{k=m}^{\infty} q_k^{-1} \le C p_m^{-1},$$

then $\{c_n\} \in RBVS$. But if

$$p_m^{-1} = o\left(\sum_{k=m}^{\infty} q_k^{-1}\right), \tag{1.14}$$

then $\{c_n\} \notin RBVS$. Further, for N large,

$$\sum_{k=\lfloor \log_2 N \rfloor}^{\infty} \frac{1}{q_k} \le \sum_{k=N}^{\infty} |c_{k+1} - c_k| \le C\left(|c_N| + \sum_{k=N}^{\infty} \frac{|c_k|}{k}\right) \le C \sum_{k=\lfloor \log_2 N \rfloor}^{\infty} \frac{1}{p_k}.$$

If, for example, $p_k \asymp q_k$ and (1.14) holds, then

$$\sum_{k=N}^{\infty} |c_{k+1} - c_k| \asymp \sum_{k=N}^{\infty} \frac{|c_k|}{k} \not\asymp |c_N|.$$

Example 1.12 ($GMS \backslash \big(QMS^* \cup RBVS\big) \ne \emptyset$) Let $u_n := n!$ and $k(n) \in \mathbb{N}$ be such that $2^{k(n)} \le n + 1 < 2^{k(n)+1}$ for any integer n. Define

$$d_s = \begin{cases} e^{-s}, & \text{if } u_n \le s < 2u_n \text{ and } s \text{ is even}, \\ e^{-s-2}, & \text{if } u_n \le s \le 2u_n - 1 \text{ and } s \text{ is odd}, \\ 2^i d_{2u_n}, & \text{if } 2^{i+1}u_n \le s < 2^{i+2}u_n \text{ and } 0 \le i \le k(n) - 2, \\ 2^{k(n)} d_{2u_n}, & \text{if } 2^{k(n)}u_n \le s < u_{n+1}. \end{cases}$$

Then

$$|\Delta d_s| \asymp \begin{cases} e^{-s}, & \text{if } u_n \le s < 2u_n, \\ 0, & \text{if } 2^{i+1}u_n \le s < 2^{i+2}u_n - 1, 0 \le i \le k(n) - 2, \\ 2^i d_{2u_n}, & \text{if } s = 2^{i+2}u_n - 1, 0 \le i \le k(n) - 2, \\ 0, & \text{if } 2^{k(n)}u_n \le s \le u_{n+1} - 2, \\ 2^{k(n)} d_{2u_n} - e^{-u_{n+1}}, & \text{if } s = u_{n+1} - 1, \end{cases}$$

and, therefore, $\{d_n\} \in GMS$. On the other hand, since $\{d_n\}$ vanishes at infinity, we have

$$\frac{1}{d_{2u_n}} \sum_{s=2u_n}^{\infty} |\Delta d_s| \ge \frac{1}{d_{2u_n}} \left| \sum_{s=u_{n+1}-1}^{\infty} \Delta d_s \right| = \frac{d_{u_{n+1}-1}}{d_{2u_n}} = 2^{k(n)} \to \infty,$$

as $n \to \infty$, and hence $\{d_n\} \notin RBVS$. We observe that this sequence is not from the QMS^* class either. Indeed, assume that there exists an increasing sequence $\{\lambda_n\}$ such that $\{\frac{d_n}{\lambda_n}\}$ is decreasing. In this case we have

$$\frac{\lambda_{n+1}}{\lambda_n} \geq \frac{d_{n+1}}{d_n} \quad \text{for all } n,$$

and since $\{\lambda_n\}$ is increasing, we also have $\frac{\lambda_{n+1}}{\lambda_n} \geq 1$ for all n. Hence, for $m \in \mathbb{N}$,

$$\frac{\lambda_{2u_m}}{\lambda_{u_m}} = \prod_{k=u_m}^{2u_m-1} \frac{\lambda_{k+1}}{\lambda_k} \geq \prod_{\substack{k=u_m \\ k \text{ is odd}}}^{2u_m-1} \frac{\lambda_{k+1}}{\lambda_k} \geq \prod_{\substack{k=u_m \\ k \text{ is odd}}}^{2u_m-1} \frac{d_{k+1}}{d_k} = \prod_{\substack{k=u_m \\ k \text{ is odd}}}^{2u_m-1} e = e^{\frac{u_m}{2}} \to \infty,$$

as $m \to \infty$. Thus, $\{d_n\} \notin QMS^*$, since $\lambda_{2n} \leq C\lambda_n$ does not hold.

Let us mention that sequences in the class *GMS* may have much less regular structure than monotone ones. The following simple observation is quite demonstrative. It is well known that a monotone sequence converges if and only if it is bounded and tends to $\pm\infty$ otherwise. This is no longer the case for general monotone sequences; just take the sequence $\{a_k\}$ with

$$a_k = \begin{cases} 1, & \text{if } k \neq 2^n,\ n \in \mathbb{N}, \\ 2, & \text{if } k = 2^n,\ n \in \mathbb{N}, \end{cases}$$

which is definitely in *GMS* and has no limit.

For $\beta_n = |a_n|$, we have $(a, \beta) \in \mathcal{GMS}$ if and only if $a \in GMS$, just by definition. Another important example of β is the following.

Example 1.13 Let $a = \{a_n\} \subset \mathbb{C}$ and $\bar{\beta} = \{\bar{\beta}_n\}$ be defined as

$$\bar{\beta}_n = \sum_{k=\lfloor n/c \rfloor}^{\lfloor cn \rfloor} \frac{|a_k|}{k} \quad \text{for some } c > 1. \tag{1.15}$$

We then define the class $\overline{GMS}$ as that of all sequences $a = \{a_k\} \subset \mathbb{C}$ satisfying $(a, \bar{\beta}) \in \mathcal{GMS}$, that is,

$$\sum_{k=n}^{2n-1} |\Delta a_k| \leq C \sum_{k=\lfloor n/c \rfloor}^{\lfloor cn \rfloor} \frac{|a_k|}{k}.$$

One has $GMS \subsetneq \overline{GMS}$. Indeed, in order to prove the embedding $GMS \subseteq \overline{GMS}$, we simply establish the inequalities

$$n \sum_{k=n}^{2n-1} |\Delta a_k| \leq C \sum_{k=\lfloor n/c \rfloor}^{\lfloor cn \rfloor} \sum_{\ell=n}^{2n-1} |\Delta a_\ell| \lesssim \sum_{k=\lfloor n/c \rfloor}^{\lfloor cn \rfloor} |a_k|,$$

where in the last step the *GMS* condition is used. To prove the claim, it now suffices to divide both sides of the inequality by n. To show the properness of the embedding

$GMS \subsetneq \overline{GMS}$, we just consider a positive sequence $\{a_k\} \in GMS$. For a fixed $m \in \mathbb{N}$, assume that $\sum_{k=m}^{2m-1} |\Delta a_k| > 0$. Then, the sequence $\{\bar{a}_n\}$ defined by

$$\bar{a}_n = \begin{cases} a_k, & \text{if } k \neq m, \\ 0, & \text{if } k = m, \end{cases}$$

belongs to the class $\overline{GMS}$, but not to the class GMS.

The estimate

$$|a_\ell| \le C \sum_{k=\lfloor n/c \rfloor}^{\lfloor cn \rfloor} \frac{|a_k|}{k}, \qquad \ell \in [n, 2n],\ n \in \mathbb{N}, \tag{1.16}$$

is a useful property of $\overline{GMS}$. This is an estimate that trivially holds for decreasing sequences (it is equivalent to the one referred to as monotonicity type property, mentioned in Section 1.1). In the case of $\overline{GMS}$ the proof is rather simple. Indeed, for any $\ell, m \in [n, 2n]$,

$$|a_\ell| - |a_m| \le |a_\ell - a_m| \le \sum_{k=n}^{2n-1} |a_k - a_{k+1}| \le C \sum_{k=\lfloor n/c \rfloor}^{\lfloor cn \rfloor} \frac{|a_k|}{k}. \tag{1.17}$$

Summing up in m from n to $2n-1$ on both sides of the inequality yields

$$|a_\ell| \le \frac{1}{n} \sum_{m=n}^{2n-1} |a_m| + C \sum_{k=\lfloor n/c \rfloor}^{\lfloor cn \rfloor} \frac{|a_k|}{k} \le C' \sum_{k=\lfloor n/c \rfloor}^{\lfloor cn \rfloor} \frac{|a_k|}{k}.$$

We also mention that general monotone sequences possess interesting multiplier properties.

Proposition 1.14 (i) *If $\{a_n\}, \{b_n\} \in GMS$, then $\{a_n b_n\} \in GMS$.*
(ii) *If $\{a_n\} \in \overline{GMS}$, $\{b_n\} \in GMS$ is nonnegative, and $b_n \asymp b_m$ for $n \asymp m$, then $\{a_n b_n\} \in \overline{GMS}$.*

Proof We have

$$\sum_{k=n}^{2n-1} |\Delta(a_k b_k)| \le \max_{n \le m \le 2n} |b_m| \sum_{k=n}^{2n-1} |\Delta a_k| + \max_{n \le m \le 2n} |a_m| \sum_{k=n}^{2n-1} |\Delta b_k| \le C |a_n b_n|,$$

where in the last inequality the GMS condition and (1.11) are used. This establishes the first part. As for the second part, we have

$$\max_{n \le m \le 2n} |b_m| \sum_{k=n}^{2n-1} |\Delta a_k| \le C \max_{n \le m \le 2n} b_m \sum_{k=\lfloor n/c \rfloor}^{\lfloor cn \rfloor} \frac{|a_k|}{k} \asymp \sum_{k=\lfloor n/c \rfloor}^{\lfloor cn \rfloor} \frac{|a_k b_k|}{k}.$$

Finally, by (1.16), we get

$$\max_{n \le m \le 2n} |a_m| \sum_{k=n}^{2n-1} |\Delta b_k| \le C b_n \sum_{k=\lfloor n/c \rfloor}^{\lfloor cn \rfloor} \frac{|a_k|}{k} \asymp \sum_{k=\lfloor n/c \rfloor}^{\lfloor cn \rfloor} \frac{|a_k b_k|}{k}.$$

Combining all the estimates we derive the desired $\overline{GMS}$ inequality. □

A straightforward generalization of the class $\overline{GMS}$ is defined as follows: a sequence $\{a_k\} \subset \mathbb{C}$ is said to belong to the class GMS^∞ if there exist constants $c, C > 1$ such that

$$\sum_{k=n}^{\infty} |\Delta a_k| \leq C \sum_{k=\lfloor n/c \rfloor}^{\infty} \frac{|a_k|}{k},$$

for every n. For such a class, the relation $\overline{GMS} \subsetneq GMS^\infty$ holds, provided that $\sum_{k=\lfloor n/c \rfloor}^{\infty} \frac{|a_k|}{k} < \infty$. We will not study this class in detail, but rather mention it for completeness. For a reference, see Notes-11.

We now give another extension of $\overline{GMS}$. For $1 \leq p < \infty$, we say that $\{a_k\} \in \overline{GMS}(p)$ if

$$\sum_{k=n}^{2n-1} |\Delta a_k| \leq C \left(\sum_{k=\lfloor n/c \rfloor}^{\lfloor cn \rfloor} \frac{|a_k|^p}{k} \right)^{\frac{1}{p}}, \tag{1.18}$$

for some constants $c, C > 1$. Note that $\overline{GMS} = \overline{GMS}(1)$. Sequences from such classes have desirable properties similar to $\overline{GMS}$. We now list several properties of $\overline{GMS}(p)$ without proofs, which are given in the references in Notes-12, along with further discussion.

We begin with the embedding relations between $\overline{GMS}(p)$ classes.

Proposition 1.15 *For $1 \leq p_1 < p_2$, there holds $\overline{GMS}(p_1) \subsetneq \overline{GMS}(p_2)$.*

The following is a counterpart of Proposition 1.14.

Proposition 1.16 *For any $p_0, p_1 \geq 1$, if $\{a_n\} \in \overline{GMS}(p_0)$ and $\{b_n\} \in \overline{GMS}(p_1)$ is such that $|b_k| \asymp |b_n|$ for $k \asymp n$, then $\{a_n b_n\} \in \overline{GMS}(p_0)$. In particular, $\{a_n n^\gamma\} \in \overline{GMS}(p_0)$ for every $\gamma \in \mathbb{R}$.*

For special choices of $\{b_n\}$, no further assumptions for $\{a_n b_n\} \in \overline{GMS}(p_0)$ are needed.

Proposition 1.17 (i) *Let $p, \gamma \geq 1$ and $\{a_n\} \in \overline{GMS}(p)$. Then*

$$\left\{ a_n |a_n|^{\gamma-1} \right\} \in \overline{GMS}(p).$$

(ii) *In particular, for any nonnegative sequence $\{a_n\} \in \overline{GMS}(p)$, $p \geq 1$, the sequence $\{a_n^\gamma\}$ belongs to $\overline{GMS}(p)$ if and only if $\gamma \geq 1$.*

We continue with a characterization of $\overline{GMS}(p)$.

Theorem 1.18 *A sequence $\{a_n\} \subset \mathbb{C}$ belongs to $\overline{GMS}(p)$, $p \geq 1$, if and only if for some positive integer ν, the conditions*

$$\max_{k=2^n,\ldots,2^{n+1}} |a_k| \lesssim \left(2^{-n} \sum_{k=2^{n-\nu}}^{2^{n+\nu}} |a_k|^p \right)^{\frac{1}{p}}, \quad n \geq \nu, \tag{1.19}$$

and

$$\sum_{k=2^n}^{2^{n+1}} |a_k - a_{k+1}| \lesssim \max_{k=2^{n-\nu},\ldots,2^{n+\nu}} |a_k|, \quad n \geq \nu, \tag{1.20}$$

hold.

We emphasize that conditions (1.19) and (1.20) are independent. Also note that (1.20) easily follows from (1.18).

The sequences in the class $\overline{GMS}(p)$ satisfy an estimate similar to (1.16).

Proposition 1.19 *Let $\{a_n\} \in \overline{GMS}(p)$, $p \geq 1$. Then, for some positive integer ν, the estimate*

$$|a_m| \lesssim \left(\sum_{k=\lfloor 2^{-\nu} m \rfloor + 1}^{2^\nu m} \frac{|a_k|^p}{k} \right)^{\frac{1}{p}}, \quad m \geq 2^\nu,$$

holds.

We conclude with further properties of $\overline{GMS}(p)$. To this end, we introduce appropriate notation. For a null sequence $\{a_k\} \subset \mathbb{C}$, we define

$$a_n^{\#} := \max_{k \geq n} |a_k| \quad \text{and} \quad M_n := \max_{k=2^n,\ldots,2^{n+1}} |a_k|.$$

Furthermore, we define the decreasing rearrangement $\{a_n^*\}$ of $\{a_n\}$ as the sequence whose terms are the same as those of $\{|a_n|\}$, rearranged in decreasing order.

Proposition 1.20 *Let $p_0 \geq 1$, $\alpha > 0$, and $0 < p < q < \infty$. If a sequence $\{a_n\} \in \overline{GMS}(p_0)$ is such that $\sum_{n=1}^{\infty} |a_n|^p n^{\alpha-1} < \infty$, then*

(i) $a_n n^{\frac{\alpha}{p}} \to 0$ *as* $n \to \infty$,
(ii) $\sum_{n=1}^{\infty} |a_n|^q n^{\frac{\alpha q}{p}-1} < \infty$, *and*
(iii) *the equivalences*

$$\sum_{n=1}^{\infty} |a_n|^p n^{\alpha-1} \asymp \sum_{n=1}^{\infty} (a_n^{\#})^p n^{\alpha-1} \asymp \sum_{n=1}^{\infty} (a_n^*)^p n^{\alpha-1} \asymp \sum_{n=1}^{\infty} 2^{n\alpha} M_n^p$$

are satisfied.

We note here that part (i) is the Abel-Olivier test (see Theorem 2.98), part (ii) is related to the $\ell^p - \ell^q$ inequalities in Section 2.4.8, and part (iii) contains the Cauchy condensation test (see Section 2.4.3).

1.3 General monotone functions

Let us turn to monotone type functions on $\mathbb{R}_+$. We cannot restrict ourselves only to sequences, since many important problems for series with terms $\{a_n\}$ are studied as those for functions f with $f(n) = a_n$. Of course, there are more interesting relations between these two settings.

Before proceeding further, it is worth settling the meaning we will give to integrals throughout the book. All the involved functions, defined in $U \subset \mathbb{R}^n$, will be measurable and locally integrable, or in other words, such that $\int_K |f(x)|\,dx < \infty$ for every compact set $K \subset U$, where the integral is understood in the sense of Lebesgue. Integrals over unbounded sets will not be taken in the sense of Lebesgue but rather as the *limit of Lebesgue integrals*. Precise definitions will be given in the corresponding parts. For a complete account of the theory of Lebesgue integrals, refer, for instance, to [B-73, Ch. 11].

Recall that the class of quasi-monotone functions QM is the class of functions $f\colon \mathbb{R}_+ \to \mathbb{R}_+$ for which there exists $\alpha > 0$ such that $t^{-\alpha}f(t)$ is decreasing.

Remark 1.21 We observe that if f is locally absolutely continuous on $(0, \infty)$, then $f \in QM$ if and only if

$$f'(t) < \frac{\gamma f(t)}{t} \tag{1.21}$$

almost everywhere, for some $\gamma > 0$. Indeed, this follows immediately from the fact that $t^{-\alpha}f(t)$ is decreasing, which implies

$$\left(t^{-\alpha}f(t)\right)' = -\alpha t^{-\alpha-1}f(t) + t^{-\alpha}f'(t) \le 0.$$

This is equivalent to $f'(t) \le \dfrac{\alpha f(t)}{t}$, and (1.21) follows with any $\gamma > \alpha$.

The converse direction follows from the fact that (1.21) is equivalent to

$$f'(t)t^{-\gamma} < \gamma t^{-\gamma-1}f(t),$$

which in turn yields the negativity of the derivative of $\dfrac{f(t)}{t^\gamma}$ (almost everywhere) and the monotone decrease of this fraction.

What is significant is that just this direction shows the importance of the absolute continuity assumption. It excludes the case of singular functions that may happen if only continuity and differentiability almost everywhere are assumed. In the latter case, the negativity of the derivative need not provide the monotone decrease of the involved fraction. For what this can cause, one may consider $\varphi_0(t) = 2\psi(t) - t$ on $[0, 1]$, where ψ is the Cantor function (we will use well-known properties of the Cantor function; see Notes-4 for appropriate references). We have $\varphi_0'(x) = -1 < 0$ almost everywhere on $[0, 1]$, though $\varphi_0(0) = 0 < 1 = \varphi_0(1)$, which gives no decrease. Further, let us define a function φ on $[0, \infty)$ as follows. Let $\varphi(t) = \varphi_0(t)$ for $t \in [0, 1)$, while

$$\varphi(t) = 2\psi(t-k) - t + 2k, \qquad t \in [k, k+1),\ k \in \mathbb{N}.$$

Of course, φ is continuous and differentiable almost everywhere, with $\varphi'(t) = -1$. Finally, define $f(t) = 2^{\varphi(t)}$. Then,

$$f'(t) = -2^{\varphi(t)} \ln 2 < 0$$

almost everywhere in $t \in [0,\infty)$. Thus, f trivially satisfies (1.21), with any $\gamma > 0$. However, though, for every $\alpha > 0$, the derivative of $\frac{f(t)}{t^\alpha}$ is almost everywhere negative, $\frac{f(t)}{t^\alpha}$ is not monotone decreasing, at least, for large values of t, and thus f is not quasi-monotone.

We now use (1.21) to relate the classes *QM* and *QMS*. Let $\{f_n\}$ be a nonnegative sequence vanishing at infinity. We construct a related function $f(t)$ by means of linear interpolation as follows: for $t \in [k, k+1)$, let

$$f(t) = f_{k+1} - (k+1-t)(f_{k+1} - f_k), \tag{1.22}$$

and $f(t) = f_1$ for $t \in [0,1]$. By this, $f(t) \in QM$ if and only if $\{f_n\} \in QMS$. Here only part "if" deserves certain discussion. It suffices to show that

$$s^{-\beta} f(s) \le t^{-\beta} f(t), \quad k \le t \le s \le k+1,$$

for some $\beta > 0$. Note that for $t \in [k, k+1)$ one has $f'(t) = f_{k+1} - f_k$. We now have two cases. If $f_k > f_{k+1}$, then $f'(t) < 0$ and (1.21) is trivial. On the other hand, if $f_k \le f_{k+1}$, we first note that $\{f_n\} \in QMS$ is equivalent to

$$f_{k+1} - f_k \le \frac{\alpha f_k}{k}$$

for some $\alpha > 0$ (see (1.8)). Then

$$f'(t) = f_{k+1} - f_k \le \alpha \frac{f_k}{k} < \gamma \frac{f(t)}{t}$$

for some $\gamma > 0$ and $k \le t < k+1$.

Further, an analogue of the *RBVS* class is given by all nonnegative functions f satisfying

$$\int_x^\infty |df(t)| \le Cf(x), \qquad x > 0. \tag{1.23}$$

We denote the collection of all such functions by *RBV*.

Remark 1.22 For any nonnegative sequence $\{f_n\}$ and for the function $f(t)$ given by (1.22), we get $f(t) \in RBV$ if and only if $\{f_n\} \in RBVS$. It then follows that the classes *RBV* and *QM* are incomparable.

Let us go on to the *GM* class (see Definition 1.4 with $\beta(x) = |f(x)|$). Similar to the *GMS* case, when comparing the *GM* class with other classes consisting of nonnegative functions, such as *QM* or *RBV*, we will assume that the functions from the *GM* class (or from any other class of the type $\mathcal{GM}$) are nonnegative. See also Remark 1.10. It follows from (1.21) and (1.23) that $QM \cup RBV \subseteq GM$. From Example 1.11, Remarks 1.21 and 1.22, and the fact that $GM \backslash QM \neq \emptyset$, we get $QM \cup RBV \subsetneq GM$. In order to prove the properness of these inclusions, it suffices to consider the sequences $\{a_n\}$ in the aforementioned examples and define

$$f(t) = a_n, \qquad t \in [n, n+1), \quad n \in \mathbb{N}.$$

We now present some properties of the *GM* class.

Proposition 1.23 *A function f defined on $\mathbb{R}_+$ belongs to the class GM if and only if*

(i) *for any $x \le u \le 2x$,*

$$|f(u)| \lesssim |f(x)|,$$

(ii) *and for any $x < y$,*

$$\int_x^y |df(t)| \lesssim \left(|f(x)| + \int_x^y \frac{|f(t)|}{t}\, dt \right). \tag{1.24}$$

The proof is omitted, as it goes exactly along the same lines as that of Proposition 1.9.

The next result is a consequence of (1.24).

Lemma 1.24 *If $f \in GM$ and f' is integrable near the origin, then f' is integrable over $(0,\infty)$, provided that in addition*

$$\int_1^\infty \frac{|f(t)|}{t}\, dt < \infty.$$

In other words, f is of bounded variation on $(0,\infty)$.

Remark 1.25 Note that assuming $\int_0^\infty t^{-1}|f(t)|\, dt < \infty$ for complex-valued $f \in GM$ rather than the conditions of Lemma 1.24 definitely ensures the boundedness of the total variation of f. The point is that the last integral cannot be convergent for such an important class as monotone functions. However, for functions vanishing at the origin the finiteness of this integral is quite natural.

Finally, the results for multipliers of general monotone functions similar to those in Proposition 1.14 can be obtained. To this end, we denote by $\overline{GM}$ the class analogous to that from Example 1.13, that is, the class of functions f such that

$$\int_x^{2x} |df(t)| \le C \int_{x/c}^{cx} \frac{|f(t)|}{t}\, dt, \tag{1.25}$$

for some constants $c, C > 1$. It is worth noting that the functions in the class $\overline{GM}$ satisfy the estimate

$$|f(u)| \lesssim \int_{x/c}^{cx} \frac{|f(t)|}{t}\, dt, \qquad x > 0,\ u \in [x, 2x]. \tag{1.26}$$

This can easily be proved as in (1.17), taking into account that for any $u, v \in [x, 2x]$ we have $|f(u)| - |f(v)| \le |f(u) - f(v)| \le \int_x^{2x} |df(t)|$.

Proposition 1.26 (i) *If $f, g \in GM$, then $fg \in GM$.*
(ii) *If $f \in \overline{GM}$, $g \in GM$ is nonnegative, and $g(x) \asymp g(y)$ for $x \asymp y$, then $fg \in \overline{GM}$.*

We conclude this section by stating useful facts on real-valued $f \in \overline{GM}$. In particular, the following statements provide certain control on the size of localized level sets of f.

In what follows we assume without loss of generality that the constant c in the definition of the class $\overline{GM}$ (cf. (1.25)) is of the form $c = 2^\nu$, with $\nu \in \mathbb{N}$. Let us define, for any function $f\colon \mathbb{R}_+ \to \mathbb{R}$ and any $n \in \mathbb{N}$,

$$A_n := \sup_{2^n \le t \le 2^{n+1}} |f(t)|,$$

$$B_n := \sup_{2^{n-2\nu} \le t \le 2^{n+2\nu}} |f(t)|.$$

For $n \in \mathbb{N} \cup \{0\}$, we say that n is a *good* number if either $n = 0$ or $B_n \le 2^{4\nu} A_n$. The rest of natural numbers consists of *bad* numbers. Note that this definition depends on the choice of f. We emphasize that the parameter ν comes from the $\overline{GM}$ condition.

To illustrate such definitions, we give a couple of examples. On the one hand, if $f(t) = \frac{1}{t^2}$ for $t \ge 1$, since

$$A_n = \frac{1}{2^{2n}}, \qquad B_n = \frac{1}{2^{2n-4\nu}},$$

we have $B_n = 2^{4\nu} A_n$, and all natural numbers n (associated with f) are good. On the other hand, if $f(t) = \frac{1}{t^3}$ for $t \ge 1$, since

$$A_n = \frac{1}{2^{3n}}, \qquad B_n = \frac{1}{2^{3n-6\nu}},$$

we have $B_n = 2^{6\nu} A_n \not\le 2^{4\nu} A_n$; thus all natural numbers n are bad. More generally, if f decreases rapidly enough (as, for instance, $1/t^3$ or e^{-t}), then all numbers $n \ne 0$ associated with f are bad.

Lemma 1.27 *Let $f \in \overline{GM}$ be real-valued. For any good number $n > 0$, there holds*

$$|E_n| := \left|\left\{x \in [2^{n-\nu}, 2^{n+\nu}] : |f(x)| \asymp A_n\right\}\right| \asymp 2^n,$$

where $|E_n|$ denotes the Lebesgue measure of E_n.

Proof We prove the statement for a specific choice of the involved constants, namely

$$|E_n| := \left|\left\{x \in [2^{n-\nu}, 2^{n+\nu}] : |f(x)| > \frac{A_n}{8C2^{2\nu}}\right\}\right| \ge \frac{2^n}{8C2^{5\nu}}, \tag{1.27}$$

where C is the constant in the $\overline{GM}$ condition.

Assume that (1.27) does not hold for some good number $n > 0$. Let us define $D_n := [2^{n-\nu}, 2^{n+\nu}] \backslash E_n$. Then, since n is good,

$$\int_{2^{n-\nu}}^{2^{n+\nu}} \frac{|f(x)|}{x}\,dx = \int_{D_n} \frac{|f(x)|}{x}\,dx + \int_{E_n} \frac{|f(x)|}{x}\,dx$$
$$\le \frac{2^{n+\nu} A_n}{8C2^{2\nu}2^{n-\nu}} + \frac{2^n B_n}{8C2^{5\nu}2^{n-\nu}} = \frac{A_n}{8C} + \frac{B_n}{8C2^{4\nu}} \le \frac{A_n}{4C}.$$

The $\overline{GM}$ condition implies that for any $x \in [2^n, 2^{n+1}]$,

$$|f(x)| \geq A_n - \int_{2^n}^{2^{n+1}} |df(t)| \geq A_n - C \int_{2^{n-\nu}}^{2^{n+\nu}} \frac{|f(t)|}{t}\, dt \geq A_n - \frac{A_n}{4} > \frac{A_n}{2},$$

which is a contradiction. □

Before stating the following lemma, let us introduce additional notation in order to distinguish the sign of f within the set E_n in Lemma 1.27:

$$E_n^+ := \{x \in E_n : f(x) > 0\}, \qquad E_n^- := \{x \in E_n : f(x) \leq 0\}.$$

Lemma 1.28 *Let $f \in \overline{GM}$ be real-valued. For any good number $n > 0$, there is an interval $(\ell_n, m_n) \subset [2^{n-\nu}, 2^{n+\nu}]$ such that at least one of the following assertions holds:*

(i) *For any $x \in (\ell_n, m_n)$, there holds $f(x) \geq 0$ and*

$$|E_n^+ \cap (\ell_n, m_n)| \asymp 2^n;$$

(ii) *For any $x \in (\ell_n, m_n)$, there holds $f(x) \leq 0$ and*

$$|E_n^- \cap (\ell_n, m_n)| \asymp 2^n.$$

Proof First of all, note that by Lemma 1.27 (or rather the explicit bound (1.27)), either $|E_n^+| \geq \frac{2^n}{16C2^{5\nu}}$ or $|E_n^-| \geq \frac{2^n}{16C2^{5\nu}}$. We assume the former and prove that (i) holds. In particular, we will prove the explicit estimate

$$|E_n^+ \cap (\ell_n, m_n)| \geq \frac{2^n}{256C^3 2^{15\nu}},$$

where C is the constant in the $\overline{GM}$ condition.

We construct a system of disjoint intervals $\{I_j = [s_j, t_j]\}_{j=1}^{p_n}$ in $\big[2^{n-\nu}, 2^{n+\nu} + \frac{2^n}{256C^3 2^{15\nu}}\big]$ as follows: let $s_1 = \inf E_n^+$, and

$$\tau_1 = \inf\{x \in [s_1, 2^{n+\nu}] : f(x) \leq 0\}.$$

If such τ_1 does not exist (i.e., if the set $\{x \in [s_1, 2^{n+\nu}] : f(x) \leq 0\}$ is empty), then we simply let $t_1 = 2^{n+\nu}$ and finish the process. Contrarily, we define

$$t_1 = \tau_1 + \frac{2^n}{256C^3 2^{15\nu}}.$$

Once we have the first interval $I_1 = [s_1, t_1]$, if $|E_n^+ \backslash I_1| > 0$, we let $s_2 = \inf E_n^+ \backslash I_1$ and define t_2 and τ_2 as above, thus obtaining a new interval $I_2 = [s_2, t_2]$. We continue this process until our collection of intervals is such that

$$|E_n^+ \backslash (I_1 \cup I_2 \cup \cdots \cup I_{p_n})| = 0,$$

where p_n is finite (see (1.28) below). By construction, for any $1 \le j \le p_n - 1$, we can find $y_j \in I_j$ such that $y_j \in E_n^+$, and $z_j \in [\tau_j, t_j]$ such that $f(z_j) \le 0$. Thus,

$$\int_{I_j} |df(t)| = \int_{s_j}^{t_j} |df(t)| \ge f(y_j) - f(z_j) \ge f(y_j) > \frac{A_n}{8C2^{2\nu}}.$$

Hence,

$$\int_{2^{n-\nu}}^{2^{n+\nu}} |df(t)| \ge \sum_{j=1}^{p_n-1} \int_{I_j} |df(t)| \ge (p_n - 1)\frac{A_n}{8C2^{2\nu}}.$$

On the other hand, the $\overline{GM}$ property and the fact that n is good imply

$$\begin{aligned}\int_{2^{n-\nu}}^{2^{n+\nu}} |df(t)| &= \sum_{k=-\nu}^{\nu-1} \int_{2^{n+k}}^{2^{n+k+1}} |df(t)| \le C \sum_{k=-\nu}^{\nu-1} \int_{2^{n+k-\nu}}^{2^{n+k+\nu}} \frac{|f(x)|}{x}\, dx \\ &\le C2\nu \int_{2^{n-2\nu}}^{2^{n+2\nu}} \frac{|f(x)|}{x}\, dx \le C2\nu B_n \int_{2^{n-2\nu}}^{2^{n+2\nu}} \frac{1}{x}\, dx = C2\nu B_n \log 2^{4\nu} \\ &\le C2^{4\nu} 8\nu^2 A_n \log 2 \le C2^{7\nu} A_n.\end{aligned}$$

We can deduce from the above estimates that

$$p_n \le 8C^2 2^{9\nu} + 1 \le 8C^2 2^{10\nu}. \qquad (1.28)$$

By the pigeonhole principle (or Dirichlet's box principle), there is an integer $j \in \{1, \ldots, p_n\}$ such that

$$|E_n^+ \cap I_j| \ge \frac{2^n}{128C^3 2^{15\nu}}.$$

Given this j, the desired interval is $(\ell_n, m_n) = (s_j, t_j - \frac{2^n}{256C^3 2^{15\nu}}) = (s_j, \tau_j) \subset [2^{n-\nu}, 2^{n+\nu}]$. □

We emphasize that Lemmas 1.27 and 1.28 can be used to derive similar statements for $\overline{GMS}$. For a number of references and applications of these results see Notes-8–14.

1.4 Weak monotone sequences and functions

Of course, much can be achieved by slightly extending the usual monotonicity to the aforementioned almost monotonicity,

$$a_n \le C a_m, \qquad m \le n.$$

This is formally a different class, but it is hardly believable that its use can fructify. In order to study number series, it is not hard to see that the above general monotonicity is too restrictive in order to give considerable generalizations of the results that are known for monotone sequences. A much more general concept is the so-called *weak monotonicity*. It is defined as follows.

Definition 1.29 We call a nonnegative sequence $\{a_k\}$ *weak monotone*, written $\{a_k\} \in WMS$, if for some constant $C \geq 1$ we have

$$a_k \leq Ca_n \quad \text{for any} \quad k \in [n, 2n]. \tag{1.29}$$

Clearly, if $\{a_k\}$ is almost monotone (cf. (1.2)), then $\{a_k\} \in WMS$.

Its counterpart for functions taken from the same set of functions where general monotone functions are defined (i.e., on $(0, \infty)$ and locally of bounded variation) is introduced in a similar manner (one does not need to require the bounded variation property when defining weak monotonicity, in general, but it is necessary in order to compare it with general monotonicity).

Definition 1.30 We say that a nonnegative function f defined on $(0, \infty)$ is *weak monotone*, written $f \in WM$, if for some absolute constant $C \geq 1$ we have

$$f(t) \leq Cf(x) \quad \text{for any} \quad t \in [x, 2x].$$

Note that trivially if $f \in WM$ and we define $a_k = f(k)$, then $\{a_k\} \in WMS$.

Clearly, in these definitions $2n$ and $2x$ can be replaced by $\lfloor cn \rfloor$ and cx, respectively, with some $c > 1$ and a different constant C. Of course, the class of weak monotone sequences/functions is wider than that of almost monotone ones. Note that we could also define those classes for complex-valued sequences (or functions), incorporating the complex modulus in the corresponding inequalities. We will omit this context, as in almost all cases the results we study in the present book are valid only for nonnegative weak monotone sequences or functions.

Here we note that general monotonicity is not alien to number series. Certain results that we will present show that general monotonicity with $\beta_n = |a_n|$, that is, *GMS* for sequences and *GM* for functions (in the complex case) are the classes that inevitably appear in the study of number series in parallel to weak monotonicity classes, specifically in those series where the differences Δa_k of a sequence $\{a_k\}$ are involved, rather than series with general term (equivalent to) a_k, where only the magnitude of the given sequence plays a role.

What is of importance is that weak monotonicity and general monotonicity with $\mathcal{GMS}$ in general are not comparable in total scope. However, the class *GMS* (recall that this is the one with $\beta_n = |a_n|$) can be compared with *WMS*. More precisely, the weak monotone class is wider than the *GMS* class (by Proposition 1.9). The latter comes into play when the former turns out to be too wide but not in the cases where monotonicity cannot be relaxed at all. Similarly, the class of *GM* functions can be compared with that of *WM* functions.

A few properties of *WMS* are in order.

Proposition 1.31 *Let* $\{a_k\}, \{b_k\} \in WMS$.

(i) *If there exists* $n \in \mathbb{N}$ *such that* $a_n = 0$, *then* $a_k = 0$ *for all* $k \geq n$.

(ii) *For every* $m < n \in \mathbb{N} \cup \{0\}$ *and every* $2^m \le k \le 2^{m+1}$ *and* $2^n \le \ell \le 2^{n+1}$ *with* $k \le \ell$,

$$a_\ell \le C a_{2^n} \le \cdots \le C^{n-m} a_{2^{m+1}} \le C^{n-m+1} a_k.$$

In particular,

$$a_\ell \le C^{\log_2 \frac{\ell}{k}+2} a_k.$$

(iii) *The sequences* $\{a_k + b_k\}$ *and* $\{a_k b_k\}$ *are WMS.*
(iv) *The sequences* $\{\min\{a_k, b_k\}\}$ *and* $\{\max\{a_k, b_k\}\}$ *are WMS.*

Proof Property (i) follows trivially from the *WMS* condition, while property (ii) is simply an iteration. Property (iii) also follows readily from the *WMS* condition.

Let us now prove (iv). Fix $n \in \mathbb{N}$ and note that for any $n \le k \le 2n$,

$$\max\{a_k, b_k\} \le \max\{Ca_n, Cb_n\} = C \max\{a_n, b_n\}.$$

Similarly,

$$\min\{a_k, b_k\} \le \min\{Ca_n, Cb_n\} = C \min\{a_n, b_n\},$$

as desired. □

Finally, we show that *WMS* cannot increase to infinity faster than any power. Moreover, it is interesting that, when restricting our attention to number series, the classes *WMS* and *QMS* are "equivalent" (see Theorem 1.33 below for a precise statement).

Lemma 1.32 *Let* $\{a_n\} \in$ *WMS with constant* $C > 0$. *Then the sequence*

$$\{n^{-\log_2 C} a_n\}$$

is bounded.

Proof For any $k \in \mathbb{N}$ and $\ell \in \mathbb{N}$ such that $2^{\ell-1} \le k < 2^\ell$, we have

$$a_k \le C a_{2^{\ell-1}} \le \cdots \le C^\ell a_1.$$

Therefore,

$$C^{-\ell} a_k = \left(2^{\log_2 C}\right)^{-\ell} a_k = \left(2^\ell\right)^{-\log_2 C} a_k \le a_1,$$

as desired. □

Theorem 1.33 *Let* $\{a_n\} \in$ *WMS with constant* $C > 1$. *Then there exists* $\{b_n\} \in$ *QMS*$(\log_2 C)$ *such that*

$$a_n \le b_n \le 2C^2 a_n, \qquad \text{for every } n \in \mathbb{N}. \tag{1.30}$$

Conversely, if $C \ge 1$ *and* $\{b_n\} \in$ *QMS*$(\log_2 C)$, *then any sequence* $\{a_n\}$ *satisfying* (1.30) *is a WMS with constant* $2C^3$.

Proof Let $s = \log_2 C$, and assume $s > 0$ (if $s = 0$, then $C = 1$ and the conclusion still holds since in this case $\{a_n\}$ is decreasing). Define

$$c_n = \sup_{k \geq n} k^{-s} a_k.$$

By Lemma 1.32, we have that $\{c_n\}$ is bounded, and, moreover, it is nonincreasing (by definition). Therefore, putting $b_n = n^s c_n$, we have that $\{b_n\} \in QMS(s)$ and $b_n \geq a_n$. Now, for any $n \in \mathbb{N}$, let ℓ be such that $n \in \{2^\ell, 2^\ell + 1, \ldots, 2^{\ell+1}\}$ and $k_0 = k_0(n) \geq n$ be such that $k_0^{-s} a_{k_0} \geq \frac{c_n}{2}$ (note that such a k_0 exists, by the properties of the supremum). Finally, let t be such that $k_0 \in \{2^t, 2^t + 1, \ldots, 2^{t+1}\}$. Applying the *WMS* condition, we have

$$\begin{aligned} b_n = n^s c_n &\leq 2^{s+1} 2^{\ell s} k_0^{-s} a_{k_0} \leq C 2^{s+1} 2^{\ell s} 2^{-ts} a_{2^t} \leq 2C^{t-\ell+1} \left(2^s\right)^{\ell-t+1} a_n \\ &= 2C^{t-\ell+1} C^{\ell-t+1} a_n = 2C^2 a_n, \end{aligned}$$

which completes the first part of the proof.

For the converse, assume that $\{b_n\} \in QMS(\log_2 C)$ with $C \geq 1$. Since (1.30) holds, we have, for any n,

$$\frac{a_n}{n^{\log_2 C}} \leq \frac{b_n}{n^{\log_2 C}} \leq 2C^2 \frac{a_n}{n^{\log_2 C}}.$$

Therefore, using that $\{n^{-\log_2 C} b_n\}$ is nonincreasing, we have, for any fixed n and $k \in [n, 2n]$,

$$\frac{a_k}{k^{\log_2 C}} \leq \frac{b_k}{k^{\log_2 C}} \leq \frac{b_n}{n^{\log_2 C}} \leq 2C^2 \frac{a_n}{n^{\log_2 C}},$$

from which we deduce

$$a_k \leq 2C^2 \frac{k^{\log_2 C}}{n^{\log_2 C}} a_n \leq 2C^3 a_n,$$

as desired. □

Remark 1.34 Despite the result in Theorem 1.33, the class *WMS* is not literally equivalent to the class *QMS*. Indeed, if this were the case, then one would obtain *QMS* = *GMS* = *WMS*, since by the results in Section 1.2 (and from condition (1.29)) one could deduce

$$QMS \subset QMS^* \subset GMS \subset WMS.$$

However, we have found in Section 1.2 an example showing that $QMS^* \subsetneq GMS$, and on the other hand it is easy to produce an example of a sequence belonging to $WMS \backslash GMS$ (for instance, any oscillating sequence satisfying (1.29), such as $\{\frac{2+\sin n}{n}\}$). Hence, $WMS \backslash QMS \neq \emptyset$.

Remark 1.35 It is also worth noting that the classes $\overline{GMS}$ and WMS are not comparable (and of course the same applies to $\overline{GM}$ and WM). In order to see that $\overline{GMS}\setminus WMS \neq \emptyset$, we proceed identically as in the proof that $\overline{GMS}\setminus GMS \neq \emptyset$. Let $\{a_k\} \in GMS$ be a positive sequence, $m \in \mathbb{N}$ be large enough, and define

$$\bar{a}_n = \begin{cases} a_n, & \text{if } n \neq m, \\ 0, & \text{if } n = m. \end{cases}$$

It is clear that since $\{a_k\} \in GMS$, one has $\{\overline{a}_k\} \in \overline{GMS}$. However, $\{\overline{a}_k\} \notin WMS$ (since if a weak monotone sequence has a zero term, all the following terms must also be zero). On the other hand, it is easy to check that an oscillating sequence such as $\left\{\frac{2+\sin n}{n}\right\}$ belongs to WMS but not $\overline{GMS}$.

We could define even more general classes of sequences than WMS as follows. For the sake of brevity, we include only their definitions and show how they relate to the WMS class given in Definition 1.29, since the latter will be the main object of study in this book.

Definition 1.36 Let $m = \{m_n\}_{n=1}^{\infty}$ be a nondecreasing positive sequence with $m_1 \geq 1$. We call a nonnegative sequence $\{a_k\}$ *weak monotone with respect to* m, written as $\{a_k\} \in WMS(m)$, if for some absolute constant $C \geq 1$ we have

$$a_k \leq Ca_n, \quad \text{for all} \quad n \leq k \leq n + m_n.$$

Note that if $m_n = n$, or more generally if $m_n \asymp n$, then $WMS(m) = WMS$. It is clear from the definition that if $m_n^{(2)} \leq Cm_n^{(1)}$ for some $C > 0$, then $WMS(m^{(1)}) \subset WMS(m^{(2)})$. More detailed embeddings between these classes in terms of the sequences $\{m_n\}$ are given by the following statement.

Proposition 1.37 *1. If M denotes the class of monotone sequences, then $M \subset WMS \subset WMS(\{1\})$.*

2. *If $m_n^{(2)} \leq Cm_n^{(1)}$ for some $C > 0$, then $WMS(m^{(1)}) \subset WMS(m^{(2)})$.*
3. *If $C_1 m_n^{(1)} \leq m_n^{(2)} \leq C_2 m_n^{(1)}$ for some $C_1, C_2 > 0$, then $WMS(m^{(1)}) = WMS(m^{(2)})$.*
4. *For any sequences $m^{(1)}$ and $m^{(2)}$ satisfying $m_n^{(2)}/m_n^{(1)} \to \infty$, there exists a sequence $\{a_k\}$ depending on both sequences $m^{(1)}$ and $m^{(2)}$ such that $\{a_k\} \in WMS(m^{(1)})\setminus WMS(m^{(2)})$.*

We do not give a proof. Instead, we refer to Notes-16 for an appropriate reference and further comments.

Another possible generalization of the WMS class is the following.

Definition 1.38 Let $\{a_k\}$ be a nonnegative sequence. We say that $\{a_k\} \in \overline{WMS}$ if for some absolute constants $c, C > 1$ we have

$$a_k \leq C \sum_{n=\lfloor k/c \rfloor}^{\lfloor ck \rfloor} \frac{a_n}{n}.$$

It is clear that $WMS \subset \overline{WMS}$, and that the $\overline{WMS}$ class generalizes the WMS class in the same spirit as the class $\overline{GMS}$ generalizes the class GMS (see Example 1.15 above). Also note that any increasing or decreasing sequence belongs to $\overline{WMS}$, and that if $\{a_n\}$ is strictly positive and $\left\{\frac{1}{a_n}\right\} \in WMS$, then $\{a_n\} \in \overline{WMS}$.

In order to show that $\overline{WMS}\backslash WMS \neq \emptyset$, it suffices to consider a positive sequence $\{a_n\} \in WMS$ and any fixed $m \in \mathbb{N}$; the sequence $\{\overline{a}_n\}$ defined by

$$\bar{a}_n = \begin{cases} a_n, & \text{if } n \neq m, \\ 0, & \text{if } n = m, \end{cases}$$

belongs to $\overline{WMS}$ but not to the class WMS. It is also easy to see that $\overline{GMS} \subsetneq \overline{WMS}$. Indeed, the inclusion follows from (1.16), which is valid for $\overline{GMS}$, while its properness simply follows from the fact that $\emptyset \neq \overline{GMS}\backslash WMS \subset \overline{WMS}$ (see Remark 1.35).

To conclude, we include a diagram (see Figure 1.1) that helps visualizing different relations between the classes of sequences studied in this chapter.

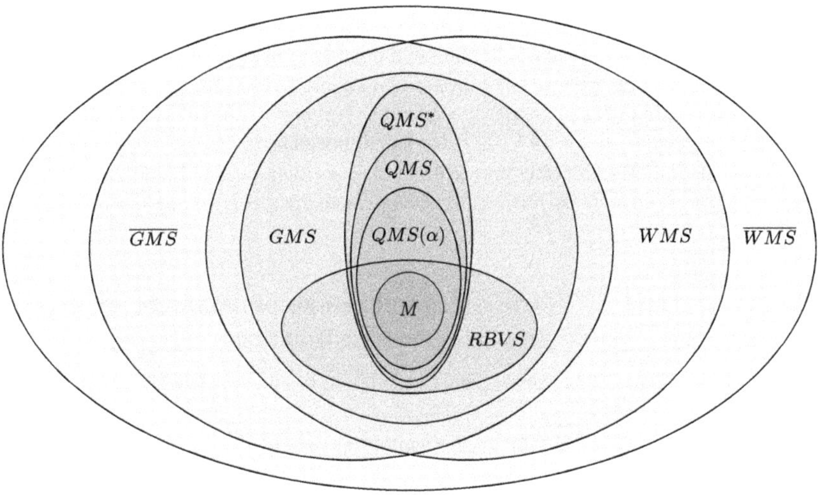

Figure 1.1 Embedding relations between sequence classes. Here M denotes the class of monotone sequences.

1.5 Notes

Widely used aforementioned extensions of monotonicity first appeared in [166], [255] in the case of sequences and [176] in the case of functions. Many generalizations of previously known results in the monotone case, especially in harmonic analysis, have been extended to these new general classes. See also Bernstein's paper [35] of 1949, where almost monotonicity is introduced.

Before proceeding further, let us give the statement of an off-topic result that is worth mentioning.

Theorem 1.39 *Let $\{a_n\}$ and $\{b_n\}$ be positive sequences and assume that the sequence $\{\frac{a_n}{b_n}\}$ is increasing. Then, the sequence*

$$\left\{\frac{\sum_{k=1}^{n} a_k}{\sum_{k=1}^{n} b_k}\right\}_{n\geq 1}$$

is also increasing.

A much stronger statement and further remarks are given in [100, Theorem 5.2].

Section 1.1 (Getting started)

1. Definition 1.3 was first given in [254, 255], as an extension of the *GMS* class (the one corresponding to $\beta_n = |a_n|$, defined in Section 1.2), which was the main tool used in that paper. In the case of functions, Definition 1.4 was first given in [176] (and again the case $\beta(t) = |f(t)|$ reduces to the *GM* case defined in Section 1.2). During the following years, the idea of Definition 1.3 attracted attention of a number of authors who studied problems of harmonic analysis in the framework of general monotonicity, by considering several different choices of β_n (or $\beta(t)$). For the case of sequences, see, e.g., [256, 94, 145, 90], and for the case of functions see [176, 89, 147, 72] (in order of appearance).

2. For the definition and further properties of Stieltjes integrals, see [B-69]. In the classical paper [47] of 1919, elementary properties that constitute the theory of Stieltjes integrals were first given.

Sections 1.2 and 1.3 (Various generalizations of monotone sequences and general monotone functions)

3. The class of quasi-monotone sequences (*QMS*) was defined independently in [236] and [250].

4. For a relatively recent survey on the main properties of the Cantor function, see [82] (and the references therein).

5. The *QMS** class was introduced in [242], under the nomenclature "*O*-regularly varying quasi-monotone sequences" (written *ORVQMS*; for simplicity, we chose a slightly different notation here). It was originally defined for complex sequences, and thus the definition we have presented here is a simplification of the original one to the case of nonnegative sequences.

6. The class of sequences with rest of bounded variation (*RBVS*) was first considered in [228] (see also [163]). Several further definitions of related sequence classes were introduced and studied in [167]. In order to prove that $RBVS \not\subset QMS$ we considered the example from Theorem 1 in [164] (see also [165] for more embedding relations between the class *RBVS* and other sequence classes).

7. The *GMS* class was first introduced in [254], and its potential was illustrated in the paper [255], where it was used to study several different problems concerning trigonometric series. Furthermore, relation (1.13) was also established in [255].

8. The class $\overline{GMS}$ defined in (1.15) was first introduced in [256]. No doubt that this class of general monotone-type sequences has attracted most attention and proved to be useful for

several applications. Perhaps, the reason for this is that, besides the fact that the condition defining such a class is very simple, the $\overline{GMS}$ class is general enough to be a substantial "improvement" of monotone sequences, but not so general to lose the desirable properties shared with nonincreasing sequences (the same reflection is valid for the class of functions $\overline{GM}$, defined in (1.25)). Some problems of harmonic analysis where these classes of sequences and/or functions have been used are the convergence [73, 93, 104, 192, 257], smoothness properties [95, 256], or weighted integrability [41, 42, 69, 74, 75, 109, 175, 176, 267, 268, 269] of Fourier series/integrals, among others.

9. The classes GM and GMS were also heavily used in [80, 81] to show the sharpness of embeddings of function spaces.

10. A different generalization of GMS was introduced by Szal in [246]. It consists of r-differences, rather than the consecutive differences (with $r \in \mathbb{N}$). More precisely, we say that $\{a_k\} \in GMS_r$ if there exists $C > 0$ such that

$$\sum_{k=n}^{2n-1} |\Delta_r a_k| := \sum_{k=n}^{2n-1} |a_k - a_{k+r}| \leq C|a_n|,$$

for all $n \in \mathbb{N}$. Applications of GMS_r are similar to those of GMS; see, e.g., [247, 248].

11. The class GMS^∞ was introduced in [94] and applied to the study of several properties of Fourier series.

12. The classes $\overline{GMS}(p)$ were first introduced in [33]. Originally, the definition was given in terms of dyadic blocks, that is, the required property is

$$\sum_{k=2^n}^{2^{n+1}} |\Delta a_k| \leq C\bigg(\sum_{k=2^{n-\nu}}^{2^{n+\nu}} \frac{|a_k|^p}{k} \bigg)^{\frac{1}{p}}, \qquad n \geq \nu,$$

for some positive integer ν and $C > 0$, instead of (1.18). However, both definitions are equivalent (cf. Theorem 1.18).

Such sequences are used in [33] in order to study several properties of functions with Fourier coefficients in the class $\overline{GMS}(p)$.

13. By taking $a = b$ in Proposition 1.14, one can build a bridge to the following general problem considered in [260]. Let $\phi\colon \mathbb{R}^+ \to \mathbb{R}^+$. Then ϕ is said to be a *GM-preserving function*, if for all sequences $a \in GMS$ we have $\phi(a) := \{\phi(a_n)\}_{n=1}^\infty \in GMS$. In order to present necessary and sufficient conditions, the following notions are introduced.

Definition 1.40 A function $\rho\colon \mathbb{R}_+ \to \mathbb{R}$ is said to be *stable* (written $\rho \in S$) if $\rho \in L^1_{\text{loc}}(\mathbb{R}_+)$ and

$$\inf_{x\in\mathbb{R}^+} \int_0^x \rho(t)\,dt > -\infty,$$

and moreover there exist $Q > 0$ and $\lambda > 1$ such that

$$\sup_{s\in[t/\lambda,\lambda t]} |\rho(s)| \leq \frac{Q}{t}\int_{t/\lambda}^t \rho(s)\,ds \qquad \text{for all } t \in \mathbb{R}_+.$$

Definition 1.41 A function $\rho\colon \mathbb{R}_+ \to \mathbb{R}$ is said to be *right stable* (written $\rho \in RS$) if $\rho \in L^1_{\text{loc}}(\mathbb{R}_+)$ and

$$\inf_{x\in\mathbb{R}^+} \int_0^x \rho(t)\,dt > -\infty,$$

and moreover there exist $Q > 0$, $D > 1$, and $\phi_0 > 0$ such that

$$\sup_{s\in[t,Dt]} |\rho(s)| \leq \frac{Q}{t}\left(\phi_0 + \int_0^t \rho(s)\,ds\right) \qquad \text{for all } t \in \mathbb{R}_+.$$

Since we consider functions in L^1, we understand the least upper bound in the L^∞ sense. It is worth mentioning that $S \subseteq RS$. We have the following necessary and sufficient conditions for a function ϕ to be GM-preserving.

Theorem 1.42 *(a) A necessary condition for $\phi\colon \mathbb{R}_+ \to \mathbb{R}_+$ to be a GM-preserving function is that $\rho = \phi' \in RS$, where ϕ' is the derivative in a weak sense.*

(b) A sufficient condition for $\phi\colon \mathbb{R}_+ \to \mathbb{R}_+$ to be a GM-preserving function is that it admits a decomposition

$$\phi(t) = \phi_0 + \int_0^t \rho(s)\,ds,$$

where $\rho\colon \mathbb{R}_+ \to \mathbb{R}$ satisfies $\rho \in S$.

An example of a GM-preserving function is $\phi(t) = t^\alpha$ with $\alpha \geq 0$ (if $\alpha < 0$, then ϕ is not a GM-preserving function any more). We refer to [260] for the proofs of these results and further discussion.

14. Lemmas 1.27 and 1.28 were obtained in [73], and although they can be used to get similar statements for $\overline{GMS}$, the latter had appeared earlier in [95] and, in fact, motivated the author of [73] to obtain the outlined version for $\overline{GM}$. In these two papers, real-valued sequences from the class $\overline{GMS}$ were used in the study of trigonometric series, in completely different problems: in [95], smoothness properties of trigonometric series are studied, while in [73], pointwise and uniform convergence of Fourier-type integrals (Hankel transforms, more precisely) are studied (see also [74] for more applications using a similar approach). A $\overline{GM}$ version of the Abel-Olivier test is also derived there (see Theorems 2.107 and 2.108 below). This shows that deeply understanding general monotone-type sequences and functions has great importance and potential in several areas of harmonic analysis.

Section 1.4 (Weak monotone sequences and functions)

15. The *WMS* class was introduced by Leindler in [162], where he obtained inequalities of Hardy–Littlewood type under such an assumption (in [169], he called such sequences "locally almost monotone"; compare the *WMS* condition with (1.2)). Several of the one-dimensional classical tests for convergence of number series with decreasing terms (Cauchy condensation test, Abel–Olivier's test, etc.) were generalized to the *WMS* class in [178].

16. Refined weak monotone classes $WMS(m)$ (cf. Definition 1.36) were introduced in [258]. In that paper, classical convergence tests, such as the Maclaurin–Cauchy integral test, the Cauchy condensation test, Schlömilch's test, and Abel–Olivier's test, were extended for the classes $WMS(m)$ (see, e.g., Notes-72 and Notes-79 of Chapter 2). It is important to mention that these tests depend on the behavior of the index sequence $\{m_n\}$.

17. Definition 1.38 was given in [178]. However, such a $\overline{WMS}$ class is too general, and classical tests for convergence of number series, such as the Cauchy condensation test, fail in such a broad context (the reader may consider finding a counterexample as an exercise). We also mention Nursultanov's papers [210, 211] considering similar classes to $\overline{WMS}$.

18. One can further extend the definition of $\overline{WMS}$ as follows: for $1 \le p < \infty$, we say that $\{a_k\} \in \overline{WMS}(p)$ if

$$|a_n| \le C\left(\sum_{k=\lfloor n/c \rfloor}^{\lfloor cn \rfloor} \frac{|a_k|^p}{k} \right)^{\frac{1}{p}},$$

and similarly, for functions, $f \in \overline{WM}(p)$ if

$$|f(x)| \le C\left(\int_{x/c}^{cx} \frac{|f(t)|^p}{t}\, dt \right)^{\frac{1}{p}};$$

see [33] and [231].

19. Further generalizations of WMS and corresponding equiconvergence results are given in [282].

2
Number series in dimension one

This chapter has multiple purposes. On the one hand, it is impossible to discuss multivariate number series without appropriate background in one-dimensional number series. Of course, such a background or essential portions of it can be found in almost any textbook on analysis. Also, more specific sources can be referred to, for instance, [B-16, B-54, B-31, B-11]. However, in order to make the presentation self-contained, we give more or less a comprehensive collection of classical tests for convergence or divergence of number series. On the other hand, since our general task is not only to reproduce multidimensional versions of these tests but also to present a solid number of improvements and refinements, we feel a need in giving an overview of recent developments in dimension one.

We now proceed to defining the notion of infinite number series in an accurate way.

Let $\{a_k\}$ be a given sequence of real or complex numbers. We form a new sequence $\{S_n\}$ as

$$S_n := \sum_{k=1}^{n} a_k.$$

Definition 2.1 The ordered pair of sequences $\big(\{a_k\}, \{S_n\}\big)$ is called an infinite series, denoted by $\sum a_k$. The number S_n is called the nth partial sum of the series. The series is said to converge (resp. diverge) if $\{S_n\}$ is convergent to a real or complex number (resp. divergent).

Throughout this chapter, the series will also be denoted as

$$\sum_{k=1}^{\infty} a_k.$$

It should also be noted that sometimes, for convenience, we may start summation with the index $k = 0$.

If the sequence $\{S_n\}$ converges to a finite number A, then A is called the sum of the series, and it is possible to write

$$A = \sum_{k=1}^{\infty} a_k.$$

Thus, for convergent series the symbol $\sum a_k$ is used to denote both the series and its sum.

The case of divergent series is traditionally more ambivalent. Typically, a divergent series is just a series that does not converge to a real or complex number. However, this could mean that the sequence $\{S_n\}$ does not have a limit, or that its limit is $\pm\infty$. We will keep the terminology "divergent" for series that are not convergent. If we wish to be more specific on the divergence of a series, we will use the terminology "divergent to infinity" if $\lim S_n = \pm\infty$, or the self-explanatory "$\{S_n\}$ does not have a limit".

It is worth emphasizing that the sum $\sum a_k$ of a convergent series is not just the sum of all the terms of the sequence $\{a_k\}$, but rather their sum in a *precise* order. Changing that order (i.e., rearranging the sequence $\{a_k\}$) may yield a different convergent series (see, e.g., (2.26)), or even a series diverging to infinity (see Theorem 2.51 below). As we will see, the order of summation is irrelevant if we consider infinite series where the sequence $\{a_k\}$ is nonnegative.

One can think about options of defining $\{S_n\}$ in a different manner. For example,

$$\sum_{k=1}^{n} \left(1 - \frac{k}{n}\right) a_k.$$

This will lead to a different notion for infinite series sometimes denoted by the same symbol $\sum a_k$. Therefore, defining infinite series by means of the ordered pair, where the second sequence is thoroughly assigned prior to the definition, presents more precise and concise information than just $\sum a_k$. By the way, the form how $\{S_n\}$ is defined in the above example is traditionally attributed to one of the methods of *summability* of infinite series, the topic we will not touch upon in this book (see Notes-2 for appropriate references).

A usual starting point in every presentation of number series is the necessary condition for the convergence of $\sum a_k$:

The sequence of the terms $\{a_k\}$ must vanish at infinity, that is,

$$\lim_{k\to\infty} a_k = 0.$$

It is unavoidably accompanied by the observation that this condition is by no means sufficient.

Another useful yet basic result to study the convergence of series is Cauchy's criterion.

Theorem 2.2 *Let $\{b_n\} \subset \mathbb{C}$ be a sequence. Then $\{b_n\}$ converges to a finite limit if and only if*

$$|b_n - b_m| \to 0 \qquad \text{as } m, n \to \infty.$$

Applying Theorem 2.2 to the sequence of partial sums $\{S_n\}$, it follows that

$$\sum_{k=1}^{\infty} a_k \qquad \text{converges if and only if} \qquad \left|\sum_{k=m}^{n} a_k\right| \to 0 \qquad \text{as } m, n \to \infty. \tag{2.1}$$

A powerful tool in the study of one-dimensional series is Abel's transformation, also known as summation by parts due to its resemblance to the method of integration by parts. Given two sequences $\{a_k\}, \{b_k\}$ and $m < n \in \mathbb{N}$, one has

$$\sum_{k=m}^{n} a_k b_k = \sum_{k=m}^{n-1} (a_k - a_{k+1}) B_m^k + a_n B_m^n, \quad \text{where } B_m^k = b_m + \cdots + b_k. \tag{2.2}$$

This relation follows from the simple observation that for every $k > m$, $b_k = B_m^k - B_m^{k-1}$. Thus, since $B_m^m = b_m$,

$$\begin{aligned} \sum_{k=m}^{n} a_k b_k &= a_m b_m + \sum_{k=m+1}^{n} a_k (B_m^k - B_m^{k-1}) \\ &= a_m b_m - a_m B_m^m + \sum_{k=m}^{n-1} (a_k - a_{k+1}) B_m^k + a_n B_m^n \\ &= \sum_{k=m}^{n-1} (a_k - a_{k+1}) B_m^k + a_n B_m^n. \end{aligned}$$

Let us introduce the corresponding notions of convergence for integrals. Given a locally integrable function $f\colon (a, \infty) \to \mathbb{C}$ and $b \in [a, \infty)$, we define the improper integral

$$\int_b^\infty f(t)\, dt = \lim_{c \to \infty} \int_b^c f(t)\, dt. \tag{2.3}$$

It is clear that one can take $a = -\infty$ and extend the integration to $\mathbb{R}$ by also taking the limit with respect to $b \to -\infty$ on the right side of (2.3). If

$$\lim_{c \to \infty} \int_b^c f(t)\, dt = A \in \mathbb{C},$$

we say that the integral $\int_b^\infty f(t)\, dt$ *converges*. Otherwise we say that it *diverges*.

It is worth mentioning that throughout the book the word *criterion* is reserved for statements that characterize a certain property (e.g., Cauchy's criterion). On the other hand, when studying the convergence character of a series, a *test* for convergence may be a statement that gives necessary/sufficient conditions for the convergence/divergence of a series (e.g., the ratio test), or also a statement that characterizes the convergence character of the given series (e.g., the Cauchy condensation test).

2.1 Series with positive terms

This section is devoted to presenting useful general facts and convergence tests for series with positive (and occasionally nonnegative) terms, which we will call *positive series* (analogously we define *nonnegative series* as those with nonnegative terms). However, we also include corresponding generalizations of the classical tests for positive series extended to series whose terms have arbitrary signs, or are even complex-valued.

We split the tests for positive series into two groups: one containing general convergence tests and the other in which certain monotonicity assumptions are posed on the sequence of terms, some of them classical as well. The former will be considered in this section, while the second one will be the subject of Section 2.4.

2.1.1 General facts around positive series

First of all, positive series are always "convergent" in a way: either to a finite limit or to infinity. This is because the sequence of partial sums is nondecreasing (while for series with arbitrary terms, a limit of its partial sums may not exist).

Theorem 2.3 *Let $\sum a_k$ be a positive (nonnegative) series.*

(i) *Then it either converges to a positive (nonnegative) number or diverges to infinity.*
(ii) *Any rearrangement of $\sum a_k$ converges or diverges as $\sum a_k$ does.*

We omit the full proof of this basic result due to its simplicity and give an idea of the proof instead. For the first part, one should use that a nondecreasing sequence converges if and only if it is bounded (see the corresponding multidimensional statement in Theorem 7.12 below), while the second part is rather trivial.

An important fact that will be used repeatedly concerns the convergence character of one of the most important families of number series, the so-called p-series. Such series and their variants will be discussed in detail in Subsection 3.1.1 below. For the moment, we will need to use that

$$\text{the series } \sum_{n=1}^{\infty} \frac{1}{n^p} \text{ converges if and only if } p > 1. \tag{2.4}$$

This can be easily proved, for instance, using Cauchy's condensation test (see Corollary 2.92 below).

The following result is a generalization of the so-called Lobachevskii's test (which was originally established for monotone sequences; see Theorem 2.96 below).

Theorem 2.4 *Let $\{a_n\}$ be a nonnegative null sequence such that $a_1 = 1$ and $a_n < 1$ for $n > 1$. For $m \in \mathbb{N} \cup \{0\}$, let*

$$p_m = \max\{n \in \mathbb{N} : a_n \geq 2^{-m}\}.$$

Then

$$\sum_{n=1}^{\infty} a_n \leq \sum_{m=0}^{\infty} \frac{p_m}{2^m}.$$

In particular, the convergence of the series $\sum_{m=0}^{\infty} p_m 2^{-m}$ implies the convergence of the series $\sum a_n$.

Proof For $m \in \mathbb{N} \cup \{0\}$, define L_m as the number of elements a_n containing 2^{-m} in their binary decomposition, namely with $\varepsilon_m = 1$ in the representation

$$a_n = \sum_{j=1}^{\infty} \varepsilon_j 2^{-j}, \qquad \text{with } \{\varepsilon_j\} \subset \{0,1\}.$$

Then,

$$\sum_{n=1}^{\infty} a_n = \sum_{m=1}^{\infty} \frac{L_m}{2^m} \tag{2.5}$$

(see Figure 2.1; this equality can be interpreted as a discrete Lebesgue integral). Since one trivially has $L_m \leq p_m$, the desired conclusion follows. □

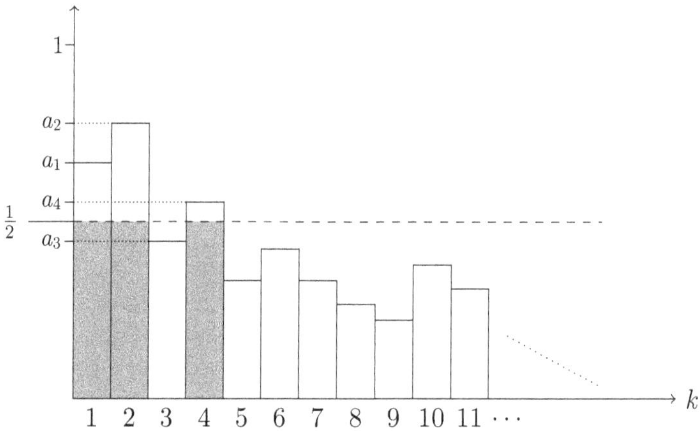

Figure 2.1 Illustration of the terms on the left-hand side series from (2.5). Here, $m = 1$ and $L_1 = 3$. The total contribution of the term $\frac{L_1}{2}$ to the series is the area highlighted in gray color. Note that for counting L_2 graphically one cannot use this picture; we should replace the terms a_1, a_2, and a_4 contributing to L_1 by $a_j - \frac{1}{2}$, $j = 1, 2, 4$, and then count how many terms lie above the horizontal line at height $\frac{1}{4}$.

Remark 2.5 First of all, note that by definition the sequence $\{p_m\}$ in Theorem 2.4 satisfies

$$p_m = \max\{n \in \mathbb{N} : a_n \geq 2^{-m}\} = \max\{n \in \mathbb{N} : a_n \geq 2^{-m} > a_{n+1}\}.$$

One cannot expect to obtain a lower bound in Theorem 2.4 that is of the form

$$C \sum_{m=0}^{\infty} \frac{q_m}{2^m} \leq \sum_{n=1}^{\infty} a_n, \tag{2.6}$$

where

$$q_m = \min\{n \in \mathbb{N} : a_n \geq 2^{-m}, a_{n+1} < 2^{-m}\},$$

with some appropriate constant $C > 0$ (independent of the sequence $\{a_n\}$). Indeed, for $M \in \mathbb{N}$, $M > 2$, consider the example

$$a_n = \begin{cases} \frac{1}{2^{n+1}}, & \text{if } n \neq M, \\ \frac{1}{2}, & \text{if } n = M. \end{cases}$$

Clearly $q_1 = M$, and the inequality

$$\frac{M}{2} = \frac{q_1}{2} \leq \sum_{m=1}^{\infty} \frac{q_m}{2^m} \leq \sum_{n=1}^{\infty} a_n < 1$$

is not possible. However, under the additional assumption that $\{a_n\}$ is decreasing, one has $q_m = p_m$ for all m, and inequality (2.6) holds with $C = \frac{1}{2}$ (see Theorem 2.96).

We now turn our attention toward averaged summability. First, we state the celebrated Cesàro summability theorem. Instead of proving it, we give a full proof of the more general Stolz–Cesàro summability theorem.

Theorem 2.6 *If a sequence $\{x_k\} \subset \mathbb{C}$ converges to a value $c \in \mathbb{C}$, then the sequence of arithmetic means,*

$$\sigma_n = \frac{x_1 + \cdots + x_n}{n},$$

also converges to c.

The Stolz–Cesàro summability theorem reads as follows.

Theorem 2.7 *Let $\{a_k\} \subset \mathbb{R}$ and let $\{b_k\}$ be a positive sequence such that $\sum b_k = \infty$. Then, the sequence defined by*

$$c_n = \frac{a_1 + \cdots + a_n}{b_1 + \cdots + b_n}$$

satisfies

$$\liminf_{n\to\infty} \frac{a_n}{b_n} \leq \liminf_{n\to\infty} c_n \leq \limsup_{n\to\infty} c_n \leq \limsup_{n\to\infty} \frac{a_n}{b_n}.$$

In particular, if $\{\frac{a_n}{b_n}\}$ converges to $c \in \mathbb{R}$, then $\{c_n\}$ also converges to c.

Note that Theorem 2.6 (with $\{a_k\} \subset \mathbb{R}$) follows from Theorem 2.7 with $b_k = 1$ for all k. Also, the extension of Theorem 2.6 to complex sequences is immediate by inspecting the real and imaginary parts of the sequence separately.

Proof of Theorem 2.7 We assume that $-\infty < \liminf \frac{a_n}{b_n}$ and $\limsup \frac{a_n}{b_n} < \infty$, since the corresponding inequality is trivial if any of such limits is infinite. Let $L := \limsup \frac{a_n}{b_n}$. By definition, for every $\varepsilon > 0$, there exists $n_0 \in \mathbb{N}$ such that

$$\frac{a_n}{b_n} \leq L + \varepsilon, \qquad \text{for every } n \geq n_0. \tag{2.7}$$

Let us denote $A_n = \sum_{k=1}^n a_k$ and $B_n = \sum_{k=1}^n b_k$. Condition (2.7) implies that for $n \geq n_0$,

$$a_1 + \cdots + a_n \leq a_1 + \cdots + a_{n_0} + (L+\varepsilon)(b_{n_0+1} + \cdots + b_n),$$

or in other words,

$$A_n \leq A_{n_0} + (L+\varepsilon)(B_n - B_{n_0}).$$

Equivalently,

$$c_n = \frac{A_n}{B_n} \leq \frac{A_{n_0}}{B_n} + (L+\varepsilon)\frac{B_n - B_{n_0}}{B_n} = (L+\varepsilon) + \frac{A_{n_0} - (L+\varepsilon)B_{n_0}}{B_n}.$$

Thus, the divergence of $\sum b_n$ implies that

$$\limsup_{n\to\infty} c_n \leq (L+\varepsilon) + \limsup_{n\to\infty} \frac{A_{n_0} - (L+\varepsilon)B_{n_0}}{B_n} = L+\varepsilon,$$

and since this holds for arbitrary $\varepsilon > 0$, we get $\limsup c_n \leq L$. The inequality concerning lim inf is proved analogously. □

The next result concerns the divergence rate of certain modified averages of the reciprocals of the terms of a convergent series.

Theorem 2.8 *In order that for every positive convergent series $\sum a_k$, the relation*

$$\lim_{n\to\infty} \frac{1}{n^\gamma} \sum_{k=1}^n \frac{1}{a_k} = \infty \tag{2.8}$$

holds, it is necessary and sufficient that $\gamma \leq 2$.

Proof To show the sufficiency, it is enough to consider $\gamma = 2$. Since $\sum a_k$ converges, for any $\varepsilon > 0$ there exists $N \in \mathbb{N}$ such that

$$\sum_{k=\lfloor n/2\rfloor+1}^n a_k < \varepsilon, \qquad n > N.$$

Using the well-known arithmetic-harmonic mean inequality

$$n\left(\sum_{k=1}^n \frac{1}{x_k}\right)^{-1} \leq \frac{1}{n}\sum_{k=1}^n x_k, \qquad x_k > 0$$

(see, e.g., [B-43]), we obtain

$$\frac{1}{n^2}\sum_{k=1}^n \frac{1}{a_k} \geq \frac{1}{n^2}\sum_{k=\lfloor n/2\rfloor+1}^n \frac{1}{a_k} \geq \frac{1}{n^2}\frac{(n-\lfloor n/2\rfloor)^2}{\sum_{k=\lfloor n/2\rfloor+1}^n a_k} \geq \frac{1}{4\varepsilon},$$

and the conclusion follows.

To verify the necessity, it suffices to find a positive convergent series for which (2.8) is not valid, provided $\gamma > 2$. We take

$$a_k = \frac{1}{k\log^2(k+1)}.$$

Then the series $\sum a_k$ converges, and

$$\frac{1}{n^\gamma}\sum_{k=1}^{n}\frac{1}{a_k} = \frac{1}{n^\gamma}\sum_{k=1}^{n} k\log^2(k+1) \asymp n^{2-\gamma}\log^2 n \to \infty,$$

if and only if $\gamma \le 2$, which completes the proof. □

Remark 2.9 Note that it is not possible to replace $\frac{1}{n^2}$ in (2.8) by $\frac{1}{n^2\log^\gamma n}$ with $\gamma > 1$. Indeed, taking

$$a_k = \frac{1}{k\log^\gamma(k+1)},$$

we have $\sum a_k < \infty$ (see, for example, Subsection 3.1.1). On the other hand,

$$\lim_{n\to\infty}\frac{1}{n^2\log^\gamma n}\sum_{k=1}^{n}\frac{1}{a_k} \le \lim_{n\to\infty}\frac{1}{n^2\log^\gamma n}n^2\log^\gamma(n+1) = 1.$$

2.1.2 Lack of limiting series

A very important observation is that for positive series neither a "biggest" convergent series nor a "smallest" divergent series exist. More precisely, if $\sum a_k$ is a convergent positive series, then there exists a positive sequence $\{b_k\}$ satisfying $b_k/a_k \to \infty$ such that $\sum b_k$ converges, and if $\sum a_k$ is a divergent positive series, then there exists a positive sequence $\{c_k\}$ satisfying $c_k/a_k \to 0$ such that $\sum c_k$ also diverges.

We begin with a theorem showing the existence of the sequences $\{b_k\}$ and $\{c_k\}$ in question.

Theorem 2.10 (i) *For any positive sequence $\{a_k\}$ such that $\sum a_k$ converges, there exists a positive decreasing sequence $\{b_k\}$ such that the series $\sum b_k$ diverges and*

$$\liminf_{k\to\infty}\frac{b_k}{a_k} = 0.$$

(ii) *For any positive sequence $\{b_k\}$ vanishing at infinity and such that the series $\sum b_k$ diverges, there exists a positive sequence $\{a_k\}$ such that $\sum a_k$ converges and*

$$\limsup_{k\to\infty}\frac{a_k}{b_k} = \infty.$$

Proof For the first part, since a_k vanishes at infinity, it is possible to find a decreasing subsequence $\{a_{n_k}\}_{k=1}^{\infty}$. In addition, the sequence of indices $\{n_k\}$ can be chosen so that $n_1 = 1$ and

$$(n_{k+1} - n_k)a_{n_k} > 1.$$

We now define

$$b_m = \frac{1}{k} a_{n_k}, \qquad n_k \le m < n_{k+1}, \quad k \in \mathbb{N}.$$

The series $\sum b_m$ diverges, since

$$\sum_{m=1}^{\infty} b_m = \sum_{k=1}^{\infty} \frac{1}{k} \sum_{m=n_k}^{n_{k+1}-1} a_{n_k} = \sum_{k=1}^{\infty} \frac{1}{k}(n_{k+1} - n_k) a_{n_k} > \sum_{k=1}^{\infty} \frac{1}{k} = \infty.$$

Moreover, for every $k \in \mathbb{N}$, we have

$$\frac{b_{n_k}}{a_{n_k}} = \frac{1}{k} \to 0,$$

and therefore the conclusion holds. To prove the second part, let $\{n_k\}$ be a sequence of indices such that $b_{n_k} < \frac{1}{k^3}$ (such a sequence exists since $b_n \to 0$) and consider a convergent positive series $\sum d_k$. We then define

$$a_m = \begin{cases} \dfrac{1}{k^2}, & \text{if } m = n_k,\ k \in \mathbb{N}, \\ d_m, & \text{otherwise.} \end{cases}$$

We have

$$\limsup_{m\to\infty} \frac{a_m}{b_m} \ge \lim_{k\to\infty} \frac{a_{n_k}}{b_{n_k}} > \lim_{k\to\infty} k = \infty,$$

and, moreover,

$$\sum_{m=1}^{\infty} a_m < \sum_{k=1}^{\infty} \frac{1}{k^2} + \sum_{m=1}^{\infty} d_m < \infty.$$ □

In what follows, we present a number of results that show different ways of constructing explicitly the aforementioned sequences $\{b_k\}$ and $\{c_k\}$.

Theorem 2.11 *Let $\sum a_k$ be a positive series.*

(i) *If $\sum a_k$ converges, then the series*

$$\sum_{n=1}^{\infty} \frac{a_n}{\left(\sum_{k=n}^{\infty} a_k\right)^{\frac{1}{2}}}$$

does as well.

(ii) *If $\sum a_k$ diverges, then the series*

$$\sum_{n=1}^{\infty} \frac{a_n}{\sum_{k=1}^{n} a_k}$$

does as well.

The proof of this theorem is omitted, since it is a classical result that can be found in several number series textbooks. Instead, we prove more general results (namely Theorems 2.12 and 2.13) that allow one to obtain Theorem 2.11 as a corollary.

Theorem 2.12 (Dini's test) *Let $\sum a_n$ be a divergent positive series and let S_n be the partial sums of $\sum a_n$, that is, $S_n = \sum_{k=1}^{n} a_k$. Then, the series*

$$\sum_{n=1}^{\infty} \frac{a_n}{S_n^{\alpha}}$$

diverges if $\alpha \leq 1$ and converges if $\alpha > 1$.

Proof We begin with the divergence instance by observing first that one can restrict oneself to the case $\alpha = 1$. This follows from the inequality $\dfrac{a_n}{S_n} \leq C\dfrac{a_n}{S_n^{\alpha}}$ for n large enough, since $S_n \to \infty$. We now have

$$\sum_{n=k}^{m} \frac{a_n}{S_n} = \sum_{n=k}^{m} \frac{S_n - S_{n-1}}{S_n} \geq \sum_{n=k}^{m} \frac{S_n - S_{n-1}}{S_m} = \frac{S_m - S_{k-1}}{S_m} = 1 - \frac{S_{k-1}}{S_m},$$

which tends to 1 as $m \to \infty$. This yields $\sum_{n=k}^{\infty} \frac{a_n}{S_n} \geq 1$ for every $k \in \mathbb{N}$, and the divergence follows for $\alpha = 1$ and thus for all $\alpha \leq 1$.

On the other hand, if $\alpha > 1$ and $n \geq 2$, we have

$$\frac{a_n}{S_n^{\alpha}} \leq \int_{S_{n-1}}^{S_n} \frac{dt}{t^{\alpha}}.$$

Therefore,

$$\sum_{n=2}^{\infty} \frac{a_n}{S_n^{\alpha}} \leq \int_{S_1}^{\infty} \frac{dt}{t^{\alpha}} < \infty,$$

which concludes the proof. □

A corresponding result holds for convergent series, where partial sums are replaced by "tails", that is, by the sums of the form

$$T_n := \sum_{k=n}^{\infty} a_k.$$

Theorem 2.13 *Let $\sum a_n$ be a convergent positive series. Then, the series*

$$\sum_{n=1}^{\infty} \frac{a_n}{T_n^{\alpha}}$$

converges if $\alpha < 1$ and diverges if $\alpha \geq 1$.

Proof Let us show the divergence case first. It is clear that it suffices to prove the statement for $\alpha = 1$, since if $\alpha > 1$ one has $\dfrac{a_n}{T_n} \leq C \cdot \dfrac{a_n}{T_n^{\alpha}}$ for all n, for some constant C independent of n. This constant comes from the inequality $T_n \leq C^{\frac{1}{\alpha-1}}$ equivalent to that in question. Writing $a_n = T_n - T_{n+1}$, we have for any $k \in \mathbb{N}$,

$$\sum_{n=k}^{m} \frac{a_n}{T_n} = \sum_{n=k}^{m} \frac{T_n - T_{n+1}}{T_n} > \sum_{n=k}^{m} \frac{T_n - T_{n+1}}{T_k} = \frac{T_k - T_{m+1}}{T_k} = 1 - \frac{T_{m+1}}{T_k} \to 1,$$

as $m \to \infty$. Hence, for sufficiently large k we have $\sum_{n=k}^{\infty} \frac{a_n}{T_n} \geq 1$, and the divergence of the corresponding series follows.

In order to show the convergence of the series $\sum_{n=1}^{\infty} \frac{a_n}{T_n^{\alpha}}$ for $\alpha < 1$, it suffices to treat the case $0 < \alpha < 1$. Note that for $n \in \mathbb{N}$ there holds

$$\frac{a_n}{T_n^{\alpha}} = \frac{T_n - T_{n+1}}{T_n^{\alpha}} \leq \int_{T_{n+1}}^{T_n} \frac{dt}{t^{\alpha}}.$$

Thus, we have

$$\sum_{n=1}^{\infty} \frac{a_n}{T_n^{\alpha}} \leq \int_0^{T_1} \frac{dt}{t^{\alpha}} < \infty,$$

as desired. □

While Theorems 2.12 and 2.13 state convergence criteria for certain series, the following statement provides a rather interesting inequality (the corresponding convergence test is trivial). The given inequality has applications in the theory of complex interpolation (see Notes-7 for an appropriate reference).

Proposition 2.14 *Let $\sum a_n$ be a convergent positive series and $0 < \alpha < 1$. Then,*

$$\sum_{k=1}^{\infty} a_k \left(\sum_{m=1}^{k} a_m \right)^{-\alpha} \leq \frac{1}{1-\alpha} \left(\sum_{k=1}^{\infty} a_k \right)^{1-\alpha}.$$

Proof Let $A = \sum_{k=1}^{\infty} a_k$. Then,

$$a_k \left(\sum_{m=1}^{k} a_m \right)^{-\alpha} = \frac{a_k}{S_k^{\alpha}} \leq \int_{S_{k-1}}^{S_k} \frac{dt}{t^{\alpha}}.$$

Summing up on k and taking into account that $S_k \to A$ as $k \to \infty$, we get

$$\sum_{k=1}^{\infty} a_k \left(\sum_{m=1}^{k} a_m \right)^{-\alpha} \leq S_1^{1-\alpha} + \sum_{k=2}^{\infty} \int_{S_{k-1}}^{S_k} \frac{dt}{t^{\alpha}} \leq \int_0^A \frac{dt}{t^{\alpha}} = \frac{A^{1-\alpha}}{1-\alpha},$$

which proves the claim. □

Another result in the spirit of Theorems 2.12 and 2.13 is the following, which, given a convergent series, provides yet more ways of constructing new convergent series with larger terms. The most interesting feature of this approach is that it depends on universal sequences of powers satisfying certain conditions.

Theorem 2.15 *Let $\{a_k\}$ be a positive sequence and $\{\gamma_k\}$ be a real-valued sequence such that $|\gamma_k| \lesssim \frac{1}{\log k}$. Then the series $\sum a_k$ and $\sum a_k^{1-\gamma_k}$ converge or diverge simultaneously.*

Proof Since $\gamma_k \to 0$, we can assume $|\gamma_k| < \frac{1}{3}$ for all k. Indeed, applying that assertion, when proven, to the sequences $\widetilde{a}_k = a_k^{1-\gamma_k}$ and $\widetilde{\gamma}_k = -\frac{\gamma_k}{1-\gamma_k}$, we see that the convergence of $\sum \widetilde{a}_k = \sum a_k^{1-\gamma_k}$ implies the convergence of $\sum \widetilde{a}_k^{1-\widetilde{\gamma}_k} = \sum a_k$.

Further, we can also assume without loss of generality that $a_k \le 1$ and $\gamma_k \ge 0$ for all k. Indeed, replacing a negative γ_k by zero would greaten the series $\sum a_k^{1-\gamma_k}$. Set $\mathcal{N} = \{k : a_k \le \frac{1}{k^2}\}$. For $k \in \mathcal{N}$, there holds

$$a_k^{1-\gamma_k} \le k^{-2(1-\gamma_k)} \le k^{-\frac{4}{3}}.$$

Hence,

$$\sum_{k\in\mathcal{N}} a_k^{1-\gamma_k} \le \sum_{k\in\mathcal{N}} \frac{1}{k^{\frac{4}{3}}} < \infty.$$

On the other hand, if $k \in \mathbb{N}\backslash\mathcal{N}$, then

$$a_k^{-\gamma_k} \le k^{2\gamma_k} = \left(e^{\log k}\right)^{2\gamma_k} \lesssim 1.$$

Therefore,

$$\sum_{k\notin\mathcal{N}} a_k^{1-\gamma_k} \lesssim \sum_{k\notin\mathcal{N}} a_k < \infty,$$

as desired. □

Theorem 2.16 *Let $\sum a_k$ be a convergent positive series. Then, the series*

$$\sum_{n=2}^{\infty} \frac{a_n}{\log n} \log \frac{1}{a_n}$$

converges.

Proof We may assume that $a_n < 1$ for all n, which yields $\log \frac{1}{a_n} > 0$. We split the series in a similar fashion as in the proof of Theorem 2.15, namely

$$\sum_{n=2}^{\infty} \frac{a_n}{\log n} \log \frac{1}{a_n} = \sum_{n:a_n>n^{-2}} \frac{a_n}{\log n} \log \frac{1}{a_n} + \sum_{n:a_n\le n^{-2}} \frac{a_n}{\log n} \log \frac{1}{a_n}.$$

We note that if $a_n > n^{-2}$, then $2\log n > \log a_n^{-1}$. Hence,

$$\sum_{n:a_n>n^{-2}} \frac{a_n}{\log n} \log \frac{1}{a_n} < 2\sum_{n=1}^{\infty} a_n.$$

On the other hand, since $a_n \to 0$, one has in particular $a_n^{\frac{1}{3}} \log \frac{1}{a_n} \to 0$ (in this case the sequence $a_n^{\frac{1}{3}} \log \frac{1}{a_n}$ is bounded by, say, $C > 0$). Therefore,

$$\sum_{n:a_n\le n^{-2}} \frac{a_n}{\log n} \log \frac{1}{a_n} = \sum_{n:a_n\le n^{-2}} \frac{a_n^{\frac{2}{3}}}{\log n} a_n^{\frac{1}{3}} \log \frac{1}{a_n} \le \sum_{n=1}^{\infty} \frac{C}{n^{\frac{4}{3}} \log n} < \infty,$$

as desired. □

To conclude, we now show that given a positive decreasing sequence $\{a_k\}$ that forms a divergent series, one may always find another positive decreasing sequence $\{b_k\}$, also forming a divergent series, for which the series of the product $\sum a_n b_n$ converges (without the monotonicity assumption, the claim would be trivial).

Proposition 2.17 *For any nonincreasing positive sequence $\{a_k\}$ vanishing at infinity such that $\sum a_k$ diverges, we can find another nonincreasing positive sequence $\{b_k\}$ that also vanishes at infinity, such that $\sum b_k$ diverges, and*

$$\sum_{k=1}^{\infty} a_k b_k < \infty.$$

Proof We may assume, without loss of generality, that $a_n \leq 1$ for all n. Since $\{a_n\}$ is decreasing, we can find an increasing sequence of natural numbers $\{n_k\}$ with $n_1 = 1$ such that

$$a_n \leq \frac{1}{k^2} \qquad \text{for all } n_k \leq n < n_{k+1}, \quad k \in \mathbb{N},$$

where, moreover, the sequence $\{n_k\}$ satisfies $n_{k+1} - n_k \leq n_{k+2} - n_{k+1}$. If we define

$$b_n = \frac{1}{n_{k+1} - n_k}, \qquad \text{for } n_k \leq n < n_{k+1}, \quad k \in \mathbb{N},$$

it is clear that $\{b_n\}$ is nonincreasing and vanishes at infinity. Moreover,

$$\sum_{n=1}^{\infty} b_n = \sum_{k=1}^{\infty} \sum_{n=n_k}^{n_{k+1}-1} \frac{1}{n_{k+1} - n_k} = \sum_{k=1}^{\infty} 1 = \infty,$$

and

$$\sum_{n=1}^{\infty} a_n b_n = \sum_{k=1}^{\infty} \sum_{n=n_k}^{n_{k+1}-1} \frac{a_n}{n_{k+1} - n_k} \leq \sum_{k=1}^{\infty} \frac{1}{k^2} < \infty,$$

as desired. □

We now proceed to giving a summary of several convergence tests for one-dimensional series.

2.1.3 Comparison-type tests

Comparison of the given positive (or even nonnegative) series, though being a very simple idea, still continues to be one of the most frequent and powerful tools for testing the convergence or divergence of the series. In fact, this basic test is used for deriving many other tests in this section.

Theorem 2.18 (Classical comparison test) *Let $\sum a_k$ and $\sum b_k$ be nonnegative series, and suppose that $a_k \leq b_k$ holds for sufficiently large k.*

(i) *If $\sum b_k$ converges, then $\sum a_k$ converges as well.*
(ii) *If $\sum a_k$ diverges, then $\sum b_k$ diverges as well.*

An obvious but more general comparison test for series with real-valued terms is the following:

Theorem 2.19 *Let $\sum a_n$, $\sum b_n$, and $\sum c_n$ be series with real-valued terms such that $a_n \leq b_n \leq c_n$ for sufficiently large n.*

(i) *If the series $\sum a_n$ and $\sum c_n$ converge, then $\sum b_n$ does as well.*
(ii) *If the series $\sum c_n$ diverges to $-\infty$, then $\sum b_n$ does as well.*
(iii) *If the series $\sum a_n$ diverges to $+\infty$, then $\sum b_n$ does as well.*

The proof of this result is rather simple and therefore omitted.

We continue with the well-known asymptotic version of the comparison test.

Theorem 2.20 *Let $\{a_k\}$ and $\{b_k\}$ be positive sequences. If*

$$\lim_{n\to\infty} \frac{a_n}{b_n} = 1, \tag{2.9}$$

then the series $\sum a_k$ and $\sum b_k$ converge or diverge simultaneously.

Proof If (2.9) holds, there exists $n_0 \in \mathbb{N}$ such that if $n \geq n_0$ then

$$\frac{1}{2}a_n \leq b_n \leq 2a_n.$$

The conclusion then follows by the classical comparison test (Theorem 2.18). □

Notice that in Theorem 2.20 the positivity assumption on the sequences $\{a_k\}$ and $\{b_k\}$ cannot be dropped. Indeed, taking

$$a_n = \frac{(-1)^n}{\sqrt{n}} + \frac{1}{n}, \qquad b_n = \frac{(-1)^n}{\sqrt{n}},$$

we have $\lim_{n\to\infty} \frac{a_n}{b_n} = 1$, and the series $\sum a_k$ diverges while the series $\sum b_k$ converges. However, one can still deduce convergence properties for series of general terms, which is the main point of the following result. Before proceeding, we introduce further terminology.

A series $\sum a_k$ is said to be *absolutely convergent* if $\sum |a_k|$ converges, and *conditionally convergent* if $\sum a_k$ converges while $\sum |a_k|$ diverges. Although the following two results involve conditional and/or absolute convergence, which we will not discuss thoroughly until Section 2.2, it is natural to include them in this section, as the conditional convergence of the involved series appears just as a consequence of some of the hypotheses, while the main feature of these statements is the comparison-type assumption.

Theorem 2.21 *Let $\{a_n\}$ and $\{b_n\}$ be complex-valued sequences such that*

$$\lim_{n\to\infty} \frac{a_n}{b_n} = 1.$$

If the series $\sum b_k$ *converges and the series* $\sum a_k$ *diverges, then* $\sum b_k$ *converges conditionally. Moreover, if we write* $a_n = b_n + c_n$*, then* $\sum c_n$ *diverges and*

$$\lim_{n\to\infty} \frac{c_n}{b_n} = 0.$$

Proof We will prove the divergence of $\sum |b_k|$ by contradiction. Assume that $\sum b_k$ converges absolutely. Since $\lim a_n/b_n = 1$, we also have

$$\lim_{n\to\infty} \frac{|a_n|}{|b_n|} = 1.$$

By Theorem 2.18, $\sum a_k$ converges absolutely, which implies the convergence of $\sum a_k$, a contradiction. For each n, define now $c_n := a_n - b_n$. Then the convergence of $\sum c_n$ implies the convergence of $\sum a_k$, and therefore $\sum c_n$ must diverge. Moreover,

$$\lim_{n\to\infty} \frac{c_n}{b_n} = \lim_{n\to\infty} \frac{a_n - b_n}{b_n} = \lim_{n\to\infty} \frac{a_n}{b_n} - 1 = 0. \qquad \square$$

Another way of obtaining extensions of Theorem 2.20 to general series is to impose an extra assumption, in addition to (2.9). This is given in the next result (in contrast with Theorem 2.21, which, with the same assumption as in the ratio comparison test, gives convergence properties of the involved series).

Theorem 2.22 *Let* $\sum a_n$ *and* $\sum b_n$ *be series of complex-valued terms. Assume that*

$$\lim_{n\to\infty} \frac{a_n}{b_n} = c \in \mathbb{C},$$

and that the sequence $\left\{\frac{a_n}{b_n}\right\}$ *is of bounded variation.*

(i) *If* $\sum b_n$ *converges, then* $\sum a_n$ *converges.*
(ii) *If* $c \neq 0$*, then the series* $\sum a_n$ *and* $\sum b_n$ *are either both divergent, both conditionally convergent, or both absolutely convergent.*

Proof We note that (i) is just du Bois-Reymond's test (Theorem 2.63 below), rewritten in a different way. In order to prove (ii), if one of the series is absolutely convergent, it follows that

$$\lim_{n\to\infty} \frac{|a_n|}{|b_n|} = \left|\lim_{n\to\infty} \frac{a_n}{b_n}\right| = |c|,$$

and therefore both series converge absolutely, by Theorem 2.18. In fact, in that result the assumption is $c = 1$, but it is enough to assume that $0 < |c| < \infty$ and rescale one of the sequences. For example, multiplying every term of the sequence $\{b_n\}$ by c does not change the convergence character. Let us now consider the case where neither of the series is absolutely convergent. Assume that $\sum b_n$ converges and define $c_n := \frac{b_n}{a_n}$. Since $c \neq 0$,

$$\lim_{n\to\infty} \frac{b_n}{a_n} = \frac{1}{c}.$$

Now, $\left\{\frac{a_n}{b_n}\right\}$ is of bounded variation, and hence, applying Theorem 2.63, we derive the convergence of the series $\sum a_n$, given that

$$\sum_{n=1}^{\infty} a_n = \sum_{n=1}^{\infty} b_n \frac{a_n}{b_n}.$$

Finally, the fact that if one of the series diverges then the other does follows as the contraposition to the statements we have just proved. □

The asymptotic comparison test (Theorem 2.20) can be refined as follows.

Theorem 2.23 *Consider two series with real-valued nonzero terms $\sum a_n$ and $\sum b_n$, and assume that*

$$\frac{b_{n+1}}{b_n} = \frac{a_{n+1}}{a_n}(1 + c_n), \qquad \textit{for } n \geq 1, \tag{2.10}$$

where the series $\sum c_n$ converges absolutely (i.e., $\sum |c_n|$ converges). Then, both series $\sum a_n$ and $\sum b_n$ converge or diverge simultaneously.

Proof Relation (2.10) implies that for every $n > 1$,

$$\frac{b_n}{a_n} = \frac{b_1}{a_1} \prod_{k=1}^{n-1} (1 + c_k).$$

We now write $b_n = a_n p_n$, with

$$p_n = \frac{b_1}{a_1} \prod_{k=1}^{n-1} (1 + c_k).$$

Note that $p_k - p_{k+1} = -c_k p_k$. Moreover, the absolute convergence of $\sum c_n$ implies that the sequence $\{p_n\}$ is bounded. Indeed, using the inequality $\log(1 + x) < x$ for $x > 0$, we get

$$|p_n| \leq \frac{b_1}{a_1} \prod_{k=1}^{n-1} (1 + |c_k|) = \frac{b_1}{a_1} \exp\left(\sum_{k=1}^{n-1} \log(1 + |c_k|) \right) < \frac{b_1}{a_1} e^C,$$

where $C = \sum_{n=1}^{\infty} |c_n|$.

Let us now assume that $\sum a_n$ converges. Thus, for all $\varepsilon > 0$, we have

$$\left| \sum_{k=m}^{n} a_k \right| < \varepsilon, \qquad \text{for all } m_0 \leq m < n,$$

for some $m_0 \in \mathbb{N}$ by Cauchy's criterion. Using Abel's transformation, for $m_0 \leq m < n$,

$$\begin{aligned} \left| \sum_{k=m}^{n} b_k \right| &= \left| \sum_{k=m}^{n} a_k p_k \right| = \left| \sum_{k=m}^{n-1} (p_k - p_{k+1}) \sum_{\ell=m}^{k} a_\ell + p_n \sum_{\ell=m}^{n} a_\ell \right| \\ &\leq \sum_{k=m}^{n-1} \left(|p_k c_k| \left| \sum_{\ell=m}^{k} a_\ell \right| \right) + |p_n| \varepsilon \leq \varepsilon \sup_n |p_n| \left(1 + \sum_{k=1}^{\infty} |c_k| \right), \end{aligned}$$

that is, $\sum b_n$ converges, again by Cauchy's criterion.

In order to prove that the convergence of $\sum b_n$ implies that of $\sum a_n$, it is enough to apply the above argument with the roles of $\{a_n\}$ and $\{b_n\}$ interchanged. However, we should ensure that when writing

$$\frac{a_{n+1}}{a_n} = \frac{b_{n+1}}{b_n}(1+d_n),$$

the series $\sum d_n$ converges absolutely. First, we observe that (2.10) implies $d_n = -\frac{c_n}{c_n+1}$. Secondly, since we are assuming that $\{a_n\}$ and $\{b_n\}$ consist of nonzero terms, there holds $c_n \neq -1$ for every n. Finally, since $c_n \to 0$, there exist finitely many indices n for which $|1+c_n| \leq \frac{1}{2}$ or, in other words, $|1+c_n| > \frac{1}{2}$ for every $n \geq n_0$. Combining these facts, we get

$$\sum_{n=1}^{\infty} |d_n| = \sum_{n=1}^{\infty} \frac{|c_n|}{|1+c_n|} < \sum_{n=1}^{n_0-1} \frac{|c_n|}{|1+c_n|} + 2\sum_{n=n_0}^{\infty} |c_n| < \infty,$$

which shows the absolute convergence of $\sum d_n$ and concludes the proof. □

An interesting feature of Theorem 2.23 is that it is sharp with respect to the assumption that $\sum c_n$ converges absolutely. More precisely, given an arbitrary sequence $\{c_n\}$ and two sequences $\{a_n\}$ and $\{b_n\}$ so that (2.10) holds, the absolute convergence of $\sum c_n$ is a necessary and sufficient condition for the equiconvergence of both series $\sum a_n$ and $\sum b_n$. For appropriate references, see Notes-13.

2.1.4 Further classical tests and their variants

We now present other tests that are among the most widely known ones (e.g., the ratio test, the root test, etc.) and their generalized variants. It is worth mentioning that these tests still have a certain flavor of comparison type, since their proofs are based on the comparison with series whose convergence character is known.

We start with the classical ratio test, which is attributed to d'Alembert.

Theorem 2.24 (ratio test; d'Alembert's test) *Let $\sum a_k$ be a positive series. Let us assume that*

$$q = \lim_{k\to\infty} \frac{a_{k+1}}{a_k} \tag{2.11}$$

exists.

(i) *If $q < 1$, then $\sum a_k$ converges.*
(ii) *If $q > 1$, then $\sum a_k$ diverges.*
(iii) *If $q = 1$, then the test is inconclusive.*

The proof of this result is rather simple and hence omitted. Besides, we give the full proof of Theorem 2.25 below, which is much more general. However, we stress that the proof of the ratio test in essence reduces to passing to the basic comparison test (Theorem 2.18), where we compare the series $\sum a_k$ to the *geometric series* $\sum q^k$ ($q > 0$). The latter converges if and only if $q < 1$, which can be proved by the following standard argument. We consider the sequence of partial sums $S_n = \sum_{k=1}^{n} q^k$, and it is clear that

$$S_{n+1} - S_n = q^{n+1},$$
$$S_{n+1} - qS_n = q.$$

Subtracting the first equality from the second, we obtain

$$S_n = \frac{q - q^{n+1}}{1-q},$$

for $q \neq 1$, which tends to a finite limit as $n \to \infty$ if and only if $q < 1$. Moreover, in this case

$$\sum_{k=1}^{\infty} q^k = \frac{q}{1-q}.$$

On the one hand, we observe that there are more general forms of Theorem 2.24 in terms of the lower and upper limits. On the other hand, one may wonder whether such a test can be refined, in the sense that some (partial) answer may be given in the case $q = 1$. Both of these observations are taken into account in the following result, which, apart from considering lower and upper limits, is based on studying ratios of the form

$$\frac{a_{mk+j}}{a_k}, \qquad j = 1, \ldots, m,$$

and comparing them with $\frac{1}{m}$ rather than analysing the ratio $\frac{a_{k+1}}{a_k}$ and comparing it with 1.

Theorem 2.25 (*m*th ratio test) *Let $\{a_n\}$ be a positive sequence. For a fixed $m \in \mathbb{N}$ and $j = 1, \ldots, m$, define*

$$L_j = \limsup_{n\to\infty} \frac{a_{mn+j}}{a_n}, \qquad l_j = \liminf_{n\to\infty} \frac{a_{mn+j}}{a_n}.$$

Let $L = \max\{L_1, L_2, \ldots, L_m\}$ and $l = \min\{l_1, l_2, \cdots, l_m\}$.

(i) *If $L < 1/m$, then $\sum a_k$ converges.*
(ii) *If $l > 1/m$, then $\sum a_k$ diverges.*
(iii) *If $l \leq 1/m \leq L$, the test is inconclusive.*

Proof If $L < 1/m$, there exists $L < r < 1/m$ and $N \in \mathbb{N}$ such that for every $n \geq N$,

$$\frac{a_{mn+j}}{a_n} \leq r, \qquad j = 1, \ldots, m. \tag{2.12}$$

Set $n_1 := 0$,

$$n_k := \sum_{s=2}^{k} m^s, \qquad k \geq 2,$$

and define

$$A_0 := \sum_{\ell=1}^{m} a_\ell, \qquad A_k := \sum_{\ell=1}^{m^k} \sum_{n=n_k+m\ell+1}^{n_k+m(\ell+1)} a_n, \qquad k \geq 1.$$

By this, $\sum_{n=1}^{\infty} a_n = \sum_{k=0}^{\infty} A_k$. Let $k_0 \geq 2$ be such that $n_k + m \geq N$ for every $k > k_0$. For such choices of k, it follows from (2.12) that

$$\begin{aligned} A_k &= \sum_{\ell=1}^{m^k} \sum_{n=m(n_{k-1}+m+\ell)+1}^{m(n_{k-1}+m+\ell+1)} a_n \\ &\leq \sum_{\ell=1}^{m^k} \sum_{n=m(n_{k-1}+m+\ell)+1}^{m(n_{k-1}+m+\ell+1)} r \cdot a_{n_{k-1}+m+\ell} \\ &= mr \sum_{\ell=1}^{m^k} a_{n_{k-1}+m+\ell} = mr \sum_{\ell=1}^{m^{k-1}} \sum_{n=n_{k-1}+m\ell+1}^{n_{k-1}+m(\ell+1)} a_n = mrA_{k-1}, \end{aligned}$$

while for $k = 1$,

$$\begin{aligned} A_1 &= \sum_{\ell=0}^{(m-1)N-1} \sum_{n=mN+\ell m}^{mN+m(\ell+1)-1} a_n \leq \sum_{\ell=0}^{(m-1)N-1} \sum_{n=mN+\ell m}^{mN+m(\ell+1)-1} r \cdot a_{N+\ell} \\ &= mr \sum_{\ell=0}^{(m-1)N-1} a_{N+\ell} = mrA_0. \end{aligned}$$

Hence, for every $k > k_0$, we have $A_k \leq (mr)^{k-k_0} A_{k_0}$. We conclude that

$$\sum_{k=k_0}^{\infty} A_k \leq A_{k_0} \sum_{k=0}^{\infty} (mr)^k < \infty,$$

since $mr < 1$, and thus derive the convergence of $\sum_{n=1}^{\infty} a_n$.

Let now $1/m < l$. Then we can find $1/m < r < l$ and $N \in \mathbb{N}$ such that, for every $n \geq N$,

$$\frac{a_{mn+j}}{a_n} \geq r, \qquad j = 1, \ldots, m.$$

Similar to the above, one may prove that $A_k \geq (mr)^{k-k_0} A_{k_0}$ for all $k \geq k_0$ (one should use exactly the same approach, but with reverse inequalities). Therefore,

$$\sum_{k=k_0}^{\infty} A_k \geq A_{k_0} \sum_{k=0}^{\infty} (mr)^k = \infty,$$

since $mr > 1$, and the divergence of $\sum_{n=1}^{\infty} a_n$ follows.

Finally, we show that the test is inconclusive if $l \leq 1/m \leq L$. Consider the example

$$a_n = \frac{1}{n(\log n)^p}, \qquad p > 0.$$

For any $m \in \mathbb{N}$ and any $j = 1, \ldots, m$, we get

$$\frac{n(\log n)^p}{m(n+1)\big(\log(m(n+1))\big)^p} \le \frac{n(\log n)^p}{(mn+j)\big(\log(mn+j)\big)^p} = \frac{a_{mn+j}}{a_n}$$
$$\le \frac{n(\log n)^p}{mn(\log n)^p} = \frac{1}{m}.$$

Since

$$\frac{n(\log n)^p}{m(n+1)\big(\log(m(n+1))\big)^p} \to \frac{1}{m}, \qquad \text{as } n \to \infty,$$

we obtain $l = L = \frac{1}{m}$, for any $p > 0$. However, this series converges if $p > 1$ and diverges if $p \le 1$; see (2.4). □

Note that Theorem 2.25 with $m=1$ generalizes the classical ratio test (Theorem 2.24). The following is an example for which the classical ratio test is inconclusive, while the "mth ratio test" (Theorem 2.25) gives an effective conclusion. Put

$$a_n = \frac{1 \cdot 3 \cdot 5 \cdots (2n-1)}{2^n(n+1)!}.$$

It is clear that

$$\frac{a_{n+1}}{a_n} = \frac{2n+1}{2(n+2)} \to 1 \quad \text{as } n \to \infty.$$

Hence the classical ratio test is inconclusive in this case. However, noting that

$$\frac{a_{2n+2}}{a_n} < \frac{a_{2n+1}}{a_n} = \frac{(2n+1)(2n+3)\cdots(4n)(4n+1)}{2^{n+1}(n+2)(n+3)\cdots(2n+1)(2n+2)}$$
$$= \left(\frac{2n+1}{2n+4}\right)\left(\frac{2n+3}{2n+6}\right)\cdots\left(\frac{4n+1}{4n+4}\right) < \left(\frac{4n+1}{4n+4}\right)^{n+1},$$

since

$$\lim_{n\to\infty}\left(\frac{4n+1}{4n+4}\right)^{n+1} = \lim_{n\to\infty}\left(\frac{4n-3}{4n}\right)^n = \lim_{n\to\infty}\left(1-\frac{3}{4n}\right)^n = e^{-\frac{3}{4}},$$

we obtain

$$\limsup_{n\to\infty}\frac{a_{2n+2}}{a_n} \le \limsup_{n\to\infty}\frac{a_{2n+1}}{a_n} \le e^{-\frac{3}{4}} < \frac{1}{2}.$$

Therefore, by Theorem 2.25, the series

$$\sum_{n=1}^{\infty}\frac{1 \cdot 3 \cdot 5 \cdots (2n-1)}{2^n(n+1)!}$$

converges.

Yet, another different version of d'Alembert's test can be obtained by inspecting the quotients

$$\frac{a_{n+1}-a_n}{a_{n+\ell}},$$

where $\ell \in \mathbb{N}$ is fixed.

Theorem 2.26 *Let $\{a_n\}$ be a positive sequence, and for $\ell \in \mathbb{N} \cup \{0\}$, define*

$$\overline{A}(\ell) = \limsup_{n\to\infty} \frac{a_{n+1} - a_n}{a_{n+\ell}},$$
$$\underline{A}(\ell) = \liminf_{n\to\infty} \frac{a_{n+1} - a_n}{a_{n+\ell}}.$$

Then, if $\overline{A}(\ell) < 0$ the series $\sum a_n$ converges, while if $\underline{A}(\ell) > 0$, such a series diverges.

Proof If $\underline{A}(\ell) > 0$, then there exists N such that for $n \geq N$ we have

$$A_n(\ell) = \frac{a_{n+1} - a_n}{a_{n+\ell}} > 0,$$

and, in particular, $a_{n+1} > a_n$. Thus, the series $\sum a_n$ diverges.

On the other hand, if $\overline{A}(\ell) < 0$, then there exists N such that for $n \geq N$ we have $A_n(\ell) < 0$, and, in particular, $b_n = a_{n+1} - a_n < 0$. Therefore, for m sufficiently large, the series $B_m := \sum_{n=m}^{\infty} b_n$ either converges to a negative number or diverges to $-\infty$. Since $B_m = \lim\limits_{n\to\infty} a_n - a_m$ and $a_n > 0$, we obtain $B_m \geq -a_m > -\infty$.

Now, let $\varepsilon > 0$ be such that $\overline{A}(\ell) + \varepsilon < 0$. For k sufficiently large, we have

$$A_k(\ell) \leq \overline{A}(\ell) + \varepsilon.$$

Multiplying this inequality by $a_{k+\ell}$ and summing up over k from m to n, we obtain

$$\sum_{k=m}^{n} b_k \leq (\overline{A}(\ell) + \varepsilon) \sum_{k=m}^{n} a_{k+\ell}.$$

Taking the limit as $n \to \infty$, we get

$$B_m \leq (\overline{A}(\ell) + \varepsilon) \sum_{k=m+\ell}^{\infty} a_k.$$

Since the term $\overline{A}(\ell) + \varepsilon$ is negative, it follows from the last inequality that

$$\sum_{k=m+\ell}^{\infty} a_k \leq \frac{B_m}{\overline{A}(\ell) + \varepsilon} < \infty,$$

that is, the series $\sum a_n$ converges. □

Remark 2.27 Note that by letting $\ell = 0$ in Theorem 2.26, we recover the classical d'Alembert's test. Indeed, in this case,

$$\overline{A}(0) = \limsup_{n\to\infty} \frac{a_{n+1}}{a_n} - 1, \qquad \underline{A}(0) = \liminf_{n\to\infty} \frac{a_{n+1}}{a_n} - 1$$

(compare it with Theorem 2.25 for $m = 1$).

As a corollary of Theorem 2.26, we can obtain the following simpler statement, in the case of ratios $\frac{a_{n+1}}{a_n}$ tending to a finite limit.

Corollary 2.28 *Let $\{a_n\}$ be a positive sequence, and for $\ell \in \mathbb{N}$, assume that the limit*

$$A(\ell) := \lim_{n\to\infty} \frac{a_{n+1} - a_n}{a_{n+\ell}}$$

exists. Then, if $A(\ell) < 0$, the series $\sum a_n$ converges, while if $A(\ell) > 0$, such a series diverges.

We continue with another classical test, the kth root test (of course, depending on the notation, it can be called mth, or nth root test).

Theorem 2.29 (kth root test) *Let $\{a_k\}$ be a nonnegative sequence. Let us assume that*

$$q = \lim_{k\to\infty} a_k^{\frac{1}{k}} \tag{2.13}$$

exists.

(i) *If $q < 1$, then $\sum a_k$ converges.*
(ii) *If $q > 1$, then $\sum a_k$ diverges.*
(iii) *If $q = 1$, then the test is inconclusive.*

As for the ratio test, one can refine the root test by involving the lower and upper limits. In fact, we omit the proof of the classical root test and prove the following more general one.

Theorem 2.30 *Let $\sum a_k$ be a nonnegative series and define*

$$L = \limsup_{k\to\infty} a_k^{\frac{1}{k}}.$$

(i) *If $L < 1$, then $\sum a_k$ converges.*
(ii) *If $L > 1$, then $\sum a_k$ diverges.*
(iii) *If $L = 1$, the test is inconclusive.*

Proof First of all, if $L < 1$, let $\varepsilon > 0$ be such that $L + \varepsilon < 1$. We observe that there are only finitely many indices a_k satisfying $a_k^{1/k} > L + \varepsilon$, or, equivalently, $a_k > (L+\varepsilon)^k$. That is, there exists $n_0 \in \mathbb{N}$ such that $a_k \le (L+\varepsilon)^k$ for every $k \ge n_0$. Since $L + \varepsilon < 1$, the series

$$\sum_{k=n_0}^{\infty} (L+\varepsilon)^k$$

converges, and so does $\sum a_k$, by the comparison test (Theorem 2.18). On the other hand, for the case $L > 1$, since $a_k^{\frac{1}{k}} > 1$ for infinitely many k, the sequence a_k fails to converge to 0, hence the series is divergent.

In order to prove that this test is inconclusive if $L = 1$, consider the example $a_n = n^{-p}$, $p > 0$. In this case

$$L = \lim_{n\to\infty} \left(n^{-p}\right)^{\frac{1}{n}} = \lim_{n\to\infty} \left(n^{\frac{1}{n}}\right)^{-p} = 1,$$

and the series $\sum n^{-p}$ converges if $p > 1$ and diverges if $p \leq 1$ (this follows immediately from Cauchy's condensation test or from the Maclaurin–Cauchy integral test; cf. Theorem 2.78). □

It is well known that the root test is stronger than the ratio test, although both of them reduce to the basic comparison test (Theorem 2.18). There are cases in which the root test ensures convergence while the ratio test does not. On the other hand, the ratio test implies the root test, since if the limit (2.11) exists, then the limit (2.13) also exists, and moreover

$$\lim_{k\to\infty} a_k^{\frac{1}{k}} = \lim_{k\to\infty} \frac{a_{k+1}}{a_k}. \tag{2.14}$$

Indeed, we first recall that for any sequence $\{c_k\}$, if the limit $\lim\limits_{k\to\infty} c_k = c$ exists, then for its geometric means,

$$\lim_{k\to\infty} \left(\prod_{n=1}^{k} c_n\right)^{\frac{1}{k}} = c.$$

For completeness, we present a brief proof of this well-known relation. It follows from

$$\log c = \lim_{k\to\infty} \log c_k = \lim_{k\to\infty} \frac{1}{k}\left(\sum_{n=1}^{k} \log c_n\right) = \lim_{k\to\infty} \log\left(\prod_{n=1}^{k} c_n\right)^{\frac{1}{k}},$$

where the second equality is justified by Theorem 2.6. The geometric means of the sequence $\left\{\frac{a_{k+1}}{a_k}\right\}$ are

$$\left(\frac{a_2}{a_1}\cdots\frac{a_{k+1}}{a_k}\right)^{\frac{1}{k}} = \left(\frac{a_{k+1}}{a_1}\right)^{\frac{1}{k}}. \tag{2.15}$$

Taking into account that $a_1^{\frac{1}{k}} \to 1$ as $k \to \infty$ and that $\lim a_{k+1} = \lim a_k$, we arrive at (2.14).

On the other hand, the root test does not imply the ratio test. To this end we consider the rearranged geometric series

$$\frac{1}{2} + 1 + \frac{1}{8} + \frac{1}{4} + \frac{1}{32} + \frac{1}{16} + \cdots, \tag{2.16}$$

where the ratios of consecutive terms alternate between the values $\frac{1}{8}$ and 2. However,

$$\limsup_{n\to\infty} a_n^{\frac{1}{n}} \leq \lim_{n\to\infty} 2^{\frac{2-n}{n}} = \frac{1}{2},$$

and thus the series converges by the root test.

Another interesting remark is that in view of (2.14) one can interpret the root test as a test on the geometric average of the consecutive ratios of the sequence $\{a_n\}$. We can deduce a simple yet interesting test based on various arithmetic means.

Theorem 2.31 *Let $\{a_k\}$ be a positive sequence.*

(i) *If*

$$\limsup_{n\to\infty} \frac{1}{n}\sum_{k=1}^{n} \frac{a_{k+1}}{a_k} < 1,$$

then the series $\sum a_k$ converges.

(ii) *If*

$$\limsup_{n\to\infty} \left(\frac{1}{n}\sum_{k=1}^{n} \frac{a_k}{a_{k+1}}\right)^{-1} > 1,$$

then the series $\sum a_k$ diverges.

Proof Part (i) follows immediately by combining the arithmetic-geometric mean inequality

$$\left(\prod_{k=1}^{n} x_k\right)^{\frac{1}{n}} \leq \frac{1}{n}\sum_{k=1}^{n} x_k, \qquad x_k > 0,$$

with $x_k = \frac{a_{k+1}}{a_k}$, and (i) of Theorem 2.30 (see also (2.15)). Part (ii) follows from the harmonic-geometric mean inequality

$$\left(\frac{1}{n}\sum_{k=1}^{n} \frac{1}{x_k}\right)^{-1} \leq \left(\prod_{k=1}^{n} x_k\right)^{\frac{1}{n}}, \qquad x_k > 0,$$

with $x_k = \frac{a_{k+1}}{a_k}$, and (ii) of Theorem 2.30 (combined with (2.15)). □

Note that (i) of Theorem 2.31 is stronger than the ratio test but weaker than the root test. Indeed, this follows from the Cesàro summability theorem (Theorem 2.6) and the following two examples: the first one is (2.16), and the second one is

$$\frac{2}{3} + 1 + \frac{8}{27} + \frac{4}{9} + \cdots + \left(\frac{2}{3}\right)^{2k-1} + \left(\frac{2}{3}\right)^{2k-2} + \cdots.$$

Although Theorem 2.31 is weaker than Theorem 2.29, in some cases it may be easier to compute the arithmetic average of the consecutive ratios than their geometric average. To illustrate this fact, we consider the following example. Let

$$d_k = \begin{cases} 1, & \text{if } k = 5m,\ m \in \mathbb{N}, \\ -1, & \text{otherwise,} \end{cases}$$

and

$$a_1 = 1, \qquad a_n = \prod_{k=1}^{n} \left(1 + \frac{d_k}{k}\right)^k.$$

Since

$$\lim_{n\to\infty}\left(1+\frac{1}{n}\right)^n = e, \qquad \lim_{n\to\infty}\left(1-\frac{1}{n}\right)^n = \frac{1}{e},$$

the limit of the consecutive ratios does not exist. In order to apply the root test, one has to evaluate the limit

$$\lim_{n\to\infty}\prod_{k=1}^{n}\left(1+\frac{d_k}{k}\right)^{\frac{k}{n}},$$

which is not an easy task. On the other hand, the arithmetic average test gives a simpler approach in this case. Indeed, by Theorem 2.6, we have

$$\begin{aligned}\lim_{n\to\infty}\frac{1}{5n}\sum_{k=1}^{5n}\frac{a_{k+1}}{a_k} &\le \frac{1}{5}\lim_{n\to\infty}\frac{1}{n}\sum_{k=1}^{n}\left(1+\frac{1}{5k}\right)^{5k} + \lim_{n\to\infty}\frac{1}{5n}\sum_{k=1}^{5n}\left(1-\frac{1}{k}\right)^{k}\\ &= \frac{1}{5}\lim_{n\to\infty}\left(1+\frac{1}{5n}\right)^{5n} + \lim_{n\to\infty}\left(1-\frac{1}{n}\right)^{n} = \frac{e}{5}+\frac{1}{e} < 1.\end{aligned}$$

This implies

$$\limsup_{n\to\infty}\frac{1}{n}\sum_{k=1}^{n}\frac{a_{k+1}}{a_k} < 1,$$

and therefore the series $\sum a_k$ converges by Theorem 2.31.

Another test of comparison type is the so-called Jamet's test. This is a less popular test, although it has been long known. As in Theorems 2.21 and 2.22, the following result involves conditional and absolute convergence, which we will discuss only in Section 2.2.

Theorem 2.32 (Jamet's test) *Let $\{a_k\}$ be a real-valued sequence of nonzero terms and assume that the limit*

$$L = \lim_{n\to\infty}\frac{-\log|a_n|}{\log n}$$

exists.

(i) *If $L > 1$, then $\sum a_k$ converges absolutely.*
(ii) *If $L < 0$, then $\sum a_k$ diverges.*
(iii) *If $0 \le L < 1$, then $\sum a_k$ either converges conditionally or diverges.*
(iv) *If $L = 1$, the test is inconclusive.*

Proof First, if $L > 1$, let $\varepsilon > 0$ be such that

$$\frac{-\log|a_n|}{\log n} > L-\varepsilon > 1,$$

for every $n \ge n_0$. This inequality is equivalent to $|a_n| < \frac{1}{n^{L-\varepsilon}}$, and by the classical comparison test (Theorem 2.18) and the convergence of the series $\sum\frac{1}{n^p}$ for $p > 1$ (see, e.g., Theorem 2.54), the absolute convergence of $\sum a_k$ follows.

Secondly, let $L < 0$, $\varepsilon > 0$ be such that

$$\frac{-\log|a_n|}{\log n} < L + \varepsilon < 0,$$

for every $n \geq n_0$. This is equivalent to $|a_n| > n^{-(L+\varepsilon)}$, or in other words, $\{a_k\}$ is unbounded, and the series $\sum a_k$ diverges.

For $0 \leq L < 1$, we analogously obtain

$$|a_n| > \frac{1}{n^{L+\varepsilon}}$$

for some $\varepsilon > 0$ such that $L+\varepsilon < 1$. By the diveregence of the series $\sum \frac{1}{n^p}$ for $p < 1$, the series $\sum a_k$ cannot converge absolutely. Thus, it either converges conditionally or diverges (take, for instance, $a_n = \frac{(-1)^n}{\log(n+1)}$, in which case $L = 0$ and $\sum a_n$ converges by Theorem 2.66; for $a_n = \frac{1}{\log(n+1)}$, the series $\sum a_n$ diverges).

Finally, for $L = 1$, we give two examples showing the inconclusivity of the test. For the sequence defined by $a_n = \frac{1}{n}$, one clearly has $L = 1$, and the series $\sum \frac{1}{n}$ diverges. On the other hand, for the sequence defined by

$$a_n = \frac{1}{n \log^2 n},$$

the series $\sum \frac{1}{n \log^2 n}$ converges, as observed in the proof of Theorem 2.25, and

$$\lim_{n\to\infty} -\frac{\log \frac{1}{n\log^2 n}}{\log n} = \lim_{n\to\infty} \frac{\log n + 2\log\log n}{\log n} = 1. \qquad \square$$

We now give a sort of condensation method applicable to any nonnegative series $\sum a_n$, even when $\{a_n\}$ fails to be monotone, in contrast with the classical condensation test (cf. Theorem 2.92 below), which requires (weak) monotonicity of $\{a_n\}$. For $A \subset \mathbb{N}$, we denote by $\#A$ the cardinality of A.

Theorem 2.33 *Let $\{a_n\}$ be a nonnegative sequence.*

(i) *The series $\sum a_n$ converges if $\{na_n\}$ is bounded and there exist $\rho, \theta \geq 0$, $\varepsilon \in (0, 1]$, and a positive convergent series $\sum c_n$ such that*

$$\#\big\{p \subset \{1, \ldots, m\} : a_{k+p} > c_{k+p}\big\} \leq \rho m^{1 \quad \varepsilon},$$

for every m large enough and every $k > m^\theta$.

(ii) *The series $\sum a_n$ diverges if there exist $\omega > 0$ and $\lambda \geq 0$ such that the inequality*

$$\max_{p\in\{1,\ldots,m\}} a_{qm+p} \geq \frac{\omega}{(qm)^{1+\frac{\lambda}{m}}} \tag{2.17}$$

holds for infinitely many m and every $q \in \mathbb{N}$.

Proof For part (i), without loss of generality, we can assume that

$$a_k \leq \frac{1}{k}, \quad \rho = 1, \quad \varepsilon < 1, \quad \theta \in \mathbb{N}.$$

Then, for every large m and every $q \geq m^{\theta-1}$, the quantities

$$A(q,m) = \sum_{p=1}^{m} a_{qm+p} \quad \text{and} \quad C(q,m) = \sum_{p=1}^{m} c_{qm+p}$$

satisfy

$$A(q,m) \leq C(q,m) + \sum_{p=1}^{\lfloor m^{1-\varepsilon} \rfloor} \frac{1}{qm+p} \leq C(q,m) + \frac{1}{qm^{\varepsilon}}. \tag{2.18}$$

We consider $m_1 \in \mathbb{N}$ and define recursively

$$m_{\ell+1} = m_\ell \left\lfloor \frac{1}{m_\ell} e^{m_\ell^{\frac{\varepsilon}{2}}} \right\rfloor,$$

which is clearly strictly increasing, provided m_1 is large enough. Furthermore, m_1 can be chosen so that for any $\ell \in \mathbb{N}$ there holds $m_\ell > \ell^{\frac{4}{\varepsilon}}$. Indeed, for any fixed $\ell_0 \in \mathbb{N}$, the above inequality is clearly true for every $\ell < \ell_0$ if m_1 is taken to be suitably large. Besides this, if $m_\ell > \ell^{\frac{4}{\varepsilon}}$ for some $\ell \geq \ell_1$ large enough,

$$m_{\ell+1} \geq e^{m_\ell^{\frac{\varepsilon}{2}}} - m_\ell > e^{\ell^2} - \ell^{\frac{4}{\varepsilon}} > (\ell+1)^{\frac{4}{\varepsilon}},$$

where we have used that the function $f(x) = e^{x^{\frac{\varepsilon}{2}}} - x$ is strictly increasing for x large enough. Therefore, defining $s = m_1^\theta + 1$ and considering the intervals

$$I_\ell = \left\{ m_\ell^{\theta-1}, m_\ell^{\theta-1} + 1, \ldots, \frac{m_{\ell+1}^\theta}{m_\ell} - 1 \right\},$$

we can write

$$\sum_{k=1}^{\infty} a_k = \sum_{k=1}^{s-1} a_k + \sum_{k=s}^{\infty} a_k = \sum_{k=1}^{s-1} a_k + \sum_{\ell=1}^{\infty} \sum_{q \in I_\ell} A(q, m_\ell).$$

By (2.18), the last series can be estimated as

$$\sum_{\ell=1}^{\infty} \sum_{q \in I_\ell} A(q, m_\ell) \leq \sum_{\ell=1}^{\infty} \sum_{q \in I_\ell} C(q, m_\ell) + \sum_{\ell=1}^{\infty} \sum_{q \in I_\ell} \frac{1}{q m_\ell^{\varepsilon}}.$$

The first series on the right side satisfies

$$\sum_{\ell=1}^{\infty} \sum_{q \in I_\ell} C(q, m_\ell) = \sum_{n=s}^{\infty} c_n < \infty.$$

Concerning the second one, estimating the inner sum

$$\sum_{q \in I_\ell} \frac{1}{q m_\ell^{\varepsilon}} \leq \frac{1}{m_\ell^{\varepsilon}} \sum_{q=m_\ell^{\theta-1}}^{m_{\ell+1}^{\theta}} \frac{1}{q} \leq \theta \frac{\log m_{\ell+1}}{m_\ell^{\varepsilon}},$$

and observing that $\log m_{\ell+1} \le m_\ell^{\frac{\varepsilon}{2}}$, we conclude that

$$\sum_{\ell=1}^{\infty} \sum_{q \in I_\ell} \frac{1}{q m_\ell^{\varepsilon}} \le \sum_{\ell=1}^{\infty} \frac{\theta}{m_\ell^{\frac{\varepsilon}{2}}} \le \sum_{\ell=1}^{\infty} \frac{\theta}{\ell^2} < \infty.$$

For part (ii), it suffices to assume that $\lambda > 0$. Indeed, if (2.17) holds for some $\lambda \ge 0$, then it also holds for $\lambda' > \lambda$. For any m for which (2.17) is fulfilled, the tail of the series

$$T_{m+1} = \sum_{k=m+1}^{\infty} a_k = \sum_{q=1}^{\infty} \sum_{p=1}^{m} a_{qm+p}$$

satisfies

$$\begin{aligned} T_{m+1} &\ge \sum_{q=1}^{\infty} \max_{p \in \{1,\ldots,m\}} a_{qm+p} \ge \frac{\omega}{m^{1+\frac{\lambda}{m}}} \sum_{q=1}^{\infty} \frac{1}{q^{1+\frac{\lambda}{m}}} \\ &\ge \frac{\omega}{m^{1+\frac{\lambda}{m}}} \int_1^{\infty} x^{-1-\frac{\lambda}{m}}\, dx = \frac{\omega}{\lambda m^{\frac{\lambda}{m}}}. \end{aligned}$$

Since such an estimate holds for infinitely many m, we conclude that

$$\liminf_{m \to \infty} T_m \ge \frac{\omega}{\lambda} > 0,$$

which implies the divergence of $\sum a_k$. □

This theorem has been used in the literature for studying intricate series such as

$$\sum_{k=1}^{\infty} \frac{1}{k^{2+\cos k}} = \infty \qquad \text{and} \qquad \sum_{k=1}^{\infty} \frac{1}{k} \left(\frac{2+\cos k}{3} \right)^k < \infty. \tag{2.19}$$

We do not include the details for these claims. Instead we give appropriate references and further discussion in Notes-20. It is also worth mentioning that the convergence character of both series in (2.19) was already established before Theorem 2.33 appeared, although the methods used were extremely sophisticated.

2.1.5 Kummer's and related tests

Kummer's test, discovered as early as 1835 by the German mathematician E. Kummer, is optimal from the theoretical point of view, since it gives a criterion for the convergence or divergence of positive series. However, checking whether a given positive series meets the hypothesis of the mentioned criterion is often a difficult task, and thus such a test may be of little help. Nevertheless, it can serve as a source for developing other tests that are simpler, in the sense that their hypotheses are easier to check. Classical examples of this are d'Alembert's, Raabe's, or Gauss' tests.

We first present Kummer's test and its proof and then discuss its consequences.

Theorem 2.34 (Strong Kummer's test) *Let $\{a_k\}$ be a positive sequence.*

(i) *The series $\sum a_k$ converges if and only if there exists a positive sequence $\{d_k\}$ and a real number $c > 0$ such that*

$$d_k \frac{a_k}{a_{k+1}} - d_{k+1} \geq c, \quad \textit{for all } k \geq 1.$$

(ii) *The series $\sum a_k$ diverges if and only if there exists a positive sequence $\{d_k\}$ such that $\sum \frac{1}{d_k}$ diverges and*

$$d_k \frac{a_k}{a_{k+1}} - d_{k+1} \leq 0, \quad \textit{for all } k \geq 1. \tag{2.20}$$

Proof (i) For the sufficiency part, note that the hypothesis implies

$$a_k \leq \frac{1}{c}\left(d_{k-1}\frac{a_{k-1}}{a_k}a_k - d_k a_k\right) = \frac{d_{k-1}a_{k-1} - d_k a_k}{c}, \qquad k \geq 2.$$

Hence,

$$\begin{aligned}\sum_{k=1}^{N} a_k &= a_1 + \sum_{k=2}^{N} a_k \leq a_1 + \frac{1}{c}\left(\sum_{k=1}^{N-1} d_k a_k - \sum_{k=2}^{N} d_k a_k\right)\\ &= a_1 + \frac{d_1 a_1}{c} - \frac{d_N a_N}{c} \leq a_1 + \frac{d_1 a_1}{c},\end{aligned}$$

since $d_k a_k > 0$ for all k.

For the necessity part, we assume that $\sum a_k$ converges and denote by A its sum. Define

$$d_k = \frac{A - \sum_{n=1}^{k} a_n}{a_k}.$$

It is clear that $d_n > 0$ for all $n \geq 1$. Moreover,

$$d_k \frac{a_k}{a_{k+1}} - d_{k+1} = \frac{a_k}{a_{k+1}} \cdot \frac{A - \sum_{n=1}^{k} a_n}{a_k} - \frac{A - \sum_{n=1}^{k+1} a_n}{a_{k+1}} = \frac{a_{k+1}}{a_{k+1}} = 1 > 0.$$

(ii) If $\{d_k\}$ is a positive sequence such that $\sum \frac{1}{d_k}$ diverges and (2.20) holds, we have

$$a_{k+1} \geq \frac{a_k d_k}{d_{k+1}} \geq \frac{a_{k-1} d_{k-1}}{d_{k+1}} \geq \cdots \geq \frac{a_1 d_1}{d_{k+1}},$$

which implies

$$\sum_{k=1}^{\infty} a_k = a_1 + \sum_{k=2}^{\infty} a_k \geq a_1 + a_1 d_1 \sum_{k=2}^{\infty} \frac{1}{d_k},$$

and therefore $\sum a_k$ diverges. As for necessity, we define

$$d_k = \frac{1}{a_k}\sum_{n=1}^{k} a_n.$$

Obviously, $d_k > 0$. Furthermore,

$$d_k \frac{a_k}{a_{k+1}} - d_{k+1} = \frac{1}{a_{k+1}} \sum_{n=1}^{k} a_n - \frac{1}{a_{k+1}} \sum_{n=1}^{k+1} a_n = -\frac{a_{k+1}}{a_{k+1}} = -1 \leq 0.$$

It remains to prove that the series $\sum \frac{1}{d_k}$ diverges. Note that since $\sum a_k$ diverges, for any $m \in \mathbb{N}$, we can find $n > m$ such that $\sum_{k=m}^{n} a_k > \sum_{k=1}^{m-1} a_k$. Therefore,

$$\sum_{k=m}^{n} \frac{1}{d_k} = \sum_{k=m}^{n} \frac{a_k}{\sum_{j=1}^{k} a_j} > \frac{1}{\sum_{k=1}^{n} a_k} \sum_{k=m}^{n} a_k = \frac{1}{1 + \frac{\sum_{k=1}^{m-1} a_k}{\sum_{k=m}^{n} a_k}} > \frac{1}{2},$$

and by Cauchy's criterion (2.1), the series $\sum \frac{1}{d_k}$ diverges. □

Note that one of the main features of Kummer's test is that the sequence $\{a_k d_k\}$ is strictly decreasing (respectively nondecreasing) for the convergence (respectively divergence) criterion. Indeed, it readily follows from

$$d_k \frac{a_k}{a_{k+1}} - d_{k+1} \geq c > 0 \qquad \text{(respectively } d_k \frac{a_k}{a_{k+1}} - d_{k+1} \leq 0\text{)}$$

that $d_k a_k - d_{k+1} a_{k+1} \geq c a_{k+1} > 0$ (respectively $d_k a_k - d_{k+1} a_{k+1} \leq 0$).

We now state a weaker form of Kummer's test that is more useful in order to derive other convergence tests. Such a test is "weaker" because, contrary to Theorem 2.34, it gives sufficient conditions for convergence or divergence of a given series rather than necessary and sufficient conditions.

Theorem 2.35 (Weak Kummer's test) *Let $\sum a_k$ be a positive series, and let $\{d_n\}$ be a positive sequence. Put*

$$l = \liminf_{n\to\infty} \left(d_n \frac{a_n}{a_{n+1}} - d_{n+1} \right), \qquad L = \limsup_{n\to\infty} \left(d_n \frac{a_n}{a_{n+1}} - d_{n+1} \right).$$

(i) *If $l > 0$, then the series $\sum a_k$ converges;*
(ii) *If $L < 0$ and $\sum \frac{1}{d_n}$ diverges, then the series $\sum a_k$ diverges;*
(iii) *If $l \leq 0 \leq L$, then the test is inconclusive.*

Proof Note that (i) is a direct consequence of Theorem 2.34. As for (ii), we can find $N \in \mathbb{N}$ such that

$$d_n \frac{a_n}{a_{n+1}} - d_{n+1} < 0, \qquad \text{for all } n \geq N,$$

or, equivalently, $d_n a_n < d_{n+1} a_{n+1}$, which in turn implies that $d_n a_n > d_N a_N$ for all $n \geq N$. Hence,

$$\sum_{n=N}^{\infty} a_n > d_N a_N \sum_{n=N}^{\infty} \frac{1}{d_n} = \infty,$$

since $\sum \frac{1}{d_n}$ diverges.

Finally, we show that this test is inconclusive if $l \le 0 \le L$. Let $a_n = \frac{1}{n(\log n)^p}$ with $p > 0$ and $d_n = n$ (so that the series $\sum a_n$ converges for $p > 1$ and diverges for $p \le 1$, and the series $\sum \frac{1}{d_n}$ diverges; cf. Section 3.1). Then

$$d_n \frac{a_n}{a_{n+1}} - d_{n+1} = (n+1)\left(\frac{(\log(n+1))^p}{(\log n)^p} - 1\right) \to 0, \qquad \text{for any } p > 0,$$

but the convergence character of $\sum a_k$ depends on the choice of p. □

We now proceed to stating two more classical tests that follow from Kummer's test.

Theorem 2.36 (Raabe's test) *Given a positive sequence $\{a_k\}$, define*

$$l = \liminf_{n\to\infty} n\left(\frac{a_n}{a_{n+1}} - 1\right), \qquad L = \limsup_{n\to\infty} n\left(\frac{a_n}{a_{n+1}} - 1\right).$$

(i) *If $l > 1$, then the series $\sum a_k$ converges;*
(ii) *If $L < 1$, then the series $\sum a_k$ diverges;*
(iii) *If $l \le 1 \le L$, then the test is inconclusive.*

This test follows directly from weak Kummer's test (Theorem 2.35) by taking $d_n = n$.

If the limit in Raabe's test actually exists, in the case $L \neq 1$ one may conclude not only the convergence/divergence of the given series but also asymptotic estimates for the tails/partial sums, respectively.

Theorem 2.37 *Consider a positive sequence $\{a_k\}$ and assume that*

$$L = \lim_n n\left(\frac{a_n}{a_{n+1}} - 1\right)$$

exists.

(i) *If $L > 1$, then the series $\sum a_k$ converges, and, moreover,*

$$\sum_{k=n}^{\infty} a_k \asymp \frac{na_n}{L-1};$$

(ii) *If $L < 1$, then the series $\sum a_k$ diverges, and, moreover,*

$$\sum_{k=1}^{n} a_k \asymp \frac{na_n}{1-L};$$

(iii) *If $L = 1$, then the test is inconclusive.*

We skip the proof of this result. See Notes-29 for appropriate references and related remarks.

A refinement of Raabe's test is known as Bertrand's test (or De Morgan–Bertrand's test).

Theorem 2.38 (Bertrand's test) *Given a positive sequence $\{a_k\}$, define*

$$l = \liminf_{n\to\infty} (\log n)\left(n\left(\frac{a_n}{a_{n+1}} - 1\right) - 1\right)$$

and

$$L = \limsup_{n\to\infty} (\log n)\left(n\left(\frac{a_n}{a_{n+1}} - 1\right) - 1\right).$$

(i) *If $l > 1$, then the series $\sum a_k$ converges;*
(ii) *If $L < 1$, then the series $\sum a_k$ diverges;*
(iii) *If $l \le 1 \le L$, then the test is inconclusive.*

Again, this test follows from weak Kummer's one by an appropriate choice of the sequence $\{d_n\}$, which is $d_n = n \log n$.

Let us now illustrate how Bertrand's test improves Raabe's test in some cases, in the sense that Bertrand's test may give satisfactory answers whenever Raabe's test is inconclusive. Consider the sequence

$$a_n = \frac{1}{n \log^\beta n}, \qquad \beta > 0.$$

The series $\sum a_n$ converges if and only if $\beta > 1$ (cf. Section 3.1). First of all we note that

$$n\left(\frac{a_n}{a_{n+1}} - 1\right) = \frac{\log^\beta(n+1)}{\log^\beta n} + n\frac{\log^\beta(n+1)}{\log^\beta n} - n.$$

Further,

$$\begin{aligned}\frac{\log^\beta(n+1)}{\log^\beta n} &= \left(1 + \frac{\log(n+1)}{\log n} - 1\right)^\beta = \left(1 + \frac{\log(n+1) - \log n}{\log n}\right)^\beta \\ &\le \left(1 + \frac{1}{n \log n}\right)^\beta,\end{aligned}$$

where the last equality follows from the mean value theorem. Since for small x we have $(1+x)^\beta = 1 + \beta x + O(x^2)$, there holds

$$\begin{aligned}n\left(\frac{a_n}{a_{n+1}} - 1\right) &\le (n+1)\left(1 + \frac{1}{n\log n}\right)^\beta - n \\ &= (n+1)\left(1 + \frac{\beta}{n \log n} + O\left(\frac{1}{n^2 \log^2 n}\right)\right) - n \\ &= 1 + \frac{\beta}{\log n} + O\left(\frac{1}{n \log n}\right),\end{aligned}$$

for any $\beta > 0$. On the other hand, in order to apply Bertrand's test, we have, for $\beta > 1$ and n large enough,

$$n\left(\frac{a_n}{a_{n+1}} - 1\right) > 1 + \frac{\beta^*}{\log n},$$

for some β^* such that $\beta > \beta^* > 1$, and if $\beta < 1$,

$$n\left(\frac{a_n}{a_{n+1}} - 1\right) < 1 + \frac{\beta^*}{\log n},$$

for some β^* such that $\beta^* < \beta < 1$. Therefore, by Bertrand's test, we obtain, for $\beta > 1$ and n large enough,

$$(\log n)\left(n\left(\frac{a_n}{a_{n+1}} - 1\right) - 1\right) > \beta > \beta^* > 1,$$

and the series $\sum \frac{1}{n \log^\beta n}$ converges. Finally, for $\beta < 1$,

$$(\log n)\left(n\left(\frac{a_n}{a_{n+1}} - 1\right) - 1\right) < \beta^* < \beta < 1,$$

and therefore the series diverges. It can be checked similarly that for $\beta = 1$, Bertrand's test is still inconclusive.

Remark 2.39 Note that Bertrand's test is very similar to Raabe's test. The only difference is that in Bertrand's test we examine, roughly speaking, a sequence with an additional logarithm. Further generalizations of Bertrand's test may be obtained by examining sequences with additional iterated logarithms. For instance, a successive test after Bertrand's one would be obtained by considering the sequence

$$(\log\log n)\left(\log n\left(n\left(\frac{a_n}{a_{n+1}} - 1\right) - 1\right) - 1\right),$$

and so on for successive iterations. All of these successive tests may be derived from Kummer's test by considering $\frac{1}{d_n}$ to be the terms of an iterated harmonic series (see Subsection 3.1.1).

Another test that somehow resembles the idea of Kummer's test is Gauss' test.

Theorem 2.40 (Gauss' test) *Let $\{a_k\}$ be a positive sequence and assume that*

$$\frac{a_k}{a_{k+1}} = 1 + \frac{\alpha}{k} + \frac{\beta_k}{k^{1+\varepsilon}}, \tag{2.21}$$

where $\alpha \in \mathbb{R}$ and $\varepsilon > 0$ are constants, and $\{\beta_k\}$ is a bounded sequence. Then the series $\sum a_k$ converges if $\alpha > 1$ and diverges if $\alpha \leq 1$.

Proof Let us first assume that $\alpha > 1$. Since β_k is bounded, by (2.21) we have, for sufficiently large k,

$$k\left(\frac{a_k}{a_{k+1}} - 1\right) = \alpha + \frac{\beta_k}{k^\varepsilon} > 1.$$

Thus, the result follows by Raabe's test (Theorem 2.36). On the other hand, if $\alpha < 1$, we have

$$k\left(\frac{a_k}{a_{k+1}} - 1\right) < 1,$$

for sufficiently large k, and the result again follows from Raabe's test. For the case $\alpha = 1$, (2.21) implies that

$$(\log k)\left(k\left(\frac{a_k}{a_{k+1}} - 1\right) - 1\right) = \frac{(\log k)\beta_k}{k^{\varepsilon}} \to 0 < 1,$$

and the conclusion follows from Bertrand's test (Theorem 2.38). □

Remark 2.41 Let us compare Gauss' and Raabe's tests. Note that Raabe's test is equivalent to asserting that, upon being able to find an expression of the form

$$\frac{a_k}{a_{k+1}} = 1 + \frac{\alpha_k}{k}, \tag{2.22}$$

the series $\sum a_k$ converges if $\liminf_k \alpha_k > 1$ and diverges if $\limsup_k \alpha_k < 1$. On the other hand, Gauss' test asserts that if

$$\frac{a_k}{a_{k+1}} = 1 + \frac{\alpha}{k} + \frac{\beta_k}{k^{1+\varepsilon}},$$

then the series $\sum a_k$ converges if $\alpha > 1$ and diverges if $\alpha \leq 1$. Hence, Gauss' test is covered by Raabe's test whenever $\alpha \neq 1$ (this is evident by the proof of Theorem 2.40) by taking $\alpha_k = \alpha + \frac{\beta_k}{k^{\varepsilon}}$ in (2.22). On the other hand, the case $\alpha = 1$ yields divergence in Gauss' test. Thus, Gauss' test can be seen as a refinement of Raabe's test by not having inconclusive cases.

Remark 2.42 Note that in contrast with the (weak) Kummer's test (or also Raabe's or Bertrand's test), Gauss' test is never inconclusive, as we have just observed. However, applying it amounts to being able to find an expression for the consecutive ratios of $\{a_k\}$ as in (2.21), which in some cases may be extremely difficult.

In the same spirit as Bertrand's test generalizes Raabe's test, one can obtain generalizations for Gauss' test as follows.

Theorem 2.43 *Let $\{a_k\}$ be a positive sequence and assume that*

$$\frac{a_k}{a_{k+1}} = 1 + \frac{1}{k} + \frac{\alpha}{k \log k} + \frac{\beta_k}{k \log^{1+\varepsilon} k},$$

where $\alpha \in \mathbb{R}$ and $\varepsilon > 0$ are constants, and $\{\beta_k\}$ is a bounded sequence. Then the series $\sum a_k$ converges if $\alpha > 1$ and diverges if $\alpha \leq 1$.

In fact, using the idea of Theorem 2.43 (that is, by adding more iterated logarithms to the original decomposition of Gauss' test), one may derive further generalized tests. To state such a generalization, we introduce specific notation for iterated logarithms. For $x > 0$ large enough and $j \in \mathbb{N}$, we denote by $\log_{(j)} x$ the jth iteration of the logarithm applied to x, that is,

$$\log_{(j)} x = \begin{cases} \log x, & \text{if } j = 1, \\ \log_{(j-1)}(\log x) & \text{if } j \geq 2. \end{cases}$$

Further (and different) notation regarding iterated logarithms will be introduced later. The generalized Gauss' test thus reads as follows.

Theorem 2.44 *Let $\{a_k\}$ be a positive sequence and assume that for k large enough, the ratio a_k/a_{k+1} can be represented as*

$$\frac{a_k}{a_{k+1}} = 1 + \frac{1}{k} + \frac{1}{k}\sum_{j=1}^{\ell-1} \frac{1}{\prod_{m=1}^{j} \log_{(m)} k} + \frac{\alpha}{k\prod_{m=1}^{\ell} \log_{(m)} k}$$
$$+ \frac{\beta_k}{k\Big(\prod_{m=1}^{\ell-1} \log_{(m)} k\Big)\big(\log_{(\ell)} k\big)^{1+\varepsilon}},$$

where $\ell \in \mathbb{N}$, $\alpha \in \mathbb{R}$, and $\varepsilon > 0$ are constants, and $\{\beta_k\}$ is a bounded sequence. Then the series $\sum a_k$ converges if $\alpha > 1$ and diverges if $\alpha \le 1$.

The proof of this result is omitted, since it runs along the same lines as that of Theorem 2.40. Note that Theorem 2.43 is just Theorem 2.44 with $\ell = 1$.

2.2 Absolute and conditional convergence

At the beginning of the present chapter we defined absolutely convergent series. Recall that $\sum a_k$ is *absolutely convergent* if the series

$$\sum_{k=1}^{\infty} |a_k| \tag{2.23}$$

converges, and $\sum a_k$ is *conditionally convergent* if $\sum a_k$ converges while $\sum |a_k|$ diverges.

The notion of absolute convergence is in a sense intermediate between the convergence of arbitrary series and the convergence of nonnegative series. Of course, absolutely convergent series are closer to nonnegative series, and in many respects there is no much difference from the nonnegative case. Indeed, we can apply the scrutiny in Section 2.1 to (2.23) and call these tests to be the tests for absolute convergence.

The following two facts are trivial yet basic results on absolute convergence.

Theorem 2.45 *If the terms of a series $\sum a_k$ satisfy*

$$|a_k| \le Cb_k$$

for all $k \ge 1$, where $\sum b_k$ is a convergent nonnegative series, then the series $\sum a_k$ converges absolutely.

Theorem 2.46 *If a series $\sum a_k$ converges absolutely, then it converges, and, moreover, we have*

$$\left|\sum_{k=1}^{\infty} a_k\right| \le \sum_{k=1}^{\infty} |a_k|.$$

In the same spirit as part (ii) of Theorem 2.11 (although the latter is exclusively about positive series), we have the following result concerning conditionally convergent series.

Theorem 2.47 *Let $\{a_k\}$ be a real-valued sequence and assume that $\sum a_k$ is conditionally convergent. Then there exists a real-valued sequence $\{\delta_k\}$ such that*

$$\lim_{k\to\infty} \delta_k = 0, \tag{2.24}$$

and

$$\sum_{k=1}^{\infty} a_k\delta_k = \infty.$$

Proof Since the series $\sum |a_k|$ diverges, by part (ii) of Theorem 2.11, there exists a sequence $\{b_k\}$ vanishing at infinity such that $\sum |a_k|b_k = \infty$ (in fact, Theorem 2.11 even gives an explicit choice of b_k). Defining $\delta_k = b_k \operatorname{sgn} a_k$, (2.24) is clearly satisfied. Moreover, we have $a_k\delta_k = |a_k|b_k$, and the assertion follows. □

Another important property of absolutely convergent series is that they are invariant under rearrangements.

Theorem 2.48 *Let $\sum a_k$ be an absolutely convergent series and let $\sum b_k$ be a rearrangement of $\sum a_k$. Then the two series converge to the same sum.*

Proof For a rearrangement $\sum b_k$ of the series $\sum a_k$, let $\sigma\colon \mathbb{N} \to \mathbb{N}$ be the permutation of the natural numbers so that $b_n = a_{\sigma(n)}$ for each n. Since $\sum a_k$ converges absolutely for $\varepsilon > 0$, we can find $n_0 \in \mathbb{N}$ so that

$$\sum_{k=n_0}^{\infty} |a_k| \le \varepsilon.$$

Let now $N \in \mathbb{N}$ be such that $\{1, \ldots, n_0\} \subset \{\sigma(1), \ldots, \sigma(N)\}$. Clearly $N \ge n_0$, and, moreover,

$$\left|\sum_{k=1}^{\infty} a_k - \sum_{k=1}^{N} b_k\right| = \left|\sum_{k=1}^{\infty} a_k - \sum_{k=1}^{N} a_{\sigma(k)}\right| = \left|\sum_{k\in\mathbb{N}\setminus\{\sigma(1),\ldots,\sigma(N)\}} a_k\right|$$
$$\le \sum_{k=n_0+1}^{\infty} |a_k| < \varepsilon. \qquad \square$$

As a complement to Theorem 2.48, the following result states that given an absolutely convergent series, we may group the terms arbitrarily into subseries and obtain a new absolutely convergent series with the same sum as the original one.

Theorem 2.49 *Let $\sum a_k$ be an absolutely convergent series. Consider an arbitrary partition of $\mathbb{N}$, that is,*

$$\mathbb{N} = \bigcup_{\omega=1}^{\Omega} \left\{n_k^{(\omega)}\right\}_{k=1}^{K_\omega}, \tag{2.25}$$

where each natural number belongs to only one of the sets on the right-hand side of (2.25) (here Ω and each of the K_ω's may be finite or infinite). Let, for each ω, $A_\omega = \left\{n_k^{(\omega)}\right\}_{k=1}^{K_\omega}$. Then, for any ω, the series

$$S^{\omega} := \sum_{k \in A_\omega} a_k$$

is absolutely convergent. Furthermore, the (possibly finite) series

$$\sum_{\omega=1}^{\Omega} S^{\omega}$$

is absolutely convergent, and its sum is $\sum a_k$.

Proof For any ω such that $K_\omega < \infty$, $\sum_{k \in A_\omega} a_k$ is just a finite sum, and it trivially converges absolutely. For $K_\omega = \infty$, the absolute convergence of $\sum_{k \in A_\omega} a_k$ follows from Theorem 2.46 and the absolute convergence of $\sum a_k$. A similar argument applies to the series (or finite sum) $\sum_{\omega=1}^{\Omega} S^{\omega}$.

Put $S := \sum a_k$. For any $\varepsilon > 0$, let $\ell \in \mathbb{N}$ be such that

$$\sum_{k=\ell+1}^{\infty} |a_k| < \varepsilon.$$

For this ε, there exist $N \in \mathbb{N}$ and $\{\omega_1, \dots, \omega_N\} \subset \mathbb{N}$ such that

$$\{1, \dots, \ell\} \subset \bigcup_{j=1}^{N} A_{\omega_j}.$$

Thus,

$$\begin{aligned} \left| S - \sum_{j=1}^{N} S^{\omega_j} \right| &= \left| S - \sum_{j=1}^{N} S^{\omega_j} + \sum_{k=1}^{\ell} a_k - \sum_{k=1}^{\ell} a_k \right| \\ &\le \left| S - \sum_{k=1}^{\ell} a_k \right| + \left| \sum_{j=1}^{N} S^{\omega_j} - \sum_{k=1}^{\ell} a_k \right| < \varepsilon + \sum_{k=\ell+1}^{\infty} |a_k| < 2\varepsilon. \end{aligned}$$

Since $\varepsilon > 0$ is arbitrary, the conclusion follows. Notice that we have implicitly used the absolute convergence of both $\sum a_k$ and $\sum S^{\omega}$ in order to rearrange their terms (cf. Theorem 2.48). □

Conditionally convergent series do not enjoy the same nice properties as absolutely convergent series when it comes to rearranging the terms. More precisely, it may happen that the rearrangement of a conditionally convergent series is also convergent, but to a different limit, or it may even diverge. For example, the series

$$1+\frac{1}{2}+\frac{1}{2}-1+\frac{1}{4}+\frac{1}{4}-\frac{1}{2}+\frac{1}{4}+\frac{1}{4}-\frac{1}{2}$$
$$+\frac{1}{8}+\frac{1}{8}-\frac{1}{4}+\frac{1}{8}+\frac{1}{8}-\frac{1}{4}+\frac{1}{8}+\frac{1}{8}-\frac{1}{4}+\frac{1}{8}+\frac{1}{8}-\frac{1}{4}+\cdots \qquad (2.26)$$

converges to 1, but its rearrangement,

$$1-1+\frac{1}{2}-\frac{1}{2}+\frac{1}{2}-\frac{1}{2}+\frac{1}{4}-\frac{1}{4}+\frac{1}{4}-\frac{1}{4}+\frac{1}{4}-\frac{1}{4}+\frac{1}{4}-\frac{1}{4}+\cdots,$$

converges to 0. As we will see later on, Leibniz's test (Theorem 2.66) yields a simple scheme for constructing many examples of conditionally convergent series.

The positive and negative parts of the sequence of the terms of a conditionally convergent series form divergent series. More precisely, if we denote

$$a_n^+ = \begin{cases} a_n, & \text{if } a_n \geq 0, \\ 0, & \text{otherwise,} \end{cases} \qquad a_n^- = \begin{cases} -a_n, & \text{if } a_n \leq 0, \\ 0, & \text{otherwise,} \end{cases} \qquad (2.27)$$

we have the following:

Theorem 2.50 *Let $\{a_k\}$ be a real-valued sequence and assume that $\sum a_k$ is conditionally convergent. Then, both series $\sum a_k^+$ and $\sum a_k^-$ are divergent.*

Proof Clearly, both series $\sum a_k^+$ and $\sum a_k^-$ cannot be convergent, since otherwise $\sum a_k$ would converge absolutely. Let us prove that the case where only one of those series is convergent is not possible. Assume, for instance, that $\sum a_k^+$ converges (to $A > 0$, say) and $\sum a_k^-$ diverges. Then we have

$$\sum_{k=1}^{n} a_k = \sum_{k=1}^{n} a_k^+ - \sum_{k=1}^{n} a_k^- \leq A - \sum_{k=1}^{n} a_k^- \to -\infty, \qquad \text{as } n \to \infty,$$

that is, $\sum a_k$ does not converge conditionally. Therefore, both series $\sum a_k^+$ and $\sum a_k^-$ necessarily diverge. □

In view of Theorem 2.50, the following observation is in order: the sum of an absolutely convergent series $\sum a_k$ is the difference of two nonnegative convergent series, precisely $\sum a_n^+ - \sum a_n^-$.

We conclude this section by presenting some results concerning rearrangements of conditionally convergent series. The first of them is the classical *Riemann's series theorem*, or *Riemann's rearrangement theorem*, which is a consequence of Theorem 2.50.

Theorem 2.51 *Let $\{a_k\}$ be a real-valued sequence and assume that $\sum a_k$ is conditionally convergent. Then, for every $\alpha \in \mathbb{R}$ (or also $\alpha = \pm\infty$), there exists a rearrangement of $\{a_k\}$, say $\{b_k\}$, such that*

$$\sum_{k=1}^{\infty} b_k = \alpha.$$

Proof We prove only the case $\alpha \in \mathbb{R}$, which can be further reduced to $\alpha = 0$ without loss of generality by replacing the first element of the series. Define the sequences $\{c_n\}$ and $\{d_n\}$ to be the rearrangements of the sequences $\{a_k^+\}$ and $\{a_k^-\}$ (as defined in (2.27)) in decreasing order, respectively. Let now

$$m_1 = \min\left\{n \in \mathbb{N} : \sum_{k=1}^{n} c_k \geq 0\right\}.$$

We note that m_1 exists, since $\sum c_k$ diverges. For $1 \leq n \leq m_1$, define $b_n := c_n$, and let $\left\{c_n^{(1)}\right\}$ be the sequence $\{c_k\}$ without the first m_1 terms. Now let

$$m_2 = \min\left\{n \in \mathbb{N} : \sum_{k=1}^{m_1} b_k - \sum_{k=1}^{m_2} d_k \leq 0\right\},$$

which, again, exists since $\sum d_k$ diverges. Further, let, for $m_1 + 1 \leq n \leq m_1 + m_2$, $b_n := d_{n-m_1}$, and let $\left\{b_n^{(1)}\right\}$ be the sequence $\{b_n\}$ without the first m_2 terms. For $n \in \mathbb{N}$, we set $M_n := \sum_{k=1}^{n} m_k$, where the numbers m_k are defined recursively as follows. On the one hand, for $\ell \in \mathbb{N}$, let

$$m_{2\ell+1} = \min\left\{n \in \mathbb{N} : \sum_{k=1}^{M_{2\ell}} b_k + \sum_{k=1}^{n} c_k^{(\ell-1)} \geq 0\right\}.$$

For $M_{2\ell} + 1 \leq n \leq M_{2\ell+1}$, we then define $b_n := c_{n-M_{2\ell}}^{(\ell-1)}$ and $\left\{c_n^{(\ell)}\right\}$ as the sequence $\left\{c_n^{(\ell-1)}\right\}$ without the first $m_{2\ell+1}$ terms. Each $m_{2\ell+1}$ is well defined due to the divergence of $\sum c_k^{(\ell-1)}$. On the other hand, let

$$m_{2\ell+2} = \min\left\{n \in \mathbb{N} : \sum_{k=1}^{M_{2\ell+1}} b_k - \sum_{k=1}^{n} d_k^{(\ell-1)} \leq 0\right\}.$$

Now, for $M_{2\ell+1} + 1 \leq n \leq M_{2\ell+2}$, we set $b_n := d_{n-M_{2\ell+1}}^{(\ell-1)}$, and let $\left\{d_n^{(\ell)}\right\}$ be the sequence $\left\{d_n^{(\ell-1)}\right\}$ without the first $m_{2\ell+2}$ terms. As above, $m_{2\ell+2}$ is well defined. The construction of this sequence $\{b_k\}$ is indeed a rearrangement of $\{a_k\}$, as it contains each term exactly once, and no additional terms. In order to prove that $\sum b_k$ converges to 0, note that for every $\varepsilon > 0$ there exists $n_0 \in \mathbb{N}$ such that

$$\max\{c_n, d_n\} < \varepsilon, \qquad \text{for all } n \geq n_0.$$

Let $\ell_0 \in \mathbb{N}$ be such that the sequences $\left\{c_n^{(\ell_0)}\right\}$ and $\left\{d_n^{(\ell_0)}\right\}$ do not contain the first n_0 terms of the original sequences $\{c_n\}$ and $\{d_n\}$, respectively. Then, due to the choice of ℓ_0, construction of the sequence $\{b_k\}$, and definition of the numbers m_k, one has, for every $M \geq M_{2\ell_0+2}$,

$$\left|\sum_{k=1}^{M} b_k\right| < \varepsilon. \qquad \square$$

It is a natural question whether it is necessary in Theorem 2.51 that the described rearrangements have the property that infinitely many terms are shifted by arbitrarily long distances from their original locations. The following result answers such a question affirmatively.

Theorem 2.52 *Let $\sum a_k$ be a conditionally convergent series. Suppose that $\{b_k\}$ is a rearrangement of the sequence $\{a_k\}$, where each term shifted forward is shifted by at most ℓ positions, where $\ell \in \mathbb{N}$ is fixed. Then $\sum b_k$ converges to the same sum as $\sum a_k$ does.*

Proof Let

$$S_n = \sum_{k=1}^{n} a_k, \qquad \text{and} \qquad S'_n = \sum_{k=1}^{n} b_k.$$

Note that the partial sum S'_n consists of the same terms as S_n, except possibly 2ℓ terms. More precisely, for n large enough, some of the last ℓ terms from S_n could be excluded in the sum S'_n, in which case they would be replaced by at most ℓ other terms $a_{n_1}, \ldots, a_{n_\ell}$, $n < n_1 < \cdots < n_\ell$, moved backwards (possibly from arbitrarily large distances). Hence,

$$|S'_n - S_n| \leq \sum_{k=n-\ell+1}^{n} |a_k| + |a_{n_1}| + \cdots + |a_{n_\ell}| \leq 2\ell \max_{k \geq n-\ell+1} |a_k|.$$

Since $\sum a_k$ converges, we have $a_n \to 0$. Moreover, since ℓ is fixed, we conclude that $2\ell \max_{k \geq n-\ell+1} |a_k| \to 0$ as $n \to \infty$, and therefore $\sum b_k$ converges to the same sum as $\sum a_k$ does. □

The following result is about the existence of very special permutations that satisfy interesting properties concerning rearrangements of series. A permutation $\sigma : \mathbb{N} \to \mathbb{N}$ is said to have Property λ if it satisfies the following:

1 If $\sum a_k$ converges, then $\sum a_{\sigma(k)}$ converges as well;
2 There exists a divergent series $\sum b_k$ such that $\sum b_{\sigma(k)}$ converges.

Theorem 2.53 *There exists a permutation $\sigma : \mathbb{N} \to \mathbb{N}$ with Property λ.*

Proof Let us first construct a permutation σ such that if $\sum a_k$ converges, then so does $\sum a_{\sigma(k)}$. We split the natural numbers in blocks as follows:

1   2 3 4   5 6 7 8 9   10 11 12 13 14 15 16   17 18 19 20 21 22 23 24 25   26 27 . . .

The kth block, written A_k, is composed by the $2k - 1$ numbers:

$$\{(k-1)^2 + 1, (k-1)^2 + 2, \ldots, k^2\}.$$

We now construct σ by permuting each block A_k according to the following rule: the permuted blocks (which we call A'_k) will begin with the middle element from

A_k, followed by the symmetric pairs in the original block (starting from the middle), which are then transposed. The new blocks A'_k are

1 3 4 2 7 8 6 9 5 13 14 12 15 11 16 10 21 22 20 23 19 24 18 25 17 31 32 . . .

In general, if the block A_k consists of the elements

$$a_{(k-1)^2+1},\, a_{(k-1)^2+2},\, \ldots,\, a_{k^2},$$

then block A'_k has the form

$$a_{\tilde{k}},\, a_{\tilde{k}+1},\, a_{\tilde{k}-1},\, a_{\tilde{k}+2},\, a_{\tilde{k}-2},\, \ldots,\, a_{k^2},\, a_{(k-1)^2+1},$$

where

$$\tilde{k} = (k-1)^2 + 1 + \left\lfloor \frac{2k-1}{2} \right\rfloor = (k-1)^2 + k.$$

Denoting

$$S_n = \sum_{k=1}^{n} a_k \quad \text{and} \quad S'_n = \sum_{k=1}^{n} a_{\sigma(k)},$$

we get S'_n as a linear combination of at most three other S_n's. For instance,

$$S'_1 = S_1, \qquad S'_3 = S_1 + (S_4 - S_2), \qquad S'_7 = S_4 + (S_8 - S_5),$$

and more generally

$$S'_n = S_{q^2} + (S_a - S_b),$$

where q^2 is the largest perfect square less than or equal to n, and where a and b can be chosen appropriately and satisfy $\max\{a, b\} \geq q^2$. Since q^2 increases with n, the first property of σ follows from Cauchy's criterion for convergence of series (2.1). Indeed, the previous identity and the convergence of $\sum a_k$ imply $S_a - S_b \to 0$ as $n \to \infty$. Thus, if $\sum a_k$ converges, then $\sum a_{\sigma(k)}$ converges as well (even to the same limit).

We have to find now a divergent series $\sum b_k$ with the property that $\sum b_{\sigma(k)}$ converges. Let $b_1 = 1$ and

$$b_k = \begin{cases} \frac{(-1)^n}{n+1}, & k \in \{n^2+1, n^2+2, \ldots, n^2+n\},\ n \geq 1, \\ \frac{(-1)^{n+1}}{n+1}, & k \in \{n^2+n+1, n^2+n+2, \ldots, (n+1)^2\},\ n \geq 1, \end{cases}$$

that is, the series $\sum b_k$ is

$$\begin{aligned} &1 - \frac{1}{2} - \frac{1}{2} + \frac{1}{2} \\ &+ \frac{1}{3} + \frac{1}{3} + \frac{1}{3} - \frac{1}{3} - \frac{1}{3} \\ &- \frac{1}{4} - \frac{1}{4} - \frac{1}{4} - \frac{1}{4} + \frac{1}{4} + \frac{1}{4} + \frac{1}{4} \\ &+ \frac{1}{5} + \frac{1}{5} + \frac{1}{5} + \frac{1}{5} + \frac{1}{5} - \frac{1}{5} - \frac{1}{5} - \frac{1}{5} - \frac{1}{5} + \cdots \end{aligned}$$

It is clear that for any $\ell \geq 2$, one has

$$\sum_{k=1}^{\ell^2+\ell} b_k = \sum_{k=1}^{\ell^2} b_k \pm 1,$$

and therefore the series $\sum b_k$ does not converge. On the other hand, rearranging the terms of this series according to the permutation σ constructed above, we arrive at the rearranged series

$$1 - \frac{1}{2} + \frac{1}{2} - \frac{1}{2} + \frac{1}{3} - \frac{1}{3} + \frac{1}{3} - \frac{1}{3} + \frac{1}{3} - \frac{1}{4} + \frac{1}{4} - \frac{1}{4} + \frac{1}{4} - \frac{1}{4} + \frac{1}{4} - \frac{1}{4}$$
$$+ \frac{1}{5} - \frac{1}{5} + \frac{1}{5} - \frac{1}{5} + \frac{1}{5} - \frac{1}{5} + \frac{1}{5} - \frac{1}{5} + \frac{1}{5} - \cdots = \sum_{k=1}^{\infty} \frac{(-1)^{k+1}}{k} = \log 2,$$

or, in other words, the series $\sum b_{\sigma(k)}$ converges. Thus, we have established that the previously constructed permutation σ satisfies Property λ. □

A simple nontrivial condition for a permutation σ to preserve the convergence is $|\sigma(n) - n| < C$ for some positive C and all $n \in \mathbb{N}$. A general criterion for permutations preserving convergence is the following:

Theorem 2.54 *A permutation $\sigma : \mathbb{N} \to \mathbb{N}$ preserves the convergence of any series $\sum a_k$ if and only if there exists $M \in \mathbb{N}$ such that for every j the set $\{\sigma(1), \ldots, \sigma(j)\}$ is the union of M or fewer blocks of consecutive integers.*

We do not give a proof of this statement. See Notes-41 for references instead.

An often overlooked result for conditionally convergent series is Cesàro's theorem. Before stating it, we introduce relevant notation. Given a real-valued sequence $\{a_k\}$, let p_n denote the number of positive terms and q_n the number of negative terms among the first n terms of $\{a_k\}$.

Theorem 2.55 (Cesàro's theorem) *If $\sum a_k$ is a conditionally convergent series, $\{|a_k|\}$ is decreasing, and the limit $\lim_{n\to\infty} \frac{p_n}{q_n}$ exists, then*

$$\lim_{n\to\infty} \frac{p_n}{q_n} = 1.$$

Proof We first write $a_k = \delta_k |a_k|$. Then $p_n - q_n = \sum_{k=1}^{n} \delta_k$ for every n. Writing

$$p_1|a_1| + \sum_{k=1}^{n-1} (p_{k+1} - p_k)|a_{k+1}| = p_n|a_n| + \sum_{k=1}^{n-1} p_k(|a_k| - |a_{k+1}|),$$

$$q_1|a_1| + \sum_{k=1}^{n-1} (q_{k+1} - q_k)|a_{k+1}| = q_n|a_n| + \sum_{k=1}^{n-1} q_k(|a_k| - |a_{k+1}|),$$

and subtracting the second equality from the first one, we obtain

$$\sum_{k=1}^{n} a_k = \sum_{k=1}^{n} \delta_k |a_k| = |a_1|(p_1 - q_1) + \sum_{k=1}^{n-1} (p_{k+1} - q_{k+1} - (p_k - q_k))|a_{k+1}|$$

$$= |a_n|(p_n - q_n) + \sum_{k=1}^{n-1} (p_k - q_k)(|a_k| - |a_{k+1}|).$$

Let us now assume $\frac{p_n - q_n}{n}$ tends to a positive limit L and arrive at a contradiction. Let n_0 be such that

$$\frac{p_n - q_n}{n} > \frac{L}{2},$$

for every $n \geq n_0$. Since $\{|a_k|\}$ is decreasing,

$$\sum_{k=n_0}^{n-1} (p_k - q_k)(|a_k| - |a_{k+1}|) + (p_n - q_n)|a_n| > \frac{L}{2}\left(\sum_{k=n_0}^{n-1} k(|a_k| - |a_{k+1}|) + n|a_n| \right)$$

$$= \frac{L}{2}\left(n_0 |a_{n_0}| + \sum_{k=n_0+1}^{n} |a_k| \right).$$

Thus,

$$\sum_{k=1}^{n} a_k > \frac{L}{2}\left(n_0 |a_{n_0}| + \sum_{k=n_0+1}^{n} |a_k| \right) - \sum_{k=1}^{n_0-1} |p_k - q_k|(|a_k| + |a_{k+1}|),$$

and since the series $\sum |a_k|$ diverges, we conclude that the latter tends to infinity as $n \to \infty$. However, this contradicts the fact that $\sum a_k$ converges, which implies that $\frac{p_n - q_n}{n}$ cannot tend to a positive limit. In an analogous way, it cannot tend to a negative limit either. Thus, if the sequence $\left\{ \frac{p_n - q_n}{n} \right\}$ tends to a finite limit, this limit must be zero. Since $n = p_n + q_n$, we obtain

$$\frac{p_n - q_n}{p_n + q_n} = \frac{1 - \frac{q_n}{p_n}}{1 + \frac{q_n}{p_n}}.$$

It is now clear that if $\frac{p_n}{q_n}$ tends to a finite limit, it must be 1. □

Theorem 2.56 *If $\{|a_n|\}$ is a decreasing sequence tending to zero and $\sum a_k$ converges, then $(p_n - q_n)a_n \to 0$ as $n \to \infty$.*

The proof is omitted, as it follows readily from Theorem 2.102 below.

A simple converse of Cesàro's theorem reads as follows.

Theorem 2.57 *If the sequence $\{|a_k|\}$ is decreasing and $p_n - q_n$ is bounded, then the series $\sum a_k$ converges.*

Proof Writing $a_k = \delta_k |a_k|$, we get $\sum_{k=1}^{n} a_k = \sum_{k=1}^{n} \delta_k |a_k|$, and since $\sum_{k=1}^{n} \delta_k = p_n - q_n$, the conclusion follows from Dirichlet's test (see Theorem 2.64 below). □

Rademacher gave a more general version of Theorem 2.57.

Theorem 2.58 *Let $\{a_n\}$ be such that $\{|a_n|\}$ is decreasing. Assume that there exists $0 < \alpha \le 1$ such that $|p_n - q_n||a_n|^{1-\alpha} \le C$ for every n. Then, the series $\sum a_k$ converges.*

Proof As in the proof of Theorem 2.55, we have

$$\sum_{k=m}^{n} a_k = \sum_{k=m}^{n} \delta_k |a_k|$$

$$= -|a_m|(p_{m-1} - q_{m-1}) + |a_n|(p_n - q_n) + \sum_{k=m}^{n-1}(p_k - q_k)(|a_k| - |a_{k+1}|).$$

On the one hand,

$$|a_m||p_{m-1} - q_{m-1}| \le |a_{m-1}||p_{m-1} - q_{m-1}| = O(|a_{m-1}|^{\alpha}) = o(1),$$

as $m \to \infty$, by assumption. On the other hand,

$$|p_n - q_n||a_n| = O(|a_n|^{\alpha}) = o(1),$$

as $n \to \infty$. Finally, for all sufficiently large $n \ge m$,

$$\sum_{k=m}^{n-1} |p_k - q_k|(|a_k| - |a_{k+1}|) \le C \sum_{k=m}^{n-1}(|a_k| - |a_{k+1}|)|a_k|^{1-\alpha}.$$

Writing $|a_k| = \frac{1}{D_k}$, where $\{D_k\}$ is an increasing sequence of positive numbers tending to infinity, the last sum equals

$$\sum_{k=m}^{n-1}\left(\frac{1}{D_k} - \frac{1}{D_{k+1}}\right)D_k^{1-\alpha} = \sum_{k=m}^{n-1}\frac{D_{k+1} - D_k}{D_{k+1}D_k^{\alpha}}.$$

Since the corresponding series converges (see Theorem 2.128 in Notes-6), the last sum vanishes as $n \ge m \to \infty$ by Cauchy's criterion for convergence of series. Combining the above facts, we get

$$\sum_{k=m}^{n} a_k \to 0 \qquad \text{as } n \ge m \to \infty,$$

as desired. □

Note that if $\alpha = 1$, Theorem 2.58 reduces to Theorem 2.55. Theorem 2.58 immediately implies the following.

Corollary 2.59 *Suppose that for some $0 < \gamma < \beta$,*

$$|p_n - q_n| \le C_1 n^{\gamma} \quad \text{and} \quad |a_n| \le \frac{C_2}{n^{\beta}} \quad \text{for all } n.$$

Then, the series $\sum a_k$ converges.

Proof Since $0 < \gamma < \beta$, there exists $0 < \alpha < 1$ such that $\gamma - \beta(1-\alpha) \le 0$. Thus,

$$|p_n - q_n||a_n|^{1-\alpha} \le C_1 C_2 n^{\gamma - \beta(1-\alpha)} \le C_1 C_2,$$

and the series $\sum a_k$ converges by Theorem 2.58. □

Despite the conclusion of Corollary 2.59, if the sequence $\{|p_n - q_n|\}$ grows faster than a certain power of n, one may always find a sequence $\{a_n\}$ such that $\{|a_n|\}$ is decreasing and $\sum a_k$ diverges (here $a_n = \delta_n |a_n|$ with $\delta_n = p_n - q_n - (p_{n-1} - q_{n-1})$). Indeed, if $|p_n - q_n| \ge cn^\beta$ for some $\beta, c > 0$ and infinitely many n, for $|a_n| = n^{-\beta}$, the sequence $(p_n - q_n)a_n$ does not vanish at infinity, and hence $\sum a_k$ diverges by Theorem 2.56.

To conclude this section, we give a test for absolute convergence based solely on the differentiation of functions generating the corresponding sequences.

Theorem 2.60 *Let $\{a_k\}$ be a real-valued sequence. Let $f\colon \mathbb{R} \to \mathbb{R}$ be such that $f(\frac{1}{k}) = a_k$ for all positive integers k. Assume that $f''(0)$ exists and is finite. Then the series $\sum k^p a_k$ converges absolutely for any $0 \le p < 1$ if and only if $f(0) = f'(0) = 0$.*

Proof Let us first prove the necessity of the conditions $f(0) = f'(0) = 0$. It suffices to prove this for $p = 0$. Since $f''(0)$ exists and is finite, it follows that f is differentiable (and hence continuous) in a neighborhood of $x = 0$. This shows the necessity of the condition $f(0) = 0$ for the (absolute) convergence of the given series, since

$$\lim_{k\to\infty} a_k = \lim_{k\to\infty} f\left(\frac{1}{k}\right) = f(0).$$

In order to prove the necessity of the condition $f'(0) = 0$, we assume on the contrary that $f'(0) = a \ne 0$. Then,

$$\lim_{k\to\infty} \frac{|a_k|}{\frac{1}{k}} = \lim_{x\to 0} \left|\frac{f(x)}{x}\right| = \lim_{x\to 0} \left|\frac{f(x) - f(0)}{x - 0}\right| = |a|.$$

Therefore, by the ratio comparison test (Theorem 2.20), the series $\sum |a_k|$ diverges, since the series $\sum \frac{1}{k}$ diverges.

Let us now prove the sufficiency part. Let $\alpha > 0$ be such that $p + \alpha < 1$. Since $f(0) = f'(0) = 0$, we have, by l'Hôpital's rule,

$$\begin{aligned}
\lim_{x\to 0^+} \frac{f(x)}{x^{1+p+\alpha}} &= \lim_{x\to 0^+} \frac{f'(x)}{(1+p+\alpha)x^{p+\alpha}} \\
&= \frac{1}{1+p+\alpha} \lim_{x\to 0^+} \left(\frac{f'(x) - f'(0)}{x - 0}\right) x^{1-p-\alpha} \\
&= \frac{f''(0)}{1+p+\alpha} \lim_{x\to 0^+} x^{1-p-\alpha} = 0.
\end{aligned}$$

Therefore,

$$\lim_{k\to\infty} \frac{k^p |a_k|}{\frac{1}{k^{1+\alpha}}} = \lim_{x\to 0^+} \left|\frac{f(x)}{x^{1+p+\alpha}}\right| = 0,$$

and since the series $\sum \frac{1}{k^{1+\alpha}}$ converges, so does the series $\sum k^p |a_k|$ by the classical comparison test (Theorem 2.18). □

As an application of Theorem 2.60, consider the series

$$\sum_{k=1}^{\infty} k^p \int_0^{\frac{1}{k}} f(x)\,dx,$$

where $0 \le p < 1$ and $f\colon \mathbb{R} \to \mathbb{R}$ is any function differentiable at $x = 0$. It is immediately verified that this series converges absolutely if and only if $f(0) = 0$.

2.3 Tests involving sequences of bounded variation

As a prelude to studying series with decreasing terms, we first present several classical convergence tests that involve the more general bounded variation (*BVS*) condition (defined in Section 1.1). Classical tests in this context are named after Abel, Dedekind, Dirichlet, du Bois-Reymond, and Leibniz.

Throughout this section, in addition to sequences of bounded variation $\{a_k\}$, we will consider auxiliary sequences $\{b_k\}$ satisfying certain properties and then obtain convergence tests for the series

$$\sum_{k=1}^{\infty} a_k b_k.$$

We emphasize that throughout this section, complex-valued sequences (rather than just nonnegative sequences) will be considered in general.

Before proceeding further, we give the following basic characterization of *BVS*.

Theorem 2.61 *The following two conditions are equivalent:*

(i) $\sum |b_n - b_{n+1}| < +\infty$, *or, in other words,* $\{b_n\} \in BVS$;

(ii) *For every convergent series* $\sum a_n$, *the series* $\sum a_n b_n$ *converges as well.*

Proof (i)⇒(ii) is the statement of du Bois-Reymond's test (see Theorem 2.63 below).

In order to prove the reverse direction, we first note that (ii) implies the boundedness of $\{b_n\}$. Indeed, suppose, on the contrary, that $\{b_n\}$ is unbounded (in this case $\{|b_n|\}$ is unbounded as well). Then there exists a subsequence $\{b_{n_j}\}$ such that $|b_{n_j}| > j^2$ for every $j \in \mathbb{N}$. Setting

$$a_n = \begin{cases} \frac{1}{j^3} & \text{if } n = n_j,\ j \in \mathbb{N}, \\ 0, & \text{otherwise,} \end{cases}$$

it is clear that $\sum a_n$ converges but $\sum a_n|b_n|$ diverges, which is a contradiction. We now have, by Abel's transformation (2.2),

$$\sum_{k=1}^{N} a_k b_k = S_N b_N + \sum_{k=1}^{N-1} S_k(b_k - b_{k+1}),$$

where $S_n = \sum_{k=1}^{n} a_k$.

Let us now consider a convergent series $\sum a_k$ with $\sum a_k = 0$. Since $\{b_k\}$ is bounded, we have $S_N b_N \to 0$. Therefore,

$$\sum_{k=1}^{\infty} a_k b_k = \sum_{k=1}^{\infty} S_k(b_k - b_{k+1}).$$

The last equality and (ii) imply that $\sum d_k(b_k - b_{k+1})$ converges for every null sequence $\{d_k\}$. The latter is in the sequence space

$$c_0 = \big\{\{d_k\} \subset \mathbb{C} : d_k \to 0\big\}.$$

Since

$$\ell^1 = \left\{\{h_k\} \subset \mathbb{C} : \sum_{k=1}^{\infty} |h_k| < \infty\right\}$$

is the β-dual space[1] of c_0 (cf. [B-13, Theorem 2.3.3 (a)], it follows that $\{b_k - b_{k+1}\} \in \ell^1$, or equivalently $\{b_k\} \in BVS$. □

2.3.1 Dedekind's and du Bois-Reymond's tests

We first present Dedekind's and du Bois-Reymond's tests, which are the most general ones involving the bounded variation property. Several other tests, such as the well-known Leibniz's test for alternating series, will be then deduced from them.

Theorem 2.62 (Dedekind's test) *Let $\{a_k\}$ and $\{b_k\}$ be two complex-valued sequences and assume $\{a_k\} \in BVS$ and $a_k \to 0$. If the sequence of partial sums $B_n = \sum_{k=1}^{n} b_k$ is bounded, then $\sum a_k b_k$ converges.*

Proof By Abel's transformation (2.2), if $M, N \in \mathbb{N}$, and $N > M + 1$, then

$$\left|\sum_{k=M+1}^{N} a_k b_k\right| = \left|a_N B_N - a_M B_M + \sum_{k=M+1}^{N-1} B_k(a_k - a_{k+1})\right|$$

$$\leq 2 \sup_{n\geq M} |a_n B_n| + \Big(\sup_{n\geq M} |B_n|\Big) \sum_{k=M+1}^{N-1} |a_k - a_{k+1}|.$$

[1] The β-dual of a sequence space X is defined as the space of all sequences $\{y_k\}$ such that the series $\sum_k x_k y_k$ converges for every $\{x_k\} \in X$.

Since a_n vanishes at infinity and B_n is uniformly bounded, the first term tends to zero as $N > M+1 \to \infty$. Also, since $\{a_k\} \in BVS$,

$$\sum_{k=M+1}^{N-1} |a_k - a_{k+1}| \to 0 \quad \text{as } N > M+1 \to \infty.$$

Thus,

$$\left| \sum_{k=M+1}^{N} a_k b_k \right| \to 0 \qquad \text{as } N > M+1 \to \infty.$$

This yields the convergence of $\sum a_k b_k$. □

We stress the importance of the assumption that $\{a_k\}$ is a null sequence in Dedekind's test. Taking a constant sequence $a_k = c \in \mathbb{C}\backslash\{0\}$ (so that $\{a_k\} \in BVS$ trivially) and $b_k = (-1)^k$ (whose partial sums $\sum_{k=1}^n b_k$ are bounded), we clearly get a divergent series $\sum a_k b_k$, which lacks only the assumption that a_k is a null sequence. This situation can be avoided if, rather than assuming the boundedness of the partial sums $\sum_{k=1}^n b_k$, we assume that they converge instead (or, in other words, $\sum b_k$ is a convergent series). This is precisely du Bois-Reymond's test.

Theorem 2.63 (du Bois-Reymond's test) *Let $\{a_k\}$ and $\{b_k\}$ be two complex-valued sequences. If $\{a_k\} \in BVS$ and $\sum b_k$ converges, then $\sum a_k b_k$ does as well.*

Proof By Abel's transformation (2.2),

$$\sum_{k=1}^{N} a_k b_k = a_N B_N + \sum_{k=1}^{N-1} B_k (a_k - a_{k+1}).$$

On the one hand, since $\{a_k\} \in BVS$, we have that $\lim_{n\to\infty} a_n$ exists and is finite, and by the convergence of B_n, the first term $a_N B_N$ tends to a finite limit (and, in particular, $B_n(a_n - a_{n+1}) \to 0$). On the other hand, for the second term we have

$$\left| \sum_{k=1}^{N-1} B_k (a_k - a_{k+1}) \right| \le \left(\sup_{n\in\mathbb{N}} |B_n| \right) \sum_{k=1}^{\infty} |a_k - a_{k+1}| < \infty,$$

and since $B_n(a_n - a_{n+1}) \to 0$, the convergence of $\sum B_k(a_k - a_{k+1})$ and, in turn, that of $\sum a_k b_k$, follows. □

2.3.2 Dirichlet's, Abel's, and Leibniz's tests

The following tests come as consequences of the tests by Dedekind and du Bois-Reymond by assuming the more restrictive property of monotonicity for $\{a_k\}$, rather than the bounded variation property. Although being trivial subcases of the mentioned tests, Dirichlet's, Abel's, and Leibniz's tests are a part of the classical tests, and so it is worth presenting them separately for the sake of completeness. Also, recall that every sequence of bounded variation is a difference of two nonincreasing sequences (cf. (1.6)).

Theorem 2.64 (Dirichlet's test) *Let $\{a_k\}$ be a monotone null sequence and $\{b_k\}$ a sequence such that the partial sums $\sum_{k=1}^{N} b_k$ are bounded. Then the series $\sum a_k b_k$ converges.*

Theorem 2.65 (Abel's test) *Let $\{a_k\}$ be a bounded monotone (either increasing or decreasing) sequence and let $\sum b_k$ be a convergent series. Then the series $\sum a_k b_k$ converges.*

Though there is a strong wish to try to somehow relax the monotonicity assumption, it cannot be extended more than to the boundedness of variation in the considered circumstances. We now give a counterexample showing that Abel's test is not valid for general monotone sequences (nor for *WMS*, since $GMS \subset WMS$). In fact, this counterexample also shows that Dedekind's test is false without the *BVS* assumption. Let

$$a_n = \begin{cases} 1 & \text{if } n \neq 2^k,\ k \in \mathbb{N}, \\ 2 & \text{if } n = 2^k,\ k \in \mathbb{N}, \end{cases}$$

so that $\{a_k\} \notin BVS$ (however, it is easy to show that $\{a_k\} \in GMS$). Taking $b_n = \frac{(-1)^n}{\ln n}$ $(n \geq 2)$, we get that

$$\sum_{k=2}^{\infty} \frac{(-1)^k}{\ln k} \tag{2.28}$$

converges by Dirichlet's test (or simply by Leibniz's test below). However, the series $\sum_{k=2}^{\infty} a_k b_k$ is the sum of the convergent series (2.28) and the divergent series

$$\sum_{k=1}^{\infty} \frac{1}{\ln 2^k} = \frac{1}{\ln 2} \sum_{k=1}^{\infty} \frac{1}{k},$$

and thus $\sum_{k=2}^{\infty} a_k b_k$ diverges.

2.3.3 Alternating series

Among the series that fall into the scope of Dirichlet's test, a very special class is that of alternating series, those of the type $\sum (-1)^k a_k$, where $\{a_k\}$ is a positive sequence satisfying certain hypotheses. One of the most immediate corollaries of Dirichlet's test is the celebrated *Leibniz's test*, also called *alternating series test*:

Theorem 2.66 (Leibniz's test) *Let $\{a_n\}$ be a monotone sequence that vanishes at infinity. Then the series*

$$\sum_{k=1}^{\infty} (-1)^k a_k \tag{2.29}$$

is convergent.

Note that Theorem 2.57 is a direct generalization of Leibniz's test.

The Leibniz test does not hold if, rather than monotonicity of the sequence $\{a_k\}$, we assume the weaker $\{a_k\} \in BVS$. To see this, just consider

$$a_k = \begin{cases} \frac{2}{\ln k}, & k = 2^n,\ n \in \mathbb{N}, \\ \frac{1}{\ln k}, & \text{otherwise}. \end{cases}$$

Alternating series with decreasing coefficients a_k, such as (2.29), are interesting for many reasons. First of all, they appear naturally in many representations of functions or numbers. For instance,

$$\log(1+x) = \sum_{n=1}^{\infty} \frac{(-1)^{n+1}}{n} x^n, \quad x \in [0,1],$$

and

$$\sin x = \sum_{n=0}^{\infty} \frac{(-1)^n}{(2n+1)!} x^{2n+1}, \quad x \in \mathbb{R}.$$

Also,

$$\frac{\pi}{4} = \sum_{n=1}^{\infty} \frac{(-1)^{n+1}}{2n-1}, \quad \text{and} \quad \frac{1}{e} = \sum_{n=0}^{\infty} \frac{(-1)^n}{n!},$$

among many others.

Secondly, every convergent alternating series has its limit $\sum_{n=1}^{\infty}(-1)^n a_n$ lying between any two successive sums S_{n-1} and S_n. Furthermore, it is very easy to obtain an estimate for the error of the partial sums. Indeed, for any decreasing sequence $\{a_k\}$, one clearly has

$$\left| \sum_{n=1}^{\infty}(-1)^n a_n - \sum_{n=1}^{N}(-1)^n a_n \right| \le a_{N+1}. \tag{2.30}$$

In particular, this estimate can be used to obtain an elementary proof of the *irrationality* of e. Indeed, assume that $e = \frac{a}{b} \in \mathbb{Q}$, with $a, b \in \mathbb{N}$. Then

$$\frac{1}{e} = \frac{b}{a} = \sum_{n=0}^{\infty} \frac{(-1)^n}{n!},$$

or equivalently

$$b(a-1)! - a! \sum_{n=0}^{a} \frac{(-1)^n}{n!} = a! \sum_{n=a+1}^{\infty} \frac{(-1)^n}{n!}.$$

Now, by (2.30), the right-hand side is a number whose absolute value is in $(0,1)$, while the left side is an integer, which is a contradiction. Thus, $e \notin \mathbb{Q}$.

In view of (2.30), alternating series are a good source for simple numerical approximations. However, in some cases the general term of the series decreases so slowly that we need many terms in order to guarantee a certain rate of approximation. For example, if we take the approximation

$$\frac{\pi}{4} = \sum_{n=1}^{N} \frac{(-1)^{n+1}}{2n-1} + \varepsilon_N,$$

then, according to (2.30), a precision of four decimal digits (i.e., $\varepsilon_N \le 5 \times 10^{-5}$) is guaranteed at the 10 001th partial sum. However, it is natural to ask if the estimate given in (2.30) can be further improved, since sometimes it may be rather rough (for instance, for the previous approximation of $\pi/4$, Theorem 2.67 below shows that the 5 000th partial sum already gives four decimal digits of precision).

In order to obtain stronger estimates than (2.30), one should assume more restrictive hypotheses on the sequence $\{a_n\}$. For example, if such a sequence is convex (i.e., if the sequence of differences $\{\Delta a_n\}$ is monotonically decreasing), the estimate (2.30) can be improved, and, moreover, a lower estimate for the error can also be obtained.

Theorem 2.67 *Let $\{a_n\}$ be a positive convex null sequence. Then*

$$\frac{a_{n+1}}{2} \le \left| \sum_{k=1}^{\infty} (-1)^k a_k - \sum_{k=1}^{n} (-1)^k a_k \right| \le \frac{a_n}{2}. \tag{2.31}$$

Proof Let $A = \sum_{k=1}^{\infty} (-1)^k a_k$ (the series clearly converges by Leibniz's test) and $S_n = \sum_{k=1}^{n} (-1)^k a_k$. Note that we can write the alternating series as

$$\sum_{n=1}^{\infty} (-1)^n a_n = S_n + (-1)^{n+1} \sum_{k=0}^{\infty} \Delta a_{n+1+2k},$$

or

$$\sum_{n=1}^{\infty} (-1)^n a_n = S_{n-1} + (-1)^n \sum_{k=0}^{\infty} \Delta a_{n+2k}.$$

Since $\{a_n\}$ is convex, we have

$$|S_n - A| = \sum_{k=0}^{\infty} \Delta a_{n+2k+1} \le \sum_{k=0}^{\infty} \Delta a_{n+2k} = |S_{n-1} - A|.$$

Now, in every convergent alternating series the limit A must lie between any two successive sums S_{n-1} and S_n, and since $a_n = |S_n - S_{n-1}|$, it follows that $a_n = |S_n - A| + |S_{n-1} - A|$. Using the fact that $|S_n - A|$ is decreasing in n, we finally obtain, for any n,

$$2|S_n - A| \le |S_n - A| + |S_{n-1} - A| = a_n \le 2|S_{n-1} - A|,$$

which is equivalent to (2.31). □

With Theorem 2.67 in hand, we can easily find that in order to approximate

$$\log 2 = \sum_{n=1}^{N} \frac{(-1)^{n+1}}{n} + \varepsilon_N$$

with four decimal places of accuracy, one needs to take the 5 000th partial sum (and no less terms).

One can obtain sharper estimates than (2.31) by imposing monotonicity conditions on the higher order differences of $\{a_k\}$. More precisely, by denoting, for $r \geq 1$,

$$\Delta^r a_k = \Delta^{r-1} a_k - \Delta^{r-1} a_{k+1}, \qquad \Delta^0 a_k = a_k,$$

we arrive at the following extension of Theorem 2.67.

Theorem 2.68 *Let $\{a_k\}$ be a decreasing null sequence. Assume that for some $r \in \mathbb{N}$, $r \geq 2$, the sequences $\{\Delta^\ell a_k\}$, $\ell = 1, \ldots, r$, decrease to zero. Then, for any $n \in \mathbb{N}$,*

$$\frac{a_{n+1}}{2} + \frac{\Delta a_{n+1}}{2^2} + \cdots + \frac{\Delta^{r-1} a_{n+1}}{2^r} \leq \left| \sum_{k=1}^{\infty} (-1)^k a_k - \sum_{k=1}^{n} (-1)^k a_k \right|$$
$$\leq \frac{a_n}{2} - \left(\frac{\Delta a_n}{2^2} + \cdots + \frac{\Delta^{r-1} a_n}{2^r} \right).$$

Note that Theorem 2.68 gives a sharper estimate than Theorem 2.67 for convex sequences. Indeed if $\{a_n\}$ is convex, then one has

$$\frac{a_{n+1}}{2} \leq \frac{a_{n+1}}{2} + \frac{\Delta a_{n+1}}{4} \quad \text{and} \quad \frac{a_n}{2} - \frac{\Delta a_n}{4} \leq \frac{a_n}{2}.$$

Proof of Theorem 2.68 Define $A = \sum_{n=1}^{\infty} (-1)^n a_n$ and $R_n = A - \sum_{k=1}^{n} (-1)^k a_k$. In what follows we also take as a convention $\sum_{k=1}^{0} b_k = 0$ for any sequence $\{b_k\}$. We start by noting that the identities

$$\sum_{k=1}^{n} (-1)^k a_k = \left(-\frac{a_1}{2} + \frac{1}{2} \sum_{k=1}^{n-1} (-1)^k \Delta a_k \right) + (-1)^n \frac{a_n}{2} \tag{2.32}$$

and

$$\sum_{k=1}^{n} (-1)^k a_k = \left(-\frac{a_1}{2} + \frac{1}{2} \sum_{k=1}^{n} (-1)^k \Delta a_k \right) + (-1)^n \frac{a_{n+1}}{2} \tag{2.33}$$

hold for any $n \geq 1$. This can easily be seen by comparing the coefficients of a_k on the left and right sides of the equalities.

Subtracting A and then multiplying by $(-1)^n$ on both sides of (2.32), we get

$$|R_n| = (-1)^{n+1} \left(\sum_{k=n+1}^{\infty} (-1)^k a_k \right)$$
$$= (-1)^n \left(-\frac{a_1}{2} + \frac{1}{2} \sum_{k=1}^{n-1} (-1)^k \Delta a_k - A \right) + \frac{a_n}{2}. \tag{2.34}$$

Now, applying identity (2.33) with $n-1$ in place of n and Δa_k in place of a_k, we obtain

$$\sum_{k=1}^{n-1}(-1)^k \Delta a_k = \left(-\frac{\Delta a_1}{2} + \frac{1}{2}\sum_{k=1}^{n-1}(-1)^k \Delta^2 a_k\right) + (-1)^{n-1}\frac{\Delta a_n}{2}.$$

Substituting the right-hand side for the left-hand one within equation (2.34), we get

$$|R_n| = (-1)^n\left(-\frac{a_1}{2} - \frac{\Delta a_1}{4} + \frac{1}{4}\sum_{k=1}^{n-1}(-1)^k \Delta^2 a_k - A\right) + \frac{a_n}{2} - \frac{\Delta a_n}{4}.$$

Repeating this procedure up to r times (where each time after the first one we apply (2.33)), we obtain

$$|R_n| = (-1)^n\left(-\sum_{j=1}^{r}\frac{\Delta^{j-1}a_1}{2^j} + \frac{1}{2^r}\sum_{k=1}^{n-1}(-1)^k \Delta^r a_k - A\right) + \frac{a_n}{2} - \sum_{j=2}^{r}\frac{\Delta^{j-1}a_n}{2^j},$$

where we have used the convention $\Delta^0 a_k = a_k$. We now notice that the term

$$(-1)^n\left(-\sum_{j=1}^{r}\frac{\Delta^{j-1}a_1}{2^j} + \frac{1}{2^r}\sum_{k=1}^{n-1}(-1)^k \Delta^r a_k - A\right) \tag{2.35}$$

is negative for all n. Indeed, since $\{\Delta^r a_k\}$ is decreasing for $j \leq r$, it follows from $|R_n| \to 0$ and the above identities that the series

$$\frac{1}{2^r}\sum_{k=1}^{\infty}(-1)^k \Delta^r a_k, \qquad n \geq 2,$$

is an alternating series itself, converging to the finite limit

$$A + \sum_{j=1}^{r}\frac{\Delta^{j-1}a_1}{2^j}.$$

Thus, the quantity

$$-\sum_{j=1}^{r}\frac{\Delta^{j-1}a_1}{2^j} + \frac{1}{2^r}\sum_{k=1}^{n-1}(-1)^k \Delta^r a_k - A$$

alternates its sign for each consecutive n. Since this quantity is positive for $n = 1$, the claim follows (the factor $(-1)^n$ in (2.35) makes the whole expression negative for each n). Hence, we establish the right inequality

$$|R_n| \leq \frac{a_n}{2} - \sum_{j=2}^{r}\frac{\Delta^{j-1}a_n}{2^j}, \qquad n \geq 1.$$

The left inequality is derived similarly beginning with the identity

$$|R_n| = (-1)^n\left(-\frac{a_1}{2} + \frac{1}{2}\sum_{k=1}^{n}(-1)^k \Delta a_k - A\right) + \frac{a_{n+1}}{2},$$

rather than (2.34). □

Consider a real-valued sequence $\{a_k\}_{k=0}^{\infty}$ and the series

$$\sum_{k=0}^{\infty}(-1)^k a_k.$$

The *Euler transform* of $\sum_{k=0}^{\infty}(-1)^k a_k$ is defined to be the series

$$\sum_{k=0}^{\infty}\frac{\Delta^k a_0}{2^{k+1}}.$$

Theorem 2.69 *For a given real-valued sequence $\{a_k\}$ such that the series $\sum_{k=0}^{\infty}(-1)^k a_k$ converges, its Euler transform also converges, and, moreover,*

$$\sum_{k=0}^{\infty}(-1)^k a_k = \sum_{k=0}^{\infty}\frac{\Delta^k a_0}{2^{k+1}}.$$

Note that the terms of the Euler transform of a series are similar to those in the estimates of Theorem 2.68. The motivation for introducing the Euler transform is that, for a given slowly convergent series $\sum_{k=1}^{\infty}(-1)^k a_k$, the speed of convergence is greatly increased in the transformed series. We also emphasize that in Theorem 2.69, the sequence $\{a_k\}$ is not assumed to be positive and decreasing. We do not give its proof (an appropriate reference and further historical remarks are given in Notes-49). Instead, we give a proof along with an error estimate under some additional assumptions on $\{a_k\}$.

Theorem 2.70 *Let $\{a_k\}$ be a decreasing null sequence. Assume that, for every $r \in \mathbb{N}$, the sequences $\{\Delta^r a_k\}$ also decrease to zero. Then*

$$A = \sum_{k=1}^{\infty}(-1)^k a_k = -\sum_{k=1}^{\infty}\frac{\Delta^{k-1} a_1}{2^k}, \tag{2.36}$$

and moreover

$$-\frac{\Delta^n a_1}{2^n} \le A - \left(-\sum_{k=1}^{n}\frac{\Delta^{k-1} a_1}{2^k}\right) \le 0.$$

Proof Evidently, (2.36) holds if and only if the formula

$$B := -A = \sum_{k=1}^{\infty}(-1)^{k+1} a_k = \sum_{k=1}^{\infty}\frac{\Delta^{k-1} a_1}{2^k}$$

holds. Let S_n denote the nth partial sum of the series $\sum(-1)^{k+1}a_k$. Define

$$U_n = U_n^{(0)} = S_n + (-1)^n\frac{a_{n+1}}{2},$$

and recursively, for $k \ge 1$,

$$U_n^{(k)} = U_n^{(k-1)} + (-1)^n\frac{\Delta^k a_{n+1}}{2^{k+1}}.$$

Noting that $U^{(0)}_{n+1} - U^{(0)}_n = \frac{(-1)^n}{2}\Delta a_{n+1}$, we derive inductively that for every k,

$$U^{(k)}_{n+1} - U^{(k)}_n = (-1)^n \frac{\Delta^{k+1} a_{n+1}}{2^{k+1}}. \tag{2.37}$$

Now, for any $r \in \mathbb{N}$ and any $k \geq 0$,

$$S_{2r} \leq U^{(0)}_{2r} \leq U^{(1)}_{2r} \leq \cdots \leq U^{(k)}_{2r} \leq U^{(k)}_{2r-1} \leq \cdots \leq U^{(1)}_{2r-1} \leq U^{(0)}_{2r-1} \leq S_{2r-1}.$$

Note also that for any $r \in \mathbb{N}$ and $k \geq 0$ we have

$$U^{(k)}_2 \leq U^{(k)}_4 \leq \cdots \leq U^{(k)}_{2r} \leq U^{(k)}_{2r-1} \leq \cdots \leq U^{(k)}_3 \leq U^{(k)}_1,$$

or equivalently

$$[U^{(k)}_2, U^{(k)}_1] \supseteq \left[U^{(k)}_4, U^{(k)}_3\right] \supseteq \cdots \supseteq \left[U^{(k)}_{2r}, U^{(k)}_{2r-1}\right].$$

Since $U^{(k)}_{2r} - U^{(k)}_{2r-1} = -\frac{\Delta^{k+1} a_{2r}}{2^{k+1}}$ decreases to zero as $r \to \infty$, there exists a number $U^{(k)}$ common to all the intervals $\left[U^{(k)}_{2r}, U^{(k)}_{2r-1}\right]$ (for all $r \in \mathbb{N}$ and $k \geq 0$) such that $U^{(k)} = \lim_{n\to\infty} U^{(k)}_n$ for any fixed k. Since $U^{(k)}_{2n} \in [S_{2n}, S_{2n-1}]$ for all n, one must have $U^{(k)} = B$ for all k. Moreover, by (2.37),

$$\left|B - U^{(k)}_n\right| \leq \frac{\Delta^{k+1} a_{n+1}}{2^{k+1}}. \tag{2.38}$$

By definition, $U^{(0)}_n - S_n = (-1)^n \frac{a_{n+1}}{2}$, and again, inductively, it is easy to see that

$$U^{(k)}_n - S_n = (-1)^n \left(\frac{a_{n+1}}{2} + \frac{\Delta a_{n+1}}{2^2} + \cdots + \frac{\Delta^k a_{n+1}}{2^{k+1}}\right).$$

Taking $n = 0$ in the last identity and defining $S_0 = 0$, we obtain

$$U^{(k)}_0 = \frac{a_1}{2} + \frac{\Delta a_1}{2^2} + \cdots + \frac{\Delta^k a_1}{2^{k+1}}.$$

Substituting in (2.38) (with $n = 0$ and arbitrary k), we have

$$\left|B - \left(\frac{a_1}{2} + \frac{\Delta a_1}{2^2} + \cdots + \frac{\Delta^k a_1}{2^{k+1}}\right)\right| \leq \frac{\Delta^{k+1} a_1}{2^{k+1}}.$$

Note that the last inequality implies

$$0 \leq B - \left(\frac{a_1}{2} + \frac{\Delta a_1}{2^2} + \cdots + \frac{\Delta^k a_1}{2^{k+1}}\right) \leq \frac{\Delta^{k+1} a_1}{2^{k+1}}.$$

Indeed, $B \geq 0$, and $U^{(k)}_0$ is the kth partial sum of a series of positive terms converging to B (due to the previous estimate). Thus, $U^{(k)}_0 \leq B$ for every k, which completes the proof. □

A typical example for which Theorem 2.70 is applied is the series with $a_n = \frac{1}{2n-1}$ (see (2.39) below).

Remark 2.71 The Euler transform, as defined above, is a series of negative terms whenever $\{\Delta^r a_k\}$ decreases to zero for each r. This is because of the sign of the original alternating series $\sum(-1)^k a_k$. The fact that the first term a_1 carries a factor of -1 in the alternating series makes the limit negative, and so all the terms of the Euler transform are negative. If the original alternating series were defined as $\sum(-1)^{k+1} a_k$, we would have the converse situation, namely with positivity instead of negativity.

We conclude the analysis of series with higher-order differences by examining the approximation of $\frac{\pi}{4}$ via the Euler transform of the series

$$\sum_{n=1}^{\infty} \frac{(-1)^{n+1}}{2n-1} = \frac{\pi}{4}. \tag{2.39}$$

The higher-order differences of the general term of this series satisfy the relation

$$\frac{\Delta^n a_1}{2^n} = \frac{n!}{1 \cdot 3 \cdot 5 \cdots (2n+1)}, \qquad \text{for any } n.$$

As mentioned above, if we want to approximate $\frac{\pi}{4}$ by means of the alternate series (2.39) with a precision of four decimal digits, we need to take the 5 000th partial sum (by Theorem 2.67). However, if we approximate $\frac{\pi}{4}$ by the Euler transform series of (2.39) with the same precision, we must take N so large that

$$\frac{\Delta^N a_1}{2^N} \leq 5 \times 10^{-5}$$

(note that the term on the left side is the error in the approximation through the Euler transform, by Theorem 2.70). The smallest N for which this estimate holds is $N = 13$. Thus, the speed of convergence of the Euler transform is significantly faster than that of the original alternating series (2.39).

In relation to alternating series, we have the following two statements.

Proposition 2.72 *Let $\{a_k\}$ be a positive sequence increasing to $+\infty$, and let $\{b_k\}$ be a real-valued sequence such that $\sum b_k = \infty$. Then $\sum a_k b_k = \infty$.*

Proof Assuming that $|\sum_{k=1}^{n} a_k b_k| < C$ for every n, we arrive at a contradiction by Dirichlet's test applied to the series $\sum b_k = \sum a_k b_k \frac{1}{a_k}$. □

An interesting statement related to Proposition 2.72 is a more counterintuitive result, where $\{a_k\}$ is not assumed to be increasing.

Proposition 2.73 *There exist a positive sequence $\{b_k\}$ and a real-valued sequence $\{a_k\}$ such that $\sum b_k = \infty$, $|a_n| \to \infty$ as $n \to \infty$, but $\sum a_k b_k$ converges.*

Proof Let $b_n = \frac{1}{n}$, which is positive, and, moreover, $\sum b_n = \infty$. For $a_n = (-1)^n \log n$, there holds $|a_n| \to \infty$, but

$$\sum_{n=1}^{\infty} a_n b_n = \sum_{n=1}^{\infty} \frac{(-1)^n \log n}{n},$$

which converges by Leibniz's test (Theorem 2.66). □

To conclude this section, we study an inequality involving alternating power series, namely power series of the form

$$\sum_{k=0}^{\infty}(-1)^{k+1}a_k x^k, \qquad x \in [0,1],$$

where $\{a_n\}$ is decreasing to zero.

Theorem 2.74 *Let $\{a_k\}_{k=0}^{\infty}$ be a convex sequence that vanishes at infinity. Then, the following inequality holds for all $x \in [0,1]$ and all $n \in \mathbb{N} \cup \{0\}$:*

$$\sum_{k=2n}^{\infty}(-1)^{k+1}a_k x^k \leq x^{2n}\sum_{k=2n}^{\infty}(-1)^{k+1}a_k.$$

Proof Rather than the inequality in the statement, we prove that for any convex sequence $\{a_k\}$,

$$\sum_{k=0}^{\infty}(-1)^{k+1}a_k x^k \leq \sum_{k=0}^{\infty}(-1)^{k+1}a_k,$$

and the result will follow just by multiplying both sides of the inequality by x^{2n} (with different $\{a_k\}$, namely with a_{k-2n} in place of a_k). We will need the following alternative form of Abel's transformation:

$$\sum_{k=0}^{m} b_k c_k = c_0 B_0^m + \sum_{k=0}^{m-1}(c_{k+1} - c_k)B_{k+1}^m,$$

where $B_\ell^n = b_\ell + \cdots + b_n$ (this equality is proved in a similar way as (2.2); here "backwards" differences $c_{k+1} - c_k$ appear, in contrast with (2.2), where "forward" differences $c_k - c_{k+1}$ appear). Putting $b_k = (-1)^k a_k$ and $c_k = x^k$, we get, for any m,

$$\sum_{k=0}^{m}(-1)^k a_k x^k = \sum_{k=0}^{m}(-1)^k a_k + \sum_{k=0}^{m-1}\left(x^k - x^{k+1}\right)\left(\sum_{j=k+1}^{m}(-1)^{j+1}a_j\right).$$

Note that in the last sum a forward difference appears, since we have added the -1 factor in the inner sum. Now, the first two series in this equality converge (because $0 \leq x \leq 1$, and hence both are alternating series with decreasing terms, which converge by Theorem 2.66), thus the last series also converges and we may write

$$\sum_{k=0}^{\infty}(-1)^k a_k x^k = \sum_{k=0}^{\infty}(-1)^k a_k + \sum_{k=0}^{\infty}\left(x^k - x^{k+1}\right)\left(\sum_{j=k+1}^{\infty}(-1)^{j+1}a_j\right).$$

It suffices to show that the last series (containing the differences $x^k - x^{k+1}$) is nonnegative (the desired inequality will then follow by multiplying the whole expression by -1; notice that our goal is to prove the inequality containing $(-1)^{k+1}$ as a factor in the general term rather than $(-1)^k$). First, grouping the terms of the last series in consecutive terms, we obtain

$$\begin{aligned}
&\sum_{k=0}^{\infty}\left(x^k - x^{k+1}\right)\left(\sum_{j=k+1}^{\infty}(-1)^{j+1}a_j\right)\\
&=\sum_{k=0}^{\infty}\Bigg[\left(x^{2k} - x^{2k+1}\right)\left(\sum_{j=2k+1}^{\infty}(-1)^{j+1}a_j\right)\\
&\qquad + \left(x^{2k+1} - x^{2k+2}\right)\left(\sum_{j=2k+2}^{\infty}(-1)^{j+1}a_j\right)\Bigg]\\
&=\sum_{k=0}^{\infty}x^{2k}(1-x)\left(\sum_{j=2k+1}^{\infty}(-1)^{j+1}a_j + x\sum_{j=2k+2}^{\infty}(-1)^{j+1}a_j\right).
\end{aligned}$$

The last expression between parentheses is nonnegative. Indeed, for $k \in \mathbb{N}$,

$$\begin{aligned}
\sum_{j=2k+1}^{\infty}(-1)^{j+1}a_j + x\sum_{j=2k+2}^{\infty}(-1)^{j+1}a_j &\geq \sum_{j=2k+1}^{\infty}(-1)^{j+1}a_j + \sum_{j=2k+2}^{\infty}(-1)^{j+1}a_j\\
&= \sum_{j=2k+1}^{\infty}(-1)^{j+1}\Delta a_j \geq 0,
\end{aligned}$$

since Δa_j is decreasing, and therefore the latter is an alternating series where the first term is nonnegative. □

We note that the conclusion of Theorem 2.74 does not hold if $\{a_k\}$ is not convex. Indeed, to see this, consider any positive decreasing sequence $\{b_k\}$ with $a_0 = b_0$ and define $a_{2k+1} = a_{2k+2} = b_k$ for $k \in \mathbb{N}$. Then,

$$\sum_{k=0}^{\infty}(-1)^{k+1}a_k x^k = -a_0 + \sum_{k=0}^{\infty}b_k\left(x^{2k+1} - x^{2k+2}\right) > -a_0,$$

for $0 < x < 1$, but

$$\sum_{k=0}^{\infty}(-1)^{k+1}a_k = -a_0.$$

We now give an application of Theorem 2.74. It allows to deduce the inequality

$$x - \frac{x^2}{2} + \frac{x^3}{3} - \frac{x^4}{4} + \frac{x^5}{5} - \cdots \leq x - \frac{x^2}{2} + \frac{x^2}{3} - \frac{x^2}{4} + \frac{x^2}{5} - \cdots,$$

which, at first glance, does not necessarily seem to be true, or even seems to be incorrect (in fact the estimate given in Theorem 2.74 may seem counterintuitive). This example (where $a_k = 1/k$) is useful for obtaining approximations of the logarithm that yield better error estimates than the usual Taylor approximations. Consider the remainder $r_{2n}(x)$ in the Taylor expansion

$$\log(1+x) = \sum_{k=1}^{\infty}\frac{(-1)^{k+1}}{k}x^k, \qquad x \in [0,1],$$

namely

$$r_{2n}(x) = \log(1+x) - \sum_{k=1}^{2n} \frac{(-1)^{k+1}}{k} x^k = \sum_{k=2n+1}^{\infty} \frac{(-1)^{k+1}}{k} x^k.$$

Note that inequality (2.30) allows us to obtain the straightforward uniform estimate

$$r_{2n}(x) \le \frac{x^{2n+1}}{2n+1} \le \frac{1}{2n+1}, \qquad x \in [0,1].$$

On the other hand, since $a_k = \frac{1}{k}$ is convex, using Theorem 2.74 we obtain

$$\begin{aligned} r_{2n}(x) &\le \frac{x^{2n+1}}{2n+1} + \sum_{k=2n+2}^{\infty} \frac{(-1)^{k+1}}{k} x^k \le \frac{x^{2n+1}}{2n+1} + x^{2n+2} \sum_{k=2n+2}^{\infty} \frac{(-1)^{k+1}}{k} \\ &\le \frac{x^{2n+1}}{2n+1} - \frac{x^{2n+2}}{2(2n+3)}. \end{aligned}$$

where the last inequality follows from Theorem 2.67. Hence,

$$r_{2n}(x) \le \frac{x^{2n+1}}{2n+1} - \frac{x^{2n+2}}{2(2n+3)},$$

and since the latter is increasing for $x \in [0,1]$ and all n, we deduce

$$r_{2n}(x) \le \frac{1}{2n+1} - \frac{1}{2(2n+3)} \le \frac{1}{4n-2}, \qquad x \in [0,1].$$

Thus, Theorem 2.74 allows one to find an absolute upper bound asymptotically twice as good as the usual one that is deduced from the Taylor expansion (together with inequality (2.30)). Note that the power series expansions of some other elementary functions on $[0,1]$ satisfy the hypotheses of Theorem 2.74. For example,

$$e^{-x} = \sum_{n=0}^{\infty} (-1)^n \frac{x^n}{n!}, \quad \text{and} \quad \arctan x = \sum_{n=0}^{\infty} (-1)^n \frac{x^{2n+1}}{2n+1}.$$

2.3.4 Segmentally alternating series

Another type of series that generalizes the alternating series is called a *segmentally alternating series.*

Definition 2.75 A series $\sum a_k$ with real-valued terms is said to be segmentally alternating of Leibniz type (or just segmentally alternating) if $\{|a_n|\}$ is decreasing to zero, and there exists a sequence $\{n_j\}_{j=1}^{\infty}$ such that in each *segment*

$$a_{n_j}, a_{n_j+1}, \ldots, a_{n_{j+1}-1}$$

half of the terms have positive sign and half have negative sign. The number $n_{j+1} - n_j$ is the *length* of the jth segment. In this case we say that $\{a_k\}$ is a segmentally alternating sequence.

Of course, alternating series are the simplest example of segmentally alternating series. A more sophisticated example of a segmentally alternating sequence is given by

$$a_n = b_n \sin\left(\frac{\pi n^2}{M}\right), \qquad \text{with } b_n \text{ decreasing to } 0, \tag{2.40}$$

where $M > 3$ is odd and has no repeated prime factors. In this case one may find segments of the form $\Xi_n = \Omega_{2n} \cup \Omega_{2n+1}$, where $\Omega_\ell = \{\ell M + 1, \ell M + 2, \ldots, \ell M + M - 1\}$, with $\ell \in \mathbb{N} \cup \{0\}$. Indeed, for each $k \in \Omega_{2n+1}$, $n \in \mathbb{N} \cup \{0\}$, one has

$$a_k = b_k \sin\left(\frac{\pi k^2}{M}\right) = -b_k \sin\left(\frac{\pi (k-M)^2}{M}\right),$$

or, in other words, for each positive term a_k with $k \in \Omega_{2n+1}$, the term a_{k-M} is negative and $k - M \in \Omega_{2n}$. Note that in this case all the segments Ξ_n have length $2(M-1)$.

Segmentally alternating series also admit a convergence test that extends the classical alternating series test (Theorem 2.66).

Theorem 2.76 *If a segmentally alternating series $\sum a_n$ has all segments with bounded lengths, then it converges.*

Proof Let $b_k = |a_k|$ and $c_k = \operatorname{sgn} a_k$. Since $b_k \to 0$, we have

$$\sum_{k=1}^{\infty} a_k = \sum_{k=1}^{\infty} b_k c_k < \infty,$$

provided that $|\sum_{k=1}^{n} c_k| < C$ for every n and some $C > 0$ by Theorem 2.64. Denote by $L \in \mathbb{N}$ the length of the largest segment of $\{a_k\}$. By hypothesis, L is finite, and therefore $|\sum_{k=1}^{n} c_k| \leq L$, which concludes the proof. □

Note that it follows from Theorem 2.76 that the series with general term (2.40) converges. On the other hand, the assertion of Theorem 2.76 is not true for segmentally alternating series with unbounded segment length. This is easily seen by considering the series

$$\left(1 - \frac{1}{2}\right) + \left(\frac{1}{2} + \frac{1}{2} - \frac{1}{4} - \frac{1}{4}\right) + \left(\frac{1}{4} + \frac{1}{4} + \frac{1}{4} + \frac{1}{4} - \frac{1}{8} - \frac{1}{8} - \frac{1}{8} - \frac{1}{8}\right) + \cdots$$

Here each segment consists of the terms inside each couple of parentheses. It is clear that the sum of all terms of an arbitrary segment equals $\frac{1}{2}$, and therefore the series is divergent.

We also note that Theorem 2.76 is weaker than the converse of Cesàro's theorem (cf. Theorem 2.57). Indeed, the terms of any segmentally alternating series satisfy the hypotheses of Theorem 2.57.

2.4 Tests for monotone sequences

In various well-known tests for convergence/divergence of number series $\sum a_k$ with positive a_k, *monotonicity* of the sequence $\{a_k\}$ used to be the basic assumption. Such series are frequently called *monotone series*. As examples, we mention tests by Cauchy, de la Vallée Poussin, Ermakov, Maclaurin, Abel–Olivier, Sapogov, Schlömilch; several such tests were named after Abel and Cauchy.

Everywhere, if possible, we present the tests in the newly obtained form, where monotonicity is replaced by weak monotonicity. It can be said that each time the sequence or function is assumed to be weak monotone, the same test would be classical if usual monotonicity were considered instead.

In some cases, we may also include generalizations of the given results where monotonicity is not involved, just as we did in Section 2.1. The latter contains results for positive series, but also some generalizations of those for series with arbitrary terms.

2.4.1 Maclaurin–Cauchy integral test and related ones

Sometimes, and this is the case for the Maclaurin–Cauchy integral test, instead of $\sum a_k$, it is more convenient to study the series

$$\sum_{k=1}^{\infty} f(k), \tag{2.41}$$

where $f\colon [1,\infty) \to [0,\infty)$. In parallel, the integral

$$\int_1^{\infty} f(t)\,dt \tag{2.42}$$

is drawn in.

The Maclaurin–Cauchy integral test reads as follows.

Theorem 2.77 *Let $f\colon [1,\infty) \to [0,\infty)$ be a decreasing function. Then, the series* (2.41) *converges if and only if the integral* (2.42) *does.*

This is probably one of the most applicable tests. Unlike many others, it does not directly infer convergence or divergence of a series by examining its terms, but it only transfers the calculations from those for sums to more convenient operations with integrals. As already mentioned, in the extended versions of the convergence tests the *WM* property substitutes for monotonicity. The following is the extended Maclaurin–Cauchy integral test.

Theorem 2.78 *Let f be a WM function. Then the series* (2.41) *and the integral* (2.42) *converge or diverge simultaneously.*

Proof On the one hand, we have

$$\sum_{n=1}^{\infty} f(n) \leq f(1) + \sum_{n=1}^{\infty} C \int_{n}^{n+1} f(x)\, dx = f(1) + C \int_{1}^{\infty} f(x)\, dx,$$

that is, the convergence of (2.42) implies the convergence of (2.41). For the reverse part, we have likewise

$$\int_{1}^{\infty} f(x)\, dx = \sum_{n=1}^{\infty} \int_{n}^{n+1} f(x)\, dx \leq C \sum_{n=1}^{\infty} f(n),$$

as desired. □

As an immediate corollary of the Maclaurin–Cauchy integral test we have the following:

Theorem 2.79 *Let $f\colon [0,\infty) \to [0,\infty)$ be a function decreasing to zero. Let $\{a_n\}$ be a nonnegative bounded sequence such that the series $\sum a_k$ diverges. Then both series*

$$\sum_{n=1}^{\infty} f(n) \quad \textit{and} \quad \sum_{n=1}^{\infty} a_n f(S_n)$$

converge or diverge simultaneously, where $S_n = \sum_{k=1}^{n} a_k$.

Proof Assume first that $\sum f(n)$ converges. By the preceeding theorem, we have that $\int_0^\infty f(x)\, dx$ converges. Consequently, setting $S_0 = 0$, we have

$$\sum_{n=1}^{\infty} a_n f(S_n) \leq \sum_{n=1}^{\infty} \int_{S_{n-1}}^{S_n} f(x)\, dx = \int_{0}^{\infty} f(x)\, dx < \infty.$$

On the other hand, if $\sum f(n)$ diverges, then so does $\int_0^\infty f(x)\, dx$. Therefore,

$$\int_{0}^{\infty} f(x)\, dx = \sum_{n=1}^{\infty} \int_{S_{n-1}}^{S_n} f(x)\, dx \leq \sum_{n=1}^{\infty} a_n f(S_{n-1}),$$

which implies that the last series diverges. To show that the series $\sum a_n f(S_n)$ diverges, it suffices to observe that the series

$$\sum_{n=1}^{\infty} a_n\big(f(S_{n-1}) - f(S_n)\big)$$

converges. Indeed, this follows from the fact that $\{a_n\}$ is bounded:

$$\sum_{n=1}^{\infty} |a_n\big(f(S_{n-1}) - f(S_n)\big)| \leq \sup_{k\in\mathbb{N}} a_k \sum_{n=1}^{\infty} |f(S_{n-1}) - f(S_n)|$$

$$= \sup_{k\in\mathbb{N}} a_k \sum_{n=1}^{\infty} (f(S_{n-1}) - f(S_n)) = f(S_0) \sup_{k\in\mathbb{N}} a_k = f(0) \sup_{k\in\mathbb{N}} a_k.$$

Note that the identity $|f(S_{n-1}) - f(S_n)| = f(S_{n-1}) - f(S_n)$ follows from the fact that f is decreasing. □

The Maclaurin–Cauchy integral test has frequently been discussed and refined, not only because of its popularity and usefulness in various estimates but also since it resembles certain operations having applications in numerical calculations (numerous versions of the Euler–Maclaurin formula come to mind), number theory, harmonic analysis, and more.

We first proceed to the very old generalizations of the integral tests given by Bromwich and Hardy. One of the features of their works is that their results apply to series containing terms of both positive and negative terms rather than to nonnegative series. The following result is due to Bromwich.

Theorem 2.80 *Let $f, g\colon [1,\infty) \to (0,\infty)$ be differentiable and satisfy the following hypotheses:*

(i) *f is decreasing to zero;*
(ii) *g is increasing to infinity;*
(iii) *g' is decreasing to zero;*
(iv) *$\displaystyle\int_1^\infty f(x)g'(x)\,dx$ converges.*

Then, the difference

$$\int_1^N f(x)e^{ig(x)}\,dx - \sum_{n=1}^N f(n)e^{ig(n)}$$

tends to a finite limit as $N \to \infty$.

Proof Let us denote $F(x) := f(x)e^{ig(x)}$, and for $n \geq 1$,

$$I_n := \int_n^{n+1} (F(x) - F(n))\,dx = \int_0^1 (F(n+t) - F(n))\,dt.$$

By the fundamental theorem of calculus, one has

$$F(n+t) - F(n) = \int_0^t F'(n+s)\,ds. \tag{2.43}$$

Let us now estimate the difference $F(n+t) - F(n)$. On the one hand,

$$|F'(x)| = |f'(x)e^{ig(x)} + if(x)g'(x)e^{ig(x)}| \leq |f'(x)| + |f(x)g'(x)|.$$

It follows from conditions (i) and (iii) that

$$|f'(x+s)| = -f'(x+s) \qquad \text{and} \qquad |f(x+s)g'(x+s)| \leq |f(x)g'(x)|,$$

where $x \geq 1$ and $0 \leq s \leq 1$. Thus,

$$|F'(n+s)| \leq f(n)g'(n) - f'(n+s), \qquad 0 \leq s \leq 1.$$

Combining this estimate with (2.43), we get

$$\begin{aligned}|F(n+t)-F(n)| &\leq tf(n)g'(n)+f(n)-f(n+t)\\ &\leq f(n)g'(n)+f(n)-f(n+1),\end{aligned}$$

for $0\leq t\leq 1$. It readily follows that

$$I_n \leq f(n)g'(n)+f(n)-f(n+1),$$

for every n. We now note that condition (iv), together with the assumptions on f and g, implies that the series

$$\sum_{n=1}^{\infty} f(n)g'(n)$$

is convergent due to the classical Maclaurin–Cauchy integral test. Furthermore, for $n_0 \leq n_1$,

$$\sum_{n=n_0}^{n_1} |f(n)-f(n+1)| = f(n_0)-f(n_1+1) \to f(n_0), \qquad \text{as } n_1 \to \infty,$$

since f is decreasing. In other words, the series $\sum |f(n)-f(n+1)|$ also converges. In view of the above inequalities, we get

$$\sum_{n=1}^{\infty} |I_n| \leq \sum_{n=1}^{\infty} |f(n)g'(n)| + \sum_{n=1}^{\infty} |f(n)-f(n+1)| < \infty,$$

and therefore the series $\sum I_n$ converges. However, noting that

$$\sum_{n=1}^{N} I_n = \int_1^N F(x)\,dx - \sum_{n=1}^{N} F(n) + \int_N^{N+1} F(x)\,dx,$$

and that the last integral vanishes as $N\to\infty$ (since f vanishes at infinity), we get

$$\left|\int_1^N F(x)\,dx - \sum_{n=1}^{N} F(n)\right| \to \left|\sum_{n=1}^{\infty} I_n\right| =: A,$$

as $N\to\infty$. □

The assumptions of Theorem 2.80 imply that the function $f(x)e^{ig(x)}$ is oscillating and decreasing in absolute value. However, note that the function g could not be the identity (thus not including the case of trigonometric series), since in that case g' would fail to decrease to zero. An example of a series for which Theorem 2.80 applies is

$$\sum_{n=1}^{\infty} \frac{(\log n)^p}{n^{1-ci}}, \qquad c>0, \quad p>0.$$

Indeed, taking $f(x) = \frac{\log^p x}{x}$ and $g(x) = c \log x$, we see that the hypotheses of Theorem 2.80 are satisfied. On the other hand, we note that the series $\sum F(n)$ generated by

$$F(x) = \frac{e^{ix^b \cos(\log x)}}{x^c}, \qquad 0 < b < c < 1, \tag{2.44}$$

is an example where Theorem 2.80 does not apply, as $g(x) = x^b \cos(\log x)$ does not increase to infinity. To overcome this obstacle, it seems natural to inquire about whether or not, in the Maclaurin–Cauchy integral test, it is sufficient to assume that the function in question is of bounded variation instead of positive and decreasing (as in some parts of Section 2.3 above). Such a condition is indeed sufficient, provided that the function possesses a continuous derivative. More precisely, Hardy obtained the following generalization of both the Maclaurin–Cauchy test and Theorem 2.80.

Theorem 2.81 *Let $f\colon [1, \infty) \to \mathbb{C}$ satisfy the following hypotheses:*

(i) *f has a continuous derivative f';*
(ii) *$f(x) \to 0$ as $x \to \infty$;*
(iii) *$\int_1^\infty |f'(x)|\, dx$ converges.*

Then, the difference

$$\int_1^N f(x)\, dx - \sum_{n=1}^N f(n)$$

tends to a finite limit as $N \to \infty$.

Proof The proof goes along the same lines as that of Theorem 2.80, although it is simpler. Just as above, we define

$$I_n = \int_n^{n+1} (f(x) - f(n))\, dx = \int_n^{n+1} (f(x) - f(n)) \frac{d}{dx}(x - n - 1)\, dx.$$

Integrating by parts, we obtain

$$\begin{aligned} I_n &= -\int_n^{n+1} (x - n - 1) f'(x)\, dx = \int_n^{n+1} (n + 1 - x) f'(x)\, dx \\ &\leq \int_n^{n+1} |f'(x)|\, dx. \end{aligned}$$

Thus, one has $\sum |I_n| < \infty$, and the result follows just in the same way as in the proof of Theorem 2.80. □

Theorem 2.81 allows one to deal, among others, with the series $\sum f(n)$ generated by (2.44). However, the last two results are still not general enough to deal with series of the type

$$\sum_{n=1}^\infty \frac{e^{in^b}}{n^c}, \qquad 0 < b < 1, \quad c > 0. \tag{2.45}$$

This series converges if $b + c > 1$, but one needs to assume $b > c$ in order to apply either Theorem 2.80 or 2.81. This can be overcome by using the following generalization of Theorems 2.80 and 2.81.

Theorem 2.82 *Let $f\colon (0,\infty) \to \mathbb{C}$ be such that for some integer $k \geq 1$ the integral*

$$\int_1^\infty |f^{(k)}(x)|\,dx$$

is convergent, and such that for every $0 \leq \ell \leq k-1$,

$$f^{(\ell)}(x) \to 0, \qquad \text{as } x \to \infty.$$

Then, the difference

$$\int_1^N f(x)\,dx - \sum_{n=1}^N f(n)$$

tends to a finite limit as $N \to \infty$.

This result is also due to Hardy. We are not presenting the original proof, but refer to the more general Theorem 2.85 below instead. It implies Theorem 2.82 as a consequence and, moreover, contains an explicit bound for the difference

$$\int_1^N f(x)\,dx - \sum_{n=1}^N f(n).$$

Theorem 2.82 allows one to deal with the series of type (2.45). Indeed, it can be checked that if $f(x) = x^{-c}e^{ix^b}$ and k is chosen so that

$$k > \frac{1-b}{1-c},$$

then the integral $\int_1^\infty |f^{(k)}(x)|\,dx$ converges and the derivatives of f vanish at infinity. Therefore, we can deduce from Theorem 2.82 that both

$$\int_1^\infty \frac{e^{ix^b}}{x^c}\,dx \quad \text{and} \quad \sum_{n=1}^\infty \frac{e^{in^b}}{n^c}$$

converge if $b + c > 1$.

We now present extensions of the Maclaurin–Cauchy integral test to the setting of functions of bounded variation. Contrary to the previous statements, in the following results we give explicit bounds for the error of approximation of the integral through the series (nonetheless, note that it is not difficult to infer error estimates from the proofs of the previous results).

The following two theorems are due to Trigub (an earlier version is due to Belinsky) and come from the needs of harmonic analysis and approximation theory. The first and basic theorem reads as follows.

Theorem 2.83 *For a function f vanishing at $\pm\infty$ and of bounded variation on $\mathbb{R}$, we have*

$$\sup_{|x|\le\pi}\left|\int_{\mathbb{R}} f(t)e^{ixt}\,dt-\sum_{k\in\mathbb{Z}} f(k)e^{ikx}\right|\le\pi\int_{\mathbb{R}}|df(t)|. \tag{2.46}$$

Proof Denoting

$$s(x)=\frac{x}{2\sin\frac{x}{2}},$$

and using a standard trick, which consists in identifying

$$e^{ixk}=s(x)\int_{k-\frac12}^{k+\frac12}e^{ixt}\,dt$$

(see, e.g., the proof of Theorem 2.29 in [B-86, Ch. 5, §2]), we get

$$\sum_{k\in\mathbb{Z}} f(k)e^{ixk}=s(x)\sum_{k\in\mathbb{Z}} f(k)\int_{k-\frac12}^{k+\frac12}e^{ixt}\,dt.$$

Since

$$\int_{\mathbb{R}} f(t)e^{ixt}\,dt=\sum_{k\in\mathbb{Z}}\int_{k-\frac12}^{k+\frac12}f(t)e^{ixt}\,dt,$$

we have

$$\begin{aligned}&\int_{\mathbb{R}} f(t)e^{ixt}\,dt-\sum_{k\in\mathbb{Z}} f(k)e^{ixk}\\&=s(x)\sum_{k\in\mathbb{Z}}\int_{k-\frac12}^{k+\frac12}\big(f(t)-f(k)\big)e^{ixt}\,dt+(1-s(x))\int_{\mathbb{R}} f(t)e^{ixt}\,dt.\end{aligned}$$

Observe that

$$1\le|s(x)|\le\frac{\pi}{2}$$

for $0<|x|\le\pi$. By this, the sum on the right-hand side is immediately bounded by $\frac{\pi}{2}\int_{\mathbb{R}}|df(t)|$. Taking into account that, also for $0<|x|\le\pi$, one has

$$\left|\frac{s(x)-1}{x}\right|\le 1, \tag{2.47}$$

there holds

$$|1-s(x)|\left|\int_{\mathbb{R}} f(t)e^{ixt}\,dt\right|\le\left|x\int_{\mathbb{R}} f(t)e^{ixt}\,dt\right|.$$

Integrating by parts in the Stieltjes sense, we see that this last integral (without the modulus) equals

$$i\int_{\mathbb{R}} e^{ixt}\,df(t).$$

Hence, its modulus is bounded from above by $\int_{\mathbb{R}}|df(t)|$, as required. □

Remark 2.84 The constant π in (2.46) is by no means the best possible. In fact, the proof itself gives a better constant, namely $\frac{\pi}{2}+1$, and at the same time, this constant can easily be improved to $\frac{\pi}{2}+c$ with some $c<1$ by obtaining a sharper bound in (2.47).

To conclude this section, we give a generalized Euler–Maclaurin formula that refines (2.46) by involving higher-order derivatives of f.

Theorem 2.85 *Suppose that M is a nonnegative integer, and that for some $r \in \mathbb{N}$ the functions f and $f^{(r)}$ are of bounded variation on $[M,\infty)$. Suppose, moreover, that $\lim_{t\to\infty} f^{(\nu)}(t) = 0$ for $0 \le \nu \le r$. Then, for $x \in [-\pi,\pi]\setminus\{0\}$, the following relation holds:*

$$\sum_{k=M}^{\infty} f(k)e^{ixk} - \int_M^{\infty} f(t)e^{ixt}\,dt = \frac{1}{2}f(M)e^{ixM} + e^{iMx}\sum_{\nu=0}^{r}\frac{(-i)^{\nu-1}}{\nu!}f^{(\nu)}(M)h^{(\nu)}(x) + \varepsilon_{r,M}(f;x),$$

where

$$\varepsilon_{r,M}(f;x) \le \frac{4}{\pi^{r+1}}\int_M^{\infty} |df^{(r)}(t)|$$

uniformly in x, and

$$h(x) = \frac{1}{x} - \frac{1}{2}\cot\frac{x}{2}.$$

In order to prove this result, we need an analogue of the Poisson summation formula, which relates infinite sums of pointwise evaluations of functions and their Fourier transforms. This formula involves the use of Fourier transforms, which are yet to be defined. For $f\colon \mathbb{R} \to \mathbb{C}$ vanishing at infinity and $y \in \mathbb{R}$, the Fourier transform of f at y (denoted by $\widehat{f}(y)$) is

$$\widehat{f}(y) = \lim_{N\to\infty}\int_{-N}^{N} f(x)e^{-ixy}\,dx =: \int_{\mathbb{R}} f(x)e^{-ixy}\,dx.$$

A priori it is not clear that $\widehat{f}(y)$ should be well defined for all $y \in \mathbb{R}$. However, a simple integration by parts shows that if $f \in BV(\mathbb{R})$, then the above limit exists for all $y \in \mathbb{R}$ except possibly for $y = 0$.

Poisson's summation type formula reads as follows.

Proposition 2.86 *If F is a function of bounded variation on $\mathbb{R}$, for all $k \in \mathbb{Z}$, there holds*

$$2F(k) = F(k+0) + F(k-0) := \lim_{x\to k^+} F(x) + \lim_{x\to k^-} F(x),$$

and if $F(x) \to 0$ as $|x| \to \infty$, then for all $x \neq 2n\pi$, $n \in \mathbb{Z}$, there holds

$$\sum_{k=-\infty}^{\infty} F(k)e^{ikx} = \sum_{k=-\infty}^{\infty} \widehat{F}(2k\pi - x).$$

We do not include the proof here. Instead, we refer to Notes-64–65, where appropriate references are given along with further discussion around the Fourier transform.

Proof of Theorem 2.85 Applying Proposition 2.86 to the function

$$F(x) = \begin{cases} 0, & \text{if } x < M, \\ \frac{1}{2}f(M), & \text{if } x = M, \\ f(x), & \text{if } x > M, \end{cases}$$

where $x \not\equiv 0 \pmod{2\pi}$, we obtain

$$\begin{aligned} &\frac{1}{2}f(M)e^{iMx} + \sum_{k=M+1}^{\infty} f(k)e^{ikx} \\ &= \int_M^{\infty} f(t)e^{itx}\,dt + \sum_{1\le|k|<\infty} \int_M^{\infty} f(t)e^{-it(2k\pi - x)}\,dt, \end{aligned}$$

or equivalently

$$\begin{aligned} &\sum_{k=M}^{\infty} f(k)e^{ikx} - \int_M^{\infty} f(t)e^{itx}\,dt \\ &= \frac{1}{2}f(M)e^{iMx} + \sum_{1\le|k|<\infty} \int_M^{\infty} f(t)e^{-it(2k\pi - x)}\,dt \end{aligned} \tag{2.48}$$

(note that the summand corresponding to $k = M$ has been added on the left side). We now examine, for each k, the integral

$$\int_M^{\infty} f(t)e^{-it(2k\pi - x)}\,dt = \int_M^{\infty} f(t)e^{it(x-2k\pi)}\,dt.$$

Integrating by parts $r + 1$ times, we get

$$\begin{aligned} &\int_M^{\infty} f(t)e^{it(x-2k\pi)}\,dt \\ &= \sum_{\nu=0}^{r} \frac{(-1)^{\nu} f^{(\nu)}(M)}{(i(x - 2k\pi))^{\nu+1}} e^{iM(x-2k\pi)} + \frac{(-1)^{r+1}}{(i(x - 2k\pi))^{r+1}} \int_M^{\infty} df^{(r)}(t)e^{it(x-2k\pi)}\,dt \\ &= e^{iMx} \sum_{\nu=0}^{r} i^{\nu-1} \frac{f^{(\nu)}(M)}{(x - 2k\pi)^{\nu+1}} + i^{r+1} \frac{1}{(x - 2k\pi)^{r+1}} \int_M^{\infty} df^{(r)}(t)e^{it(x-2k\pi)}\,dt. \end{aligned}$$

On the one hand,

$$\sum_{1\le|k|<\infty} e^{iMx}\sum_{\nu=0}^{r} i^{\nu-1}\frac{f^{(\nu)}(M)}{(x-2k\pi)^{\nu+1}}$$
$$= e^{iMx}\sum_{\nu=0}^{r} i^{\nu-1}f^{(\nu)}(M)\sum_{1\le|k|<\infty}\frac{1}{(x-2k\pi)^{\nu+1}}.$$

Since

$$\sum_{1\le|k|<\infty}\frac{1}{(x-2k\pi)^{\nu+1}} = (-1)^{\nu-1}\sum_{1\le|k|<\infty}\frac{1}{(2k\pi-x)^{\nu+1}},$$

we obtain

$$e^{iMx}\sum_{\nu=0}^{r} i^{\nu-1}f^{(\nu)}(M)\sum_{1\le|k|<\infty}\frac{1}{(x-2k\pi)^{\nu+1}}$$
$$= e^{iMx}\sum_{\nu=0}^{r}(-i)^{\nu-1}f^{(\nu)}(M)\sum_{1\le|k|<\infty}\frac{1}{(2k\pi-x)^{\nu+1}}.$$

Now, the Mittag–Leffler expansion for the cotangent function

$$\pi\cot(\alpha\pi) = \frac{1}{\alpha} - \sum_{k\neq 0}\frac{1}{k-\alpha} \qquad \alpha\notin\mathbb{Z} \tag{2.49}$$

implies that

$$h(x) = \sum_{k\neq 0}\frac{1}{2k\pi - x} \quad \text{and} \quad h^{(p)}(x) = p!\sum_{k\neq 0}\frac{1}{(2k\pi-x)^{p+1}},$$

for $x \not\equiv 0 \pmod{2\pi}$. Hence,

$$\sum_{1\le|k|<\infty} e^{iMx}\sum_{\nu=0}^{r} i^{\nu-1}\frac{f^{(\nu)}(M)}{(x-2k\pi)^{\nu+1}} = e^{iMx}\sum_{\nu=0}^{r}\frac{(-i)^{\nu-1}}{\nu!}f^{(\nu)}(M)h^{(\nu)}(x).$$

On the other hand,

$$\left|\sum_{1\le|k|<\infty} i^{r+1}\frac{1}{(x-2k\pi)^{r+1}}\int_M^{\infty} df^{(r)}(t)e^{it(2k\pi-x)}\,dt\right|$$
$$\le \left(\int_M^{\infty}|df^{(r)}(t)|\right)\sum_{1\le|k|<\infty}\frac{1}{|x-2k\pi|^{r+1}}.$$

Let us now estimate the last sum. We split it in two sums, each containing the positive and negative indices k, respectively. Taking into account that $x \in [-\pi, \pi]$, we have

$$\sum_{k=1}^{\infty} \frac{1}{|x-2k\pi|^{r+1}} \leq \sum_{k=1}^{\infty} \frac{1}{(2k\pi-\pi)^{r+1}} = \frac{1}{(2\pi)^{r+1}} \sum_{k=1}^{\infty} \frac{1}{(k-1/2)^{r+1}}$$
$$= \frac{1}{(2\pi)^{r+1}} \left(2^{r+1} + \sum_{k=2}^{\infty} \frac{1}{(k-1/2)^{r+1}} \right).$$

In the same spirit as in the Maclaurin–Cauchy test, we can estimate the last sum by the corresponding integral, namely

$$\sum_{k=2}^{\infty} \frac{1}{(k-1/2)^{r+1}} \leq \int_1^{\infty} \frac{1}{(x-1/2)^{r+1}}\, dx = \int_{1/2}^{\infty} \frac{1}{x^{r+1}}\, dx = \frac{2^r}{r}.$$

Thus,

$$\sum_{k=1}^{\infty} \frac{1}{|x-2k\pi|^{r+1}} \leq \frac{1}{(2\pi)^{r+1}} \left(2^{r+1} + \frac{2^r}{r} \right) \leq \frac{2^{r+2}}{(2\pi)^{r+1}} = \frac{2}{\pi^{r+1}}.$$

As for the series with negative indices k, we have

$$\sum_{k=-\infty}^{-1} \frac{1}{|x-2k\pi|^{r+1}} = \sum_{k=1}^{\infty} \frac{1}{(x+2k\pi)^{r+1}} \leq \sum_{k=1}^{\infty} \frac{1}{(2k\pi-\pi)^{r+1}} \leq \frac{2}{\pi^{r+1}}.$$

Thus,

$$\sum_{1 \leq |k| < \infty} \int_M^{\infty} f(t) e^{-it(2k\pi - x)} = e^{iMx} \sum_{\nu=0}^{r} \frac{(-i)^{\nu-1}}{\nu!} f^{(\nu)}(M) h^{(\nu)}(x) + \varepsilon_{r,M}(f;x),$$

where $\varepsilon_{r,M}(f;x) \leq \dfrac{4}{\pi^{r+1}} \displaystyle\int_M^{\infty} |df^{(r)}(t)|$. Combining the last equality with (2.48), the conclusion follows. □

Note that the proved formula remains valid at $x = 0$, provided that $f \in L^1(\mathbb{R})$, that is, if $\int_M^{\infty} |f(x)|\, dx$ converges (or if the series $\sum |f(k)|$ converges).

2.4.2 Ermakov-type integral tests

Ermakov's test is related to the integral test. Although such a relation may not be evident by looking at the statement of Ermakov's test, it is in its proof where Maclaurin–Cauchy's test comes into play. Throughout this section we will state several versions of Ermakov's test, namely the original test and some of its generalizations. We will only give the proof of the most general result (Theorem 2.90), which will imply the previous ones as particular cases.

In its original form, Ermakov's test reads as follows.

Theorem 2.87 (Ermakov's test) *Let $f\colon [1,\infty) \to (0,\infty)$ be a decreasing function. If for t large enough $\frac{f(e^t)e^t}{f(t)} \le q < 1$, then the series $\sum_{k=1}^{\infty} f(k)$ converges, while if $\frac{f(e^t)e^t}{f(t)} \ge 1$, then the series $\sum_{k=1}^{\infty} f(k)$ diverges.*

One can get a wider family of tests by replacing e^t with a positive increasing function $\varphi(t)$ that satisfies certain properties. Furthermore, as in most of the classical tests involving decreasing sequences/functions, Ermakov's test is also valid for *WM* functions. Taking these two observations into account, we present the corresponding generalization of Ermakov's test.

Theorem 2.88 *Let f be a WM function, and let $\varphi\colon [1,\infty) \to (0,\infty)$ be a monotone increasing function. Suppose that φ has a continuous derivative and, moreover, that $\varphi(t) > t$ for all $t > t_0$.*

If for t large enough

$$\frac{f(\varphi(t))\varphi'(t)}{f(t)} \le q < 1, \tag{2.50}$$

then the series $\sum_{k=1}^{\infty} f(k)$ converges, while if

$$\frac{f(\varphi(t))\varphi'(t)}{f(t)} \ge 1, \tag{2.51}$$

then the series $\sum_{k=1}^{\infty} f(k)$ diverges.

Another direction for generalizations of Ermakov's test is suggested by replacing t with a more general function $\psi(t)$ in the denominators at (2.50) and (2.51). For decreasing functions, the statement is as follows.

Theorem 2.89 (generalized Ermakov's test) *Let $f\colon (0,\infty) \to (0,\infty)$ be a decreasing function, and $\varphi, \psi\colon (0,\infty) \to (0,\infty)$ be differentiable increasing functions such that*

$$\lim_{x\to\infty} \varphi(x) = \lim_{x\to\infty} \psi(x) = \infty,$$

and, additionally,

$$\varphi(x) > \psi(x) \quad \text{and} \quad \psi'(x) > 0 \quad \text{for all } x > 0.$$

If

$$\lim_{x\to\infty} \frac{f(\varphi(t))\varphi'(t)}{f(\psi(t))\psi'(t)} = r \in (0,\infty),$$

then the series $\sum_{n=1}^{\infty} f(n)$ converges if $r < 1$ and diverges if $r > 1$.

To conclude this section, we state and prove the most general version of Ermakov's test, where all the above generalizations are incorporated.

Theorem 2.90 *The conclusion of Theorem 2.89 holds if we replace the assumption that f is decreasing by $f \in WM$.*

Before proving this result, we must recall Cauchy's criterion for convergence of integrals (which is a counterpart of (2.1)).
Given a locally integrable function $f\colon (0,\infty) \to \mathbb{C}$ and $a \geq 0$, we have that

$$\int_a^\infty f(x)\,dx \quad \textit{converges if and only if} \quad \left|\int_{y_1}^{y_2} f(x)\,dx\right| \to 0 \tag{2.52}$$

as $y_2 > y_1 \to \infty$.

Proof of Theorem 2.90 Let first $r < 1$. Then there exists $\varepsilon > 0$ such that for sufficiently large y and any $x > y$,

$$\frac{f(\varphi(x))\varphi'(x)}{f(\psi(x))\psi'(x)} < r + \varepsilon < 1.$$

Hence,

$$\int_{y_1}^{y_2} f(\varphi(x))\varphi'(x)\,dx < (r+\varepsilon)\int_{y_1}^{y_2} f(\psi(x))\psi'(x)\,dx, \qquad \text{for any } y < y_1 < y_2,$$

or equivalently

$$\int_{\varphi(y_1)}^{\varphi(y_2)} f(x)\,dx < (r+\varepsilon)\int_{\psi(y_1)}^{\psi(y_2)} f(x)\,dx. \tag{2.53}$$

By (2.53) and the fact that $\varphi(x) > \psi(x)$ for all $x > 0$, we get

$$\begin{aligned}\int_{\psi(y_1)}^{\varphi(y_1)} f(x)\,dx - \int_{\psi(y_2)}^{\varphi(y_2)} f(x)\,dx &= \int_{\psi(y_1)}^{\psi(y_2)} f(x)\,dx - \int_{\varphi(y_1)}^{\varphi(y_2)} f(x)\,dx \\ &> (1-r-\varepsilon)\int_{\psi(y_1)}^{\psi(y_2)} f(x)\,dx.\end{aligned}$$

Therefore,

$$\begin{aligned}\int_{\psi(y_1)}^{\psi(y_2)} f(x)\,dx &< \frac{1}{1-r-\varepsilon}\left(\int_{\psi(y_1)}^{\varphi(y_1)} f(x)\,dx - \int_{\psi(y_2)}^{\varphi(y_2)} f(x)\,dx\right) \\ &< \frac{1}{1-r-\varepsilon}\int_{\psi(y_1)}^{\varphi(y_1)} f(x)\,dx,\end{aligned}$$

uniformly in $y_2 > y_1 > y$. By fixing y_1 and letting $y_2 \to \infty$, it follows from the last inequality that $\int_1^\infty f(x)\,dx$ converges, and by the Maclaurin–Cauchy integral test (Theorem 2.78), the series $\sum_{n=1}^\infty f(n)$ converges as well.

Let now $r > 1$. Then there exists $\varepsilon > 0$ such that for sufficiently large y and $x > y$,

$$\frac{f(\varphi(x))\varphi'(x)}{f(\psi(x))\psi'(x)} > r - \varepsilon > 1,$$

which implies that

$$\int_{\varphi(y_1)}^{\varphi(y_2)} f(x)\,dx > (r-\varepsilon)\int_{\psi(y_1)}^{\psi(y_2)} f(x)\,dx, \qquad \text{for any } y < y_1 < y_2.$$

It then follows that

$$\begin{aligned}\int_{\psi(y_2)}^{\varphi(y_2)} f(x)\,dx - \int_{\psi(y_1)}^{\varphi(y_1)} f(x)\,dx &= \int_{\varphi(y_1)}^{\varphi(y_2)} f(x)\,dx - \int_{\psi(y_1)}^{\psi(y_2)} f(x)\,dx \\ &> (r-\varepsilon-1)\int_{\psi(y_1)}^{\psi(y_2)} f(x)\,dx > 0.\end{aligned}$$

We conclude that

$$\int_{\psi(y_2)}^{\varphi(y_2)} f(x)\,dx > (r-\varepsilon-1)\int_{\psi(y_1)}^{\psi(y_2)} f(x)\,dx + \int_{\psi(y_1)}^{\varphi(y_1)} f(x)\,dx \geq C > 0.$$

We now fix y_1 and let $y_2 \to \infty$, so that $\varphi(y_2) > \psi(y_2) \to \infty$ as $y_2 \to \infty$. By Cauchy's criterion for convergence of integrals (2.52), the divergence of $\int_1^\infty f(x)\,dx$ follows. Finally, by Theorem 2.78, the series $\sum_{n=1}^{\infty} f(n)$ diverges as well. □

2.4.3 Cauchy's condensation test and its generalizations

The well-known *Cauchy's condensation test* states that *if $\{a_k\}$ is a positive monotone decreasing null sequence, then the series $\sum a_k$ and the series*

$$\sum_{k=1}^{\infty} 2^k a_{2^k} \tag{2.54}$$

converge or diverge simultaneously.

A direct generalization of Cauchy's condensation test is obtained by replacing the index sequence $\{2^k\}$ in (2.54) by another *lacunary* sequence $\{u_k\}$ with similar properties. We say that a positive sequence $\{u_k\}$ (which we will assume to take values on $\mathbb{N}$, since it will be a sequence of indices) is lacunary if it is increasing and

$$\frac{u_{k+1}}{u_k} \geq q > 1$$

for all $k \in \mathbb{N}$. More generally, we say that $\{u_k\} \in \Lambda$ if it is increasing, takes values on $\mathbb{N}$, and can be split into finitely many lacunary sequences, or, in other words, there is $r \in \mathbb{N}$ such that

$$\frac{u_{k+r}}{u_k} \geq q > 1$$

for all $k \in \mathbb{N}$, or, equivalently, if for any n,

$$\sum_{k=1}^{n} u_k \leq C u_n. \tag{2.55}$$

The following generalization of Cauchy's condensation test involving lacunary index sequences also admits the more general *WMS* assumption rather than monotonicity of the sequence $\{a_k\}$.

Theorem 2.91 (generalized Cauchy's condensation test) *Let $\{a_k\} \in WMS$ be a null sequence and let $\{u_k\} \in \Lambda$ be such that $u_{k+1} = O(u_k)$. Then the three series*

$$\sum_{k=1}^{\infty} a_k, \qquad \sum_{k=1}^{\infty} u_k |a_{u_k} - a_{u_{k+1}}|, \qquad \textit{and} \qquad \sum_{k=1}^{\infty} u_k a_{u_k}$$

converge or diverge simultaneously.

Proof Let us begin with proving the equiconvergence of the first and third series. Since $u_{k+1} = O(u_k)$ and $\{a_k\} \in WMS$, there exists a constant M so that $a_n \le M a_{u_k}$ for all $u_k \le n \le u_{k+1}$ (note that for such values of n one also has $a_{u_{k+1}} \le M a_n$). Thus,

$$(u_{k+1} - u_k) a_{u_{k+1}} \le M \sum_{n=u_k}^{u_{k+1}-1} a_n \le M^2 (u_{k+1} - u_k) a_{u_k},$$

and by the comparison test, the result already follows in one direction, since

$$\sum_{n=u_1}^{\infty} a_n = \sum_{j=1}^{\infty} \sum_{n=u_j}^{u_{j+1}-1} a_n.$$

In order to prove the reverse direction, let $r \in \mathbb{N}$ be such that $u_{k+r}/u_k \ge q > 1$ for all k (r can be found since $\{u_k\} \in \Lambda$). Moreover, we can also find a sequence $\{\ell_k\}_{k \ge r}$ such that for every k, $0 \le k - \ell_k \le r - 1$, and

$$u_{\ell_k+1} - u_{\ell_k} \ge \frac{u_{k+1} - u_{k+1-r}}{r}. \tag{2.56}$$

Indeed, since we can write $u_{k+1} - u_{k+1-r}$ as the sum of r positive terms,

$$u_{k+1} - u_{k+1-r} = \sum_{j=0}^{r-1} (u_{k+1-j} - u_{k-j}),$$

there is at least one term in the last sum that is larger than or equal to the average of all the terms, or, in other words, for some $\ell_k \in \{k - r + 1, \dots, k\}$, (2.56) holds. Using that $\{a_k\} \in WMS$ and the estimate $u_{k+1} = O(u_k)$, we obtain, for $k \ge r$,

$$u_{k-r+1} a_{u_{\ell_k+1}} \ge C u_{\ell_k+1} a_{u_{\ell_k+1}} \ge C u_{k+1} a_{u_{\ell_k+1}} \ge C u_{k+1} a_{u_{k+1}},$$

where throughout this proof C denotes a generic constant (not the *WMS* constant of $\{a_k\}$) that may vary in different occurrences. Combining this inequality with those obtained above, we conclude

$$\sum_{k=r}^{\infty} u_{k+1} a_{u_{k+1}} \leq C \sum_{k=r}^{\infty} (u_{\ell_k+1} - u_{\ell_k}) a_{u_{\ell_k+1}} \leq C \sum_{k=1}^{\infty} (u_{k+1} - u_k) a_{u_{k+1}}$$

$$\leq C \sum_{k=u_1}^{\infty} a_k \leq C \sum_{k=1}^{\infty} (u_{k+1} - u_k) a_{u_k} \leq C \sum_{k=1}^{\infty} u_k a_{u_k},$$

which yields the equiconvergence of the first and third series in the statement.

Let us now study the equiconvergence of the second and third series. Applying Abel's transformation, we get

$$\sum_{k=1}^{n} u_k a_{u_k} = \left(\sum_{j=1}^{n} u_j \right) a_{u_n} + \sum_{k=1}^{n-1} \left(\sum_{j=1}^{k} u_j \right) (a_{u_k} - a_{u_{k+1}})$$

$$\leq C u_n \sum_{k=n}^{\infty} |a_{u_k} - a_{u_{k+1}}| + C \sum_{k=1}^{n-1} u_k |a_{u_k} - a_{u_{k+1}}|$$

$$\leq C \sum_{k=1}^{\infty} u_k |a_{u_k} - a_{u_{k+1}}|,$$

where we have used property (2.55). On the other hand, since $\{a_k\} \in WMS$ and $u_{k+1} = O(u_k)$, we easily derive that

$$|a_{u_k} - a_{u_{k+1}}| \leq C a_{u_k},$$

and hence

$$\sum_{k=1}^{\infty} u_k |a_{u_k} - a_{u_{k+1}}| \leq C \sum_{k=1}^{\infty} u_k a_{u_k},$$

as desired. □

As an immediate consequence of Theorem 2.91, we obtain Cauchy's condensation test for *WMS*.

Corollary 2.92 *Let $\{a_k\} \in WMS$. Then the three series*

$$\sum_{k=1}^{\infty} a_k, \qquad \sum_{k=0}^{\infty} 2^k |a_{2^k} - a_{2^{k+1}}|, \qquad \text{and} \qquad \sum_{k=0}^{\infty} 2^k a_{2^k}$$

converge or diverge simultaneously.

In relation to Corollary 2.92, we give the following straightforward observation:

Theorem 2.93 *Let $\sum a_k$ be a positive series with nonincreasing terms. Then the series $\sum a_k$ converges if and only if there exists a sequence of positive integers $u_0 < u_1 < u_2 < \cdots$ such that the series*

$$\sum_{k=0}^{\infty} (u_{k+1} - u_k) a_{u_k}$$

converges.

Proof For the "only if" part we simply use Corollary 2.92 (i.e., with $u_k = 2^k$). For the "if" part, for any increasing sequence of positive integers $\{u_k\}$ we have, since a_k is nonincreasing,

$$\sum_{k=u_0}^{\infty} a_k = \sum_{k=0}^{\infty} \sum_{n=u_k}^{u_{k+1}-1} a_n \le \sum_{k=0}^{\infty} \sum_{n=u_k}^{u_{k+1}-1} a_{u_k} = \sum_{k=0}^{\infty} (u_{k+1} - u_k) a_{u_k}.$$

If there exists a sequence $\{u_k\}$ such that the last series is finite, the convergence of $\sum a_k$ follows. □

Note that in Theorem 2.93 one may not replace the wording "there exists an increasing sequence $\{u_k\}$" with "for every increasing sequence $\{u_k\}$". Indeed, one may easily check that for $a_k = \frac{1}{k^2}$ and $u_k = 2^{k!}$, the series $\sum a_k$ converges, but $(u_{k+1} - u_k) a_{u_k} \not\to 0$, and so the series $\sum (u_{k+1} - u_k) a_{u_k}$ diverges.

There is another generalization of Cauchy's condensation test, where differences of the sequence $\{u_k\}$ are considered rather than differences of the sequence $\{a_{u_k}\}$. This assertion is due to *Schlömilch*. Let

$$\overleftarrow{\Delta}\, u_k := u_{k+1} - u_k$$

denote the "backward" difference. Note that $\overleftarrow{\Delta}\, u_k = -\Delta u_k$. The statement is as follows.

Theorem 2.94 (Schlömilch's test) *Let $\sum a_k$ be a positive series with nonincreasing terms, and let $u_0 < u_1 < u_2 < \cdots$ be a sequence of positive integers such that*

$$\frac{\Delta u_k}{\Delta u_{k-1}} = \frac{\overleftarrow{\Delta}\, u_k}{\overleftarrow{\Delta}\, u_{k-1}} \le C.$$

Then the series $\sum a_k$ converges if and only if the series

$$\sum_{k=1}^{\infty} \overleftarrow{\Delta}\, u_k a_{u_k} = \sum_{k=1}^{\infty} (u_{k+1} - u_k) a_{u_k}$$

does.

Schlömilch's test in turn was generalized by *de la Vallée Poussin*. We now give an extension of the latter generalization. As above, it is just the same assertion but with monotonicity foregoing the *WMS* condition. For convenience, we state this result in terms of a function $f \in WM$ and the convergence of the integral $\int_1^\infty f(t)\,dt$ rather than of a sequence $\{a_k = f(k)\} \in WMS$ and the convergence of the series $\sum a_k = \sum f(k)$ (by the Maclaurin–Cauchy integral test we may always restrict ourselves to such a setting).

Theorem 2.95 *Let $\{u_k\}$ be an increasing sequence of positive numbers such that $u_{k+1} = O(u_k)$ and $u_k \to \infty$. Let f be a WM function. Then, the series*

$$\sum_{k=1}^{\infty} f(u_k)\, \overleftarrow{\Delta}\, u_k \tag{2.57}$$

and

$$\sum_{k=1}^{\infty} f(u_{k+1}) \overleftarrow{\Delta} u_k \tag{2.58}$$

both converge (or diverge) with $\int_1^\infty f(t)\,dt$.

Proof Given $k \in \mathbb{N}$, let $N_k \in \mathbb{N}$ be the number such that $2^{N_k} \leqslant u_k < 2^{N_k+1}$. Since

$$2^{N_{k+1}-N_k-1} = \frac{2^{N_{k+1}}}{2^{N_k+1}} < \frac{u_{k+1}}{u_k} \le C_1,$$

it follows that $N_{k+1} - N_k \le C_2$.

For $t \in [u_k, u_{k+1}]$, let n be the integer such that $2^n \le t < 2^{n+1}$. Then $n \in \{N_k, \ldots, N_{k+1}\}$, and by the *WM* property we obtain

$$f(t) \le Cf(2^n) \le C^2 f\left(2^{n-1}\right) \le \cdots \le C^{n-N_k} f\left(2^{N_k+1}\right) \le C^{n-N_k+1} f(u_k),$$

for $n > N_k$. Now, since $n - N_k + 1 \le N_{k+1} - N_k + 1 \le C_2 + 1$, it follows that for every $t \in [u_k, u_{k+1}]$,

$$f(t) \lesssim f(u_k).$$

Therefore,

$$(u_{k+1} - u_k) f(u_{k+1}) \lesssim \int_{u_k}^{u_{k+1}} f(t)\,dt \lesssim (u_{k+1} - u_k) f(u_k).$$

Hence,

$$\sum_{k=1}^{\infty} f(u_{k+1}) \overleftarrow{\Delta} u_k \lesssim \int_{u_1}^{\infty} f(t)\,dt \lesssim \sum_{k=1}^{\infty} f(u_k) \overleftarrow{\Delta} u_k, \tag{2.59}$$

which proves only one direction of the convergence/divergence claim. It remains to prove that the divergence of the integral $\int_1^\infty f(t)\,dt$ implies the divergence of (2.58), and that the convergence of the same integral implies the convergence of (2.57). Let us proceed to the divergence instance first. Let $\{r_k\} \subset \mathbb{N}$ be such that $2^{r_k-1} u_k \le u_{k+1} < 2^{r_k} u_k$ for every k. Since $u_{k+1} = O(u_k)$, it follows that $r_k \le r$ for some $r \in \mathbb{N}$. Applying the right inequality of (2.59) to the sequence $\{2^r u_k\}$ in place of $\{u_k\}$, we obtain

$$\int_{2^r u_1}^{\infty} f(t)\,dt \lesssim \sum_{k=1}^{\infty} f(2^r u_k) \overleftarrow{\Delta} (2^r u_k) = 2^r \sum_{k=1}^{\infty} f(2^r u_k) \overleftarrow{\Delta} u_k,$$

and hence the divergence of $\int_1^\infty f(t)\,dt$ implies the divergence of the last series. By the *WM* condition, we have

$$f(u_{k+1}) \ge C^{-1} f\left(2^{r_k} u_k\right) \ge C^{-(r-r_k+1)} f\left(2^r u_k\right) \ge C^{-r} f\left(2^r u_k\right),$$

which proves the divergence of the series on the left side of (2.59). As for the remaining convergence claim, we have

$$\int_{2^{-r}u_1}^{\infty} f(t)\,dt = \sum_{k=1}^{\infty}\int_{2^{-r}u_k}^{2^{-r}u_{k+1}} f(t)\,dt \geq C^{-r}\sum_{k=1}^{\infty} f(2^{-r}u_{k+1})\overleftarrow{\Delta}(2^{-r}u_k)$$

$$= 2^{-r}C^{-r}\sum_{k=1}^{\infty} f(2^{-r}u_{k+1})\overleftarrow{\Delta}u_k.$$

The *WM* condition and the fact that $u_{k+1} < 2^r u_k$ imply

$$C^{-r}f(u_k) \leq f(2^{-r}u_{k+1}),$$

which proves the convergence of the series on the right side of (2.59) (provided that $\int_{u_1}^{\infty} f(t)\,dt$ converges) and completes the proof. □

The following test, due to Lobachevskii, can be seen as a dual of Cauchy's condensation test.

Theorem 2.96 *Let $\{a_n\}$ be a nonincreasing sequence such that $1 = a_1 > a_2$. For $m \in \mathbb{N} \cup \{0\}$, let $p_m > 1$ be the unique natural number such that*

$$a_{p_m} \geq 2^{-m} \qquad \text{and} \qquad a_{p_m+1} < 2^{-m}.$$

Then

$$\frac{1}{2}\sum_{m=1}^{\infty}(p_m - 1)2^{-m} \leq \sum_{n=2}^{\infty} a_n \leq \sum_{m=1}^{\infty}(p_m - 1)2^{-m}.$$

In particular, the series $\sum_{m=0}^{\infty} p_m 2^{-m}$ and $\sum a_n$ converge or diverge together.

Proof The upper inequality is proved in a very similar fashion as in Theorem 2.4. For $n > 1$, since $a_n < 1$, we have

$$a_n = t_n 2^{-m_n},$$

with $m_n \in \mathbb{N} \cup \{0\}$ and $\frac{1}{2} \leq t_n < 1$ uniquely determined. Define the sets

$$E_m = \{n \in \mathbb{N} : a_n = 2^{-m}t_n\},$$

where $\frac{1}{2} \leq t_n < 1$. It follows from the definitions that $p_0 - 1 = 0$, and

$$p_m - 1 = \sum_{j=0}^{m-1} |E_j|.$$

If we now define, for $m \in \mathbb{N} \cup \{0\}$, the quantity L_m as the number of elements a_n containing 2^{-m} in their binary decomposition, namely with $\varepsilon_m = 1$ in the representation

$$a_n = \sum_{j=1}^{\infty} \varepsilon_j 2^{-j}, \qquad \text{with } \{\varepsilon_j\} \subset \{0, 1\},$$

then one has

$$L_m \le \sum_{j=0}^{m-1} |E_j| = p_m - 1$$

(see also the proof of Theorem 2.4 for the details). On the other hand, note that if $n \in E_{m-1}$, then a_n contains the term 2^{-m} in its binary decomposition, by the definition of E_{m-1}. Thus, $|E_{m-1}| \le L_m$, and since $|E_{m-1}| = p_m - p_{m-1}$, we conclude

$$p_m - p_{m-1} \le L_m \le p_m - 1.$$

Hence,

$$\sum_{m=1}^{\infty} (p_m - p_{m-1})2^{-m} \le \sum_{m=1}^{\infty} L_m 2^{-m} = \sum_{n=2}^{\infty} a_n \le \sum_{m=1}^{\infty} (p_m - 1)2^{-m}.$$

We now note that

$$\sum_{m=1}^{\infty} p_{m-1}2^{-m} = \frac{1}{2}\sum_{m=0}^{\infty} p_m 2^{-m} = \frac{1}{2} + \frac{1}{2}\sum_{m=1}^{\infty} p_m 2^{-m} = \frac{1}{2}\sum_{m=1}^{\infty} (p_m + 1)2^{-m},$$

since $p_0 = 1$. Hence,

$$\sum_{m=1}^{\infty} (p_m - p_{m-1})2^{-m} = \frac{1}{2}\sum_{m=1}^{\infty} (p_m - 1)2^{-m}.$$

Combining this identity with the previous inequalities, we finally obtain

$$\frac{1}{2}\sum_{m=1}^{\infty} (p_m - 1)2^{-m} \le \sum_{n=2}^{\infty} a_n \le \sum_{m=1}^{\infty} (p_m - 1)2^{-m}.$$ □

2.4.4 Sapogov's test

This test not only looks special, it indeed is such in the sense that, contrary to the other tests for monotone positive series, it does not allow refinements toward weak monotonicity or general monotonicity.

Theorem 2.97 (Sapogov's test) *If $\{b_k\}$ is a positive increasing sequence, then the series*

$$\sum_{k=1}^{\infty} \left(1 - \frac{b_k}{b_{k+1}}\right) \tag{2.60}$$

as well as

$$\sum_{k=1}^{\infty} \left(\frac{b_{k+1}}{b_k} - 1\right) \tag{2.61}$$

converge if the sequence $\{b_k\}$ is bounded, or diverge otherwise.

Proof We may assume that

$$\lim_{k\to\infty}\frac{b_k}{b_{k+1}}=\lim_{k\to\infty}\frac{b_{k+1}}{b_k}=1,$$

since otherwise $\{b_n\}$ is unbounded, and furthermore the given series trivially diverge.

Since for k large enough one has

$$1-\frac{b_k}{b_{k+1}}\asymp\log\left(\frac{b_k}{b_{k+1}}\right),$$

the convergence of the series (2.60) is equivalent to the convergence of the series

$$\sum_{k=1}^{\infty}\log\left(\frac{b_k}{b_{k+1}}\right). \tag{2.62}$$

Since the partial sums of the latter are

$$\sum_{k=1}^{n}\log\left(\frac{b_k}{b_{k+1}}\right)=\log\left(\frac{b_1}{b_{n+1}}\right),$$

we conclude that the convergence of (2.60) is equivalent to the boundedness of $\{b_k\}$. Similarly, the boundedness of $\{b_k\}$ is equivalent to the convergence of (2.61). Indeed, since for k large enough,

$$\frac{b_{k+1}}{b_k}-1\asymp\log\left(\frac{b_{k+1}}{b_k}\right)=-\log\left(\frac{b_k}{b_{k+1}}\right),$$

the convergence of (2.61) is equivalent to that of (2.62) (and thus to the boundedness of $\{b_k\}$, as we had established before). □

Theorem 2.97 cannot be true if $\{b_k\}$ (as well as $\{b_k^{-1}\}$) is *WMS*. Taking $b_k=1$, $k\neq 2^m$ and $b_k=2$, $k=2^m$, we have that $\{b_k\}\in$ *WMS* is a bounded sequence, but in this case the series (2.60) diverges. Clearly, a necessary condition for $\sum_{k=1}^{\infty}(1-\frac{b_k}{b_{k+1}})$ to converge is that $\frac{b_k}{b_{k+1}}\to 1$, which is not satisfied in this example (note that the series (2.61) also diverges, as the necessary condition $\frac{b_{k+1}}{b_k}\to 1$ is not fulfilled).

2.4.5 Abel–Olivier's test and related topics

In the theory of series with monotone terms, statements not directly about convergence/divergence of the series but about the behavior of its terms take their deserved place. The most famous example is *Abel–Olivier's kth term test.* One of our goals is to show that this test and certain of its extensions are more substantial than it has been meant before.

Theorem 2.98 *Let $\{a_k\}$ be a nonincreasing sequence. If the series $\sum a_k$ converges, then $ka_k\to 0$ as $k\to\infty$.*

We emphasize that Abel–Olivier's test gives only a necessary condition for a series with decreasing coefficients a_n to converge (which by the way is also valid for *WMS*, as we will prove below). However, an application of Abel's transformation (2.2) yields

$$\sum_{k=1}^{n} a_k = na_n + \sum_{k=1}^{n-1} k(a_k - a_{k+1}).$$

This simple observation allows one to refine the original test as follows.

Theorem 2.99 (Extended Abel–Olivier's test) *If $\{a_k\} \in WMS$, then $\sum_{k=1}^{\infty} a_k$ converges if and only if $ka_k \to 0$ and*

$$\sum_{k=1}^{\infty} k(a_k - a_{k+1})$$

converges. Furthermore, in this case $\sum_{k=1}^{\infty} a_k = \sum_{k=1}^{\infty} k(a_k - a_{k+1})$.

Instead of proving this statement, we discuss in full detail the following generalization, where we consider the series $\sum_{k=1}^{\infty} a_k b_k$ with a positive weight $\{b_k\}$ (note that in the above classical case $b_k = 1$).

Theorem 2.100 *Let $\{a_k\} \in WMS$ and $\{b_k\}$ be a sequence of nonnegative numbers such that*

$$B_n = \sum_{k=1}^{n} b_k \le M \sum_{k=\lfloor n/2 \rfloor}^{n} b_k, \qquad M > 0.$$

Then $\sum_{k=1}^{\infty} a_k b_k$ converges if and only if

$$B_n a_n \to 0 \qquad \text{and} \qquad \sum_{k=1}^{\infty} B_k \Delta a_k \text{ converges.} \tag{2.63}$$

Furthermore, in this case $\sum_{k=1}^{\infty} a_k b_k = \sum_{k=1}^{\infty} B_k \Delta a_k$.

Proof By Abel's transformation (2.2), we have

$$\sum_{k=1}^{n} a_k b_k = a_n B_n + \sum_{k=1}^{n-1} B_k \Delta a_k. \tag{2.64}$$

If (2.63) holds, it is clear that $\sum_{k=1}^{\infty} a_k b_k$ converges. On the other hand, if the latter converges, it suffices to show that $a_n B_n \to 0$ as $n \to \infty$, by (2.64). Since $\{a_k\} \in WMS$, we have $a_n \le C^2 a_k$ for $k \in \{\lfloor n/2 \rfloor, \dots, n\}$. Thus,

$$a_n B_n \le C^2 M \sum_{k=\lfloor n/2 \rfloor}^{n} a_k b_k \to 0 \quad \text{as } n \to \infty,$$

by the Cauchy criterion for convergence of series (2.1). □

If $\alpha > -1$, the sequence $b_n = n^\alpha$ satisfies the hypothesis of Theorem 2.100. The case $\alpha = 0$ corresponds to Theorem 2.99.

Remark 2.101 It is worth noting that the Abel–Olivier test (Theorem 2.99) holds under the assumption that $\{a_k\} \in \overline{WMS}$ (see Definition 1.38). Indeed, for such a class of sequences, there holds

$$ka_k \lesssim \sum_{n=\lfloor k/c \rfloor}^{\lfloor ck \rfloor} a_n \to 0 \qquad \text{as } k \to \infty,$$

by Cauchy's criterion. What is more, the extended Abel–Olivier test is valid for $\{a_k\} \in \overline{WMS}(p)$, $p \geq 1$ (see Notes-81 and Notes-18 in Chapter 1).

The following is a version of Theorem 2.100 for nonincreasing sequences $\{a_k\}$. It differs from Theorem 2.100 in the sense that, by imposing stronger assumptions on $\{a_n\}$, one may relax the hypotheses about $\{b_n\}$.

Theorem 2.102 *Let $\{b_n\}$ be a complex-valued sequence and let $\{a_n\}$ be a nonincreasing positive sequence that vanishes at infinity. Then the series $\sum a_k b_k$ converges if and only if* (2.63) *holds.*

Proof By (2.64), one direction of the assertion is clear. For the reverse direction, if $\sum a_k b_k$ converges, it suffices to show that $a_n B_n \to 0$ as $n \to \infty$. Define

$$S_n = \sum_{k=1}^{n} a_k b_k, \qquad S_0 = 0.$$

Changing b_1, we may assume that $S_n \to 0$ as $n \to \infty$. Indeed, defining $S = \sum_{k=2}^{\infty} a_k b_k$, for $b_1 = -\frac{S}{a_1}$ it is clear that $S_n \to 0$. Now, for any $\varepsilon > 0$ there exists n_0 such that $|S_n| < \varepsilon$ for $n \geq n_0$. For such choices of n, we have, by Abel's transformation,

$$\begin{aligned} |a_n B_n| &= a_n \left| \sum_{k=1}^{n} b_k \right| = a_n \left| \sum_{k=1}^{n} \frac{S_k - S_{k-1}}{a_k} \right| = a_n \left| \frac{S_n}{a_n} + \sum_{k=1}^{n-1} S_k \left(\frac{1}{a_k} - \frac{1}{a_{k+1}} \right) \right| \\ &\leq |S_n| + a_n \max_{k \in \mathbb{N}} |S_k| \sum_{k=1}^{n_0-1} \left(\frac{1}{a_k} - \frac{1}{a_{k+1}} \right) + a_n \max_{k \geq n_0} |S_k| \sum_{k=n_0}^{n} \left(\frac{1}{a_k} - \frac{1}{a_{k+1}} \right) \\ &\leq 2\varepsilon + \frac{a_n}{a_1} \max_{k \in \mathbb{N}} |S_k|. \end{aligned}$$

Since $a_n \to 0$ as $n \to \infty$ and $\varepsilon > 0$ is arbitrary, the conclusion follows. □

By requiring in Theorem 2.102 that the sequence $\{a_k\}$ is decreasing (rather than $\{a_k\} \in WMS$), we can drop the hypothesis $B_n \lesssim \sum_{k=\lfloor n/2 \rfloor}^{n} b_k$ in Theorem 2.100 and still obtain the same conclusion. This assumption cannot be omitted if $\{a_k\}$ is not decreasing, as the following example shows: if $a_n = \sqrt{n}$ and $b_n = n^{-2}$, then $\{a_n\}$ is clearly *WMS*, and on the one hand,

$$\sum_{k=1}^{\infty} a_k b_k = \sum_{k=1}^{\infty} \frac{1}{k^{3/2}} < \infty,$$

while, on the other hand, $B_n \geq 1$ for every n. Therefore $a_n B_n \geq \sqrt{n} \not\to 0$.

A natural question is whether Theorem 2.102 holds with replacing the assumption that $\{a_k\}$ is monotone by the more general $\{a_k\} \in WMS$, and assuming, additionally, that $\{a_k\}$ vanishes at infinity (that is, we have a stronger assumption on $\{a_k\}$ and a weaker assumption on $\{b_k\}$ as compared with Theorem 2.100). In this case the conclusion is again not true, as the following example shows. Define

$$a_n = \begin{cases} \frac{1}{n}, & \text{if } n \text{ is odd}, \\ \frac{2}{n}, & \text{if } n \text{ is even}, \end{cases} \qquad b_n = \begin{cases} 1, & \text{if } n \text{ is odd}, \\ -\frac{1}{2}, & \text{if } n \text{ is even}. \end{cases}$$

It is clear that $\{a_k\} \in WMS$ and, moreover, it vanishes at infinity. Further,

$$\sum_{n=1}^{\infty} a_n b_n = \sum_{n=1}^{\infty} \frac{(-1)^{n+1}}{n} = \log 2,$$

although $a_{2n} B_{2n} = \frac{2}{2n} \cdot \frac{n}{2} = \frac{1}{2} \not\to 0$.

Another interesting observation is that the reverse statement of Theorem 2.102 does not hold. That is, for any positive decreasing sequence $\{a_n\}$, one can find a (nonnegative) sequence $\{b_n\}$ such that $a_n B_n \to 0$ but $\sum a_k b_k = \infty$. To prove this claim, given $\{a_k\}$ decreasing to zero, consider a sequence $\{n_k\} \subset \mathbb{N}$ such that $\dfrac{a_{n_{k+1}}}{a_{n_k}} \leq 1/2$. Let

$$b_n = \begin{cases} \dfrac{1}{k a_{n_k}}, & \text{if } n = n_k \text{ for some } k \in \mathbb{N}, \\ 0, & \text{otherwise}. \end{cases}$$

It is clear that $\sum a_k b_k = \sum \frac{1}{k} = \infty$. However, since $\{a_n\}$ is decreasing, we have, for $n_k \leq m < n_{k+1}$,

$$\begin{aligned} a_m B_m &\leq a_{n_k} \sum_{n=1}^{m} b_n = a_{n_k} \sum_{j=1}^{k} \frac{1}{j a_{n_j}} \leq \sum_{j=1}^{k} \frac{1}{j} \left(\frac{1}{2}\right)^{k-j} = \sum_{j=1}^{\lceil k/2 \rceil} \frac{1}{j} \left(\frac{1}{2}\right)^{k-j} \\ &+ \sum_{j=\lceil k/2 \rceil+1}^{k} \frac{1}{j} \left(\frac{1}{2}\right)^{k-j} \leq \frac{1}{2^{\frac{k}{2}-1}} \sum_{j=1}^{\lceil k/2 \rceil} \frac{1}{j} + \frac{2}{k} \sum_{j=1}^{\lceil k/2 \rceil} \frac{1}{2^{j-1}} \to 0, \end{aligned}$$

as $k \to \infty$ (here $\lceil c \rceil$ denotes the ceiling function of $c > 0$), which proves the claim.

As a consequence of Theorem 2.102, we have the following:

Corollary 2.103 *Let $\{b_n\}$ be a complex sequence such that $\sum b_k$ converges. If $\{c_n\}$ is a nondecreasing sequence of positive numbers that tends to infinity, then we have*

$$\lim_{n\to\infty} \frac{1}{c_n} \sum_{k=1}^{n} b_k c_k = 0.$$

Proof Define $a_k = \frac{1}{c_k}$ for $k \in \mathbb{N}$. By assumption, the series

$$\sum_{k=1}^{\infty} b_k = \sum_{k=1}^{\infty} \frac{a_k b_k}{a_k}$$

converges. Moreover, since $\{a_k\}$ is nonincreasing and vanishing at infinity, we can apply Theorem 2.102 to derive

$$a_n \sum_{k=1}^{n} \frac{b_k}{a_k} = \frac{1}{c_n} \sum_{k=1}^{n} b_k c_k \to 0, \quad \text{as } n \to \infty. \qquad \square$$

A natural question concerning the classical Abel–Olivier test is how much the condition of monotonicity of the sequence $a = \{a_k\}$ can be weakened so that the conclusion $\lim\limits_{k\to\infty} k a_k = 0$ still holds, at least in some weaker sense. Of course, one can suppose that the monotonicity condition holds except for a finite number of terms, but this represents a very cheap extension of the original result. We now give a statement of quantitative nature in this direction.

To this end, we introduce appropriate notation. Let $\sum a_k$ be a positive convergent series, denote its sum by A, and put $S_n = \sum_{k=1}^{n} a_k$. For a given $n \in \mathbb{N}$, let

$$P_n = P_n(\{a_k\}) = \{k \in \{1,2,...,n\} : a_k \geq a_n\},$$
$$R_n = R_n(\{a_k\}) = \{k \in \{1,2,...,n\} : a_k < a_n\},$$

and

$$p_n = \sum_{k\in P_n} (a_k - a_n) \qquad \text{and} \qquad r_n = \sum_{k\in R_n} (a_n - a_k).$$

Notice that both p_n and r_n are nonnegative. Furthermore,

$$S_n = p_n - r_n + n a_n \quad \text{and} \quad n a_n - r_n > 0. \tag{2.65}$$

Lemma 2.104 *For every positive convergent series $\sum a_k$, we have*

$$\lim_{n\to\infty} p_n = \sum_{k=1}^{\infty} a_k = A. \tag{2.66}$$

Proof First, notice that, as $p_n < S_n < A$, the sequence $\{p_n\}$ is bounded by A. Thus, to establish (2.66) it suffices to prove that for every $\varepsilon > 0$, there is an $n_0 \in \mathbb{N}$ such that for all $n > n_0$, we have $p_n > A - 2\varepsilon$. Choose $\varepsilon > 0$ and let $n_1 \in \mathbb{N}$ be such that $S_{n_1} > A - \varepsilon$. Let $n_0 > n_1$ be an integer such that $a_n < \frac{\varepsilon}{n_1}$ and $a_n \leq \min\{a_1, a_2, \ldots, a_{n_1}\}$ for all $n > n_0$ (such an n_0 exists since $\{a_k\}$ vanishes at infinity). Thus, $\{1, 2, \ldots, n_1\} \subset P_n$ for all $n > n_0$, and hence,

$$p_n \geq \sum_{k=1}^{n_1} (a_k - a_n) = S_{n_1} - n_1 a_n > A - \varepsilon - \varepsilon,$$

which completes the proof. □

The following is a straightforward consequence of (2.65) and Lemma 2.104.

Theorem 2.105 *For every positive convergent series $\sum a_k$, we have*

$$\lim_{n\to\infty} (na_n - r_n) = 0.$$

Finally, we arrive at the desired generalization of the classical Abel–Olivier test.

Corollary 2.106 *Let $\sum a_k$ be a positive convergent series. Then*

$$\lim_{n\to\infty} r_n = 0 \quad \textit{if and only if} \quad \lim_{n\to\infty} na_n = 0.$$

This readily follows from Theorem 2.105. Note that if $\{a_k\}$ is monotone, then $r_n = 0$ for every $n \in \mathbb{N}$ and we recover the classical Abel–Olivier test (Theorem 2.98).

Let us now turn our attention to generalizations of Abel–Olivier's test in other directions. We focus on relaxing the monotonicity condition. An extended Abel–Olivier test also holds for $\{a_k\} \in \overline{GMS}$ (recall that the class $\overline{GMS}$ was defined in Example 1.13; the classes $\overline{GMS}$ and WMS are incomparable, as shown in Remark 1.34, and therefore Theorem 2.100 does not apply for $\overline{GMS}$). In fact, the extended Abel–Olivier test for $\overline{GMS}$ still holds if the sequence $\{a_k\}$ is real-valued rather than nonnegative.

Theorem 2.107 *Let $\{a_k\} \in \overline{GMS}$ be a real-valued sequence. Then, $\sum_{k=1}^{\infty} a_k$ converges if and only if*

$$ka_k \to 0 \quad \text{and} \quad \sum_{k=1}^{\infty} k\Delta a_k \text{ converges.}$$

Furthermore, in this case $\sum_{k=1}^{\infty} a_k = \sum_{k=1}^{\infty} k\Delta a_k$.

Instead of proving this theorem directly, we prove a "basic" counterpart for functions (that is, without involving differences). The conclusion will then follow by Remark 1.5 and Abel's transformation.

Theorem 2.108 *Let $f \in \overline{GM}$ be real-valued. If the integral*

$$\int_0^{\infty} f(t)\, dt \tag{2.67}$$

converges, then

$$tf(t) \to 0 \qquad \text{as } t \to \infty.$$

Proof In order to prove this result, we use Lemmas 1.27 and 1.28. Recall that in these we assume without loss of generality that the constant c in the $\overline{GM}$ condition (1.25) is of the form $c = 2^\nu$, with $\nu \in \mathbb{N}$ (which we also assume here). Recall also the definitions

$$A_n := \sup_{2^n \le t \le 2^{n+1}} |f(t)|,$$
$$B_n := \sup_{2^{n-2\nu} \le t \le 2^{n+2\nu}} |f(t)|,$$

and that for $n \in \mathbb{N} \cup \{0\}$, we say that n is a *good* number if either $n = 0$ or $B_n \le 2^{4\nu} A_n$. Otherwise, n is a bad number.

Clearly, if we show that $2^n A_n \to 0$ as $n \to \infty$, the result will follow. First of all, the convergence of (2.67) is equivalent to

$$\left| \int_M^N f(t)\, dt \right| \to 0 \qquad \text{as } N > M \to \infty,$$

by (2.1) (or, more precisely, by its analogue for integrals).

We distinguish between two cases, namely if there are finitely or infinitely many good numbers. Let us first assume that there are infinitely many of them. For any good number $n > 0$, it follows from Lemma 1.28 that

$$2^n A_n \asymp \left| \int_{\ell_n}^{m_n} f(t)\, dt \right|,$$

and, moreover, $f(x) \ge 0$ (or $f(x) \le 0$) for all $x \in (\ell_n, m_n) \subset [2^{n-\nu}, 2^{n+\nu}]$. Since the integrals $\int_{\ell_n}^{m_n} f(t)\, dt$ vanish as $n \to \infty$ (by the convergence of (2.67)) we deduce that

$$2^n A_n \to 0 \qquad \text{as } n \to \infty,\ n \text{ good}. \tag{2.68}$$

We now prove that $2^n A_n$ also vanishes as $n \to \infty$ whenever n runs through bad numbers. If n is a bad number, then $A_n < 2^{-4\nu} B_n$, and $B_n = A_{s_1}$, with $|n - s_1| \le 2\nu$. Let us first suppose that $s_1 < n$ and find the largest good number m that is smaller than n. If there is any good number in the set $\{s_1, s_1 + 1, \dots, n-1\}$, we just choose m to be the largest good number from such a set and conclude the procedure. Otherwise, s_1 is a bad number. Then there exists s_2 satisfying $|s_1 - s_2| \le 2\nu$ such that $A_{s_1} < 2^{-4\nu} B_{s_1} = 2^{-4\nu} A_{s_2}$. Also, note that $s_2 < s_1$; the opposite is not possible. Indeed, if $s_2 \ge s_1$, then $|n - s_2| \le 2\nu$, and hence $A_{s_2} \le B_n$. In this case, we would have

$$A_{s_1} < 2^{-4\nu} B_{s_1} = 2^{-4\nu} A_{s_2} \le 2^{-4\nu} B_n < B_n,$$

which contradicts the fact that $A_{s_1} = B_n$. As before, if there is any good number in $\{s_2, s_2 + 1, \dots, s_1 - 1\}$, we choose m to be the largest good number from such a set and we are done.

Repeating this process, we arrive at a finite sequence $n = s_0 > s_1 > \cdots > s_{j-1} > s_j$, $j \ge 1$, where all the numbers in the set $\{s_{j-1}, s_{j-1} + 1, \dots, s_0\}$ are bad, and there exists a good number m (strictly positive, if n is large enough) satisfying

$s_j \le m < s_{j-1}$ (fix it to be the largest from $\{s_j, s_j+1, \dots, s_{j-1}-1\}$). Note that $A_{s_k} < 2^{-4\nu} A_{s_{k+1}}$ and $|s_k - s_{k+1}| \le 2\nu$ for any $0 \le k \le j-1$, thus $n \le s_j + 2j\nu$. Also, the number m obtained by this procedure tends to infinity whenever $n \to \infty$, since there are infinitely many good numbers. The fact that m is good yields

$$\begin{aligned} 2^n A_n &< 2^{n-4\nu} A_{s_1} < \cdots < 2^{n-4j\nu} A_{s_j} \le 2^{s_j - 2j\nu} A_{s_j} \\ &\le 2^{s_j - 2j\nu} B_m \le 2^{m-2j\nu+4\nu} A_m \le 2^{m+2\nu} A_m \to 0 \end{aligned}$$

as $n \to \infty$, by (2.68) (here the inequality $A_{s_j} \le B_m$ follows from the definition of B_m and the fact that $s_{j-1} < m \le s_j$ and $|s_j - s_{j-1}| \le 2\nu$).

Suppose now that $n < s_1$. Then either there is a good number m such that $n < m \le s_1$, or s_1 is bad, in which case we can find $s_1 < s_2$ such that $|s_2 - s_1| \le 2\nu$ and $A_{s_1} < 2^{-4\nu} A_{s_2}$ (note that the case $s_2 < s_1$ is not possible since it leads to a contradiction). As before, we iterate the procedure until we find a set $\{s_{j-1}, s_{j-1}+1, \dots, s_j\}$ that contains at least one good number. Since there are infinitely many good numbers, we arrive at a finite sequence $n = s_0 < s_1 < \cdots < s_{j-1} < s_j$, where the numbers in the set $\{s_0, s_0+1, \dots, s_{j-1}\}$ are bad, and there is a good number m such that $s_{j-1} < m \le s_j$. Fix m to be any good number from $\{s_{j-1}+1, s_{j-1}+2, \dots, s_j\}$. Since $n < m$, we have

$$2^n A_n < 2^{n-4\nu} A_{s_1} < \cdots < 2^{n-4j\nu} A_{s_j} < 2^{m-4j\nu} B_m \le 2^m A_m \to 0$$

as $n \to \infty$, by (2.68) (again, the inequality $A_{s_j} \le B_m$ follows from the definition of B_m and the fact that $s_{j-1} < m \le s_j$ and $|s_j - s_{j-1}| \le 2\nu$).

Assume now that there are finitely many good numbers n. Let $N \in \mathbb{N}$ be such that $m \le N$ for all good numbers m. If $n > N$, then n is a bad number, so that $A_n < 2^{-4\nu} B_n$, and $B_n = A_{s_1}$ for some s_1 satisfying $|n - s_1| \le 2\nu$. If $s_1 < n$, one can find, in a similar way as above, a sequence $n = s_0 > s_1 > \cdots > s_{j-1} > s_j$, where $s_0, s_1, \dots, s_{j-1}$ are bad and s_j is good, and moreover $n \le s_j + 2j\nu$. Since $s_j \le N$, we have

$$j \ge \frac{n - s_j}{2\nu} \ge \frac{n - N}{2\nu} \tag{2.69}$$

and deduce

$$2^n A_n < 2^{n-4\nu} A_{s_1} < \cdots < 2^{n-4j\nu} A_{s_j} \le 2^{s_j - 2j\nu} A_{s_j} \le 2^{N-2j\nu} \max_{0 \le k \le N} A_k.$$

The latter vanishes as $n \to \infty$, since in such a case $j \to \infty$, by (2.69).

Finally, it remains to investigate the case $s_1 > n$. We will actually show that this case is not possible at all. Let n be such that $A_n > 0$ (if this n does not exist, our assertion follows trivially). If $s_1 > n$, then there is an infinite sequence of bad numbers $n = s_0 < s_1 < s_2 < \cdots$ such that $A_{s_j} < 2^{-4\nu} B_{s_j} = 2^{-4\nu} A_{s_{j+1}}$ for every $j \ge 0$. Hence,

$$\frac{A_{s_{j+1}}}{A_{s_j}} > 2^{4\nu} \qquad \text{for all } j \ge 0,$$

or, in other words, the sequence A_{s_k} does not vanish as $k \to \infty$. This contradicts that f vanishes at infinity, showing the case $s_1 > n$ to be impossible. The proof is complete. □

A natural question around Abel–Olivier's test is whether it is sharp with respect to the necessary condition $na_n \to 0$, specifically if there exists $u_n \to \infty$ such that for a convergent series $\sum a_n$ with decreasing terms, one has $nu_na_n \to 0$. This is well known to be false.

Proposition 2.109 *For any positive sequence $\{u_n\}$ such that* $\limsup_n u_n = \infty$, *there exists a nonincreasing sequence $\{a_n\}$ such that $\sum a_n$ converges, but*

$$nu_na_n \nrightarrow 0 \quad \textit{as } n \to \infty.$$

Proof Since $\limsup_n u_n = \infty$, we can find an increasing sequence $\{m_k\} \subset \mathbb{N}$ such that $\{u_{m_k}\}$ is increasing, $u_{m_1} \geq 1$, and moreover

$$\sum_{k=1}^{\infty} \frac{1}{\sqrt{u_{m_k}}} < \infty.$$

Assume without loss of generality that $m_1 = 1$. Defining $a_1 = 1$ and

$$a_n = \frac{1}{m_k\sqrt{u_{m_k}}}, \qquad \text{if } m_{k-1} < n \leq m_k, \quad k \geq 2,$$

we note that a_n is nonincreasing. Observe that for $k \in \mathbb{N}$ and $n = m_k$, we have $nu_na_n = \sqrt{u_{m_k}} \to \infty$. Further,

$$\sum_{n=1}^{\infty} a_n = 1 + \sum_{k=2}^{\infty} \frac{m_k - m_{k-1}}{m_k\sqrt{u_{m_k}}} < 1 + \sum_{k=1}^{\infty} \frac{1}{\sqrt{u_{m_k}}} < \infty.$$

Thus, $\sum_n a_n$ is a convergent monotone series such that $\limsup_n nu_na_n = \infty$. □

2.4.6 Behavior of the terms of monotone series

While Abel–Olivier's test is the most relevant result for characterizing the behavior of monotone convergent series, there are further results that describe the behavior of the terms of (either convergent or divergent) series with monotone-type terms. We present a variety of such results in this section.

We begin with a generalization of the following statement:
If a series $\sum_{k=1}^{\infty} a_k$ with nonincreasing terms diverges, then the series

$$\sum_{k=1}^{\infty} \min\left\{a_k, \frac{1}{k}\right\}$$

does as well.

This statement may be proved in a fairly simple way using Cauchy's condensation test or Abel–Olivier's test. Moreover, as one may expect, it can be extended to $\{a_k\} \in WMS$.

Theorem 2.110 *Let $\{a_k\} \in WMS$ and assume that the series $\sum_{k=1}^{\infty} a_k$ is divergent. Then, the series*

$$\sum_{k=1}^{\infty} \min\left\{a_k, \frac{1}{k}\right\} \tag{2.70}$$

is also divergent.

Proof Assume first that there exists $n \in \mathbb{N}$ such that for every $k \geq n$,

$$a_k \leq \frac{1}{k}.$$

In this case the series (2.70) clearly diverges by the divergence of $\sum_{k=1}^{\infty} a_k$.

On the other hand, if for every $n \in \mathbb{N}$ we can find $k \geq n$ such that

$$a_k \geq \frac{1}{k},$$

then

$$k \min\left\{a_k, \frac{1}{k}\right\} \not\to 0, \qquad \text{as } k \to \infty,$$

and the divergence of (2.70) follows from Abel–Olivier's test. Indeed, the sequence $\min\left\{a_k, \frac{1}{k}\right\}$ is *WMS* by property (iv) in Proposition 1.31. □

A natural question that arises from Theorem 2.110 is whether, given two positive sequences $\{a_k\}, \{b_k\} \in WMS$ such that the series $\sum a_k$ and $\sum b_k$ are divergent, the series $\sum \min\{a_k, b_k\}$ is also divergent. We now show that the answer to this question is negative, even in the more restrictive case where $\{a_k\}$ and $\{b_k\}$ are nonincreasing.

Proposition 2.111 *There exist nonincreasing positive sequences $\{a_k\}$ and $\{b_k\}$ vanishing at infinity such that the series $\sum a_k$ and $\sum b_k$ are divergent, but*

$$\sum_{k=1}^{\infty} \min\{a_k, b_k\}$$

converges.

The proof of this result is based on the existence of *general* positive sequences $\{c_k\}$ and $\{d_k\}$ so that the series $\sum c_k$ and $\sum d_k$ diverge, while $\sum \min\{c_k, d_k\} < \infty$. Disregarding the monotonicity assumptions, this kind of pairs of sequences are easy to construct. Indeed, for instance, the simple examples

$$c_k = \begin{cases} \frac{1}{n}, & \text{if } k = 2^n,\ n \in \mathbb{N}, \\ \frac{1}{k^2}, & \text{if } 2^n < k < 2^{n+1},\ n \in \mathbb{N}, \end{cases}$$

$$d_k = \begin{cases} \frac{1}{n}, & \text{if } k = 2^n + 1,\ n \in \mathbb{N}, \\ \frac{1}{k^2}, & \text{if } 2^n + 1 < k \le 2^{n+1},\ n \in \mathbb{N}, \end{cases}$$

and $c_1 = d_1 = d_2 = 1$ satisfy the desired properties.

Proof of Proposition 2.111 Let $\{c_k\}$ and $\{d_k\}$ be two positive sequences such that the series $\sum c_k$ and $\sum d_k$ diverge and $\sum \min\{c_k, d_k\}$ converges. Let us construct two nonincreasing sequences $\{a_k\}$ and $\{b_k\}$ depending on $\{c_k\}$ and $\{d_k\}$, respectively. First, we define an increasing sequence of natural numbers $\{\lambda_k\}$ as follows: let $\lambda_1 = 1$, and for $k \ge 1$,

$$\lambda_{k+1} = \min\left\{m \in \mathbb{N} : c_k \ge \frac{c_{k+1}}{m},\ d_k \ge \frac{d_{k+1}}{m},\ m > \lambda_k\right\}.$$

Thus, $c_k \ge \dfrac{c_{k+1}}{\lambda_{k+1}}$ and $d_k \ge \dfrac{d_{k+1}}{\lambda_{k+1}}$ for all $k \in \mathbb{N}$. Define another increasing sequence of natural numbers $\{n_s\}$ as

$$n_1 = 1, \qquad n_s = \sum_{\ell=1}^{s}\left(\prod_{j=1}^{\ell}\lambda_j\right), \quad s \ge 2.$$

Finally, let $a_1 = c_1$, $b_1 = d_1$, and for any $s \ge 1$,

$$a_k = \frac{c_{s+1}}{\prod_{j=1}^{s+1}\lambda_j}, \qquad n_s < k \le n_{s+1},$$

$$b_k = \frac{d_{s+1}}{\prod_{j=1}^{s+1}\lambda_j}, \qquad n_s < k \le n_{s+1}.$$

We now show that $\{a_k\}$ is nonincreasing (the instance of $\{b_k\}$ is very similar and we omit it). We only need to check that $a_k \ge a_{k+1}$ at the indices $k = n_s$, for $s \in \mathbb{N}$. We have, by the construction of $\{\lambda_k\}$,

$$a_{n_s} = \frac{c_s}{\prod_{j=1}^{s}\lambda_j} \ge \frac{1}{\prod_{j=1}^{s}\lambda_j}\cdot\frac{c_{s+1}}{\lambda_{s+1}} = a_{n_s+1},$$

which shows that $\{a_k\}$ is nonincreasing. As for the divergence of $\sum a_k$, one has

$$\sum_{k=1}^{\infty} a_k = a_1 + \sum_{s=1}^{\infty}\sum_{k=n_s+1}^{n_{s+1}} a_k = c_1 + \sum_{s=1}^{\infty}\sum_{k=n_s+1}^{n_{s+1}} \frac{c_{s+1}}{\prod_{j=1}^{s+1}\lambda_j}$$

$$= c_1 + \sum_{s=1}^{\infty} c_{s+1}\frac{n_{s+1} - n_s}{\prod_{j=1}^{s+1}\lambda_j} = \sum_{k=1}^{\infty} c_k = \infty.$$

The proof that $\sum b_k$ diverges is very similar and is therefore omitted. To conclude,

$$\begin{aligned}\sum_{k=1}^{\infty} \min\{a_k, b_k\} &= \min\{a_1, b_1\} + \sum_{s=1}^{\infty} \sum_{k=n_s+1}^{n_{s+1}} \min\{a_k, b_k\} \\ &= \min\{a_1, b_1\} + \sum_{s=1}^{\infty} \min\{c_{s+1}, d_{s+1}\} \frac{n_{s+1} - n_s}{\prod_{j=1}^{s+1} \lambda_j} \\ &= \sum_{k=1}^{\infty} \min\{c_k, d_k\} < \infty,\end{aligned}$$

since $\{c_k\}$ and $\{d_k\}$ are chosen so that the last series converges. □

Note that the assertion of Proposition 2.111 would be true even if the sequences $\{a_k\}$ and $\{b_k\}$ were assumed to be not only nonincreasing but also actually strictly decreasing. To see this, it suffices to construct a somewhat modified sequence $\{\lambda_k\}$ that delivers the desired properties on $\{a_k\}$ and $\{b_k\}$ by also modifying the sequence $\{n_s\}$ slightly.

To compare the behavior of terms of weak monotone convergent and divergent series, we follow Dvoretzky, who investigated the interrelation between such series in the monotone case. Roughly speaking, Dvoretzky's test states that for the series with decreasing sequences of terms $\{a_k\}$, $\{b_k\}$, convergent and divergent, respectively, one may find infinitely many sets $\mathcal{N} = \{n, n+1, \dots, n+m-1\}$ of arbitrary length and such that

$$a_k < b_k \qquad \text{for all } k \in \mathcal{N}.$$

In fact, Dvoretzky's result is more delicate, as it specifies those sets $\mathcal{N}$. Here we state and prove its corresponding generalization for *WMS* rather than for decreasing sequences.

Theorem 2.112 (Dvoretzky's type test) *If $\{a_k\}, \{b_k\} \in WMS$, and the series $\sum a_k$ and $\sum b_k$ are convergent and divergent, respectively, then for every $M > 1$, there exists an increasing sequence $\{R_j\}$, $R_j \to \infty$, such that for all k with $R_j \le k < MR_j$, we have $a_k < b_k$.*

Proof For fixed $M > 1$ and $k \in \mathbb{N}$, put $\Sigma_k = \Sigma_k(M) = \{j \in \mathbb{N} : M^{k-1} \le j < M^k\}$ and

$$A_k = \sum_{j \in \Sigma_k} a_j, \qquad B_k = \sum_{j \in \Sigma_k} b_j.$$

Note that if $a_n \ge b_n$ for some $n \in \Sigma_k$, the *WMS* condition implies that

$$C \frac{B_{k+1}}{M^{k+1} - M^k} \le \frac{A_{k-1}}{M^{k-1} - M^{k-2}}. \tag{2.71}$$

Indeed, on the one hand, for any $n \in \Sigma_k$, we have $a_n \le C_1 a_\ell$ for every $\ell \in \Sigma_{k-1}$, and on the other hand, $b_m \le C_2 b_n$ for every $m \in \Sigma_{k+1}$. Therefore, $b_m \le C_3 a_\ell$, which in turn implies (2.71).

Note that, since (2.71) implies $A_{k-1} \geq CM^{-2}B_{k+1}$, and the series $\sum A_k$ converges while $\sum B_k$ diverges, there must be infinitely many k for which the inequality $A_{k-1} \geq CM^{-2}B_{k+1}$ does not hold. Thus, there are infinitely many k for which $a_j < b_j$ for every $j \in \Sigma_k$, and the conclusion follows. □

Remark 2.113 In the original proof of Theorem 2.112, it is actually shown that for $M > 1$ there exist infinitely many $n \in \mathbb{N}$ such that $a_k < b_k$ for all $M^n \leq k < M^{n+1}$. We omit the corresponding reformulation (see Notes-88 for references).

The following is an interesting observation concerning weak monotonicity and divergent series. It is clear that power-type sequences $a_n = n^\alpha$ are *WMS* for any $\alpha \in \mathbb{R}$. In particular, for this type of sequences one has $\{a_n\} \in WMS$ and also $\left\{\frac{1}{a_n}\right\} \in WMS$. A natural question to ask is what kind of sequences $\{a_n\} \in WMS$ are such that $\left\{\frac{1}{a_n}\right\} \notin WMS$. Perhaps the simplest example of such a sequence can be $a_n = 2^{-n}$. Thus, at a first glance, it seems that sequences $\{a_n\} \in WMS$ satisfying $\left\{\frac{1}{a_n}\right\} \notin WMS$ have nice properties in terms of fast decay and may therefore be tempted to conjecture that *if* $\{a_n\} \in WMS$ *is such that* $\left\{\frac{1}{a_n}\right\} \notin WMS$*, then* $\sum a_k$ *converges* (given that some *WMS* that decay slower than exponentials, such as n^α for $\alpha < -1$, already give examples of convergent series with $\left\{\frac{1}{a_n}\right\} \in WMS$). However, this statement is not true, as our next result shows.

Theorem 2.114 *There exists* $\{a_n\} \in WMS$ *such that* $\left\{\frac{1}{a_n}\right\} \notin WMS$ *and* $\sum a_n = \infty$.

Proof Let us introduce two increasing sequences of natural numbers $\{n_s\}$ and $\{N_s\}$ as follows: let $n_1 = 1$, and for $s \geq 1$, $n_{s+1} = 2^{n_s}$, and

$$N_s = \sum_{k=1}^{s} n_k.$$

We now define the sequence

$$a_k = \frac{1}{n_{s+1}}, \qquad N_s \leq k < N_{s+1}, \qquad s \in \mathbb{N}.$$

Note that $\{a_k\} \in WMS$ (it is actually nonincreasing), and the series $\sum a_k$ diverges, since

$$\sum_{k=1}^{\infty} a_k = \sum_{s=1}^{\infty} \sum_{k=N_s}^{N_{s+1}-1} a_k = \sum_{s=1}^{\infty} \sum_{k=N_s}^{N_{s+1}-1} \frac{1}{n_{s+1}} = \sum_{s=1}^{\infty} \frac{N_{s+1} - N_s}{n_{s+1}} = \sum_{s=1}^{\infty} 1 = \infty.$$

Moreover, $\left\{\frac{1}{a_n}\right\} \notin WMS$, which follows from the fact that for $k = N_s - 1$, one has

$$\frac{\frac{1}{a_{k+1}}}{\frac{1}{a_k}} = \frac{\frac{1}{a_{N_s}}}{\frac{1}{a_{N_s-1}}} = \frac{a_{N_s-1}}{a_{N_s}} = \frac{n_{s+1}}{n_s} = \frac{2^{n_s}}{n_s} \to \infty,$$

as $s \to \infty$. □

2.4.7 Applications to Orlicz-type classes

This subsection is devoted to applications of monotone-type series/functions to functional analysis. In particular, we investigate an old question on characterizing Orlicz-type spaces. Typical examples include $L \log L$ spaces, Lorentz–Zygmund spaces, and so on. Such results have important applications in functional analysis (for further comments and references, see Notes-89 at the end of the chapter). Their analogs in two and arbitrary dimensions are given in Subsections 6.2.3 and 9.2.3, respectively.

Theorem 2.115 *Let $\{a_n\}$ be a positive sequence and let $\varphi\colon [0,\infty) \to [0,\infty)$ be nondecreasing such that*

$$\varphi(x^2) \le C\varphi(x), \qquad x > 1, \qquad C > 1. \tag{2.72}$$

Then

(i) *The convergence of*

$$\sum_{n=1}^{\infty} \frac{a_n}{n^2}\varphi(a_n) \tag{2.73}$$

implies that of

$$\sum_{n=1}^{\infty} \frac{a_n}{n^2}\varphi(n). \tag{2.74}$$

(ii) *If there exists $C > 0$ such that $a_n \le \frac{C}{n}\sum_{k=\lfloor n/2\rfloor}^{n} a_k$ for all n, then the convergence of* (2.74) *implies that of* (2.73).

In particular, if $\{a_n\} \in WMS \cup \overline{GMS}$, both series (2.73) *and* (2.74) *converge or diverge simultaneously.*

For the second part of this theorem, we note that both WMS and (nonnegative) $\overline{GMS}$ satisfy the estimate

$$a_n \le \frac{C}{n}\sum_{k=\lfloor n/2\rfloor}^{n} a_k.$$

For WMS, this estimate is straightforward, while for $\overline{GMS}$ it is proved in (1.16) and (1.17).

An example of a function φ satisfying the hypotheses of Theorem 2.115 is

$$\varphi(x) = \log^{\gamma}(1+x), \qquad \gamma \ge 0.$$

Proof of Theorem 2.115 Assume first that the series $\sum_{n=1}^{\infty} \frac{a_n}{n^2}\varphi(a_n)$ converges. Let

$$E = \{n \in \mathbb{N} : a_n \ge \sqrt{n}\}.$$

Splitting the series into two sums, over E and $\mathbb{N}\backslash E$, we get

$$\sum_{n=1}^{\infty}\frac{a_n}{n^2}\varphi(n)\le\sum_{n\in E}\frac{a_n}{n^2}\varphi(a_n^2)+\sum_{n\in\mathbb{N}\backslash E}\frac{\sqrt{n}}{n^2}\varphi(n)$$
$$\le C\sum_{n\in E}\frac{a_n}{n^2}\varphi(a_n)+\sum_{n\in\mathbb{N}\backslash E}\frac{\varphi(n)}{n^{3/2}},$$

where the last inequality follows from (2.72) (in fact, if (2.72) holds for $x>1$, then it also holds for all $x>0$ since φ is nondecreasing, and therefore $\varphi(x^2)\le\varphi(x)$ for $0<x\le 1$). In order to conclude the proof of this part, it remains to prove that the series

$$\sum_{n\in\mathbb{N}\backslash E}\frac{\varphi(n)}{n^{3/2}}$$

converges. Condition (2.72) implies the estimate

$$\varphi(x)\le C\log^{\gamma}x,\qquad x\ge 2,$$

for some $C,\gamma>0$. Indeed, for $n\in\mathbb{N}$ and $x>1$, we have

$$\varphi(x)\le C\varphi(\sqrt{x})\le\cdots\le C^n\varphi\big(x^{2^{-n}}\big).$$

Putting $n=\log_2 s$, we obtain $\varphi(x)\le s^{\log_2 C}\varphi(x^{1/s})$. Fixing $x=2^s$, we get

$$\varphi(x)\le(\log_2 x)^{\log_2 C}\varphi(2)\asymp\log^{\gamma}x,$$

where $\gamma=\log_2 C$. Therefore, the series in question converges.

Assume now that the series $\sum_{n=1}^{\infty}\frac{a_n}{n^2}\varphi(n)$ converges. Note that the series $\sum_{n=1}^{\infty}\frac{a_n}{n^2}$ also converges, since φ is nondecreasing. Therefore,

$$\frac{a_n}{n}\le C\sum_{k=\lfloor n/2\rfloor}^{n}\frac{a_k}{k^2}\to 0\qquad\text{as } n\to\infty.$$

In particular, if $\{a_n\}\in WMS$, such a decay of $\{a_n\}$ follows from the Abel–Olivier test for WMS (Theorem 2.99). Therefore, we may find $n_0\in\mathbb{N}$ such that if $n\ge n_0$ the estimate $a_n\le n$ holds. Again, since φ is nondecreasing, we get

$$\sum_{n\ge n_0}\frac{a_n}{n^2}\varphi(a_n)\le\sum_{n\ge n_0}\frac{a_n}{n^2}\varphi(n),$$

which implies the convergence of $\sum_{n=1}^{\infty}\frac{a_n}{n^2}\varphi(a_n)$. □

The counterpart for functions is as follows. The proof is omitted due to its similarity to that of Theorem 2.115 above.

Theorem 2.116 *Let $f\colon(0,1]\to[0,\infty)$ and let $\varphi\colon[0,\infty)\to[0,\infty)$ be nondecreasing such that*

$$\varphi(x^2)\le C\varphi(x),\qquad x>1.$$

(i) *If the integral*

$$\int_0^1 f(x)\varphi(f(x))\,dx \tag{2.75}$$

converges, then the integral

$$\int_0^1 f(x)\varphi\Big(\frac{1}{x}\Big)\,dx \tag{2.76}$$

does as well.

(ii) *If there exists $C > 0$ such that the estimate $f(x) \le \frac{C}{x}\int_{x/2}^x f(t)\,dt$ holds for all $x \in (0,1]$, then the convergence of* (2.76) *implies the convergence of* (2.75).

In particular, if $f \in WM \cup \overline{GM}$, both integrals (2.75) *and* (2.76) *converge or diverge simultaneously.*

Note that the formulation of Theorem 2.116 is slightly different from that of Theorem 2.115. However, applying the change of variables $x \to 1/x$ in the integrals (2.75) and (2.76), we obtain

$$\int_1^\infty \frac{f(1/x)}{x^2}\varphi(f(1/x))\,dx \qquad \text{and} \qquad \int_1^\infty \frac{f(1/x)}{x^2}\varphi(x)\,dx,$$

respectively, which resemble the corresponding series in Theorem 2.115. Also, note that as in the case of $\overline{GMS}$, the estimate

$$f(x) \le \frac{C}{x}\int_{x/2}^x f(t)\,dt$$

holds for nonnegative $f \in \overline{GM}$ (cf. (1.26)).

2.4.8 $\ell^p - \ell^q$ inequalities and Stechkin's lemma

In this section we present $\ell^p - \ell^q$ inequalities for series involving terms that satisfy certain monotonicity conditions.

The first result is an inequality for *WMS* inspired by the well-known $\ell^q \subset \ell^p$ inequality. For a nonnegative sequence $\{a_n\}$ and $0 < q < p \le \infty$, one has

$$\Big(\sum_{k=1}^\infty a_k^p\Big)^{\frac{1}{p}} \le \Big(\sum_{k=1}^\infty a_k^q\Big)^{\frac{1}{q}}. \tag{2.77}$$

In particular, the convergence of $\sum_{k=1}^\infty a_k^q$ implies that of $\sum_{k=1}^\infty a_k^p$. It is interesting that for weak monotone sequences we can improve the latter statement as follows.

Theorem 2.117 *Let $\{a_k\} \in WMS$ and $0 < q < p < \infty$. Then, the inequality*

$$\Big(\sum_{k=1}^\infty a_k^p\Big)^{\frac{1}{p}} \le 3^{\frac{1}{q}}C^2\Big(\sum_{k=1}^\infty a_k^q k^{\frac{q}{p}-1}\Big)^{\frac{1}{q}}$$

holds, where C is the WMS constant of $\{a_k\}$. In particular, the convergence of the series

$$\sum_{k=1}^{\infty} a_k^q k^{\frac{q}{p}-1}$$

implies that of

$$\sum_{k=1}^{\infty} a_k^p.$$

For $p=\infty$ and $0<q<\infty$, we have

$$\sup_n a_n \le 2^{\frac{1}{q}} C \Big(\sum_{k=1}^{\infty} \frac{a_k^q}{k} \Big)^{\frac{1}{q}}.$$

Proof By the *WMS* condition, we have, as in Cauchy's condensation test (see Theorem 2.92),

$$\Big(\sum_{k=1}^{\infty} a_k^p \Big)^{\frac{1}{p}} \le C \Big(\sum_{k=0}^{\infty} 2^k a_{2^k}^p \Big)^{\frac{1}{p}}.$$

Further, using the inequality (2.77) (with 1 in place of p and $\frac{q}{p}$ in place of q), and the fact that

$$2^{k-1} a_{2^k}^q \le C^q \sum_{\ell=2^{k-1}}^{2^k-1} a_\ell^q,$$

we obtain

$$\Big(\sum_{k=0}^{\infty} 2^k a_{2^k}^p \Big)^{\frac{1}{p}} \le \Big(\sum_{k=0}^{\infty} 2^{k\frac{q}{p}} a_{2^k}^q \Big)^{\frac{1}{q}} = \Big(a_1^q + 2 \sum_{k=1}^{\infty} 2^{k-1} a_{2^k}^q 2^{k(\frac{q}{p}-1)} \Big)^{\frac{1}{q}}$$
$$\le \Big(a_1^q + 2C^q \sum_{k=1}^{\infty} a_k^q k^{\frac{q}{p}-1} \Big)^{\frac{1}{q}} \le 3^{\frac{1}{q}} C \Big(\sum_{k=1}^{\infty} a_k^q k^{\frac{q}{p}-1} \Big)^{\frac{1}{q}}.$$

For the case $p=\infty$, if $n>1$ then the estimate

$$\frac{n}{2} a_n^q \le C^q \sum_{k=\lfloor n/2 \rfloor}^{n} a_k^q$$

holds. Hence,

$$a_n \le \Big(2C^q \sum_{k=\lfloor n/2 \rfloor}^{n} \frac{a_k^q}{k} \Big)^{\frac{1}{q}} \le 2^{\frac{1}{q}} C \Big(\sum_{k=1}^{\infty} \frac{a_k^q}{k} \Big)^{\frac{1}{q}},$$

with the last term independent of n. Taking the supremum on the left-hand side completes the proof. □

It is easy to check that the exponent $1-\frac{q}{p}$ in Theorem 2.117 is the best possible, that is, the result does not hold if we replace $1-\frac{q}{p}$ by $1-\frac{q}{p}+\varepsilon$ whenever $\varepsilon>0$. It suffices to consider the series with $a_k=k^{-\frac{1}{p}}$ as a counterexample.

Of course, a counterpart of Theorem 2.117 also holds in the context of functions.

Theorem 2.118 *Let $f\in WM$ and $0<q<p<\infty$. Then, the inequality*

$$\left(\int_0^\infty f(x)^p\,dx\right)^{\frac{1}{p}}\le 2^{\frac{1}{q}}C^2\left(\int_0^\infty f(x)^q x^{\frac{q}{p}-1}\,dx\right)^{\frac{1}{q}}$$

holds, where C is the WM constant for f. In particular, the convergence of

$$\int_0^\infty f(x)^q x^{\frac{q}{p}-1}\,dx$$

implies that of

$$\int_0^\infty f(x)^p\,dx.$$

For $p=\infty$ and $0<q<\infty$, we have

$$\sup_{x\in(0,\infty)} f(x)\le 2^{\frac{1}{q}}C\left(\int_0^\infty \frac{f(x)^q}{x}\,dx\right)^{\frac{1}{q}}.$$

In particular, the convergence of the last integral implies that f is bounded.

Proof Since $f\in WM$,

$$\int_0^\infty f(x)^p\,dx=\sum_{k=-\infty}^{\infty}\int_{2^k}^{2^{k+1}} f(x)^p\,dx\le C^p\sum_{k=-\infty}^{\infty}2^k f(2^k)^p.$$

Applying the inequality (2.77) (with 1 in place of p and $\frac{q}{p}$ in place of q), we get

$$\begin{aligned}\left(\int_0^\infty f(x)^p\,dx\right)^{\frac{1}{p}}&\le C\left(\sum_{k=-\infty}^{\infty}2^{k\frac{q}{p}}f(2^k)^q\right)^{\frac{1}{q}}\\&=C\left(\sum_{k=-\infty}^{\infty}2^{-k+1}2^{k\frac{q}{p}}\int_{2^{k-1}}^{2^k} f(2^k)^q\,dx\right)^{\frac{1}{q}}\\&\le 2^{\frac{1}{q}}C^2\left(\sum_{k=-\infty}^{\infty}\int_{2^{k-1}}^{2^k} x^{\frac{q}{p}-1}f(x)^q\,dx\right)^{\frac{1}{q}},\end{aligned}$$

where in the last step we used the *WM* inequality.

For the case $p=\infty$, the *WM* inequality yields, for any $t>0$,

$$\frac{t}{2}f(t)^q\le C^q\int_{t/2}^t f(x)^q\,dx.$$

Thus,

$$f(t) \le \left(2C^q \int_{t/2}^{t} \frac{f(x)^q}{x}\,dx\right)^{\frac{1}{q}} \le 2^{\frac{1}{q}} C\left(\int_0^\infty \frac{f(x)^q}{x}\,dx\right)^{\frac{1}{q}},$$

where the last term is independent of t. Taking the supremum on the left-hand side concludes the proof. □

Our next result is closely related to the above $\ell^p - \ell^q$ inequalities. It is a generalization of the so-called Stechkin's lemma, which has several interesting applications in approximation theory, especially in nonlinear approximation and, more recently, the approximation of high-dimensional parametric elliptic partial differential equations (see the appropriate references in Notes-90–91). Originally, this result was proved for decreasing sequences $\{a_k\}$, although the extension to the quasi-monotone case (1.7) is straightforward.

Theorem 2.119 *Let $\alpha \ge 0$, and $\{a_k\} \in QMS(\alpha)$. For $0 < q < p < \infty$ and any $n \le N \in \mathbb{N}$, we have*

$$\left(\sum_{k=n}^{N} k^{\alpha(p-q)} a_k^p\right)^{\frac{1}{p}} \le C n^{\frac{1}{p}-\frac{1}{q}} \left(\frac{1}{n^{\alpha(p-q)}} \sum_{k=1}^{N} k^{\alpha(p-q)} a_k^q\right)^{\frac{1}{q}},$$

where

$$C = \sup_n \left(n^{-\alpha p-1} \sum_{k=1}^{n} k^{\alpha p}\right)^{-\frac{1}{q}}$$

(in particular, if $\{a_k\}$ is nonincreasing, then $C = 1$). For the case $p = \infty$ and $0 < q < \infty$, we have

$$n^{-\alpha} a_n \le n^{-\frac{1}{q}} \left(\sum_{k=1}^{N} k^{-\alpha q} a_k^q\right)^{\frac{1}{q}}.$$

Proof Since $\{k^{-\alpha q} a_k^q\}$ is decreasing, we have

$$\left(\sum_{k=1}^{N} k^{\alpha(p-q)} a_k^q\right)^{\frac{1}{q}} \ge \left(\sum_{k=1}^{n} k^{\alpha(p-q)} a_k^q\right)^{\frac{1}{q}} \ge \frac{a_n}{n^\alpha}\left(\sum_{k=1}^{n} k^{\alpha p}\right)^{\frac{1}{q}}$$

$$= a_n n^{\alpha(\frac{p}{q}-1)+\frac{1}{q}} \left(n^{-\alpha p-1} \sum_{k=1}^{n} k^{\alpha p}\right)^{\frac{1}{q}} \ge C_0 n^{\alpha(\frac{p}{q}-1)+\frac{1}{q}} a_n,$$

with

$$C_0 = \inf_n \left(n^{-\alpha p-1} \sum_{k=1}^{n} k^{\alpha p}\right)^{\frac{1}{q}}.$$

Defining

$$C = C_0^{-1} = \sup_n \left(n^{-\alpha p-1} \sum_{k=1}^{n} k^{\alpha p}\right)^{-\frac{1}{q}},$$

we can rewrite the above inequality as

$$a_n \le C n^{-\alpha(\frac{p}{q}-1)-\frac{1}{q}} \left(\sum_{k=1}^{N} k^{\alpha(p-q)} a_k^q \right)^{\frac{1}{q}}. \tag{2.78}$$

On the other hand, the fact that $\{k^{-\alpha(p-q)} a_k^{p-q}\}$ is decreasing implies

$$\sum_{k=n}^{N} a_k^p = \sum_{k=n}^{N} k^{\alpha(p-q)} \frac{a_k^p}{k^{\alpha(p-q)}} \le \frac{a_n^{p-q}}{n^{\alpha(p-q)}} \sum_{k=1}^{N} k^{\alpha(p-q)} a_k^q. \tag{2.79}$$

Applying (2.78), we get

$$\begin{aligned} \left(\frac{a_n^{p-q}}{n^{\alpha(p-q)}} \right)^{\frac{1}{p}} &\le C \left(\frac{n^{\alpha(p-q)-\frac{\alpha p}{q}(p-q)-\frac{p-q}{q}}}{n^{\alpha(p-q)}} \right)^{\frac{1}{p}} \left(\sum_{k=1}^{N} k^{\alpha(p-q)} a_k^q \right)^{\frac{1}{q}-\frac{1}{p}} \\ &= C n^{-\frac{\alpha}{q}(p-q)+\frac{1}{p}-\frac{1}{q}} \left(\sum_{k=1}^{N} k^{\alpha(p-q)} a_k^q \right)^{\frac{1}{q}-\frac{1}{p}}. \end{aligned}$$

Combining this estimate with (2.79) yields the desired inequality.

For the case $p = \infty$ and $0 < q < \infty$, using the monotonicity of $\{k^{-\alpha q} a_k^q\}$, we get, for $n \le N$,

$$n^{-\frac{1}{q}} \left(\sum_{k=1}^{N} k^{-\alpha q} a_k^q \right)^{\frac{1}{q}} \ge n^{-\frac{1}{q}} \left(\sum_{k=1}^{n} k^{-\alpha q} a_k^q \right)^{\frac{1}{q}} \ge n^{-\alpha} a_n,$$

which completes the proof. □

We emphasize that in Theorem 2.119 an explicit expression for the constant in the obtained inequality is given. For applications of Stechkin's lemma (and its variants), it is essential to control such constants. The case $C = 1$ is of special importance.

The classical Stechkin's lemma follows from Theorem 2.119.

Corollary 2.120 *Let $\{a_k\}$ be a positive decreasing sequence. For $0 < q < p < \infty$ and any $n < N \in \mathbb{N}$, we have*

$$\left(\sum_{k=n}^{N} a_k^p \right)^{\frac{1}{p}} \le n^{\frac{1}{p}-\frac{1}{q}} \left(\sum_{k=1}^{N} a_k^q \right)^{\frac{1}{q}}.$$

For the case $p = \infty$ and $0 < q < \infty$, we have

$$a_n = \max_{n \le k \le N} a_k \le n^{-\frac{1}{q}} \left(\sum_{k=1}^{N} a_k^q \right)^{\frac{1}{q}}.$$

Finally, we have another version of Stechkin's lemma for sequences satisfying an estimate of the type

$$a_n \le A \left(\frac{1}{n} \sum_{k=1}^{n} a_k^\alpha \right)^{\frac{1}{\alpha}} + B \left(\sum_{k=n+1}^{\infty} \frac{a_k^\beta}{k} \right)^{\frac{1}{\beta}}, \tag{2.80}$$

for some $0 < \alpha, \beta < \infty$ and constants $A, B \ge 0$.

In order to prove the following statement, we will make use of Hölder's inequality: *if* $\{b_k\}$ *and* $\{c_k\}$ *are nonnegative sequences, and* $1 < q, q' < \infty$ *are such that* $\frac{1}{q} + \frac{1}{q'} = 1$, *then*

$$\sum_{k=1}^{\infty} b_k c_k \le \left(\sum_{k=1}^{\infty} b_k^q\right)^{\frac{1}{q}} \left(\sum_{k=1}^{\infty} c_k^{q'}\right)^{\frac{1}{q'}}. \tag{2.81}$$

Theorem 2.121 *Let* $0 < q < p < \infty$ *and let* $\{a_k\}$ *be a nonnegative sequence satisfying the estimate* (2.80) *with some* $0 < \alpha, \beta \le q$. *Then, there exists a constant* C *depending on* A, B, p, q, *and* β, *such that for every* $n \in \mathbb{N}$,

$$\left(\sum_{k=n}^{\infty} a_k^p\right)^{\frac{1}{p}} \le C n^{\frac{1}{p}-\frac{1}{q}} \left(\sum_{k=1}^{\infty} a_k^q\right)^{\frac{1}{q}}.$$

In particular, if (2.80) *holds with* $B = 0$, *we have*

$$\left(\sum_{k=n}^{\infty} a_k^p\right)^{\frac{1}{p}} \le A n^{\frac{1}{p}-\frac{1}{q}} \left(\sum_{k=1}^{\infty} a_k^q\right)^{\frac{1}{q}}.$$

For the case $p = \infty$ *and* $0 < q < \infty$,

$$\max_{k \ge n} a_k \le C n^{-\frac{1}{q}} \left(\sum_{k=1}^{\infty} a_k^q\right)^{\frac{1}{q}}.$$

Remark 2.122 Note that decreasing sequences satisfy the estimate (2.80) with $A = 1$, $B = 0$, and any $\alpha > 0$. In this case, Theorem 2.121 reduces to Corollary 2.120 (with $N = \infty$).

Sequences in the class *WMS* also satisfy an estimate of the form (2.80) with $B = 0$. Indeed, we have, for any $\alpha > 0$,

$$a_n^\alpha \le \frac{2C^\alpha}{n} \sum_{k=\lfloor n/2 \rfloor}^{n} a_k^\alpha, \quad \{a_k\} \in WMS,$$

where C is the *WMS* constant of $\{a_k\}$ (i.e., for arbitrary $\alpha > 0$, (2.80) holds with $A = 2^{\frac{1}{\alpha}} C$; such an estimate was already used in the proof of Theorem 2.117). The inequality readily follows from the definition and the fact that $\{a_k^\alpha\} \in WMS$.

Proof of Theorem 2.121 Using Hölder's inequality, we estimate the two sums on the right-hand side of (2.80). For $\alpha < q$, we apply (2.81) with

$$b_k = \begin{cases} a_k^\alpha, & \text{if } k \le n, \\ 0, & \text{otherwise}, \end{cases} \qquad c_k = \begin{cases} 1, & \text{if } k \le n, \\ 0, & \text{otherwise}, \end{cases}$$

$q_0 = \frac{q}{\alpha}$, and $q_1 = \frac{q}{q-\alpha}$, to obtain

$$\left(\frac{1}{n}\sum_{k=1}^{n} a_k^\alpha\right)^{\frac{1}{\alpha}} \le \frac{1}{n^{\frac{1}{\alpha}}} \left(\sum_{k=1}^{n} a_k^q\right)^{\frac{1}{q}} \left(\sum_{k=1}^{n} 1\right)^{\frac{1}{\alpha}-\frac{1}{q}} = \left(\frac{1}{n}\sum_{k=1}^{n} a_k^q\right)^{\frac{1}{q}}.$$

As for the last sum in (2.80), consider first the case $\beta < q$. We have by (2.81) with

$$b_k = \begin{cases} a_k^\beta, & \text{if } k > n, \\ 0, & \text{otherwise,} \end{cases} \qquad c_k = \begin{cases} \frac{1}{k}, & \text{if } k > n, \\ 0, & \text{otherwise,} \end{cases}$$

$q_0 = \frac{q}{\beta}$, and $q_1 = \frac{q}{q-\beta}$,

$$\begin{aligned}\left(\sum_{k=n+1}^{\infty} \frac{a_k^\beta}{k}\right)^{\frac{1}{\beta}} &\le \left(\sum_{k=n+1}^{\infty} a_k^q\right)^{\frac{1}{q}} \left(\sum_{k=n+1}^{\infty} \frac{1}{k^{\frac{q}{q-\beta}}}\right)^{\frac{1}{\beta}-\frac{1}{q}} \\ &\le C_0 \left(\frac{1}{n} \sum_{k=n+1}^{\infty} a_k^q\right)^{\frac{1}{q}},\end{aligned}$$

where the constant C_0 depends on q and β. Combining both estimates with (2.80), we get

$$a_n \le A\left(\frac{1}{n}\sum_{k=1}^{n} a_k^q\right)^{\frac{1}{q}} + C_0 B\left(\frac{1}{n}\sum_{k=n+1}^{\infty} a_k^q\right)^{\frac{1}{q}} \le C\left(\frac{1}{n}\sum_{k=1}^{\infty} a_k^q\right)^{\frac{1}{q}},$$

for $\alpha, \beta < q$, where the constant C in the last inequality depends on all the previous ones.

The estimate is still valid if either $\alpha = q$ or $\beta = q$. Note also that, in particular, the last estimate already implies the corresponding inequality for $p = \infty$. The rest of the proof is very similar to that of Theorem 2.119. For $p < \infty$, we have

$$\begin{aligned}\left(\sum_{k=n}^{\infty} a_k^p\right)^{\frac{1}{p}} &= \left(\sum_{k=n}^{\infty} a_k^q a_k^{p-q}\right)^{\frac{1}{p}} \le C\left(\sum_{k=n}^{\infty} a_k^q \left(\frac{1}{k}\sum_{\ell=1}^{\infty} a_\ell^q\right)^{\frac{p-q}{q}}\right)^{\frac{1}{p}} \\ &\le C n^{\frac{1}{p}-\frac{1}{q}} \left(\sum_{k=n}^{\infty} a_k^q\right)^{\frac{1}{p}} \left(\sum_{\ell=1}^{\infty} a_\ell^q\right)^{\frac{1}{q}-\frac{1}{p}} \le C n^{\frac{1}{p}-\frac{1}{q}} \left(\sum_{k=1}^{\infty} a_k^q\right)^{\frac{1}{q}},\end{aligned}$$

as desired. □

2.4.9 Equiconvergence results involving general monotonicity

We now present some equiconvergence results for *GMS* and *GM* that do not naturally fall into any other class of tests (save, possibly, Theorem 2.127 below, which can be seen as a condensation-type test, although it also clearly belongs to a family of tests inherent to *GMS*).

We begin with a statement that easily follows from the *GMS* condition and Abel's transformation.

Theorem 2.123 *Let $\{a_k\} \in GMS$ be vanishing at infinity. Then, the series*

$$\sum_{k=1}^{\infty} |a_k| \qquad \textit{and} \qquad \sum_{k=1}^{\infty} k|\Delta a_k|$$

converge or diverge simultaneously.

Although Theorem 2.123 is a direct consequence of Theorem 2.127 below, it is instructive to include its proof, as it gives a clear idea on how to generalize it to $\mathcal{GMS}$ classes.

Proof of Theorem 2.123 On the one hand, the *GMS* condition yields

$$\sum_{k=1}^{\infty} |a_k| \gtrsim \sum_{k=1}^{\infty}\sum_{n=k}^{2k} |\Delta a_n| \asymp \sum_{k=1}^{\infty} k|\Delta a_k|. \tag{2.82}$$

Secondly, by Abel's transformation (2.2) and the fact that $a_k \to 0$, we have, for any $m \in \mathbb{N}$,

$$\sum_{k=1}^{m} |a_k| = \sum_{k=1}^{m-1} k\Delta|a_k| + m|a_m| \le \sum_{k=1}^{m-1} k|\Delta a_k| + m\left|\sum_{k=m}^{\infty} \Delta a_k\right| \le \sum_{k=1}^{\infty} k|\Delta a_k|,$$

which concludes the proof. □

We note that Theorem 2.123 does not hold for $\{a_k\} \in WMS\backslash GMS$. To see this, consider the example

$$a_k = \begin{cases} 2\cdot 4^{-m}, & \text{if } 2^m \le k < 2^{m+1} \text{ and } k \text{ is even,} \\ 4^{-m}, & \text{if } 2^m \le k < 2^{m+1} \text{ and } k \text{ is odd,} \end{cases} \qquad m \in \mathbb{N}.$$

It is clear that $\{a_k\} \in WMS\backslash GMS$. Furthermore,

$$\sum_{k=1}^{\infty} |a_k| = \sum_{m=0}^{\infty} \sum_{k=2^m}^{2^{m+1}-1} a_k \le 2\sum_{m=0}^{\infty} 2^m 4^{-m} < \infty,$$

while

$$\sum_{k=1}^{\infty} k|\Delta a_k| = \sum_{m=0}^{\infty} \sum_{k=2^m}^{2^{m+1}-1} k|\Delta a_k| \asymp \sum_{k=1}^{\infty} 2^m \sum_{k=2^m}^{2^{m+1}-1} |\Delta a_k| \asymp \sum_{k=1}^{\infty} 2^m 2^m 4^{-m} = \infty.$$

We continue with the converse of Theorem 2.123 for *GM* functions.

Theorem 2.124 *Let $f \in GM$ be vanishing at infinity. Then, the integrals*

$$\int_0^{\infty} |f(t)|\,dt \qquad \text{and} \qquad \int_0^{\infty} t|df(t)|$$

converge or diverge simultaneously.

Proof The *GM* condition implies that

$$\int_0^{\infty} |f(t)|\,dt \gtrsim \int_0^{\infty} \left(\int_t^{2t} |df(s)|\right) dt \asymp \int_0^{\infty} \frac{1}{t}\left(\int_t^{2t} s|df(s)|\right) dt$$
$$\asymp \int_0^{\infty} s|df(s)|.$$

Further, integration by parts (in the sense of Stieltjes) and the fact that f vanishes at infinity yield, for $M > 0$,

$$\int_0^M |f(t)|\,dt = M|f(M)| - \int_0^M t\,d|f|(t)$$
$$\leq M\left|\int_M^\infty df(t)\right| + \int_0^M t|df(t)| \leq \int_0^\infty t|df(t)|,$$

which eventually yields the equiconvergence of the desired integrals. □

As mentioned above, analyzing the proofs of Theorems 2.123 and 2.124, one may easily find conditions on $(a,\beta) \in \mathcal{GMS}$ (resp. $(f,\beta) \in \mathcal{GM}$), so that the corresponding conclusions hold.

Proposition 2.125 *Let $(a,\beta) \in \mathcal{GMS}$ be such that $\{a_k\}$ vanishes at infinity and*

$$\sum_{k=1}^\infty \beta_k \lesssim \sum_{k=1}^\infty |a_k|. \tag{2.83}$$

Then, the conclusion of Theorem 2.123 holds.

Proof The only difference with respect to the proof of Theorem 2.123 is that

$$\sum_{k=1}^\infty |a_k| \gtrsim \sum_{k=1}^\infty \beta_k \gtrsim \sum_{k=1}^\infty \sum_{n=k}^{2k} |\Delta a_n| \asymp \sum_{k=1}^\infty k|\Delta a_k|.$$

holds rather than (2.82). The rest of the proof (which does not make any reference to the *GMS* condition) is just the same. □

An important example of $(a,\beta) \in \mathcal{GMS}$ satisfying (2.83) is the pair that yields the $\overline{GMS}$ class (cf. Example 1.13), where

$$\beta_n = \sum_{k=\lfloor n/c\rfloor}^{\lfloor cn\rfloor} \frac{|a_k|}{k}, \quad \text{for some } c > 1.$$

Proposition 2.126 *Let $(f,\beta) \in \mathcal{GM}$ be such that f vanishes at infinity and*

$$\int_0^\infty \beta(x)\,dx \lesssim \int_0^\infty |f(x)|\,dx.$$

Then, the conclusion of Theorem 2.124 holds.

We skip the proof, as it is analogous to that of Proposition 2.125. It is worth mentioning that the $\overline{GM}$ class (see (1.25)) is an example for which Proposition 2.126 applies.

We conclude with a generalization of Theorem 2.123 that may be seen as a version of Cauchy's condensation test (Theorem 2.91) where we replace the *WMS* assumption by the more restrictive *GMS*, and where a stronger assumption is taken on the sequence $\{u_k\}$.

Theorem 2.127 *Let $\{a_k\} \in GMS$ be vanishing at infinity, and let $\{u_k\}$ be a strictly increasing sequence of natural numbers such that $\overleftarrow{\Delta} u_{k+1} = O(\overleftarrow{\Delta} u_k)$. Then, the series*

$$\sum_{k=1}^{\infty} |a_k|, \quad \sum_{k=1}^{\infty} \overleftarrow{\Delta} u_k |a_{u_k}|, \quad \text{and} \quad \sum_{k=1}^{\infty} u_k |a_{u_k} - a_{u_{k+1}}|$$

converge or diverge together.

The conclusion of this theorem is in a sense different from that of the classical Cauchy condensation test (Theorem 2.91), since the "condensing" sequences to be considered differ in general from those in Theorem 2.91. For instance, the sequence $\{u_k = k\}$ is an example of a valid "condensing" sequence for Theorem 2.127, but not for Theorem 2.91, since it is not in the class Λ.

Proof of Theorem 2.127 First of all, we note that

$$\sum_{k=1}^{\infty} u_k |a_{u_k} - a_{u_{k+1}}| \le \sum_{k=1}^{\infty} u_k \sum_{s=u_k}^{u_{k+1}-1} |\Delta a_s| \le \sum_{s=u_1}^{\infty} s|\Delta a_s|. \tag{2.84}$$

On the one hand, since $\{a_n\} \in GMS$,

$$\sum_{n=u_1}^{\infty} |a_n| \gtrsim \sum_{n=u_1}^{\infty} \sum_{k=n}^{2n} |\Delta a_k| \gtrsim \sum_{k=u_1}^{\infty} k|\Delta a_k|,$$

while on the other hand, since $\{a_k\}$ vanishes at infinity,

$$n|a_n| = n\left|\sum_{k=n}^{\infty} \Delta a_k\right| \le \sum_{k=n}^{\infty} k|\Delta a_k|.$$

We then have, by Abel's transformation and the last estimate, for $n > u_1$,

$$\sum_{k=u_1}^{n} |a_k| \le \sum_{k=u_1}^{n-1} k(|a_k| - |a_{k+1}|) + n|a_n| \le \sum_{k=u_1}^{\infty} k|\Delta a_k|,$$

which establishes the equivalence

$$\sum_{k=u_1}^{\infty} |a_k| \lesssim \sum_{k=u_1}^{\infty} k|\Delta a_k| \lesssim \sum_{k=u_1}^{\infty} |a_k|.$$

By (2.84), such an equivalence readily implies the estimate

$$\sum_{k=1}^{\infty} u_k |a_{u_k} - a_{u_{k+1}}| \lesssim \sum_{k=u_1}^{\infty} |a_k|.$$

Since $\overleftarrow{\Delta} u_{k+1} = O(\overleftarrow{\Delta} u_k)$, it follows that $u_{k+1} \lesssim u_k$. Indeed, since

$$u_{k+1} - u_k \lesssim (u_k - u_{k-1}), \qquad k \ge 2,$$

summing up in k we obtain

$$\sum_{k=2}^{n}(u_{k+1}-u_k) \lesssim \sum_{k=2}^{n}(u_k-u_{k-1}),$$

or, equivalently, $u_{n+1}-u_2 \le C(u_n-u_1)$, which readily yields

$$\frac{u_{n+1}}{u_n} \le C + \frac{u_2-Cu_1}{u_n} \lesssim 1.$$

Now, $\{a_k\} \in GMS$ implies

$$|a_s| \lesssim |a_{u_k}|,$$

for all $s \in \{u_k, \dots, u_{k+1}\}$ and all k, since $\{|a_k|\} \in GMS$, and therefore also $\{|a_k|\} \in WMS$ (cf. Proposition 1.9). Then, for any $M \in \mathbb{N}$,

$$\sum_{s=u_1}^{u_{M+1}-1} |a_s| = \sum_{k=1}^{M}\sum_{s=u_k}^{u_{k+1}-1} |a_s| \lesssim \sum_{k=1}^{M} \overleftarrow{\Delta} u_{k+1}|a_{u_k}| \lesssim \sum_{k=1}^{M} \overleftarrow{\Delta} u_k|a_{u_k}|.$$

By Abel's transformation, we have

$$\begin{aligned}\sum_{k=1}^{M} \overleftarrow{\Delta} u_k|a_{u_k}| &\le \sum_{k=1}^{M-1} u_{k+1}(|a_{u_k}|-|a_{u_{k+1}}|) + u_{M+1}|a_{u_M}| \\ &\lesssim \sum_{k=1}^{\infty} u_k|a_{u_k}-a_{u_{k+1}}|,\end{aligned}$$

which establishes

$$\sum_{k=u_1}^{\infty} |a_k| \lesssim \sum_{k=1}^{\infty} \overleftarrow{\Delta} u_k|a_{u_k}| \lesssim \sum_{k=1}^{\infty} u_k|a_{u_k}-a_{u_{k+1}}|,$$

as desired. □

The conclusion of Theorem 2.127 does not hold in general if the assumption $\overleftarrow{\Delta} u_{k+1} = O(\overleftarrow{\Delta} u_k)$ is omitted. Indeed, one may easily check that, for the choices $a_k = \frac{1}{k\log k}$ and $u_k = 2^{k!}$,

$$\sum_{k=2}^{\infty} a_k = \infty \qquad \text{and} \qquad \sum_{k=1}^{\infty} u_k|a_{u_k}-a_{u_{k+1}}| < \infty.$$

2.5 Notes

Section 2.1 (Series with positive terms)

1. Theorem 2.8 was proposed as a problem for the readers in [31] (the solution, which is the proof appearing here, was given in [32]).

2. For more general methods of summability of number series (including Cesàro or Cesàro–Stolz summability, cf. Theorems 2.6 and 2.7, respectively), see [B-1, B-4, B-7, B-15, B-20, B-27, B-44, B-46, B-61, B-67, B-68]. Euler's work (translated from Latin in [28]) laid the foundation for later developments in the theory of summability, including methods like aforementioned ones, which were developed in the nineteenth century.

3. Theorem 2.10 may be found in [B-54, Ch. IX, §41], although only the proof of the first part appears there. Note that the second part of Theorem 2.10 is quite simple to prove due to the freedom of choice of the sequence of indices $\{n_k\}$, which only needs $b_k \to 0$ as an assumption. The conclusion does not hold if $b_k \not\to 0$. Indeed, if $b_k = 1$ for all k, then $\sum b_k$ diverges, and since

$$\limsup_{k\to\infty} \frac{a_k}{b_k} = \limsup_{k\to\infty} a_k = \infty,$$

we have that $a_k \not\to 0$ (in particular, it is not bounded), and thus $\sum a_n$ cannot converge.

4. The proof of Theorem 2.11 may be found in [19]. Both statements in Theorem 2.11 are classical results: part (i) is due to du Bois-Reymond [39], while part (ii) is due to Abel [2].

5. The proofs of Theorems 2.12 and 2.13 can be found in [128]: it is noted there that Theorem 2.13 is actually easier to prove and can be used to prove Theorem 2.12 later. This last result is also rather classical and is due to Dini (cf. [128, B-54]).

6. The convergence part of Dini's test (Theorem 2.12) is often proved by means of the following theorem due to Pringsheim [220] (see also [B-54, p. 291]).

Theorem 2.128 *If $\sum a_n$ is a divergent series of positive terms and $\{S_n\}$ denotes the partial sums of $\sum a_n$, then the series*

$$\sum_{n=1}^{\infty} \frac{S_{n+1} - S_n}{S_{n+1} S_n^{\alpha}}$$

converges for any $\alpha > 0$.

7. For applications of Proposition 2.14 in the theory of function spaces, and in particular complex interpolation methods, we refer to [B-79, Ch. 2].

8. A slightly weaker version of Theorem 2.15 was given in [244]. See also [B-61, Ch. IV, Problem 3.3], where one direction of the corresponding statement is given for the particular example $\gamma_n = \frac{1}{n}$.

9. Theorem 2.16 is given as an exercise in [B-61, Ch. IV, Problem 3.3].

10. Theorem 2.19 was mentioned in [182], and although it is trivial, it served the authors as a starting point to try to obtain more general comparison tests (see also Theorem 3.22, which also appeared in [182], and Notes-24 in Chapter 3 for further discussion).

11. The ratio test (Theorem 2.20) is sometimes referred to as the *second comparison test.*

12. Theorem 2.21 appeared in [20], while Theorem 2.22 appeared in [129]. Both of them conclude the absolute and/or conditional convergence (or divergence) of certain series, and in fact it seems hopeless to find any conclusion that does not involve absolute or conditional convergence when considering comparison-type tests for series with both positive and negative terms. Examples of applications of Theorems 2.21 and 2.22 may be found in appropriate papers.

13. Theorem 2.23 was obtained by Jehle in [137]. Its sharpness with respect to the assumption that $\sum c_k$ converges absolutely is proved in [9].

14. D'Alembert's test (or the ratio test) appeared as early as 1768 in a work by d'Alembert [68]. It is worth mentioning that such a test was just a small piece of content in a volume of a series of publications (in total eight volumes) by the same author, where he studied several problems in both mathematics and physics.

15. The m-th ratio test (Theorem 2.25) was stated in [11] (slightly differently, in a way that does not recover the classical ratio test), along with several examples comparing this test with the classical one, like that presented above (an earlier simplified version of this test appeared in [207]; further remarks and generalizations of the mth ratio test may be found in [119]). However, in [11], the full proof of Theorem 2.25 is given only for the case $m = 2$. In a similar spirit, and motivated by the mth ratio test, the author of [132] develops a so-called second Raabe's test (where the terms $\frac{a_n}{a_{2n}}$ and $\frac{a_n}{a_{2n+1}}$ replace the usual ratio $\frac{a_n}{a_{n+1}}$ in Theorem 2.36), a second Gauss' test, and, finally, a second Kummer's test that includes all of them as particular cases.

16. The root test (Theorem 2.30) can be used to easily prove that for $m > 1$, any series of the type

$$\sum_{k=1}^{\infty} \frac{k^n}{m^k}, \qquad n \in \mathbb{N}, \tag{2.85}$$

converges. However, as noted in [235], in calculus courses one is often satisfied with determining the convergence of a series and does not look into what the precise value of that series is. The author of [235] shows two different methods to determine the sum of (2.85).

17. A more general relation than (2.14) holds true. For a positive sequence $\{x_k\}$, one always has

$$\liminf_n \frac{x_{n+1}}{x_n} \le \liminf_n \sqrt[n]{x_n} \le \limsup_n \sqrt[n]{x_n} \le \limsup_n \frac{x_{n+1}}{x_n},$$

see [B-78, Theorem 2.3.27].

18. The author of [66] was interested in the relation between the ratio and root tests. It was noted that the root test does not imply the ratio test, by means of the example

$$\frac{1}{2} + 1 + \frac{1}{8} + \frac{1}{4} + \frac{1}{32} + \frac{1}{16} + \cdots,$$

which served as a motivation for obtaining Theorem 2.31. It can be viewed as an "averaged ratio test". Moreover, being an improvement of the classical ratio test, it is still weaker than the root test. However, the true strength of Theorem 2.31 lies in the fact that the average of ratios of a sequence may be much easier to compute than the nth root of the general term, as shown in the examples at the end of Subsection 2.1.4 (refer to [66, 67] for further discussion).

19. There are interesting historical notes around Jamet's test (Theorem 2.32) in [117]. Such a test was already introduced by Cauchy in his lectures published in 1821 [B-20]. However, as mentioned in [117], "the name *Cauchy's test* would not be sufficiently distinctive". According to [117], the nomenclature *Jamet's test* was proposed in [B-77] (this test is also referred to as "second Cauchy's test" in [187]). Jamet's contribution may be found in [134].

Further comments on Jamet's test may be found in Goursat's book [B-38, pp. 394–395]. In particular, it is noted there that since Jamet's test is based on the comparison with the sequence $\frac{1}{n^p}$, that test may be iterated based on the comparison with the terms of the sequences

$$\frac{1}{n(\log n)^p}, \qquad \frac{1}{n \log n (\log \log n)^p}, \qquad \text{etc.}$$

In these cases, the corresponding limits to be inspected are

$$\lim_{n\to\infty} -\frac{\log(n|a_n|)}{\log\log n}, \qquad \lim_{n\to\infty} -\frac{\log(n|a_n|\log n)}{\log\log\log n}, \qquad \text{etc.}$$

These "logarithmic iterations" are analogous to the possible "logarithmic iterations" of Raabe's test (Theorem 2.36). For instance, the first iteration of Raabe's test yields Bertrand's test (Theorem 2.38), and further iterations of the latter may be obtained as discussed in Remark 2.39.

In fact, the relationship between Jamet's test and Raabe's test is in parallel to the relationship between the ratio test and the root test. Recall that the root test is stronger than the ratio test, or, in other words, the ratio test implies the root test (cf. (2.14)). Similarly, Jamet's test is stronger than Raabe's test. That is, if $\{a_k\}$ is a positive sequence and the limit

$$\lim_{n\to\infty} n\left(1 - \frac{a_{n+1}}{a_n}\right) = L$$

in Raabe's test exists (note that in Theorem 2.36 the expressions in lim inf and lim sup are written slightly differently, although simple manipulations allow to write them in this latter form), then one has

$$\lim_{n\to\infty} -\frac{\log a_n}{\log n} = L.$$

This fact is shown in [187]. A much deeper consideration and further observations on the relationship between Raabe's and Jamet's tests are given in a more recent paper [117].

Further generalizations of Raabe's and Cauchy's tests in the spirit of logarithmic iterations are given in [4]. We also mention the recent work by Slyusarchuk [238], where the aforementioned tests with logarithmic iterations are obtained in the context of operator integrals. The same author has extended several other convergence tests in that context; see the references in [238].

20. Theorem 2.33 as well as the details for deriving the divergence/convergence of the series in (2.19) (by means of Theorem 2.33 itself) appeared in the relatively recent paper [160]. The first series in (2.19) was shown to diverge in [48], using an approach that is "at the frontier between analysis and number theory". An interesting historical note concerning the second series in (2.19) is also mentioned in [160]. It was proposed in a calculus exam by mistake, and the problem remained unsolved for ten years, until a solution appeared in [240].

21. Kummer's test first appeared in [157] (1835). It is necessary to note that Kummer's test in Theorem 2.34 is not the same that appeared originally in [157], but a more modern version in [259] (the proof that we reproduce was extracted from this last reference). In fact, Kummer's original test in [157] was even weaker than what we called "weak Kummer's test" (Theorem 2.35) obtained by Dini in [78]. Weak Kummer's test may be found in classical series textbooks, as, for instance, in [B-16, Ch. II, §12].

22. The following interesting historical note regarding Kummer's test is given in [112]. It is mentioned there that Pringsheim [B-68] called it a "most remarkable criterion" of "indeed surprising generality" that stands in need of clarification. Pringsheim's opinion was by no means exceptional; many mathematicians must have been similarly puzzled and left unsatisfied by Kummer's original proof of his criterion in 1835. As Knopp notes [B-54], it was not until 1885 that O. Stolz gave an "extremely simple proof, by means of which the criterion was first rendered fully intelligible". Pringsheim also noticed that Kummer's (original) test "appeared as totally erratic" and he thus aimed at presenting it "freed from this mysterious isolation". He gave two different proofs of it, one explanatory and another one that only "proves the correctness of the criterion a posteriori in a simpler way". For further references as well as a comprehensive account of this historical note, we refer to [112].

23. There are also different tests that are formulated in the spirit of Kummer's test. In the book by Salekhov [B-75, Ch.II, §4], there is a test attributed to Giudici, which turns out to be exactly the same statement as the first part of Kummer's test (Theorem 2.34). However, this issue is somewhat mysterious, since the only reference given there is to the old book [B-71].

24. Another interesting remark around Kummer's test is that the basic ratio test (Theorem 2.24) also follows from weak Kummer's test by taking $d_n = 1$. In fact, the British mathematician Augustus De Morgan (1806–1871) proposed a hierarchy for ratio-type tests based on the examination of a sequence of the type

$$d_n \frac{a_n}{a_{n+1}} - d_{n+1},$$

involving the consecutive ratios of the sequence $\{a_k\}$. By this, the ratio test would be the first test of the hierarchy, Raabe's test the second one, and so on. Obviously, the strong Kummer's test is the most general test in such a hierarchy.

25. Recent attempts to generalize Kummer's test are taken in [22]. In that paper, the authors improved the test in two different ways. First, they showed that rather than involving consecutive terms, Kummer's test can be written in terms of nonconsecutive terms as long as the length between these terms is fixed. More precisely, the term

$$d_k \frac{a_k}{a_{k+1}} - d_{k+1}$$

in both appearances of Theorem 2.34 may be replaced by the term

$$d_k \frac{a_k}{a_{k+m}} - d_{k+m},$$

for any fixed $m \in \mathbb{N}$, and the conclusion remains the same. Another direction in which Kummer's test was generalized in [22] is in describing the relation between the sequences $\{a_k\}$ and $\{c_k\}$ in series of the form $\sum a_k c_k$. The convergence criterion (corresponding to the first part of Kummer's test) is as follows.

Theorem 2.129 *Let $\{a_k\}$ and $\{c_k\}$ be positive sequences. The series $\sum a_k c_k$ converges if and only if there exists a positive sequence $\{d_k\}$ such that*

$$d_k \frac{a_k}{a_{k+1}} - d_{k+1} \ge c_{k+1}, \quad \text{for all } k \ge 1.$$

For the proof, we refer to [22, Theorem 3.1]. As for the divergence criterion (the second part of Kummer's test), the statement is slightly more involved, and one of the implications needs additional hypotheses (cf. [22, Theorem 3.2]).

Theorem 2.130 *Let $\{a_k\}$ and $\{c_k\}$ be positive sequences.*

(a) If there exists a positive sequence $\{d_k\}$ such that $\sum \frac{1}{d_k}$ diverges and

$$d_k \frac{a_k}{a_{k+1}} - d_{k+1} \le -c_{k+1}, \quad \text{for all } k \ge 1,$$

then all series $\sum a_k$, $\sum \frac{1}{c_k}$, $\sum (d_k - c_k) a_k$, and $\sum d_k a_k$ diverge. If, additionally, the series $\sum \frac{c_k}{d_k}$ diverges, then $\sum a_k c_k$ diverges as well.

(b) Assume that the series $\sum a_k$ and $\sum a_k c_k$ diverge and that for every $m \in \mathbb{N}$, there exists $n \ge m$ such that

$$a_m + \cdots + a_n \ge a_m c_m + \cdots + a_n c_n.$$

Then, there exists a positive sequence $\{d_k\}$ such that $\sum \frac{1}{d_k}$ diverges, and

$$d_k \frac{a_k}{a_{k+1}} - d_{k+1} \le -c_{k+1}, \quad \text{for all } k \ge 1.$$

26. In [223], the author gives an interesting remark on Raabe's test (Theorem 2.36). To begin the discussion, let us observe that Raabe's test may be reformulated as follows: first, if

$$l = \liminf_{n\to\infty} n\left(\frac{a_n}{a_{n+1}} - 1\right) > 1,$$

then the series $\sum a_k$ converges. On the other hand, if $n\left(\frac{a_n}{a_{n+1}} - 1\right) < 1$ for all $n \geq n_0$, then $\sum a_k$ diverges. Intuitively, inspecting the terms

$$n\left(1 - \frac{a_{n+1}}{a_n}\right)$$

should give an analogous statement to Raabe's test. Indeed, in this case, if

$$l' = \liminf_{n\to\infty} n\left(1 - \frac{a_{n+1}}{a_n}\right) > 1,$$

then the series $\sum a_k$ converges, while if $n\left(1 - \frac{a_{n+1}}{a_n}\right) < 1$ for all $n \geq n_0$, then $\sum a_k$ diverges. One would expect that such tests are equivalent, but in fact this is not the case, as shown in [223]. The author considers the series

$$\sum_{n=1}^{\infty} \frac{(n-1)!\,n!\,4^n}{(2n)!\,\sqrt{n}}.$$

On the one hand, it is easy to see that

$$n\left(\frac{a_n}{a_{n+1}} - 1\right) = 1 + \frac{1}{8n} + o\left(\frac{1}{n^2}\right) \to 1^+,$$

which renders the above Raabe's test (Theorem 2.36) inconclusive. However, for the alternative Raabe's test, one has

$$n\left(1 - \frac{a_{n+1}}{a_n}\right) = 1 - \frac{3}{8n} + o\left(\frac{1}{n^2}\right) \to 1^-,$$

which yields the divergence of the corresponding series by using the alternative Raabe's test (since $n\left(1 - \frac{a_{n+1}}{a_n}\right) < 1$ for large n). This discrepancy motivated the author of [223] to obtain a refinement of Raabe's test (more precisely of the divergence part of the test). The precise statement is as follows.

Theorem 2.131 *If $\{a_k\}$ is a positive sequence and*

$$n\left(\frac{a_n}{a_{n+1}} - 1\right) = 1 + O\left(\frac{1}{n}\right), \qquad n \to \infty,$$

then the series $\sum a_k$ diverges.

27. A comparison between Raabe's test and "Schlömilch's test" is given in [224]. However, we emphasize that what the author calls "Schlömilch's test" in [224] does not correspond to Theorem 2.94. In particular, the test in [224] reads as follows.

Theorem 2.132 *Let $\sum a_k$ be a positive series.*

- *If $\liminf n \ln\left(\frac{a_n}{a_{n+1}}\right) > 1$, then the series $\sum a_k$ converges.*
- *If $n \ln\left(\frac{a_n}{a_{n+1}}\right) \leq 1$ for all $n \geq n_0$, then the series $\sum a_k$ diverges.*

In [224], the exact relationship between Raabe's test and "Schlömilch's test" (Theorem 2.132) is described and a useful improved version of the divergence parts of these tests is given.

28. Notably, versions of Raabe's test for complex-valued sequences or for alternating series also exist. In [118], the author obtained the following two results (for a proof, see the given reference).

Theorem 2.133 *Let $\sum a_k$ be a series consisting of nonzero terms. Assume that*

$$L = \lim_n n\left(\left|\frac{a_n}{a_{n+1}}\right| - 1\right)$$

exists and is finite.

(a) If $L > 1$, then the series $\sum a_k$ converges absolutely;
(b) If $L < 0$, then the series $\sum a_k$ diverges;
(c) If $0 \le L < 1$, then the series $\sum a_k$ either diverges or converges conditionally;
(d) If $L = 1$, then the test is inconclusive.

Theorem 2.134 *Let $\{a_k\}$ be a positive sequence and assume that*

$$L = \lim_n n\left(\frac{a_n}{a_{n+1}} - 1\right)$$

exists and is finite. If $0 < L \le 1$, then the series $\sum(-1)^n a_n$ converges, and the convergence is necessarily conditional if $0 < L < 1$.

It is also seen in [118] that the convergence character of the three series

$$\sum_{n=1}^{\infty} \frac{(-1)^{n-1}(2n-1)!}{4^n(n!)^2}, \qquad \sum_{n=1}^{\infty} \frac{(-1)^{n-1}(2n)!}{4^n(n!)^2}, \qquad \sum_{n=1}^{\infty} \frac{(-1)^{n-1}(2n+1)!}{4^n(n!)^2}$$

can be decided by means of Theorems 2.133 and 2.134. Indeed, by calculating the limit in Theorem 2.133 applied to the general terms of each of these series, one obtains the values $3/2$, $1/2$, and $-1/2$, respectively. Hence, the first series converges absolutely, the second one converges conditionally, and the last one diverges (in order to obtain the conditional convergence of the second series, Theorem 2.133 is not enough; Theorem 2.134 must also be applied).

29. The proof of Theorem 2.37 is rather standard and may be derived as a consequence of the results in [40] (see also [B-10, §1.9]). It is based on the fact that if the limit in Theorem 2.37 exists, then $\{a_k\}$ is regularly varying, written

$$\lim_n \frac{a_{\lfloor \xi n \rfloor}}{a_n} = \psi(\xi) \in (0, \infty), \qquad \text{for any } \xi \ge 1.$$

Theorem 2.37 was also extended in [6] to estimates of logarithmic type.

30. The following result in [188] readily follows from the inequality between arithmetic and geometric means and generalizes the sufficient conditions in Gauss' test (Theorem 2.40).

Theorem 2.135 *If $\{a_n\}$ is a positive sequence, $a_1 = 1$, and $a_{n+1}/a_n = 1 - d_n$, then*

$$\sum_{n=1}^{\infty} a_n \le 1 + \sum_{n=1}^{\infty}\left(1 - \frac{1}{n}\sum_{k=1}^{n} d_k\right)^n, \tag{2.86}$$

and, in particular, the convergence of the series on the right-hand side of (2.86) *implies the convergence of $\sum a_n$. Furthermore, if* (2.21) *holds with $\alpha > 1$, the series on the right of* (2.86) *converges.*

31. It is proved in [252] that in the Gauss test, in the case

$$\frac{a_{k+1}}{a_k} = 1 - \frac{1}{k} - r_k,$$

the most natural (and sharp) condition for divergence is $r_k = o\left(\frac{1}{k \log k}\right)$, contrary to $r_k = O\left(\frac{1}{k^{1+\varepsilon}}\right)$, for some $\varepsilon > 0$, given in various sources.

32. The generalized Gauss test (Theorem 2.44) can be found in [3].

33. An interesting feature of a positive convergent series is that its terms can be grouped so that the sequence of grouped sums monotonically decreases, that is, one can find an increasing sequence of integers $\{N_j\}$ with $N_0 = 1$ such that

$$\sum_{k=N_0}^{N_1} a_k \geq \sum_{k=N_1+1}^{N_2} a_k \geq \dots \tag{2.87}$$

One possible construction is as follows. Define $A_0 = \sum a_k$ and suppose that the sequence $\{a_k\}$ has infinitely many nonzero terms (otherwise the construction is trivial). Define

$$N_1 = \min\left\{n \in \mathbb{N} : \sum_{k=1}^{n} a_k \geq \frac{A_0}{2}\right\}.$$

Further, set $A_1 = \sum_{k=N_1+1}^{\infty} a_k$ and let

$$N_2 = \min\left\{n > N_1 : \sum_{k=N_1+1}^{n} a_k \geq \frac{A_1}{2}\right\}.$$

It is clear that

$$\sum_{k=1}^{N_1} a_k \geq \frac{A_0}{2} \geq \sum_{k=N_1+1}^{\infty} a_k \geq \sum_{k=N_1+1}^{N_2} a_k \geq \frac{A_1}{2}.$$

Using this procedure, we define, recursively for $j \geq 3$, $A_{j-1} = \sum_{k=N_{j-1}+1}^{\infty} a_k$ and

$$N_j = \min\left\{n > N_{j-1} : \sum_{k=N_{j-1}+1}^{n} a_k \geq \frac{A_{j-1}}{2}\right\}.$$

The sequence of indices $\{N_j\}$ defined in this way is such that (2.87) holds. Indeed, for $j \in \mathbb{N}$,

$$\sum_{k=N_{j-1}+1}^{N_j} a_k \geq \frac{A_{j-1}}{2} \geq \sum_{k=N_j+1}^{\infty} a_k \geq \sum_{k=N_j+1}^{N_{j+1}} a_k.$$

In the late 1950s, P. L. Ul'yanov asked whether this kind of grouping is possible under the weaker assumption that $\{a_k\} \in \ell^p$ with $p > 1$, or, in other words, that $\sum a_k^p$ converges. Shortly after that, Ul'yanov himself constructed an example of a sequence $\{a_k\} \in \bigcap_{p>2} \ell^p$ for which any such grouping is impossible. For the case $1 < p \leq 2$, this problem remained unsolved.

This problem, for $1 < p \leq 2$, was partially solved by Erdős, Joó, and Székely [98] in 1987, under the additional assumption that the sequence $\{a_k\}$ admits a decreasing majorant $\{b_k\} \in \ell_2$. Later on, in 1988, A. A. Lunin [184] showed that the desired grouping is possible for $\{a_k\} \in \ell^2$.

Theorem 2.136 *Let $\{a_k\}$ be a positive sequence and assume that*

$$\sum_{k=1}^{\infty} a_k^2 < \infty.$$

Then there exists an increasing sequence of integers $\{N_j\}$ with $N_0 = 1$ for which (2.87) *holds.*

Section 2.2 (Absolute and conditional convergence)

34. In [120], Hardy gave an abstract version of Theorem 2.48. In particular, he showed that for absolutely convergent series, any summation type whose corresponding ordinal belongs to the Cantor's second class ordinals may be arbitrarily rearranged, and the resulting sum remains the same (for an informal discussion on Cantor's ordinals see [B-24, Ch. 7], while for a more formal exposition we refer to [B-57, Ch. 4]). Since Cantor's second class ordinals form an enumerable set, in particular, Hardy's result applies to rearrangements of absolutely convergent series in any dimension (cf. Theorem 7.44 below).

35. Riemann's rearrangement theorem (Theorem 2.51) is often taught during the first years of undergraduate studies, and at first glance it may come as an unexpected phenomenon.

36. A problem closely connected to rearrangements of series is that of considering number series of the type

$$\sum_{n=1}^{\infty} \varepsilon_n a_n, \qquad \varepsilon_n = \pm 1. \tag{2.88}$$

In fact, in the proof of Theorem 2.47 we implicitly construct series of this type. Given any series $\sum a_n$ with real-valued terms, one may ask about the convergence character of (2.88). If the signs ε_n are chosen randomly (each with probability $\frac{1}{2}$), the problem of determining the probability that (2.88) converges was first posed by Steinhaus in 1922 [243]. The solution to this problem is as follows.

Theorem 2.137 *Let $\{a_n\}$ be a real sequence and $\{\varepsilon_n\}$ be such that each ε_n is either $+1$ or -1 with probability $\frac{1}{2}$. Then, the following are equivalent.*

(i) *The series $\sum \varepsilon_n a_n$ converges with probability 1.*
(ii) *The sereis $\sum \varepsilon_n a_n$ is bounded with probability 1.*
(iii) *The series $\sum a_n^2$ converges.*

For a proof, see [B-51, Ch. 2]. The implications and observations of this result are rather interesting. First of all, the probability that the series $\sum \varepsilon_n a_n$ converges can only be either 0 or 1 due to Kolmogorov's zero-one law. Secondly, no assumptions are made on the sequence $\{a_n\}$. In particular, by choosing $a_n = 1$, the partial sums $\sum_{n=1}^{m} \varepsilon_m$ model the one-dimensional random walk, and according to Theorem 2.137, these are unbounded. Finally, for the p-series $\sum \frac{1}{n^p}$, the corresponding series $\sum \frac{\varepsilon_n}{n^p}$ converges with probability 1 if and only if $p > \frac{1}{2}$.

Further interesting facts concerning complex-valued series of the form (2.88) are deeply studied in [136] (see also the references therein).

37. Theorem 2.52 was obtained in [97].

38. In the paper [153], the authors raise interesting questions concerning rearrangements of sequences. In particular, they introduce Property λ of a permutation and prove that permutations with such a property exist (precisely Theorem 2.53). The mere existence of a permutation with Property λ is already an interesting question, but in [153] the authors consider further problems, such as whether the set of all permutations with Property λ can easily be characterized, or how many permutations with such a property exist (they show that the answer is "uncountably many"). Further deep discussions and results are given in [153] and the references therein.

39. Cesàro's theorem (Theorem 2.55) was described by both Bromwich [B-16] and Hardy [125] as "curious" (see the quotations in [209]). Theorem 2.58 was given by Rademacher in [225] (a proof may also be found in [209]). Corollary 2.59 and further discussion are given in [133].

40. For convex sequences $\{|a_k|\}$ (i.e., decreasing such that $\{\Delta |a_k|\}$ is also decreasing), a version of Theorem 2.57 holds. In order to state it, we introduce the sequences

$$P_n = \sum_{k=1}^{n} p_k, \qquad Q_n = \sum_{k=1}^{n} q_k.$$

Theorem 2.138 *If the sequence $\{|a_n|\}$ is convex, $|P_n - Q_n| \leq Cn$ for all n, and*

$$(p_n - q_n)a_n \to 0 \quad as\ n \to \infty, \tag{2.89}$$

then the series $\sum a_k$ converges.

A converse of Theorem 2.138 also holds.

Theorem 2.139 *If the sequence $\{|a_n|\}$ is convex and the series $\sum a_k$ converges, then* (2.89) *holds.*

Note that combining Theorems 2.138 and 2.139 yields a criterion for the convergence of $\sum a_k$, provided that $|P_n - Q_n| \leq Cn$. For a proof of these statements, we refer to [209].

41. Theorem 2.54 was obtained in [10]. Equivalent characterizations were given in [172, 217], all of which are discussed in the survey [232]. See also [B-27, Ch. V, §2, Problem 2].

42. Although in the present book we do not dive really deep into the topic of rearrangements of series, it is worth giving some comments and references that study parts of this topic in detail.

One of the first papers to study the properties of permutations that preserve the convergence/divergence of *any* convergent/divergent series $\sum a_k$ is due to Kronrod [156]. In fact, such permutations are characterized in that paper. If a permutation $\sigma : \mathbb{N} \to \mathbb{N}$ does not preserve convergence of all series, or, in other words, if there exists $\{a_k\}$ such that $\sum a_k$ converges but $\sum a_{\sigma(n)}$ diverges, we say that σ is a divergent permutation. Divergent permutations are also characterized in [156]. As for permutations preserving the sum of a convergent series, they are fully characterized by Levi [172] and Agnew [10].

Subsequent papers studied interesting properties of divergent permutations. In particular, in [130], the authors show that for any divergent permutation, any closed interval $I \subset [-\infty, \infty]$ (the two-point compactification of $\mathbb{R}$), and any $A \in I \cap \mathbb{R}$, one may find a sequence $\{a_k\}$ satisfying $\sum a_k = A$ and such that every point of I is an accumulation point of the sequence of rearranged partial sums $\left\{ \sum_{k=1}^{n} a_{\sigma(k)} \right\}$. This fact was rediscovered independently in [276] (in fact, the author of the latter recognizes that the paper was about to be published and decided to withdraw it upon noticing the existence of [130]). In the end, the paper [276] was published and contains a different approach to [130], which translates into simpler proofs for the main results.

For the interested reader, further details on the theory of rearrangements of series may be found in [107, 135, 172, 199, 200, 201, 217, 218, 278] and the references therein. For a relatively recent survey, see [277].

43. Theorem 2.60 was obtained in [7]. It is also noted there that the assumption of the existence of $f''(0)$ may be relaxed. For instance, it is enough that the limit

$$\lim_{x\to 0^+} \frac{f'(x)}{x^\alpha}$$

exists for some $0 < \alpha < 1$, or that $g''(0)$ exists, where $g(x) = |x|^\alpha f(x)$ (both conditions are satisfied whenever $f''(0)$ exists).

Section 2.3 (Tests involving sequences of bounded variation)

44. Although being direct consequences of Dedekind's and du Bois-Reymond's tests, Dirichlet's and Abel's tests are elements of the classical theory of series involving sequences of bounded variation (in fact, the two latter tests appeared in 1862 (cf. [79]) and in 1826 (cf. [1]), respectively).

45. The application of the estimate (2.30) related to Leibniz's test in order to prove the irrationality of e can be found in [215].

46. A counterexample showing that Dedekind's test is not valid if we replace the condition $\{a_k\} \in BVS$ by the more general $\{a_k\} \in GMS$ is given in [178].

47. Interestingly enough, Theorem 2.67 was extracted from the short paper [55] by P. Calabrese (we do not know if this result was obtained earlier), who was at the time an undergraduate student. The result comes from a class assignment that consisted of determining the number of terms of the alternating harmonic series (i.e., the series with general term $a_n = \frac{(-1)^n}{n}$) needed for its partial sum to have four digits of decimal precision. His solution was the estimate given in Theorem 2.67 (and the answer he obtained was 10 000 terms). His instructor (and several other calculus textbooks at the time) suggested that one would need the double of that amount of terms in order to obtain the said approximation. However, that answer was based on estimate (2.30), which, in view of Theorem 2.67, is "half optimal" for convex sequences. It is also worth noting that in [55], Theorem 2.67 is stated for strictly convex sequences (i.e., $\Delta^2 a_k > 0$), although the result is still valid for just convex sequences.

48. Theorem 2.68 was obtained in [138]. In [266], motivated by the preceding work in [280], the author considered a method for proving Theorems 2.67 and 2.68 based on Cantor's theorem on nested intervals. This approach in turn allowed the author to obtain a proof for Theorem 2.70, which we reproduced here.

49. For a proof of Theorem 2.69, we refer to Knopp's book [B-54, pp. 244–246]. In that same book, interesting historical notes are given. Apparently, Euler's transformation [101] was already used in a particular example by N. Fatzius, according to a letter that J. Bernoulli wrote to Leibniz in 1704. Further remarks may be found in [B-54, pp. 244].

50. In [96], the authors investigated the best possible constants $a > 0$ and $b > -a$ such that, for fixed $\alpha > 0$ and $\beta > -\alpha$, the inequalities

$$\frac{1}{2\alpha n + a} < \left| \sum_{k=n+1}^{\infty} \frac{(-1)^{k-1}}{\alpha k + \beta} \right| < \frac{1}{2\alpha n + b}$$

hold for every $n \geq 1$ (see also (2.31)). Denoting $S(\alpha, \beta) = \sum_{k=1}^{\infty} \frac{(-1)^{k-1}}{\alpha k+\beta}$, the best constants are

$$a = \left(\frac{1}{\alpha + \beta} - S(\alpha, \beta) \right)^{-1} - 2\alpha \qquad \text{and} \qquad b = \alpha + 2\beta.$$

The authors then used this result to answer a question from [261] on the *Gregory-Leibniz series* $\sum_{k=1}^{\infty} \frac{(-1)^{k-1}}{2k-1}$ and concluded that the best constants c and d, so that

$$\frac{1}{4n+c} < \left| \sum_{k=n+1}^{\infty} \frac{(-1)^{k-1}}{2k-1} \right| < \frac{1}{4n+d}$$

holds for all $n \geq 1$, are $c = \dfrac{4}{4-\pi} - 4$ and $d = 0$.

51. An interesting problem concerning alternating series was studied recently by T. Tao in [251]. The problem consists of determining whether the series with alternating sign

$$\sum_{n=1}^{\infty} \frac{(-1)^n n}{p_n},$$

where p_n is the nth prime, converges or diverges. It is shown in [251] that the convergence of this series is equivalent to the convergence of

$$\sum_{n=2}^{\infty} \frac{(-1)^{\pi(n)}}{n \log n},$$

where $\pi(x) = \#\{p \text{ prime}, p \leq x\}$, $x \geq 2$, is the prime counting function. What is more, such series are shown to be convergent under the assumption that the Hardy–Littlewood prime tuples conjecture holds true (cf. [127] and [159, Conjecture 1.3]).

52. Further examples of series with alternating signs are considered in [57] (see also the references therein). Also, estimates of Hölder and Minkowski types for partial sums of series with alternating signs were obtained in [60].

53. Segmentally alternating series were introduced in [234], where example (2.40) was also discussed. One may expect similar estimates for the error of approximation, or the sum of alternating series in the context of segmentally alternating series. In fact, it is easy to see (cf. [234]) that if a segmentally alternating series has segments of constant length 2ℓ, then for any $n \in \mathbb{N}$,

$$|S_n| = \left| \sum_{k=1}^{n} a_k \right| \leq \sum_{k=1}^{\ell} |a_k| \leq \ell |a_1|$$

(recall that for segmentally alternating series the sequence $\{|a_k|\}$ is nonincreasing). Although the assumption of having segments of bounded length in Theorem 2.76 is essential, one can still obtain the following positive result.

Theorem 2.140 *Let $\sum a_n$ be a divergent positive series and assume that the sequence $\{a_n\}$ is decreasing. Then we may choose a sequence $\{\varepsilon_n\}$ consisting of 0's and 1's such that the series $\sum_{n=1}^{\infty}(-1)^{\varepsilon_n} a_n$ is a convergent segmentally alternating series, with segments of unbounded length.*

For a proof and deeper discussion on segmentally alternating series, we refer the reader to [234]. See also [27].

54. Theorem 2.74 as well as further discussion appeared in [65]. It was also observed in [65] that the conclusion of Theorem 2.74 is no longer true if $\{a_k\}$ is not convex.

Section 2.4 (Tests for monotone sequences)

55. The most classical tests that are mentioned at the beginning of Section 2.4 may be found in, e.g., [B-54], [B-16], [B-31], or the more recent [B-11] and [B-15], in the context of monotone sequences. Many of them were obtained in the context of *WMS* in [178].

56. It is worth mentioning that integral tests (including but not restricted to the classical Maclaurin–Cauchy test) are of special interest, not only because they provide simple tools for convergence of series but also because of the applications they have, as in problems of harmonic analysis and approximation theory, or in advanced problems of number theory. In [54, Theorem 1], a theorem is obtained about convergence of the Riemann type sums $\frac{1}{m}\sum_{k=1}^{m} f\left(\frac{k}{m}\right)$, related to the improper Riemann integral $\int_0^1 f(t)\,dt$, where f is a stepwise function. The Prime Number Theorem follows from that result immediately, and the theorem itself is studied thoroughly, with numerous relations to number series involved. Such a result was an improvement of an earlier statement in [275], about which it is noted that looking "quite innocent, (it) is actually strongly connected with the Prime Number Theorem".

Maclaurin–Cauchy integral and related tests

57. An interesting problem is to determine for which classes of functions one may derive an integral-type test (this is precisely one of our goals, by replacing monotone functions by $f \in WM$). Another simple example of a class of functions (yet more general than WM) for which the Maclaurin–Cauchy test holds is that consisting of all f such that

$$\frac{1}{C} f(k+1) \le f(x) \le Cf(k), \qquad x \in [k, k+1],$$

for some $C > 1$ independent of x. This task has also been undertaken by the authors in [83] (see also the references therein), where Cesàro summability of the series and integrals is involved. There, the multidimensional case is also considered.

58. In relation to the integral test (see Theorem 2.78), it is interesting to obtain explicit bounds for the remainder term. More precisely, if $f\colon [1,\infty) \to \mathbb{R}_+$ is a monotone function and $\sum f(k)$ converges to $S > 0$, one may ask what are the best bounds for the remainder $R_n = S - \sum_{k=1}^{n} f(k)$ of the nth partial sum. In general, if we only require that f is decreasing, the best estimate for such a remainder is

$$\int_{n+1}^{\infty} f(x)\,dx \le R_n \le \int_{n}^{\infty} f(x)\,dx.$$

However, it is shown in [196] that if we additionally require f to be convex, such estimates can be improved. These were further refined in [202], and the refined version reads as

$$\frac{1}{2} f(n+1) + \int_{n+1}^{\infty} f(x)\,dx \le R_n \le \int_{n+\frac{1}{2}}^{\infty} f(x)\,dx.$$

For smooth f, Boas [37] showed that

$$\frac{1}{8} f'\left(n+\frac{1}{2}\right) + \int_{n+\frac{1}{2}}^{\infty} f(x)\,dx < R_n < \int_{n+\frac{1}{2}}^{\infty} f(x)\,dx$$

by means of the Euler–Maclaurin formula and the second mean value theorem. For further results on approximating the sum of a convergent series, see the recent paper [5] and the references therein.

59. Theorem 2.80 was obtained by Bromwich in [51]. As Hardy explains in [124], "Roughly, it may be said that what Dr. Bromwich proved amounted to this, that the relations between

$$\sum_{k=1}^{\infty} f(k) \quad \text{and} \quad \int_1^{\infty} f(t)\, dt$$

as regards convergence or divergence, established by Maclaurin and Cauchy when $f(x)$ is positive and decreasing, still subsist when $f(x)$ is multiplied by an oscillatory factor of the type $e^{ig(x)}$ provided $g(x)$ tends to infinity more slowly than x". Hardy observed that, for continuously differentiable functions, the Maclaurin–Cauchy integral test holds under a more general assumption. More precisely, the involved function can be taken of bounded variation rather than positive and decreasing. This is precisely Theorem 2.81, which appeared in [124] (condition (iii) in that statement is precisely the boundedness of variation of f under the given assumptions).

60. In [56, Theorem 4], the smoothness requirement of Theorem 2.81 was replaced by the milder bounded variation condition. Based on the fact that any function of bounded variation may be written as a difference of two decreasing functions, the result states that

$$\left| \sum_{n=1}^{N} f(n) - \int_1^N f(x)\, dx \right|$$

tends to a finite limit whenever $N \to \infty$. Note that Theorem 2.83 with $x = 0$ gives an upper bound for such a difference:

$$\left| \sum_{n=1}^{N} f(n) - \int_1^N f(x)\, dx \right| \le \pi \int_1^N |df(x)|. \tag{2.90}$$

Since f is assumed to be of bounded variation and thus the right-hand side is finite as $N \to \infty$, we obtain the same conclusion as in Theorem 2.81, with an explicit bound for the difference. As an example of importance of relations of the type (2.90) in practice, the author of [56] refers to [204].

61. Theorem 2.82 was obtained by Hardy in [124], the same paper where Theorem 2.81 was proved. It is worth citing one more of his comments: "I have not thought it worth while to encumber this theorem (or the others...) with explicit statements as regards the continuity of such derivatives of $f(x)$ as figure in them. All such derivatives are, of course, supposed continuous: only conditions relating to their behaviour as $x \to \infty$ are in any way relevant to the problem. Sometimes, however, these conditions may be broken for $x = 0$ or some other value of x. Then the formula may need modification, but the application of the theorem to questions of convergence is not affected". See also [179, 227] for related tests.

62. There are yet some more generalizations to Theorem 2.82. Motivated by the fact that Theorem 2.82 does not apply to series of the type (2.45) with $b \ge 1$, one may aim at further generalizations of Hardy's theorem. For example, we have the following three practical corollaries obtained in [241]. In all of them $f(x)$ and its derivatives $f^{(j)}(x)$, defined for $x \ge 1$, are supposed to be continuous and tend to zero as $x \to \infty$.

Theorem 2.141 *Let $f\colon (0, \infty) \to \mathbb{C}$. The series $\sum_{n=1}^{\infty} f(n)$ converges if the integrals*

$$\int_1^{\infty} f(x)\, dx \quad \text{and} \quad \int_1^{\infty} |f^{(k)}(x)|\, dx$$

converge for some $k \ge 1$.

Theorem 2.142 *Let $f\colon (0, \infty) \to \mathbb{C}$. The series $\sum_{n=1}^{\infty} f(n)$ converges if $\int_1^{\infty} f(x)\, dx$ converges and*

$$\int_1^{\infty} f^{(k)}(x) \sin(2\pi N x)\, dx = O\left(\frac{1}{N^r}\right),$$

for some $r > 0$ and $k \ge 1$.

Finally, sufficient conditions for divergence are given in a similar way.

Theorem 2.143 *Let $f\colon (0,\infty) \to \mathbb{C}$. If the integral $\int_1^\infty f(x)\,dx$ converges and*

$$\lim_{N\to\infty} \int_1^\infty f'(x)\sin(2\pi Nx)\,dx = A \neq 0,$$

then the series $\sum_{n=1}^\infty f(n)$ diverges.

Theorem 2.142 can be applied to deduce that the series in (2.45) converges if $0 < b \le 1$ and $b+c > 1$, or if $1 < b < 2$ and $c-b+1 > 0$ (note the contrast with Theorem 2.82, which can be applied only in the case $b < 1$).

63. Theorems 2.83 and 2.85 were obtained in [262] (in fact, [262] contains an improved version of Theorem 2.83). Both of them are also available in the book [B-80, Ch. IV]. We note that the statement of Theorem 2.85 presented here differs slightly from the original one given in [262]. An extension of Theorem 2.85 to several dimensions is given in [174]. Further, an interesting version for two-dimensional polygonal sums was recently obtained in [45].

64. For a proof of such a version of Poisson's summation formula (Proposition 2.86), we refer to [B-80, 4.1.4]. The representation (2.49) may be found in [B-80, 2.5.4].

65. We note that the definition of the Fourier transform given above is probably not the most standard one. Typically, the Fourier transform $\widehat{f}$ is defined for functions $f \in L^1(\mathbb{R})$ (which implies that the integral defining $\widehat{f}$ is absolutely convergent for any $y \in \mathbb{R}$, and thus convergent and well defined everywhere). Another subtlety to be taken into account is that if $f \in L^1(\mathbb{R})$, then $\widehat{f}$ is defined as a Lebesgue integral, while the definition we gave above is through improper Riemann integrals. It is important to remark that as an improper Riemann integral, one may consider the Fourier transform of functions $f \notin L^1(\mathbb{R})$. For instance, we claimed that if $f \in BV(\mathbb{R})$ vanishes at infinity, the Fourier transform (as an improper Riemann integral) is well defined (except possibly at $y = 0$). Indeed, integrating by parts, we get

$$\int_{-N}^{N} f(x)e^{-ixy}\,dx = -\frac{1}{iy}\Big(f(N)e^{-iNy} - f(-N)e^{iNy}\Big) + \frac{1}{iy}\int_{-N}^{N} e^{-ixy}\,df(x),$$

for $y \neq 0$. Since $f \in BV(\mathbb{R})$, the integral $\int_{\mathbb{R}} e^{-ixy}\,df(x)$ converges absolutely for any $y \in \mathbb{R}$, and therefore, letting $N \to \infty$ in the above equality, we obtain

$$\widehat{f}(y) = \frac{1}{iy}\int_{\mathbb{R}} e^{-ixy}\,df(x),$$

which is well defined for all $y \neq 0$. Note also that by definition $\widehat{f}(0) = \int_{\mathbb{R}} f(x)\,dx$, where the last integral may be divergent.

Ermakov-type integral tests

66. Ermakov's test in its most basic form (Theorem 2.87), although it may or may not be a material of undergraduate courses in analysis, appears in any book on number series (see, for instance, [B-54, Ch. IX, §40, 177] or [B-31]; the original result by Ermakov is published in [99]). Some generalizations had also been studied, as, for example, Theorem 2.88, which can be found in [B-31] (in the case of decreasing functions f; for $f \in WM$, this result was obtained in [178]).

67. In [52], Bugaev suggested that the convergence results of number series are a reflection of the so-called law of conjugation, which was invented earlier by him referring also to Cauchy as the earlier author of a particular case. He mainly illustrated this by some tests related to that of Ermakov.

68. We also mention that Theorem 2.89 is stated and proved in [158].

Cauchy's condensation test

69. See [B-8, Introduction] for basic properties and applications of lacunary sequences (or more generally sequences from the class Λ).

70. The Cauchy condensation test for decreasing sequences with generalized condensing sequences was obtained in [58]. The author also relates it to the Maclaurin–Cauchy integral test.

71. Cauchy's condensation test and Schlömilch's test for weak monotone sequences (Theorems 2.91 and 2.95, respectively) were obtained in [178]. As for the original Schlömilch test, we refer to [B-54, Ch. III, §14, 77] or [B-11, p. 44]. For an original version of de la Vallée Poussin test (Theorem 2.95), which applies to decreasing sequences, see [205, Theorems 1 and 1a].

72. For an extension of Cauchy's condensation test and Schlömilch's test to the classes *WMS*(m) (cf. Definition 1.36), see [258]. We give the statement of the corresponding Schlömilch's test (see Theorem 2.94 for the *WMS* case).

Theorem 2.144 *Let* $\{a_k\} \in WM(m)$ *and let* $\{u_k\}$ *be a nondecreasing sequence of integers such that* $u_k \to \infty$ *and*

$$\overleftarrow{\Delta} u_k \le C m_{u_k}$$

for some $C \geq 1$ *and all* $k \in \mathbb{N}$. *Then,*

(i) *the convergence of the series*

$$\sum_{k=1}^{\infty} \overleftarrow{\Delta} u_k a_{u_k} \tag{2.91}$$

implies the convergence of the series $\sum a_k$.

(ii) *If, additionally,*

$$\frac{\overleftarrow{\Delta} u_{k+1}}{\overleftarrow{\Delta} u_k} \le C.$$

for some $C > 0$, *then the series* (2.91) *and* $\sum a_k$ *converge or diverge simultaneously.*

73. Besides the ones presented here, several interesting variations of the classical Cauchy's condensation test can be found in [46].

74. Lobachevskii's test (Theorem 2.96), although not being widely spread in the common literature on number series, is also a part of the most classical results in the topic. In fact, such a result was obtained by this famous Russian mathematician during the first half of the nineteenth century. Theorem 2.96 has been stated here not in its original form but rather as a slight adaptation. The original test by Lobachevskii [181, B-58] (see also [B-75]) reads as follows.

Theorem 2.145 *Let* $\{a_n\}$ *be a nonnegative null sequence such that for some* $k \in \mathbb{N}$, $a_{k+1} < a_k$ *and* $\{a_n\}_{n \geq k+1}$ *is nonincreasing. For* $m \in \mathbb{N} \cup \{0\}$, *let* $p_m > k$ *be the unique natural number such that*

$$\frac{a_{p_m}}{a_k} \geq 2^{-m} \qquad \text{and} \qquad \frac{a_{p_m+1}}{a_k} < 2^{-m}.$$

Then

$$\frac{1}{2} a_k \sum_{m=1}^{\infty} (p_m - k) 2^{-m} \le \sum_{n=k+1}^{\infty} a_n \le a_k \sum_{m=1}^{\infty} (p_m - k) 2^{-m}. \tag{2.92}$$

In particular, the series $\sum_{m=0}^{\infty} p_m 2^{-m}$ *and* $\sum a_n$ *converge or diverge together.*

In Theorem 2.96 we have simplified the statement of Theorem 2.145 by just assuming that $k = 1$ and $a_1 = 1$.

75. In [181], it is noted that the same procedure as in the proof of Theorem 2.145 can be done by replacing the factor 2^{-m} by a different one γ^{-m}, where $\gamma > 1$ is such that $a_1 = \gamma^{-1}$ (this includes the definition of p_m in terms of estimates with respect to γ^{-m}, the decomposition of the ratios $\frac{a_n}{a_k}$ as a number in base γ, etc.). In this case, one may derive the inequality

$$\frac{\gamma-1}{\gamma}\sum_{m=2}^{\infty}(p_m-k)\gamma^{-m} \le \sum_{n=k+1}^{\infty} a_n \le (\gamma-1)\sum_{m=2}^{\infty}(p_m-k)\gamma^{-m},$$

which in a sense generalizes (2.92) (note that, in contrast with (2.92), the left and right series in the above inequalities start at the index $m = 2$).

Sapogov's test

76. Sapogov's test may also be found in [B-31, Ch. 2, Problem 8] (for the "divergence" we refer to [B-16, p. 36]). It is proved in [178] that even an extension to sequences with bounded variation is possible in Sapogov's test only in its "positive" part.

Abel–Olivier's test and related topics

77. Abel–Olivier's test is a basic necessary condition for convergence of the series with decreasing terms. It may be found in several books (see, e.g., [B-54, Ch. III, §14]). Its version for *WMS* was obtained in [178] (previous results in this direction were also obtained for the smaller *QMS* class, cf. [208]). What we understand as an "optimal" version of Abel–Olivier's test is Theorem 2.99, which rather than giving a necessary condition for convergence gives a characterization in terms of two different conditions.

78. There is an interesting historical note related to the classical Abel–Olivier test (Theorem 2.98). In his original paper [214], Olivier claimed that for positive series $\sum a_k$, convergence is equivalent to the fact that $ka_k \to 0$ as $k \to \infty$ (he actually stated a more general claim involving series with real terms). One year later, a note had been published by Abel [2] in the same journal where Olivier's paper [214] was published, stating that Olivier's claim was wrong (and taking also the opportunity to publicly complain that the field of analysis was lacking rigor at that time). To prove that Olivier's result is wrong, he proved the divergence of the series $\sum \frac{1}{n\log n}$, which clearly satisfy Olivier's condition $ka_k \to 0$. He also established the classical Abel–Oliver's test for monotone series. A more extensive comment on the history of this result can be found in [108], along with some extractions from the original papers [2] and [214]. Abel also found a counterexample to a theorem by Cauchy on the continuity of the convergent series of continuous functions, which he called an "exception". See [239] for a detailed discussion.

79. A (sharp) version of the Abel–Olivier test is given for the classes *WMS*(m) (cf. Definition 1.36) in [258]. In order to state it, we first introduce the following notation. For a nondecreasing index sequence $\{m_n\}$, set

$$m'_n := \min\{n, m_n\}.$$

The test reads as follows.

Theorem 2.146 (i) *Let* $\{a_k\} \in WMS(m)$ *be such that* $\sum a_k$ *converges. If* $\{m'_n\}$ *satisfies the doubling condition*

$$m'_{2k} \lesssim m'_k,$$

then $m'_n a_n \to 0$.

(ii) *Let* $\{u_n\}$ *be a sequence with* $\limsup u_n = \infty$*. Then there exists a positive convergent series* $\sum a_k$ *with* $\{a_k\} \in WMS(m)$ *such that* $u_n m'_n a_n \not\to 0$.

80. An interesting observation concerning the classical Abel–Olivier test is made in [230]. In particular, the authors proved that the monotonicity assumption on $\{a_k\}$ can be dropped if the convergence of $\{ka_k\}$ in the conclusion of the test is weakened. For instance, convergence of the latter in density is obtained. We say that a set $A \subset \mathbb{N}$ has zero density if

$$d(A) := \lim_n \frac{|A \cap \{1, \ldots, n\}|}{n} = 0.$$

A real-valued sequence $\{b_k\}$ is said to converge to $L \in \mathbb{R}$ in density if for every $\varepsilon > 0$ the set $A(\varepsilon) = \{n \in \mathbb{N} : |b_k - L| \geq \varepsilon\}$ has zero density. In [230], it is shown that if $\sum a_k$ converges, where $\{a_k\}$ is only assumed to be a positive sequence, then $ka_k \to 0$ in density (further remarks are given in [206]).

What is more, the above convergence in density can be rewritten in terms of ideals. An admissible ideal $\mathfrak{I}$ of subsets of natural numbers is an ideal that is additive (i.e., $A, B \in \mathfrak{I}$ implies $A \cup B \in \mathfrak{I}$), hereditary (i.e., $B \subset A \in \mathfrak{I}$ implies $B \in \mathfrak{I}$), and contains all singletons and does not contain $\mathbb{N}$. A real-valued sequence $\{b_k\}$ is said to $\mathfrak{I}$-converge to $L \in \mathbb{R}$ if for each $\varepsilon > 0$ we have $A(\varepsilon) \in \mathfrak{I}$ (with $A(\varepsilon)$ as defined above). Convergence in density is precisely $\mathfrak{I}_d$-convergence with

$$\mathfrak{I}_d = \{A \in \mathbb{N} : d(A) = 0\}.$$

Such a type of convergence was first introduced in [152].

The authors of [230] characterize the ideals for which Abel–Olivier's test for $\mathfrak{I}$-convergence holds. Curiously enough, such ideals are related to the convergence of the harmonic series with deleted terms (see Subsection 3.1.2). The precise statement is as follows.

Theorem 2.147 *Let* $\mathfrak{I}$ *be an admissible ideal of subsets of* $\mathbb{N}$*. The following are equivalent.*

(i) *For any positive sequence* $\{a_k\}$*, if the series* $\sum a_k$ *converges, then the sequence* $\{ka_k\}$ $\mathfrak{I}$*-converges to 0.*
(ii) *The ideal* $\mathfrak{I}$ *contains the ideal*

$$\mathfrak{I}_h = \left\{A \subset \mathbb{N} : \sum_{n \in A} \frac{1}{n} < \infty\right\}.$$

Further generalizations of Abel–Olivier's test in this direction are given in [106].

81. The proof of the extended Abel–Olivier test for the class $\overline{WMS}(p)$, $p \geq 1$ (see Remark 2.101 and Notes-18 in Chapter 1) is given in [33, Theorem 2.9].

82. Theorem 2.102 is due to Kronecker (see also [B-5, §2.2.2] for some of its applications). We remark that the proof in [B-5] lacks certain details and is different from the one presented here.

83. It is worth noting that the weighted Abel–Olivier test (Theorem 2.102) was given for decreasing sequences $\{a_n\}$ and positive $\{b_n\}$ in [213]. However, as the author noted there, such a result was already proved by Hardy in [123] without stating it explicitly. What is more, Hardy showed in [123] that such a result is sharp in the sense of Proposition 2.109. More precisely, he showed that the condition $B_n a_n \to 0$ cannot be replaced by the more general

$$u_n B_n a_n \to 0,$$

for any positive sequence $\{u_n\}$ with $\limsup u_n = \infty$.

84. The idea of generalizing Abel–Olivier's test in the direction of Corollary 2.106 was recently considered by the authors of [103], where they also obtained the preliminary statements in Lemma 2.104 and Theorem 2.105.

85. The Abel–Olivier test for real-valued $\overline{GM}$ functions (Theorem 2.108), along with Lemmas 1.27 and 1.28, was obtained in [73] (see also Notes-14 in Chapter 1).

86. In the paper [115], the author proves the sharpness (in the sense of Proposition 2.109) of an "iterated" Abel–Olivier test. Such an iteration consists on assuming that both $\{a_n\}$ and $\{2^k a_{2^k}\}$ are decreasing. In this case, it is not hard to see that the convergence of $\sum a_n$ implies

$$n \log n a_n \to 0$$

(in fact, we will give a detailed multidimensional version of this statement in Proposition 9.48), and the statement in [115] shows that this decay cannot be improved.

Behavior of the terms of monotone series

87. To the best of our knowledge, the first version of Theorem 2.110 was obtained in [273] for decreasing sequences and was used in the study of trigonometric series. In the original proof (applied to decreasing sequences), the author makes use of Cauchy's condensation test. However, in the case where $\{a_k\} \in WMS$, it is also possible to prove Theorem 2.110 by means of Abel–Olivier's test for *WMS* (that is, Theorem 2.100 with $b_k = 1$), while this is not possible in the monotone case (see Remark 9.43 for further details).

88. Dvoretzky's test (Theorem 2.112) appeared in [85], although, as mentioned there, earlier related results were already obtained by Pringsheim and Hamming (curiously enough, Hamming's paper [116] appeared two years before Dvoretzky's paper [85] and both carry the same title). The corresponding generalization for *WMS*, as for the previous classical tests involving monotone sequences, was obtained in [178].

Applications to Orlicz-type classes

89. A continuous version of Theorem 2.115 (i.e., Theorem 2.116) was first obtained in [264] and aimed to study embedding properties for Orlicz spaces of functions. Such results were widely used in functional analysis, see, e.g., [B-9, 161, 263, 264]. For a specific choice of Orlicz function φ (characterizing $L \log L$ spaces), this result was obtained by Hardy–Littlewood, Zygmund, Lorentz, Herz, and Bennett (see [34, pp. 34–35]).

$\ell^p - \ell^q$ inequalities and Stechkin's lemma

90. The $\ell^p \subset \ell^q$ inequality (2.77) may be found in several sources containing basic theory of number series. See, for instance, [B-43, Theorems 19 and 167].

91. Corollary 2.120 and its versions are used in many papers dealing with nonlinear approximation and, more recently, the approximation of high-dimensional parametric elliptic partial differential equations; see, e.g., [62, 63, 77, 155, 253, B-28]. In some of these sources it is called Stechkin's lemma. See the historical notes in [B-28].

Equiconvergence results involving general monotonicity

92. Theorems 2.123 and 2.124 were first obtained in [178].

3
Miscellaneous tests and results

In this chapter, we collect several miscellaneous results for one-dimensional series that are less known than classical ones, yet interesting and notable. We also discuss one of the most important number series, the harmonic series. Multidimensional extensions of these miscellaneous results will not be given for the sake of both extent and simplicity.

3.1 The harmonic series

The harmonic series

$$\sum_{n=1}^{\infty} \frac{1}{n}$$

is perhaps the most popular example of a number series. In view of this fact, this section will be devoted to descriptions of several properties of the harmonic series, ranging from the classical ones, which can be found in every monograph on number series, to less known ones, though interesting.

There are hundreds of ways to show that the harmonic series diverges (in fact, many of the above tests can be applied), but among them all we highlight the following very simple yet beautiful one. Assume that the harmonic series converges and denote its sum by H. If we consider the subseries of the harmonic series consisting only of even terms, it is clear that this subseries converges to the value $\frac{H}{2}$. Formally,

$$\sum_{n=1}^{\infty} \frac{1}{2n} = \frac{1}{2} \sum_{n=1}^{\infty} \frac{1}{n} = \frac{H}{2}.$$

Therefore, the subseries of the harmonic series consisting only of the odd terms also converges to $\frac{H}{2}$. However, since for every $n \in \mathbb{N}$

$$\frac{1}{2n-1} > \frac{1}{2n},$$

we have

$$\frac{H}{2} = \sum_{n=1}^{\infty} \frac{1}{2n-1} > \sum_{n=1}^{\infty} \frac{1}{2n} = \frac{H}{2},$$

which is a contradiction. Therefore, the harmonic series diverges.

The sequence of partial sums of the harmonic series (written $H_n := \sum_{k=1}^{n} \frac{1}{k}$) increases to infinity very slowly. For instance, we have $H_{1000} < 7.5$ and $H_{10000} < 10$. On the other hand, if we denote

$$k_m = \min\{n \in \mathbb{N} : H_n \geq m\}, \qquad m \in \mathbb{N},$$

that is, the minimal number of terms needed for the partial sums H_n to exceed m, then $k_{10} = 12367$ and $k_{1000} > 1.1 \times 10^{434}$.

As for the asymptotic behavior, we have

$$H_n = \sum_{m=1}^{n} \frac{1}{m} \asymp \log n,$$

which can be proved using the Maclaurin–Cauchy integral test, for instance.

In some cases, it is possible to obtain the exact value of series related to the harmonic series in terms of the Riemann zeta function. The following result is the main tool in this respect.

Theorem 3.1 *Let $\{m_k\}$ be a real sequence vanishing at infinity and define $M_n = \sum_{k=1}^{n} m_k$ for $n \in \mathbb{N}$. Then,*

$$\sum_{k=1}^{\infty} \frac{M_k}{(k+1)(k+n+1)} = m_1\left(\frac{H_n}{n} - \frac{1}{n+1}\right) + \frac{1}{n}\sum_{j=1}^{n}\sum_{k=1}^{\infty} \frac{m_{k+1}}{j+k+1}$$

$$= \frac{1}{n}\sum_{j=1}^{n}\sum_{k=1}^{\infty} \frac{m_k}{j+k}.$$

We skip the proof of this result. Instead, we give appropriate references and further comments in Notes-2. We only mention that Theorem 3.1 (and further manipulations) allows one to obtain interesting equalities, as, for example,

$$\sum_{n=1}^{\infty} \frac{H_n^2}{n^2} = \frac{17}{4}\zeta(4), \quad \sum_{n=1}^{\infty} \frac{H_n^2}{n^4} = \frac{97}{24}\zeta(6) - 2\zeta(3)^2,$$

and

$$\sum_{n=1}^{\infty} \left(\frac{H_n}{n}\right)^3 = \frac{93}{16}\zeta(6) - \frac{5}{2}\zeta(3)^2,$$

where ζ is the Riemann zeta function:

$$\zeta(s) = \sum_{n=1}^{\infty} \frac{1}{n^s}, \qquad s \in \mathbb{C}, \qquad \operatorname{Re} s > 1.$$

3.1.1 Iterated harmonic series

Since for $p > 1$ the so-called p-series

$$\sum_{n=1}^{\infty} \frac{1}{n^p}$$

converges, one may wonder what happens with "intermediate" series between the harmonic series and all the p-series, that is, series whose general term a_n satisfies

$$\lim_{n\to\infty} na_n = 0, \qquad \text{and} \qquad \lim_{n\to\infty} n^p a_n = \infty, \tag{3.1}$$

for all $p > 1$. A natural candidate for such an intermediate series is

$$\sum_{n=1}^{\infty} \frac{1}{nH_n}, \qquad \text{or, equivalently,} \qquad \sum_{n=1}^{\infty} \frac{1}{n \log n},$$

so that (3.1) is satisfied (in fact, the general term of this new series even "passes" Abel–Olivier's test, while the general term of the harmonic series does not). Though Abel–Olivier's test does not show that this series diverges, applying Cauchy's condensation test, we easily find that the series

$$\sum_{n=1}^{\infty} \frac{1}{nH_n}$$

diverges. More generally, for $\ell \in \mathbb{N}$, one can define the ℓth iterated harmonic series whose nth partial sum is

$$H_n^{(\ell)} := \sum_{m=1}^{n} \frac{1}{m \prod_{j=0}^{\ell-1} H_m^{(j)}}, \qquad H_n^{(0)} = H_n.$$

An iterated application of Cauchy's condensation test shows that for every $\ell \in \mathbb{N}$, the sequence of partial sums $\{H_n^{(\ell)}\}$ of the ℓth iterated harmonic series is unbounded, so the corresponding series diverges for any ℓ. The reason we call those series "ℓth iterated harmonic series" is that, after applying iteratively the Cauchy condensation test to it exactly ℓ times, one recovers the harmonic series. For this same reason, any series of the form

$$\sum_{n=1}^{\infty} \frac{1}{n\left(\prod_{j=0}^{\ell-1} H_n^{(j)}\right) \cdot (H_n^{(\ell)})^p}, \qquad p > 1, \tag{3.2}$$

is convergent, since by applying iteratively the Cauchy condensation test to this series exactly $\ell + 1$ times, we recover a convergent p-series.

Remark 3.2 For $a > 1$, we denote $t_a^{(0)}(n) := \lceil \log_a n \rceil$ and, for $j \geq 1$,

$$t_a^{(j)}(n) := \begin{cases} 1, & \text{if } n = 1, 2, \ldots, \lceil a \rceil - 1, \\ \lceil \log_a t_a^{(j-1)}(n) \rceil, & \text{if } n \geq \lceil a \rceil, \end{cases}$$

(i.e., $t_a^{(j)}(n)$ is an iterated logarithm), where $\lceil\cdot\rceil$ denotes the ceiling function. It can easily be proved that for n large enough and every $j \geq 0$, one has $H_n^{(j)} \asymp t_a^{(j)}(n)$ (note that $t_a^{(j)}(n) \asymp t_b^{(j)}(n)$ for $a, b > 1$). In other words, every iteration of the harmonic series is just the addition of an iterated logarithm to the denominator of the preceding series.

In relation with iterated harmonic series, we consider another interesting example that can be seen as an infinitely iterated harmonic series. For $a \geq 2$, define $\pi_a(1) = 1$ and set, recursively,

$$\pi_a(n) := \lceil \log_a n \rceil \cdot \pi_a(\lceil \log_a n \rceil) = t_a^{(0)}(n) \cdot \pi_a(t_a^{(0)}(n)). \tag{3.3}$$

Note that for any n, $\pi_a(n)$ is a finite product of iterated logarithms, which, roughly speaking, is formed as follows: first we find the smallest natural number k so that the kth iterated (base-a) logarithm applied to n is smaller than a. Once such a k is found, we take the product of all the ith iterations of the logarithm of n, for $i = 1, \ldots, k$. Of course, some correction is needed (in the formal definition of $t_a^{(j)}(n)$ we also involve ceiling functions), but this already gives a clear intuition behind the definition of $\pi_a(n)$.

For instance, we have

$$\pi_a(2) = \lceil \log_a 2 \rceil \cdot \pi_a\big(\lceil \log_a 2 \rceil\big) = \pi_a(1) = 1,$$

and if $a < 3$,

$$\pi_a(3) = \lceil \log_a 3 \rceil \cdot \pi_a\big(\lceil \log_a 3 \rceil\big) = 2 \cdot \pi_a(2) = 2.$$

Note that we assume $a \geq 2$ in the definition of π_a, since for some numbers $a \in (1, 2)$, $\pi_a(n)$ may be not well defined in general. For instance, if $a = \frac{3}{2}$ then $\pi_{\frac{3}{2}}(2) = 2\pi_{\frac{3}{2}}(2)$ (which implies $\pi_{\frac{3}{2}}(2) = 0$), or if $a = m^{1/m}$ for $m \in \mathbb{N}$, then we would have

$$\pi_{m^{1/m}}(m) = \lceil \log_{m^{1/m}} m \rceil \cdot \pi_{m^{1/m}}\big(\lceil \log_{m^{1/m}} m \rceil\big) = m \cdot \pi_{m^{1/m}}(m),$$

which is absurd unless $\pi_{m^{1/m}}(m) = 0$.

Remark 3.3 We assert that if

$$b_n^{(a)} := \frac{1}{n \cdot \pi_a(n)},$$

then such a sequence $\{b_n^{(a)}\}$ tends to zero. This convergence is slower than that of the terms in the series (3.2), but more rapid than that in any of the iterated harmonic series. In order to prove this claim, we first write $\pi_a(n)$ as

$$\pi_a(n) = \prod_{j : t_a^{(j-1)}(n) \geq \lceil a \rceil} t_a^{(j)}(n).$$

For any n, this is a finite product that precisely captures the recursive nature of $\pi_a(n)$ (see its definition (3.3)). Showing that $b_n^{(a)}$ tends to zero more rapidly than the terms of any iterated harmonic series is rather straightforward. It suffices to observe that the terms of the ℓth iterated harmonic series are asymptotically equivalent to

$$\frac{1}{n\prod_{j=0}^{\ell-1} t_a^{(j)}(n)}, \qquad n \in \mathbb{N},$$

and therefore, for any $\ell \geq 1$ and sufficiently large n,

$$b_n^{(a)} \cdot \left(n \prod_{j=0}^{\ell-1} t_a^{(j)}(n)\right) = \frac{n\prod_{j=0}^{\ell-1} t_a^{(j)}(n)}{n \cdot \pi_a(n)} \leq \frac{1}{t_a^{(\ell)}(n)} \to 0 \qquad \text{as } n \to \infty. \tag{3.4}$$

We now show that $b_n^{(a)}$ tends to zero slower than the terms of any iterated p-series or, in other words, any series of the type (3.2).

Proposition 3.4 *For any $\ell \in \mathbb{N}$ and $p > 1$, we have*

$$\frac{n\left(\prod_{j=0}^{\ell-1} H_n^{(j)}\right) \cdot \left(H_n^{(\ell)}\right)^p}{n \cdot \pi_a(n)} \to \infty,$$

as $n \to \infty$.

Proof Note that since $H_n^{(j)} \asymp t_a^{(j)}(n)$, by inequality (3.4), for n large enough,

$$\begin{aligned}
\frac{n\left(\prod_{j=0}^{\ell-1} H_n^{(j)}\right) \cdot \left(H_n^{(\ell)}\right)^p}{n \cdot \pi_a(n)} &\asymp \frac{\left(\prod_{j=0}^{\ell-1} t_a^{(j)}(n)\right) \cdot \left(t_a^{(\ell)}(n)\right)^p}{\prod_{j:t_a^{(j-1)}(n)\geq\lceil a\rceil} t_a^{(j)}(n)} \\
&= \frac{\left(t_a^{(\ell)}(n)\right)^{p-1}}{\prod_{j>\ell:t_a^{(j-1)}(n)\geq\lceil a\rceil} t_a^{(j)}(n)} \\
&= \frac{\left(t_a^{(\ell)}(n)\right)^{p-1}}{\prod_{j>0:t_a^{(j-1)}\left(t_a^{(\ell)}(n)\right)\geq\lceil a\rceil} t_a^{(j)}\left(t_a^{(\ell)}(n)\right)} \\
&= \frac{\left(t_a^{(\ell)}(n)\right)^{p-1}}{\pi_a\left(t_a^{(\ell)}(n)\right)}.
\end{aligned}$$

Here the second-to-last equality follows from the fact that

$$t_a^{(j-1)}(n) = t_a^{(j-\ell-1)}\left(t_a^{(\ell)}(n)\right),$$

for all $j > \ell \geq 0$ and $n \in \mathbb{N}$. Thus, it now suffices to prove that $\frac{n^{p-1}}{\pi_a(n)} \to \infty$. In order to show this, for $n \in \mathbb{N}$, let $k = k(n)$ be the natural number such that

$$\pi_a(n) = \prod_{j:t_a^{(j-1)}(n)\geq\lceil a\rceil} t_a^{(j)}(n) = \prod_{j=0}^{k} t_a^{(j)}(n).$$

We have

$$\begin{aligned}\frac{n^{p-1}}{\pi_a(n)} &= \frac{n^{p-1}}{\prod_{j=0}^{k} t_a^{(j)}(n)} = \frac{n^{p-1}}{\prod_{j=0}^{k}\left(t_a^{(j)}(n)\right)^p}\prod_{j=0}^{k}\left(t_a^{(j)}(n)\right)^{p-1} \\ &\asymp \frac{n^{p-1}}{\left(\log_a n\right)^p}\frac{\prod_{j=0}^{k}\left(t_a^{(j)}(n)\right)^{p-1}}{\prod_{j=1}^{k}\left(t_a^{(j)}(n)\right)^p} \\ &= \frac{n^{p-1}}{\left(\log_a n\right)^p}\prod_{j=0}^{k-1}\frac{\left(t_a^{(j)}(n)\right)^{p-1}}{\left(t_a^{(j+1)}(n)\right)^p}\left(t_a^{(k)}(n)\right)^{p-1}.\end{aligned}$$

By the definition of $\pi_a(n)$ and k, we have $t_a^{(k)}(n) \geq 1$. Moreover, we also have

$$\frac{n^{p-1}}{\left(\log_a n\right)^p} \to \infty, \qquad \frac{\left(t_a^{(j)}(n)\right)^{p-1}}{\left(t_a^{(j+1)}(n)\right)^p} \to \infty, \quad \text{for all } 0 \leq j \leq k-1.$$

Therefore, we conclude that $\frac{n^{p-1}}{\pi_a(n)} \to \infty$. □

Another interesting feature of the series $\sum b_n^{(a)}$ is that, while the Cauchy condensation test can be used to determine the divergence of any iterated harmonic series, or the convergence of (3.2), it is inconclusive in this case. Indeed, it transforms the series into itself (whenever the condensation factor is equal to a, which we may assume to be a natural number for the purpose):

$$a^n b_{a^n}^{(a)} = \frac{1}{\pi_a(a^n)} = \frac{1}{n \cdot \pi_a(n)} = b_n^{(a)}.$$

The convergence of $\sum b_n^{(a)}$ is characterized in terms of a as follows.

Theorem 3.5 *For $a \geq 2$, the series*

$$\sum_{n=1}^{\infty}\frac{1}{n \cdot \pi_a(n)}$$

converges if and only if $a < e$.

Proof For simplicity, we pass to integrals to prove this result. We first introduce an analog of $\pi_a(n)$ defined for any real number $x \geq 1$. That is, we define $\pi_a(1) = 1$ and set, recursively,

$$\pi_a(x) := \lceil \log_a x \rceil \cdot \pi_a(\lceil \log_a x \rceil);$$

see also (3.3). It is clear that the function

$$g_a(x) = \frac{1}{x\pi_a(x)}, \qquad x \geq 1,$$

is decreasing and, moreover, $g_a(n) = b_n^{(a)}$ for every $n \in \mathbb{N}$. Thus, by the Maclaurin–Cauchy integral test (Theorem 2.78), we can equivalently study the convergence or divergence of the integral

$$\int_1^{\infty} \frac{1}{x\pi_a(x)}\, dx,$$

rather than that of the series $\sum_{n=1}^{\infty} b_n^{(a)}$. Let

$$d_n^{(a)} := \int_1^{a^n} \frac{1}{x\pi_a(x)}\, dx, \qquad n \in \mathbb{N}.$$

Taking into account that by definition for $n \geq 2$ and $x \in (a^{n-1}, a^n]$ one has $\pi_a(x) = n\pi_a(n)$, we get

$$d_n^{(a)} - d_{n-1}^{(a)} = \int_{a^{n-1}}^{a^n} \frac{1}{x\pi_a(x)}\, dx = b_n^{(a)} \int_1^a \frac{1}{x}\, dx = b_n^{(a)} \log a,$$

and hence, for $m \geq 2$,

$$d_m^{(a)} - d_1^{(a)} = \sum_{k=2}^{m} \left(d_k^{(a)} - d_{k-1}^{(a)}\right) = (\log a) \sum_{k=2}^{m} b_k^{(a)} = (\log a) \sum_{k=2}^{m} \frac{1}{k\pi_a(k)}. \tag{3.5}$$

Let us now show that $\left\{d_n^{(a)}\right\}$ is bounded for $2 \leq a < e$, which is equivalent to the convergence of the desired series. In this case, it follows from (3.5) and the monotonicity of $g_a(x)$ that for $n \geq 2$,

$$\begin{aligned} d_{\lceil a^n \rceil - 1}^{(a)} - d_1^{(a)} &= (\log a) \sum_{k=2}^{\lceil a^n \rceil - 1} \frac{1}{k\pi_a(k)} \leq (\log a) \int_1^{\lceil a^n \rceil - 1} \frac{1}{x\pi_a(x)}\, dx \\ &\leq (\log a) \int_1^{a^n} \frac{1}{x\pi_a(x)}\, dx = d_n^{(a)} \log a \leq d_{\lceil a^n \rceil - 1}^{(a)} \log a, \end{aligned}$$

where the last inequality simply follows from the fact that $\left\{d_n^{(a)}\right\}$ is increasing. These estimates imply that

$$d_n^{(a)} \leq \frac{d_1^{(a)}}{1 - \log a}$$

for $n \geq 2$ and prove the convergence of $\sum_{n=1}^{\infty} b_n^{(a)}$.

Let us now turn to the divergence part. We will equivalently show that for $a \geq e$ the sequence $\left\{d_n^{(a)}\right\}$ is unbounded. Since $a \geq e$, it follows from (3.5) and the monotonicity of $g_a(x)$ that

$$\begin{aligned} d_{\lceil a^n \rceil}^{(a)} - d_1^{(a)} &= (\log a) \sum_{k=2}^{\lceil a^n \rceil} \frac{1}{k\pi_a(k)} \geq (\log a) \int_2^{\lceil a^n \rceil} \frac{1}{x\pi_a(x)}\, dx \\ &\geq (\log a) \int_2^{a^n} \frac{1}{x\pi_a(x)}\, dx \geq d_n^{(a)} - \int_1^2 \frac{1}{x\pi_a(x)}\, dx. \end{aligned}$$

Therefore,

$$d_{\lceil a^n \rceil}^{(a)} \geq d_1^{(a)} + d_n^{(a)} - \int_1^2 \frac{1}{x\pi_a(x)}\, dx = d_n^{(a)} + \int_2^a \frac{1}{x\pi_a(x)}\, dx,$$

or, in other words, the sequence $\left\{d_n^{(a)}\right\}$ has a subsequence that is bounded from below by an arithmetic progression. Hence, $d_n^{(a)} \to \infty$, as desired. □

Remark 3.6 (i) For integer $a \geq 2$, Theorem 3.5 can be proved without introducing a continuous version of the sequence $b_n^{(a)}$ and passing to integrals (see Notes-7 for further details).

(ii) In contrast with Theorem 3.5, the choice of a is not relevant to determine the convergence or divergence of the iterated harmonic series or the iterated p-series (3.2), which, for any $a > 1$, can equivalently be written as

$$\sum_{n=1}^{\infty} \frac{1}{n \prod_{j=0}^{\ell-1} t_a^{(j)}(n)} \quad \text{and} \quad \sum_{n=1}^{\infty} \frac{1}{n\big(\prod_{j=0}^{\ell-1} t_a^{(j)}(n)\big) \cdot \big(t_a^{(\ell)}(n)\big)^p},$$

respectively, by Remark 3.2. In other words, when there are finitely many logarithmic factors, the base a of the logarithm plays no role, but when the number of logarithmic factors grows infinitely (as it happens in the denominator of $b_n^{(a)}$), the base of the logarithm does actually matter.

3.1.2 Harmonic series with deleted terms

An interesting problem arises from the divergence of the harmonic series. It suggests to consider the harmonic series without certain terms. In 1914, Kempner [142] showed that if we consider the harmonic series without the terms containing the digit "9" in their decimal representation, then the resulting series converges. Although the result itself is surprising, the argument that proves it is quite simple. For every $j \in \mathbb{N}$, there are exactly $8 \times 9^{j-1}$ integers of j digits that do not contain the digit "9". Moreover, their reciprocals are all bounded by $1/10^{j-1}$, and therefore the corresponding series is bounded from above by

$$\sum_{j=1}^{\infty} 8\left(\frac{9}{10}\right)^{j-1} = 80.$$

Some remarks are in order. First of all, note that this upper bound is rather rough, since the real value of the corresponding series is actually slightly less than 23 (see Notes-8 for references). Secondly, it is interesting to notice that as the number of digits j increases, the proportion of integers of j digits not containing any "9" (with respect to all integers of j digits) decreases dramatically, and thus most of the terms from the harmonic series are removed. For instance, among all the terms with 1 000 digits in the denominator (there are 9×10^{999}), the ones not containing any "9" are 8×9^{999}. In other words, the proportion of terms with denominator having 1 000 digits and not containing any "9" is

$$\frac{8 \times 9^{999}}{9 \times 10^{999}} = \frac{8}{9} \times \left(\frac{9}{10}\right)^{999} < 10^{-45}.$$

The same exact argument can be carried out if we substitute any other digit for "9". More generally, a similar approach can be used if we replace the role of the single digit by a string of arbitrarily many integer digits as, for instance, "32" or "719201".

Theorem 3.7 *Let X be a string of n digits written in decimal system. If we delete all the terms whose reciprocals contain the string X from the harmonic series, then the resulting series converges.*

Proof Define, for $k, n \in \mathbb{N}$, the sets

$$D_k^n = \{m \in \mathbb{N} : m \text{ has } k \text{ digits in base } 10^n\} = \{m \in \mathbb{N} : 10^{n(k-1)} \leq m < 10^{nk}\}$$

and denote the subset of D_k^n that contains the digit X in base 10^n by $\mathcal{Y}_k = \mathcal{Y}_k(n)$. Note that the amount of elements m in the set D_k^n that contain the string X is larger than the amount of elements containing the digit X in base 10^n (for example, the number 5123 contains the string "12" but does not contain the number "12" as a digit in base 100; the digits in base 100 are "51" and "23"). By this,

$$\sum_{\substack{m=1 \\ m \text{ does not contain} \\ \text{the string } X}}^{\infty} \frac{1}{m} \leq \sum_{k=1}^{\infty} \sum_{m \in D_k^n \setminus \mathcal{Y}_k} \frac{1}{m} \leq \sum_{k=1}^{\infty} \sum_{m \in D_k^n \setminus \mathcal{Y}_k} 10^{-n(k-1)}$$

$$= 10^n \sum_{k=1}^{\infty} 10^{-nk} |D_k^n \setminus \mathcal{Y}_k|. \tag{3.6}$$

Let us obtain an upper bound for $|D_k^n \setminus \mathcal{Y}_k|$. An arbitrary number $m \in D_k^n$ is written as the concatenation

$$m = \underbrace{\underline{m_{k-1}}\,\underline{m_{k-2}}\,\ldots\,\underline{m_1}\,\underline{m_0}}_{\substack{k \text{ blocks} \\ \text{of } n \text{ digits}}}, \tag{3.7}$$

with each $m_\ell \in \{0, 1, \ldots, 10^n - 1\}$, $\ell = 0, 1 \ldots, k-2$, and $m_{k-1} \in \{1, \ldots, 10^n - 1\}$. We will allow the string X to appear in the position of m_{k-1} in (3.7) so as to avoid situations of the type $X =$ "0123", with $m_{k-1} = 123$ (in this case the string X clearly does not appear in the position of m_{k-1} in (3.7)). Taking this into account, if the string X is not to appear in any but the first position of (3.7), we only exclude one possibility from each choice (recall that we exclude the case $m_{k-1} = 0$). Thus,

$$|D_k^n \setminus \mathcal{Y}_k| \leq (10^n - 1)^k,$$

and by (3.6),

$$\sum_{\substack{m=1 \\ m \text{ does not contain} \\ \text{the string } X}}^{\infty} \frac{1}{m} \leq 10^n \sum_{k=1}^{\infty} (1 - 10^{-n})^k = 10^{2n}(1 - 10^{-n}),$$

which proves the convergence of the desired series. □

Note that the bound we have obtained for the series corresponding to Theorem 3.7 is far from being optimal, since in the proof we do not delete as many terms as we could.

One can go a step further and refine Theorem 3.7 as follows. For $0 \leq \lambda \leq 1$, let A_λ be the set of all positive integers n such that the number of 9's in the decimal representation of n is less than or equal to λ times the number of digits of n. Note that $A_1 = \mathbb{N}$ and A_0 is the set of positive integers without any "9" in their decimal representation, so

$$\sum_{n\in A_0} \frac{1}{n} < \infty \qquad \text{and} \qquad \sum_{n\in A_1} \frac{1}{n} = \infty.$$

Also note that if $\lambda_1 < \lambda_2$, then $A_{\lambda_1} \subset A_{\lambda_2}$. Hence,

1. If the series $\sum_{n\in A_{\lambda_1}} \frac{1}{n}$ converges, then the series $\sum_{n\in A_\lambda} \frac{1}{n}$ converges for all $0 \leq \lambda \leq \lambda_1$.
2. If the series $\sum_{n\in A_{\lambda_1}} \frac{1}{n}$ diverges, then the series $\sum_{n\in A_\lambda} \frac{1}{n}$ diverges for all $\lambda_1 \leq \lambda \leq 1$.

In view of these observations, one may wonder what the threshold $0 \leq \lambda \leq 1$ that separates convergent and divergent series is. The answer is given by the following result:

Theorem 3.8 *The series* $\sum_{n\in A_\lambda} \frac{1}{n}$

(i) *converges if* $0 \leq \lambda < \frac{1}{10}$, *and*
(ii) *diverges if* $\frac{1}{10} \leq \lambda \leq 1$.

Roughly speaking, in order to remain with a convergent series, it is not necessary to delete as many terms from the harmonic series as in the original Kempner's example. Since the proof of Theorem 3.8 is rather lengthy, we omit it for the sake of brevity. Instead, we refer the reader to [183], where the result first appeared. We also note that in the definition of A_λ the digit "9" may be replaced by any other digit, and Theorem 3.8 would remain the same under this new definition.

Above we have discussed the convergence of the harmonic series when some of the terms are deleted. On the other hand, for any $x \geq 0$, it is possible to obtain such a number x as a subsum of the harmonic series, or, in other words, there exists a sequence $\{c_n\}$ consisting only of 0's and 1's such that

$$\sum_{n=1}^{\infty} \frac{c_n}{n} = x.$$

This kind of series are called selective sums and will be discussed in more detail in Section 3.3 (see Theorem 3.14 in particular).

A rather surprising fact on selective sums of the harmonic series is that one may obtain any rational number as a finite sum of the terms of the harmonic series.

Theorem 3.9 *Let $r > 0$ be a rational number. Then there exists a finite set $A \subset \mathbb{N}$ such that*

$$\sum_{n\in A} \frac{1}{n} = r.$$

Proof We start with the case $r < 1$ and proceed by induction on the numerator of r. If such a numerator equals 1, then r is just a term of the harmonic series. Now assume that the result is true for every $r < 1$ having numerator less than $p \in \mathbb{N}$, and let $r = p/q$. Let m be the natural number such that

$$\frac{1}{m} \leq \frac{p}{q} < \frac{1}{m-1}.$$

The second inequality implies that $pm - q < p$. By induction hypothesis, there exists a finite set $A \subset \mathbb{N}$ such that

$$\sum_{n \in A} \frac{1}{n} = \frac{pm-q}{mq},$$

and furthermore $m \notin A$, since

$$\frac{pm-q}{mq} = \frac{p}{q} - \frac{1}{m} < \frac{1}{m-1} - \frac{1}{m} = \frac{1}{m(m-1)}.$$

Therefore, it is clear that when defining $A' = A \cup \{m\}$ one has $\sum_{n \in A'} \frac{1}{n} = \frac{p}{q}$.

We now prove the case $r \geq 1$. Let $N \geq 1$ be the natural number such that

$$\sum_{n=1}^{N} \frac{1}{n} \leq r < \sum_{n=1}^{N+1} \frac{1}{n}.$$

If we put $s = r - \sum_{n=1}^{N} \frac{1}{n}$, then s is a rational number satisfying $s < \frac{1}{N+1} < 1$. Therefore, the above argument leads to a finite set $A \subset \mathbb{N}$ such that

$$\sum_{n \in A} \frac{1}{n} = s.$$

We conclude that

$$r = \sum_{n=1}^{N} \frac{1}{n} + s = \sum_{n=1}^{N} \frac{1}{n} + \sum_{n \in A} \frac{1}{n}.$$

None of the terms of the last sum is repeated in the first sum on the right, since $\min A > N$. □

3.1.3 Series of the reciprocals of the primes

While studying the harmonic series, Euler, who had already found his well-known product formula

$$\sum_{n=1}^{\infty} \frac{1}{n} = \prod_{p \text{ prime}} \left(1 + \frac{1}{p} + \frac{1}{p^2} + \cdots\right) = \prod_{p \text{ prime}} \frac{1}{1 - p^{-1}}, \tag{3.8}$$

got the idea of studying the series of reciprocals of primes:

$$\sum_{p \text{ prime}} \frac{1}{p}.$$

He showed, by means of arguments lacking the rigor of modern mathematics, that this series diverges, and, moreover,

$$\sum_{\substack{p \text{ prime} \\ p \le n}} \frac{1}{p} \asymp \log\log n, \tag{3.9}$$

for n large enough. Although his arguments were questionable, the conclusion was right, as rigorously established later. For historical completeness, we include the proof as Euler wrote it. A thorough modern proof will follow.

To begin with, we take logarithms on both sides of equality (3.8) and then use the Taylor expansion of $\log x$. This yields

$$\begin{aligned} \log\left(\sum_{n=1}^{\infty} \frac{1}{n}\right) &= -\sum_{p \text{ prime}} \log\left(1 - \frac{1}{p}\right) = \sum_{p \text{ prime}} \left(\frac{1}{p} + \frac{1}{2p^2} + \frac{1}{3p^3} + \cdots\right) \\ &= \sum_{p \text{ prime}} \frac{1}{p} + \sum_{n=2}^{\infty} \frac{1}{n} \sum_{p \text{ prime}} \frac{1}{p^n}, \end{aligned}$$

where the second series on the right-hand side is convergent. Euler then used the relation

$$\sum_{n=1}^{\infty} \frac{1}{n} = \log \infty.$$

He probably meant that the partial sums up to m increase with the rate of $\log m$ in order to conclude that

$$\log\log\infty = \log\left(\sum_{n=1}^{\infty} \frac{1}{n}\right) = C + \sum_{p \text{ prime}} \frac{1}{p},$$

which presumably stands for the relation (3.9). Rather than the first known proof (see Notes-14–15 for references), we include a much simpler one, which is just a corollary of the so-called prime number theorem.

Theorem 3.10 *Let p_n denote the nth prime number. Then, for n large enough, one has*

$$p_n \asymp n \log n. \tag{3.10}$$

This theorem was first proved by Hadamard and, independently, by de la Valleé Poussin in 1896 using methods involving the Riemann zeta function. We do not include its proof since it is rather far away from the scope of this book. Equivalence (3.10) allows one to obtain (3.9) easily. Indeed, consider the prime-counting function

$$\pi(x) = \#\{p \text{ prime}, p \le x\}, \qquad x \ge 2.$$

For n large, we have

$$\sum_{\substack{p \text{ prime} \\ p \le n}} \frac{1}{p} = \sum_{k=1}^{\pi(n)} \frac{1}{p_k} \asymp \sum_{k=2}^{\pi(n)} \frac{1}{k \log k} \asymp \log\log \pi(n),$$

for the last equivalence, see also Subsection 3.1.1. Finally, the well-known asymptotic

$$\pi(n) \asymp \frac{n}{\log n}$$

for n large yields

$$\sum_{\substack{p \text{ prime} \\ p \leq n}} \frac{1}{p} \asymp \log\log \pi(n) = \log\log n - \log\log\log n \asymp \log\log n,$$

as desired.

3.2 A test resembling l'Hôpital's rule

Recall that l'Hôpital's rule asserts that if $\frac{f(x)}{g(x)}$ is indeterminate at infinity (of the form $\frac{0}{0}$ or $\frac{\infty}{\infty}$), then

$$\lim_{x\to\infty} \frac{f(x)}{g(x)} = \lim_{x\to\infty} \frac{f'(x)}{g'(x)},$$

provided that $g'(x) \neq 0$ for large x and that the limit $\lim_{x\to\infty} \frac{f'(x)}{g'(x)}$ exists. Also, by the comparison test (Theorem 2.20), if a_n is equivalent to b_n for large n, then both series $\sum a_n$ and $\sum b_n$ converge or diverge together. In view of these ideas, one may conjecture the following:

Let $g'(x) \neq 0$ for large x and assume $\frac{f(x)}{g(x)}$ is indeterminate at infinity (of the form $\frac{0}{0}$ or $\frac{\infty}{\infty}$). If the series

$$\sum \frac{f'(n)}{g'(n)}$$

converges, then

$$\sum \frac{f(n)}{g(n)}$$

does as well.

We will see that this conjecture is not valid in general. Also, the reverse implication of this conjecture does not hold, just as in the case of l'Hôpital's rule. This is easily seen by taking

$$f(x) = \frac{\sin(\pi x)}{x}, \qquad g(x) = \frac{1}{x}.$$

In this case $\sum \frac{f(n)}{g(n)}$ converges, while $\sum \frac{f'(n)}{g'(n)}$ does not. Another observation is that the assumption that $\frac{f(x)}{g(x)}$ is indeterminate at infinity cannot be removed if the conjecture is true. This can be seen by taking

$$f(x) = 6\pi x + \sin(6\pi x), \qquad g(x) = e^{\sin 3\pi x} f(x),$$

for which such an assumption is not true. In this case $\sum \frac{f'(n)}{g'(n)}$ converges, while $\sum \frac{f(n)}{g(n)}$ diverges.

In fact, a refined version of the conjecture holds true. More precisely, stronger conditions should be assumed, and surprisingly, the interrelation between the series

$$\sum_{n=1}^{\infty} \frac{f(n)}{g(n)} \quad \text{and} \quad \sum_{n=1}^{\infty} \frac{f'(n)}{g'(n)}$$

is different in the cases where the indetermination is $\frac{0}{0}$ or $\frac{\infty}{\infty}$. For the indetermination of the form $\frac{0}{0}$, we have the following result.

Theorem 3.11 *Let f and g be differentiable functions on $(0, \infty)$ such that the ratio $\frac{f(x)}{g(x)}$ generates an indetermination of the form $\frac{0}{0}$ at infinity. Assume that there exists $x_0 > 0$ such that $g(x) \neq 0$, $g'(x) \neq 0$ for $x \geq x_0$, and that $g(n), g'(n) \neq 0$ for all n. If the series*

$$\sum_{n=1}^{\infty} \sup_{x \geq n} \left| \frac{f'(x)}{g'(x)} \right| \tag{3.11}$$

converges, then the series

$$\sum_{n=1}^{\infty} \frac{f(n)}{g(n)} \tag{3.12}$$

converges absolutely.

Proof Since f and g are differentiable and $g(x) \neq 0$ for x large enough, we can define, for n sufficiently large,

$$h_n(x) = \frac{f(n)}{g(n)} g\left(\frac{1}{x}\right) - f\left(\frac{1}{x}\right), \qquad x \in \left(0, \frac{1}{n}\right].$$

Note that h is differentiable and $h_n(1/n) = 0$. Also, since

$$\lim_{x \to \infty} f(x) = \lim_{x \to \infty} g(x) = 0,$$

$\lim_{x \to 0} h_n(x) = 0$. Thus, setting $h_n(0) = 0$, by Rolle's theorem (see Theorem 3.35 in Notes-17), there exists $b = b_n \in \left(0, \frac{1}{n}\right)$ such that $h'(b) = 0$. In other words, for any $n \geq 1$, there exists $c = c_n = \frac{1}{b_n} > n$ such that

$$\left| \frac{f(n)}{g(n)} \right| = \left| \frac{f'(c)}{g'(c)} \right| \leq \sup_{x \geq n} \left| \frac{f'(x)}{g'(x)} \right|,$$

which leads to the desired conclusion. □

We note that if there exists $N \in \mathbb{N}$ such that $\left|\frac{f'(x)}{g'(x)}\right|$ is decreasing for $x > N$, then the assumption that (3.11) converges in Theorem 3.11 can be replaced by

$$\sum_{n=1}^{\infty} \left| \frac{f'(n)}{g'(n)} \right| < \infty,$$

since in that case

$$\sup_{x \geq n} \left| \frac{f'(x)}{g'(x)} \right| = \left| \frac{f'(n)}{g'(n)} \right|$$

for sufficiently large n.

The conclusion of Theorem 3.11 is not true if the indetermination is of the form $\frac{\infty}{\infty}$. To see this, we consider the following examples:

$$f(x) = \log x, \qquad g(x) = x \log^2 x - 2x \log x + 2x.$$

Note that $\frac{f(x)}{g(x)}$ generates the indetermination $\frac{\infty}{\infty}$ as $x \to \infty$. Furthermore,

$$f'(x) = \frac{1}{x}, \qquad g'(x) = \log^2 x,$$

and $\frac{f'(x)}{g'(x)}$ is decreasing for large x. By this, the series corresponding to (3.11) is equivalent to

$$\sum_{n=2}^{\infty} \frac{f'(n)}{g'(n)} = \sum_{n=2}^{\infty} \frac{1}{n \log^2 n},$$

and therefore it is convergent. However, in this case the series (3.12) diverges, since

$$\sum_{n=2}^{\infty} \frac{f(n)}{g(n)} \asymp \sum_{n=2}^{\infty} \frac{\log n}{n \log^2 n} = \sum_{n=2}^{\infty} \frac{1}{n \log n}.$$

Finally, we state a version of Theorem 3.11 purely in the context of sequences (instead of sequences generated by pointwise evaluation of functions).

Theorem 3.12 *Let $\{a_n\}$ and $\{b_n\}$ be such that $\frac{a_n}{b_n}$ generate the indetermination $\frac{0}{0}$ as $n \to \infty$. Assume that the sequence $\{b_n\}$ is strictly decreasing to 0 and that the series*

$$\sum_{n=1}^{\infty} \sup_{m \geq n} \left| \frac{a_m - a_{m+1}}{b_m - b_{m+1}} \right| \tag{3.13}$$

converges. Then the series

$$\sum_{n=1}^{\infty} \frac{a_n}{b_n}$$

converges absolutely.

Proof Since $\{b_k\}$ is strictly decreasing, one has $\Delta b_k = b_k - b_{k+1} > 0$. Moreover, since $a_k \to 0$, we have, for every $n \in \mathbb{N}$,

$$\sum_{k=n}^{\infty} \Delta a_k = \sum_{k=n}^{\infty} (a_k - a_{k+1}) = a_n.$$

Clearly, in order to prove the desired result it suffices to show that for every n, there exists $m \geq n$ such that $\frac{a_n}{b_n} \leq \frac{\Delta a_m}{\Delta b_m}$. Assuming that such an m does not exist, we arrive at a contradiction. For some n, and all $m \geq n$, we have

$$\frac{|a_n|}{b_n} = \left| \frac{a_n}{b_n} \right| > \left| \frac{\Delta a_m}{\Delta b_m} \right| = \frac{|\Delta a_m|}{\Delta b_m},$$

and hence $|a_n| \Delta b_m > |\Delta a_m| b_n$. Summing up this inequality in m from n to infinity, we obtain

$$|a_n| b_n = |a_n| \sum_{m=n}^{\infty} \Delta b_m > b_n \sum_{m=n}^{\infty} |\Delta a_m| \geq b_n \left| \sum_{m=n}^{\infty} \Delta a_m \right| = b_n |a_n|,$$

which is impossible. Thus, the result follows. □

We note that similar to Theorem 3.11, if there exists $N \in \mathbb{N}$ such that

$$\left|\frac{a_m - a_{m+1}}{b_m - b_{m+1}}\right|$$

is decreasing for $m \geq N$, then the assumption that (3.13) converges in Theorem 3.12 becomes

$$\sum_{n=1}^{\infty}\left|\frac{a_n - a_{n+1}}{b_n - b_{n+1}}\right| = \sum_{n=1}^{\infty}\left|\frac{\Delta a_n}{\Delta b_n}\right| < \infty.$$

3.3 Selective sums of positive convergent series

Given a positive convergent series

$$\sum_{n=1}^{\infty} a_n = A,$$

it is natural to consider certain subseries such as

$$A_1 = \sum_{n=1}^{N} a_n, \qquad A_2 = \sum_{n=1}^{\infty} a_{5n}, \qquad A_3 = \sum_{n=N}^{\infty} a_n, \qquad \text{etc.}$$

and try to evaluate their sums. The "subsums" A_j, $j = 1, 2, 3$, are called *selective sums*. More precisely, we have the following definition.

Definition 3.13 Let $\sum_{n=1}^{\infty} a_n$ be a series consisting of real-valued terms. A number $x \in \mathbb{R}$ is called a selective sum of the series if there exists a sequence $\{c_n\}$ consisting only of 0's and 1's such that

$$x = \sum_{n=1}^{\infty} c_n a_n.$$

Note that the series studied in Subsection 3.1.2 (harmonic series with deleted terms) and Subsection 3.1.3 (series of the reciprocals of the primes) are special selective sums of the harmonic series.

We also note that the collection of all selective sums of a positive series $\sum a_n$ with $a_n \to 0$ is invariant under rearrangement of the sequence $\{a_n\}$. Thus, for simplicity and without loss of generality, we study selective sums of series with positive nonincreasing terms.

Before proceeding further, we state the following result on selective sums of divergent positive series, which is convenient although not within the main scope of this section.

Theorem 3.14 *Let $\{a_n\}$ be a nonincreasing positive sequence vanishing at infinity such that $\sum a_n$ diverges. Then any number $x \geq 0$ is a selective sum of $\sum a_n$.*

Proof Let $x > 0$. We define the sequence $\{c_n\}$ recursively as follows: let $c_1 = 1$ if $a_1 \le x$ and 0 otherwise, and for $n \ge 2$,

$$c_n = \begin{cases} 1, & \text{if } \sum_{k=1}^{n} c_k a_k \le x, \\ 0, & \text{otherwise.} \end{cases}$$

The sequence $\{c_n\}$ gives rise to a nonnegative series $\sum a_n c_n$ bounded by x. If the equality $\sum_{k=1}^{n} c_k a_k = x$ is attained for some finite n, the proof is then complete. If this is not the case, since $\sum a_n$ diverges and $a_n \to 0$, the partial sums $\sum_{k=1}^{n} c_k a_k$ converge to x. □

An interesting problem is to determine the set of selective sums of a given positive convergent series. Let us examine two simple examples for which we can readily determine the corresponding selective sums. First, the geometric series

$$\sum_{n=1}^{\infty} \frac{c_n}{2^n}, \tag{3.14}$$

whose sum is 1 when $c_n = 1$ for all n, has a maximal set of selective sums $S = [0, 1]$. Indeed, for any choice of the sequence $\{c_n\}$, the series (3.14) converges to a real number between 0 and 1 whose binary representation is $0.c_1c_2c_3\cdots$. Since any real number in $[0, 1]$ can be written in this form (which easily follows from Theorem 3.16 below), the claim follows.

On the other hand, we consider the series

$$\sum_{n=1}^{\infty} \frac{2c_n}{3^n}, \tag{3.15}$$

whose sum is also 1 whenever $c_n = 1$ for all n. Since for each n we have $2c_n = 0$ or 2, we can view the set S of selective sums of $\sum \frac{2}{3^n}$ as the subset of all $x \in [0, 1]$ such that the representation of x in base 3 has no digits equal to 1. In other words, the set S is the classical Cantor set (see, e.g., [B-82, Ch. 3]), which is well known to have zero measure. This contrasts with the previous example for which the set of selective sums is the whole interval $[0, A]$, where $A = \sum_{n=1}^{\infty} a_n$.

We proceed to the description of the sets of selective sums of certain series with nonnegative and nonincreasing terms. First, we have the following lemma yielding a necessary and a (different) sufficient condition for a number x to be a selective sum.

Theorem 3.15 *Let $\sum a_n$ be a convergent positive series of nonincreasing terms, with its sum denoted by A. If for $x \in (0, A]$ we define*

$$n_x = \min\{m \in \mathbb{N} : a_m \le x\},$$

then

(i) *A necessary condition for x to be a selective sum of the series* $\sum a_n$ *is that*

$$\sum_{k=n_x}^{\infty} a_k \geq x. \tag{3.16}$$

(ii) *A sufficient condition for x to be a selective sum of the series* $\sum a_n$ *is that*

$$\sum_{k=n_y}^{\infty} a_k \geq y,$$

for every $y \in [0, x]$.

Before proving this result, we note that since the series $\sum a_n$ is assumed to be convergent, the number $n_x = \min\{m \in \mathbb{N} : a_m \leq x\}$ exists for any given $x > 0$. Secondly, we note that condition (3.16) alone is not sufficient to guarantee that x is a selective sum of the given series. This is illustrated by a simple example given by the sequence

$$a_n = \begin{cases} 6, & \text{if } n = 1, \\ 4, & \text{if } n = 2, \\ \dfrac{1}{2^{n-2}}, & \text{if } n \geq 3. \end{cases}$$

It is clear that the series $\sum a_n$ is convergent with sum $6 + 4 + 1 = 11$. Moreover, if $x = 8$, we have that $n_x = 1$, since $a_1 < 8$ and $\sum_{n=1}^{\infty} a_n \geq 8$. However, it is clear that the number 8 is not a selective sum of the series $\sum a_n$.

Proof of Theorem 3.15 In order to prove the necessity part, note that if (3.16) does not hold, it precisely means that after excluding all the terms of the series that are greater than x (which obviously cannot be a part of a selective sum that sums up to x), the sum of the remaining terms is strictly smaller than x. Since all the terms a_n are positive, the claim follows.

In order to prove sufficiency, we assume that x is not equal to the sum of finitely many terms of the series (otherwise there is nothing to prove). By assumption (3.16), we have

$$\sum_{k=n_x}^{\infty} a_k \geq x.$$

If the equality holds, then the proof is complete. Assume that we have

$$\sum_{k=n_x}^{\infty} a_k > x$$

instead. Let N_1 be the integer number such that

$$\sum_{k=n_x}^{N_1} a_k < x < \sum_{k=n_x}^{N_1+1} a_k. \tag{3.17}$$

Put $x_1 := x$. Since, by assumption, $x_1 > x_2 := x_1 - \sum_{k=n_{x_1}}^{N_1} a_k > 0$ satisfies property (3.16) (with x_2 in place of x), we find that

$$\sum_{k=n_{x_2}}^{\infty} a_k \geq x_2 = x_1 - \sum_{k=n_{x_1}}^{N_1} a_k \geq a_{n_{x_2}}. \tag{3.18}$$

By (3.17), we deduce that $x_2 < a_{N_1+1}$. Moreover, we note that $n_{x_2} > N_1$. Indeed, if this is not the case, it follows from (3.17), (3.18), and the monotonicity of $\{a_k\}$, that

$$a_{N_1} \leq a_{n_{x_2}} \leq x_2 < a_{N_1+1},$$

which is a contradiction. As before, if we have equality on the left-hand side of (3.18), the proof is complete. We assume that the inequality is strict and we can repeat the previous process infinitely. At the $(\ell-1)$th step, $\ell \geq 3$, we find n_{x_ℓ} such that

$$\sum_{k=n_{x_\ell}}^{\infty} a_k \geq x_\ell := x - \sum_{m=1}^{\ell-1} \sum_{k=n_{x_m}}^{N_m} a_k \geq a_{n_{x_\ell}}.$$

This completes the proof if we have equality on the left-hand side. If it is a strict inequality, we define N_ℓ as the integer number such that

$$\sum_{k=n_{x_\ell}}^{N_\ell} a_k < x_\ell < \sum_{k=n_{x_\ell}}^{N_\ell+1} a_k.$$

Now

$$0 < x_\ell - \sum_{k=n_{x_\ell}}^{N_\ell} a_k = x - \sum_{m=1}^{\ell} \sum_{k=n_{x_m}}^{N_m} a_k < a_{N_\ell+1}.$$

Since $a_n \to 0$, we conclude that the sums $\sum_{m=1}^{\ell} \sum_{k=n_{x_m}}^{N_m} a_k$ converge to x as $\ell \to \infty$. □

The following is a characterization of those series for which the set of selective sums is maximal.

Theorem 3.16 *Let $\sum a_n$ be a convergent positive series with sum A such that $\{a_n\}$ is nonincreasing. Then the set of selective sums of this series is equal to $[0, A]$ if and only if*

$$a_n \leq \sum_{k=n+1}^{\infty} a_k, \tag{3.19}$$

for every $n \geq 1$.

Proof For the "if" part, we show that for every $x \in [0, A]$ we have

$$\sum_{k=n_x}^{\infty} a_k \geq x,$$

where, as above, $n_x = \min\{m \in \mathbb{N} : a_m \leq x\}$. The conclusion will then follow by Theorem 3.15. Let us then fix $x \in [0, A]$, thus obtaining a natural number n_x. If $n_x = 1$, then $\sum_{k=1}^{\infty} a_k \geq x$. On the other hand, if $n_x > 1$, then $a_{n_x} \leq x < a_{n_x-1}$, and by (3.19) we obtain

$$\sum_{k=n_x}^{\infty} a_k \geq a_{n_x-1} > x,$$

as desired.

For the "only if" part, assume that there exists $n \in \mathbb{N}$ such that (3.19) does not hold. If x is a number such that $\sum_{k=n+1}^{\infty} a_k < x < a_n$, the same argument as in the necessity part of Theorem 3.15 shows that x cannot be a selective sum of the series $\sum a_n$. □

For instance, the set of selective sums of the series $\sum_{n=1}^{\infty} 2^{-n}$ is the interval $[0, 1]$, since

$$\frac{1}{2^n} = \sum_{k=n+1}^{\infty} 2^{-k}.$$

On the opposite side of Theorem 3.16, the set of selective sums of series for which (3.19) does not hold for any n is also described easily.

Theorem 3.17 *Let $\sum a_n$ be a convergent positive series with sum A such that $\{a_n\}$ is nonincreasing. If*

$$a_n > \sum_{k=n+1}^{\infty} a_k$$

for every $n \in \mathbb{N}$, then the set of selective sums S of the series is of the form

$$S = \bigcap_{n=1}^{\infty} F_n,$$

where F_n is the union of the 2^n disjoint closed intervals of the form

$$\left[\sum_{k=1}^{n} c_k a_k, \sum_{k=1}^{n} c_k a_k + \sum_{k=n+1}^{\infty} a_k\right],$$

with each c_k equal to 0 or 1.

We omit the proof of this result. The general idea is that the intervals F_n are built in a way analogous to the construction of the Cantor set (that is, starting with the full interval $[0, A]$ and removing central open intervals iteratively, according to some rule), and that any $x \in S$ may easily be obtained as a selective sum of the given series related to the aforementioned construction.

3.4 "Change of variables" for sums

When computing integrals, a substitution technique, or change of variables, is of constant use. Very often it is convenient to replace calculations with sums by those with integrals just for this reason.

Given a continuously differentiable bijective function $\varphi\colon [a,b] \to I$, where $I \subset \mathbb{R}$ is an interval, and a continuous function $f\colon I \to \mathbb{C}$, we have

$$\int_a^b f(\varphi(x))\varphi'(x)\,dx = \int_{\varphi(a)}^{\varphi(b)} f(y)\,dy.$$

In the case of series (or sums), we do not have such a nice equality. However, one could still write simple identities of the type

$$\sum_{n=5M}^{5N} a_{\lfloor n/5 \rfloor} = 5\sum_{n=M}^{N} a_n, \tag{3.20}$$

but in general there is no formula for $\sum a_{\varphi(n)}$ whenever $\varphi\colon \mathbb{N} \to \mathbb{R}$. Actually, equality (3.20) is not even an equality if we take limits of summation that are not multiples of 5 in the left-side sum. In this case we only have the upper estimate

$$\sum_{n=M}^{N} a_{\lfloor n/5 \rfloor} \le 5 \sum_{n=\lfloor M/5 \rfloor}^{\lfloor N/5 \rfloor} a_n.$$

This is the main idea of the following result: an upper estimate for series with transformed indices in a way that resembles the change of variables for integrals.

Theorem 3.18 *Let $\{a_n\}_{n=0}^{\infty} \in$ WMS be a positive sequence and assume that $\sum a_n$ converges. Then,*

$$\sum_{k=0}^{\infty} 2^k a_{\lfloor 2^k x \rfloor} \le \phi(x) \sum_{n=0}^{\infty} a_n, \qquad \textit{for any } x > 0,$$

where

$$\phi(x) = \frac{4C^2}{x}\big(\chi_{[1,\infty)}(x) - (1 - \log_2 x)\chi_{(0,1)}(x)\big),$$

and C is the WMS constant for $\{a_n\}$.

Proof For $x \ge 1$, let $m \in \mathbb{N} \cup \{0\}$ and assume that $2^m \le x < 2^{m+1}$. Then

$$\sum_{k=0}^{\infty} 2^k a_{\lfloor 2^k x \rfloor} \le C \sum_{k=0}^{\infty} 2^k a_{2^{k+m}} = C2^{-m} \sum_{k=m}^{\infty} 2^k a_{2^k} = C2^{-m} \sum_{k=m}^{\infty} \sum_{\ell=2^k}^{2^{k+1}-1} a_{2^k}$$

$$\le \frac{4C^2}{x} \sum_{k=m}^{\infty} \sum_{\ell=2^{k-1}}^{2^k-1} a_\ell \le \frac{4C^2}{x} \sum_{n=0}^{\infty} a_n.$$

On the other hand, if $2^{-(m+1)} \le x < 2^{-m}$,

$$\begin{aligned}\sum_{k=0}^{\infty} 2^k a_{\lfloor 2^k x\rfloor} &\le C\sum_{k=0}^{\infty} 2^k a_{\lfloor 2^{k-m-1}\rfloor} = C2^{m+1}\sum_{k=-m-1}^{\infty} 2^k a_{\lfloor 2^k\rfloor}\\ &\le C2^{m+1}(m+1)\left(\sum_{k=0}^{\infty} 2^k a_{2^k} + a_0\right)\\ &\le C\frac{2}{x}\log_2\frac{2}{x}\left(\sum_{k=0}^{\infty} 2^k a_{2^k} + a_0\right) \le \frac{4C^2}{x}(1-\log_2 x)\sum_{n=0}^{\infty} a_n,\end{aligned}$$

which completes the proof. □

One may also obtain an identity for the series $\sum a_k$ involving integrals of the terms $a_{\lfloor 2^k x\rfloor}$.

Theorem 3.19 *For any nonnegative sequence $\{a_n\}$, we have*

$$\sum_{n=1}^{\infty} a_n = \int_1^2 \sum_{k=0}^{\infty} 2^k a_{\lfloor 2^k x\rfloor}\, dx.$$

Proof Indeed, since

$$\frac{1}{2^n}\sum_{k=2^n}^{2^{n+1}-1} a_k = \frac{1}{2^n}\int_{2^n}^{2^{n+1}} a_{\lfloor x\rfloor}\, dx = \frac{1}{2^n}\int_1^2 2^n a_{\lfloor 2^n y\rfloor}\, dy = \int_1^2 a_{\lfloor 2^n y\rfloor}\, dy,$$

we conclude that

$$\sum_{n=1}^{\infty} a_n = \sum_{n=0}^{\infty}\sum_{k=2^n}^{2^{n+1}-1} a_k = \sum_{n=0}^{\infty} 2^n \int_1^2 a_{\lfloor 2^n y\rfloor}\, dy = \int_1^2 \sum_{n=0}^{\infty} 2^n a_{\lfloor 2^n y\rfloor}\, dy,$$

where the last equality is justified by Tonelli's theorem (i.e., we can change the order of summation and integration at any time since the involved functions are all nonnegative). □

To conclude this section, we present an equiconvergence result of two series that are not exactly related through a change of variables, but in a way that reminds of it. We found this section to be the most appropriate to contain it due to its nature.

Theorem 3.20 *Let $f\colon [1,\infty) \to (0,\infty)$ be a strictly increasing function such that $f(x) \to \infty$ as $x \to \infty$. Then, the series*

$$\sum_{n=1}^{\infty} \frac{1}{f(n)} \quad \text{and} \quad \sum_{n=\lceil f(1)\rceil}^{\infty} \frac{f^{-1}(n)}{n^2},$$

where f^{-1} denotes the inverse of f, converge or diverge simultaneously.

Proof First of all we note that since the inverse f^{-1} is also strictly increasing,

$$\sum_{n=\lceil f(1)\rceil}^{\infty} \frac{1}{n^2} f^{-1}(n) = \sum_{k=\lceil f(1)\rceil}^{\infty} \left(\sum_{n:k<f^{-1}(n)\leq k+1} \frac{1}{n^2} f^{-1}(n) \right)$$
$$= \sum_{k=\lceil f(1)\rceil}^{\infty} \sum_{n=\lceil f(k)\rceil}^{\lfloor f(k+1)\rfloor} \frac{1}{n^2} f^{-1}(n).$$

We proceed to obtaining estimates of the inner sum from above and from below. On the one hand,

$$\sum_{n=\lceil f(k)\rceil}^{\lfloor f(k+1)\rfloor} \frac{1}{n^2} f^{-1}(n) \geq k \int_{\lceil f(k)\rceil}^{\lfloor f(k+1)\rfloor+1} \frac{1}{x^2}\, dx$$
$$\geq k\left(\frac{1}{\lfloor f(k)\rfloor + 1} - \frac{1}{\lfloor f(k+1)\rfloor + 1} \right).$$

On the other hand, for the upper estimate, we distinguish between two cases. First, if $\lceil f(k)\rceil = 1$, then

$$\sum_{n=\lceil f(k)\rceil}^{\lfloor f(k+1)\rfloor} \frac{1}{n^2} f^{-1}(n) \leq 2k\left(1 + \sum_{n=2}^{\lfloor f(k+1)\rfloor} \frac{1}{n^2}\right) \leq 2k\left(1 + \int_1^{\lfloor f(k+1)\rfloor} \frac{1}{x^2}\, dx\right)$$
$$\leq 8k\left(\frac{1}{\lfloor f(k)\rfloor + 1} - \frac{1}{\lfloor f(k+1)\rfloor + 1} \right).$$

Secondly, if $\lceil f(k)\rceil \neq 1$, we have

$$\sum_{n=\lceil f(k)\rceil}^{\lfloor f(k+1)\rfloor} \frac{1}{n^2} f^{-1}(n) \leq 2k \int_{\lceil f(k)\rceil-1}^{\lfloor f(k+1)\rfloor} \frac{1}{x^2}\, dx$$
$$\leq 2k\left(\frac{1}{\lceil f(k)\rceil - 1} - \frac{1}{\lfloor f(k+1)\rfloor + 1} \right)$$
$$\leq 6k\left(\frac{1}{\lfloor f(k)\rfloor + 1} - \frac{1}{\lfloor f(k+1)\rfloor + 1} \right).$$

Combining the above estimates and applying the comparison test (Theorem 2.18), we derive that the series

$$\sum_{n=\lceil f(1)\rceil}^{\infty} \frac{1}{n^2} f^{-1}(n) \quad \text{and} \quad \sum_{k=1}^{\infty} k\left(\frac{1}{1+\lfloor f(k)\rfloor} - \frac{1}{1+\lfloor f(k+1)\rfloor} \right)$$

converge or diverge simultaneously. Applying Abel's transformation (2.2) to the last series, we get

$$\sum_{k=1}^{m} k\left(\frac{1}{1+\lfloor f(k)\rfloor}-\frac{1}{1+\lfloor f(k+1)\rfloor}\right)=m\left(\frac{1}{1+\lfloor f(1)\rfloor}-\frac{1}{1+\lfloor f(m+1)\rfloor}\right)$$
$$+\sum_{k=1}^{m-1}\left(\frac{1}{1+\lfloor f(k+1)\rfloor}-\frac{1}{1+\lfloor f(1)\rfloor}\right)$$
$$=\frac{1}{1+\lfloor f(1)\rfloor}-\frac{m}{1+\lfloor f(m+1)\rfloor}+\sum_{k=1}^{m-1}\frac{1}{1+\lfloor f(k+1)\rfloor}.$$

If the series

$$\sum_{n=1}^{\infty}\frac{1}{f(n)} \tag{3.21}$$

converges, then by Abel–Olivier's test (Theorem 2.98), we have $\frac{m}{f(m)}\to 0$ as $m\to\infty$. This, together with the above estimates, implies the convergence of

$$\sum_{n=\lceil f(1)\rceil}^{\infty}\frac{f^{-1}(n)}{n^2}.$$

On the other hand, if the latter converges, the fact that f^{-1} is increasing implies that $\left\{\frac{f^{-1}(n)}{n^2}\right\}\in\overline{WMS}$ (see Definition 1.38). In this case, we can apply Abel–Olivier's test for $\overline{WMS}$ (cf. Remark 2.101) and derive that

$$\frac{f^{-1}(n)}{n}\to 0 \quad \text{as } n\to\infty, \qquad \text{or, equivalently,} \qquad \frac{x}{f(x)}\to 0 \quad \text{as } x\to\infty,$$

where we made the substitution $n=f(x)$. The convergence of the series $\sum_{n=1}^{\infty}\frac{1}{f(n)}$ readily follows from the condition $\frac{m}{f(m)}\to 0$ and the above identities. □

3.5 When does a transformed series inherit the convergence?

In this section we consider convergence tests involving series of the type $\sum f(a_n)$, where f and $\{a_n\}$ are real-valued. The following result essentially characterizes the functions f for which the transformed series $\sum f(a_n)$ inherits the convergence of the original series $\sum a_n$ in terms of the asymptotic behaviour of f near the origin.

Theorem 3.21 *Let f be a function defined on $\mathbb{R}$. The following are equivalent:*

(i) *$|f(x)|\le C|x|$ for x in a neighborhood of zero.*
(ii) *For any absolutely convergent series $\sum a_k$, the series $\sum f(a_k)$ converges absolutely.*

Proof The fact that (i) implies (ii) is clear. In order to prove that (ii) implies (i), we argue by contradiction. We assume that there exists a sequence $\{c_k\}$ vanishing at infinity, such that $|f(c_k)/c_k|\to\infty$ as $k\to\infty$. We now find a sequence $\{a_n\}=\{c_{k_n}\}$ so that $\sum a_n$ converges absolutely but $\sum f(a_n)$ diverges. We choose a_n so that for any

n we have $0 < |a_n| < 1/n^2$, the sequences $\{a_n\}$ and $\{f(a_n)\}$ have constant sign, and $\lambda_n = f(a_n)/a_n > n$ (we can assume without loss of generality that $a_n, f(a_n) > 0$). Consider the series obtained from $\sum a_n$ by repeating each term P_n times, where $P_n = \lfloor 1/(n^2 a_n) \rfloor$. Then

$$\sum_{n=1}^{\infty}\Bigg(\sum_{1\le m\le P_n} a_n\Bigg) = \sum_{n=1}^{\infty} P_n a_n \le \sum_{n=1}^{\infty}\frac{1}{n^2},$$

that is, the considered series converges. However,

$$\sum_{n=1}^{\infty}\Bigg(\sum_{1\le m\le P_n} f(a_n)\Bigg) = \sum_{n=1}^{\infty} P_n f(a_n) = \sum_{n=1}^{\infty} P_n a_n \lambda_n \ge \frac{1}{2}\sum_{n=1}^{\infty}\frac{\lambda_n}{n^2}$$
$$> \frac{1}{2}\sum_{n=1}^{\infty}\frac{1}{n} = \infty.$$

Arriving at a contradiction, we conclude that if (ii) holds, (i) follows. □

In the same spirit, the following result also involves functions that have a certain asymptotic expansion near the origin, although in this case we start with a convergent series of real-valued terms and characterize the convergence of $\sum f(a_n)$ in terms of the higher order terms of the asymptotic expansion of f near the origin.

Theorem 3.22 *Let $\sum a_n$ be a convergent series with real-valued terms. If f is a real-valued function such that, in a neighborhood of zero,*

$$f(x) = \alpha x + \beta x^{2k} + o(x^{2k}), \qquad \beta \ne 0,\ k \in \mathbb{N}, \tag{3.22}$$

then $\sum f(a_n)$ converges if and only if $\sum a_n^{2k}$ does.

Proof By assumption, there exists $\varepsilon > 0$ such that for $|x| < \varepsilon$ one has

$$\alpha x + \left(\beta - \frac{|\beta|}{2}\right)x^{2k} \le f(x) \le \alpha x + \left(\beta + \frac{|\beta|}{2}\right)x^{2k}.$$

Since $\sum a_n$ converges, $a_n \to 0$. Therefore, there exists $N = N(\varepsilon)$ such that

$$\alpha a_n + \left(\beta - \frac{|\beta|}{2}\right)a_n^{2k} \le f(a_n) \le \alpha a_n + \left(\beta + \frac{|\beta|}{2}\right)a_n^{2k},$$

for all $n \ge N$. The conclusion now follows from the last inequality and Theorem 2.19. □

Remark 3.23 Theorem 3.22 cannot be extended to the case of f having the asymptotic expansion

$$f(x) = \alpha x + \beta x^{2k+1} + o(x^{2k+1}), \qquad \beta \ne 0,\ k \in \mathbb{N}, \tag{3.23}$$

near the origin. In fact, there are counterexamples for both directions of the result (i.e., for the necessity and sufficiency). Indeed, if on the one hand we consider

$$a_n = \frac{(-1)^n}{\sqrt[4]{n}},$$

it is clear that both series $\sum a_n$ and $\sum a_n^3$ converge by the alternating series test (Theorem 2.66). However, if we choose $f(x) = x + x^3 + x^4$, then $\sum f(a_n)$ is the sum of two convergent series and the harmonic series, and therefore it diverges.

On the other hand, if we define the sequence

$$a_{2n} = \frac{(-1)^n}{\sqrt[4]{2n}} + \frac{1}{3\sqrt{2n}}, \qquad a_{2n+1} = -\frac{1}{3\sqrt{2n}}, \qquad n \in \mathbb{N},$$

and the function $f(x) = x + x^3 - x^4$, it is easy to see that both $\sum a_n$ and $\sum f(a_n)$ converge, but the series $\sum a_n^3$ and $\sum a_n^4$ are both divergent.

It is also worth noting that if (3.23) holds and $\sum |a_n|^{2k+1}$ converges, then, provided that $\sum a_n$ converges as well, the convergence of $\sum f(a_n)$ follows. Indeed, a simple sufficient condition for $\sum |a_n|^{2k+1}$ to converge is that $\sum a_n^{2\ell}$ converges for some $\ell \in \{1, 2, \ldots, k\}$. The converse, however, is not true, as shown by the example of the function

$$f(x) = \sin x = x + \frac{x^3}{6} + o(x^3), \qquad x \to 0,$$

and the sequence $a_n = \frac{(-1)^n}{\log n}$.

3.6 An integral-ratio test

The following test is a combination of the ratio and the integral tests, in which, roughly speaking, the ratios of successive terms of the sequence form a sequence of bounded variation.

Theorem 3.24 *Let $\{a_n\}$ be a positive sequence. Let $m \in \mathbb{N}$, and $r\colon (0,\infty) \to (0,\infty)$ be a differentiable function satisfying the following properties:*

(i) $r(n) = \dfrac{a_{n+1}}{a_n}$,

(ii) $0 < A \le r(x) \le B$ *for all* $x > m$, *and*

(iii) $\int_m^M |r'(t)|\,dt < D < \infty$ *for* $M > m$.

Then the convergence of the integral

$$\int_m^\infty e^{R_m(x)}\,dx, \qquad \text{where } R_m(x) = \int_m^x (r(t) - 1)\,dt,$$

implies that of the series $\sum a_n$, and the divergence of the integral

$$\int_m^\infty e^{\rho_m(x)}\,dx, \qquad \text{where } \rho_m(x) = \int_m^x \left(1 - \frac{1}{r(t)}\right)dt,$$

implies that of the series $\sum a_n$.

Proof Let us start with the convergence test. Define

$$d_n = \int_n^{n+1} (r(t) - 1)\,dt - (r(n+1) - 1) = \int_n^{n+1} (r(t) - r(n+1))\,dt.$$

Integrating by parts, we get

$$|d_n| = \left| - \int_n^{n+1} (t-n)r'(t)\,dt \right| \le \int_n^{n+1} |r'(t)|\,dt.$$

By assumption, we have

$$\sum_{k=m}^{M} |d_k| \le \int_m^{M+1} |r'(t)|\,dt < D,$$

for $M > m$, and therefore

$$\sum_{k=m}^{M} d_k > -D, \qquad M > m.$$

We now have, for any $n \le x < n+1$,

$$\begin{aligned} R_m(x) &= \sum_{k=m}^{n-1} \int_k^{k+1} (r(t) - 1)\,dt + \int_n^x (r(t) - 1)\,dt \\ &= \sum_{k=m}^{n-1} d_k + \sum_{k=m}^{n-1} (r(k+1) - 1) + \int_n^x (r(t) - 1)\,dt \\ &> -D - 1 + \sum_{k=m}^{n-1} (r(k+1) - 1). \end{aligned}$$

Thus,

$$\int_n^{n+1} e^{R_m(x)}\,dx > e^{-D-1} \prod_{k=m}^{n-1} e^{r(k+1)-1}.$$

Finally, using $e^{y-1} \ge y$ for any $y \in \mathbb{R}$, we obtain

$$\int_n^{n+1} e^{R_m(x)}\,dx > e^{-D-1} \prod_{k=m+1}^{n} r(k) = \frac{e^{-D-1}}{a_{m+1}} a_{n+1}, \qquad n \ge m.$$

Hence,

$$\infty > \int_m^{\infty} e^{R_m(x)}\,dx > \frac{e^{-D-1}}{a_{m+1}} \sum_{n=m}^{\infty} a_{n+1}.$$

The divergence test is proved in a similar manner by defining

$$d_n = \int_n^{n+1} \left(1 - \frac{1}{r(t)}\right) dt - \left(1 - \frac{1}{r(n+1)}\right)$$

and by using the inequality

$$e^{1-\frac{1}{y}} \leq y, \qquad y > 0.$$

The rest of the details are omitted as they are analogous to those in the proof of the convergence part. □

Remark 3.25 Note that assumptions (i) and (iii) from Theorem 3.24 imply that the sequence of ratios $\left\{\frac{a_{n+1}}{a_n}\right\}_{n=1}^{\infty}$ is of bounded variation, since

$$\int_m^{\infty} |r'(x)|\, dx \geq \sum_{n=m}^{\infty} |r(n+1) - r(n)| = \sum_{n=m}^{\infty} \left| \frac{a_{n+2}}{a_{n+1}} - \frac{a_{n+1}}{a_n} \right|.$$

An example of a series for which Theorem 3.24 easily answers whether it converges or not is

$$\sum_{n=2}^{\infty} \prod_{k=1}^{n-1} \left(1 - \frac{\alpha}{k}\right), \qquad \alpha > 0. \tag{3.24}$$

The function given by $r(x) = 1 - \frac{\alpha}{x}$ satisfies the hypotheses of Theorem 3.24. On the one hand, if $\alpha > 1$, we have

$$R_1(x) = -\int_1^x \frac{\alpha}{t}\, dt = \alpha \log \frac{1}{x}.$$

Thus, the integral

$$\int_1^{\infty} e^{R_1(x)}\, dx = \int_1^{\infty} \frac{1}{x^{\alpha}}\, dx$$

converges, which implies the convergence of series (3.24). On the other hand, if $\alpha \leq 1$, we similarly have

$$c \int_{\alpha+1}^{\infty} \frac{1}{(x-\alpha)^{\alpha}}\, dx \leq \int_{\alpha+1}^{\infty} e^{\rho_{\alpha+1}(x)}\, dx \leq C \int_{\alpha+1}^{\infty} \frac{1}{(x-\alpha)^{\alpha}}\, dx,$$

for some constants $0 < c < C$, and the corresponding integral diverges. Thus, concluding, series (3.24) converges if and only if $\alpha > 1$.

3.7 A divergence test for positive series

The following result is based on the comparison of the general terms a_n with those whose indices are given by a certain increasing function $f(n)$. The latter is in turn assumed to have at least a certain rate of increase.

Theorem 3.26 *Let $\{\lambda_n\}$ be a sequence of nonnegative real numbers such that*

$$\lambda_n \leq 1, \qquad n \in \mathbb{N}. \tag{3.25}$$

Let $f\colon \mathbb{N} \to \mathbb{N}$ be a function satisfying

$$f(n+1) - f(n) \geq n\lambda_n + 1, \qquad n \in \mathbb{N}, \tag{3.26}$$

and let $\{a_n\}$ be a positive null sequence satisfying

$$a_n \le a_{n+1} + \lambda_n a_{f(n)}, \qquad n \in \mathbb{N}. \tag{3.27}$$

Then the series $\sum a_n$ diverges.

Proof We may assume that $\lambda_n \not\equiv 0$, otherwise $a_n \ge a_1 > 0$ for every n, and the conclusion is trivial. Let now $N > 1$ be such that $\lambda_{N-1} \ne 0$. This yields $f(n) > n$ for $n > N$ by applying (3.26) iteratively. Note also that (3.26) implies that f is strictly increasing. Applying (3.27) starting from $f(n)$ until $m-1$, we derive that for $n \ge 1$ and each m such that $f(n) < m < f(n+1)$,

$$b_n := \max\left\{0, a_{f(n)} - \sum_{k=f(n)}^{f(n+1)-1} \lambda_k a_{f(k)}\right\} \le a_m.$$

By this inequality and (3.26), we have

$$\sum_{m=f(n)}^{f(n+1)-1} a_m \ge a_{f(n)} + (f(n+1) - f(n) - 1)b_n \ge a_{f(n)} + n\lambda_n b_n$$
$$\ge a_{f(n)} + \sum_{k=N}^{n} \lambda_n b_n.$$

Assuming now that $\sum a_n < \infty$, we are going to arrive at a contradiction. Summing up this last inequality in $n \ge N$, we get

$$\sum_{m=f(N)}^{\infty} a_m = \sum_{n=N}^{\infty} \sum_{m=f(n)}^{f(n+1)-1} a_m \ge \sum_{n=N}^{\infty} a_{f(n)} + \sum_{n=N}^{\infty}\sum_{k=N}^{n} \lambda_n b_n$$
$$= \sum_{n=N}^{\infty} a_{f(n)} + \sum_{k=N}^{\infty}\sum_{n=k}^{\infty} \lambda_n b_n.$$

Since $\lambda_n \le 1$, it follows from the definition of b_n that

$$\sum_{n=k}^{\infty} \lambda_n b_n \ge \sum_{n=k}^{\infty} \lambda_n a_{f(n)} - \sum_{n=f(k)}^{\infty} \lambda_n a_{f(n)} = \sum_{n=k}^{f(k)-1} \lambda_n a_{f(n)}.$$

Finally, it follows from (3.27) that

$$\sum_{n=k}^{f(k)-1} \lambda_n a_{f(n)} \ge \sum_{n=k}^{f(k)-1} (a_n - a_{n+1}) = a_k - a_{f(k)}.$$

Therefore, combining this with the previous inequalities, we obtain

$$\sum_{m=f(N)}^{\infty} a_m = \sum_{m=f(N)}^{\infty} a_m - \sum_{k=N}^{\infty} a_{f(k)} + \sum_{k=N}^{\infty} a_{f(k)} \ge \sum_{k=N}^{\infty}\sum_{n=k}^{\infty} \lambda_n b_n + \sum_{k=N}^{\infty} a_{f(k)}$$
$$\ge \sum_{k=N}^{\infty}\sum_{n=k}^{f(k)-1} \lambda_n a_{f(n)} + \sum_{k=N}^{\infty} a_{f(k)} \ge \sum_{k=N}^{\infty} a_k.$$

Since $f(N) > N$ and $a_N > 0$, the inequality

$$\sum_{k=N}^{\infty} a_k \le \sum_{m=f(N)}^{\infty} a_m$$

is impossible. Here we arrive at a contradiction, and hence $\sum a_k$ diverges. □

Remark 3.27 We note that if only two of the three conditions (3.25), (3.26), and (3.27) hold, the conclusion of Theorem 3.26 is no longer true. Let us examine each case carefully. We first consider the case where (3.25) does not hold. Let $\lambda_n \equiv \lambda > 1$ be constant, and let f be any function that satisfies (3.26). Define recursively $f^0(m) = m$ and $f^\ell(m) = f\big(f^{\ell-1}(m)\big)$, $\ell \ge 1$. For $n \ge 1$, let $k = k_n$ be the largest nonnegative integer for which there exists $r \in \mathbb{N}$ satisfying $f^k(r) = n$ (clearly if n does not belong to the image of f, then $k = 0$). This in turn uniquely determines $r = r_n$, since f is strictly increasing. If we define $a_n = \lambda^{-k_n} 2^{-r_n}$, then (3.27) holds, since $f(n) = f(f^k(r)) = f^{k+1}(r)$. Thus, $a_{f(n)} = \lambda^{-k_n-1} 2^{-r_n}$, and

$$a_n = 2^{-r_n}\lambda^{-k_n} = \lambda 2^{-r_n}\lambda^{-k_n-1} = \lambda a_{f(n)} \le a_{n+1} + \lambda a_{f(n)}.$$

However, it is clear that $\sum a_n$ converges. Indeed,

$$\sum_{n=1}^{\infty} a_n \le \sum_{r=1}^{\infty} 2^{-r} \sum_{k=0}^{\infty} \lambda^{-k} < \infty.$$

Secondly, let us consider the case where (3.26) is not satisfied. Let us fix $\lambda_n \equiv 1$ and consider a strictly increasing sequence of natural numbers $\{n_k\}_{k=1}^{\infty}$. We define

$$a_n = \begin{cases} 2^{-k}, & \text{if } n = n_k,\ k \in \mathbb{N}, \\ 2^{-k-1}, & \text{if } n = n_k + 1,\ k \in \mathbb{N}, \\ 0, & \text{otherwise.} \end{cases}$$

The series $\sum a_n$ converges, and if we define $f(n) = n_{k+1}$ for $k \ge 2$ and $n_{k-1} + 1 < n \le n_k + 1$, we see that (3.27) holds. However, (3.26) does not hold, since f is not strictly increasing.

Finally, we consider the case where (3.27) is not satisfied. We fix $\lambda_n \equiv 1$. In this case it is rather easy to find a counterexample, since, as we will observe in Remark 3.28 below, this condition controls in a way the speed of convergence of the sequence a_n to zero. Therefore, for any function f satisfying (3.26), if a_n is such that $a_n > 2a_{n+1}$ for all n, then (3.27) does not hold, and $\sum a_n < \sum 2^{-n} a_1 < \infty$.

Remark 3.28 Note that condition (3.27) in Theorem 3.26 can be rewritten as

$$a_n - a_{n+1} \le \lambda_n a_{f(n)}. \tag{3.28}$$

This condition in a sense does not allow a_n to approach zero too fast, which results in the divergence of the series $\sum a_n$. For instance, if $\lambda_n \equiv 1, f(n) = n^2$ and $a_n = n^{-2}$

(which, despite the convergence of $\sum a_k$, does not tend to zero as fast as other sequences such as $b_n = 2^{-n}$), the above condition reads as

$$n^{-2} - (n+1)^{-2} \leq n^{-4},$$

which is clearly not true for $n \geq 2$.

Example 3.29 Let us present some examples of possible choices for $\{\lambda_n\}$ and f in Theorem 3.26.

(i) For fast increasing functions f, for instance, $f(n) = 2^n$, the best choice for the coefficients λ_n is $\lambda_n \equiv 1$, since in this case (3.26) is always fulfilled. On the other hand, for a given sequence $\{a_n\}$, assumption (3.27) is more likely to be satisfied for higher values of λ_n.

(ii) If $\lambda_n \equiv 1$, it is known [44, 219] that if $f(n) \leq cn^2$ for some $0 < c < \frac{1}{2}$, then (3.27) does not imply the divergence of the series $\sum a_n$ (particularly in this case there exist values of n for which (3.26) is not satisfied). For this choice of f, one may always find $\delta = \delta(c) > 0$ so small that

$$a_n = \frac{1}{n \log n (\log\log n)^{1+\delta}}$$

satisfies (3.27), while $\sum a_n$ converges (see, e.g., Section 3.1).

(iii) In relation with (ii), if $0 < c < \frac{1}{2}$ and $f(n) = \lfloor cn^2 \rfloor$, we may always find a sequence $\{\lambda_n\}$, with $\lambda_n \leq 1$, such that

$$\lfloor c(n+1)^2 \rfloor - \lfloor cn^2 \rfloor \geq n\lambda_n + 1,$$

for every n. This example illustrates how the coefficients λ_n can compensate the fact that f increases to infinity not fast enough in order to still derive the divergence test from Theorem 3.26 (compare with (i)).

(iv) The balance between the choices of f and the sequence $\{\lambda_n\}$ mentioned in (iii) plays a crucial role in condition (3.27) (or equivalently (3.28)). As observed in Remark 3.28, in order to guarantee the divergence of a series $\sum a_n$ by means of Theorem 3.26 (we assume that $a_n \to 0$, since otherwise the conclusion is trivial), the term $\lambda_n a_{f(n)}$ on the right-hand side of (3.28) ought to be small. This is achieved if either of the terms λ_n or $a_{f(n)}$ are small. In order for $a_{f(n)}$ to be small enough (with respect to the difference $a_n - a_{n+1}$), f must increase fast to infinity (since a_n vanishes at infinity). If f does not increase to infinity fast enough (so that $a_{f(n)}$ is not small enough), the condition

$$f(n+1) - f(n) \geq n\lambda_n + 1$$

will be satisfied only whenever λ_n is small enough and so will be the product $\lambda_n a_{f(n)}$.

It can be proved that if the sequence $\{a_n\}$ in Theorem 3.26 is assumed to be decreasing, condition (3.25) is redundant whenever a slight modification of (3.26) holds. The precise statement is as follows.

Theorem 3.30 *Let $\{\lambda_n\}$ be a sequence of nonnegative real numbers and let $f\colon \mathbb{N} \to \mathbb{N}$ be a function satisfying*

$$f(n+1) - f(n) \geq n\lambda_{n+1} + 1, \qquad n \in \mathbb{N}. \tag{3.29}$$

Let also $\{a_n\}$ be a decreasing positive sequence satisfying (3.27). Then the series $\sum a_n$ diverges.

Proof As in the proof of Theorem 3.26, we may assume that there is $N > 1$ such that $\lambda_{N-1} \neq 0$, and hence $f(N) > N$. For $N \leq n \leq M$, it follows from (3.27) that

$$a_{f(n)} \leq a_{f(M)+1} + \sum_{k=f(n)}^{f(M)} \lambda_k a_{f(k)}.$$

Multiplying both sides by $f(n) - f(n-1)$ and summing up in n, $N \leq n \leq M$, we obtain

$$\begin{aligned}
&\sum_{n=N}^{M} (f(n) - f(n-1)) a_{f(n)} \\
&\leq (f(M) - f(N-1)) a_{f(M)+1} + \sum_{n=N}^{M} (f(n) - f(n-1)) \sum_{k=f(n)}^{f(M)} \lambda_k a_{f(k)} \\
&\leq f(M) a_{f(M)+1} + \sum_{k=f(N)}^{f(M)} \lambda_k a_{f(k)} \sum_{n:N\leq n, f(n)\leq k} (f(n) - f(n-1)) \\
&\leq f(M) a_{f(M)+1} + \sum_{k=f(N)}^{f(M)} \lambda_k a_{f(k)} (k - f(N-1)).
\end{aligned}$$

Using (3.29), we get

$$\lambda_k (k - f(N-1)) \leq \lambda_k (k-1) < f(k) - f(k-1).$$

Therefore,

$$\sum_{n=N}^{M} (f(n) - f(n-1)) a_{f(n)} < f(M) a_{f(M)+1} + \sum_{k=f(N)}^{f(M)} (f(k) - f(k-1)) a_{f(k)}. \tag{3.30}$$

We now assume that $\sum a_k$ converges and arrive at a contradiction. Since $\{a_k\}$ is decreasing, we have

$$\sum_{k=f(N)}^{\infty} (f(k) - f(k-1)) a_{f(k)} \leq \sum_{k=f(N)}^{\infty} \sum_{\ell=f(k-1)}^{f(k)-1} a_\ell \leq \sum_{n=1}^{\infty} a_n < \infty. \tag{3.31}$$

Taking into account that the convergence of $\sum a_n$ implies $na_n \to 0$ (by Abel–Olivier's test, Theorem 2.98), letting $M \to \infty$ in (3.30) and using the convergence of the first series in (3.31), we arrive at

$$\sum_{n=N}^{\infty}(f(n)-f(n-1))a_{f(n)} \leq \sum_{k=f(N)}^{\infty}(f(k)-f(k-1))a_{f(k)},$$

which is impossible since $f(N) > N$ and $\{a_n\}$ is a positive sequence. Therefore, we obtain a contradiction and conclude that $\sum a_n$ diverges. □

By examining the proof of Theorem 3.30, where the monotonicity of $\{a_k\}$ is used, one can easily obtain a generalization for $\{a_k\} \in WMS$.

Theorem 3.31 *The conclusion of Theorem 3.30 holds for $\{a_k\} \in WMS$ if the function f is additionally assumed to satisfy*

$$\frac{f(n+1)}{f(n)} = O(1).$$

Proof The structure of the proof is the same as that of Theorem 3.30. The only difference is that we must show how the *WMS* condition replaces the monotonicity assumption. On the one hand, Abel–Olivier's test for decreasing sequences (Theorem 2.98) is used to derive that $na_n \to 0$ for a convergent series $\sum a_n$. As we saw in Chapter 2, this test is valid if $\{a_k\} \in WMS$ (cf. Theorem 2.99). On the other hand, the monotonicity condition is also used to derive the inequality

$$\sum_{k=f(N)}^{f(M)}(f(k)-f(k-1))a_{f(k)} \leq \sum_{k=f(N)}^{f(M)}\sum_{\ell=f(k-1)}^{f(k)-1} a_\ell$$

in (3.31). Since we assume that $f(n+1) \leq Cf(n)$ for some absolute constant $C > 1$, the *WMS* condition yields $a_{f(n+1)} \leq C'a_\ell$ for every $f(n) \leq \ell \leq f(n+1) \leq Cf(n)$. By this,

$$\sum_{k=f(N)}^{f(M)}(f(k)-f(k-1))a_{f(k)} = \sum_{k=f(N)}^{f(M)}\sum_{\ell=f(k-1)}^{f(k)-1} a_{f(k)} \leq C'\sum_{k=f(N)}^{f(M)}\sum_{\ell=f(k-1)}^{f(k)-1} a_\ell,$$

that is, (3.31) also holds (up to the constant C') in this case, and the rest follows. □

We observe that the only difference between conditions (3.29) and (3.27) is the index of the sequence $\{\lambda_n\}$. Though the difference seems to be negligible, it is in fact shown in [44] that condition (3.29) cannot be replaced by (3.27) in Theorem 3.30.

3.8 Notes

Section 3.1 (The harmonic series)

1. Interesting historical remarks on the first published proof of the divergence of the harmonic series can be found in [30]. We briefly discuss them and then refer to that paper for further details. The first dated publication of the divergence of the harmonic series is due to Pietro Mengoli, in 1650 [B-63]. However, this result was already established as early as 1350 by Nicole Oresme in a series of questions (*Questiones super geometriam Euclidis*, cf. [B-39, pp. 131–135]) that were copied as a manuscript and not published until the twentieth century.

Apparently, Mengoli did not know about the existence of Oresme's work. Interestingly enough, Mengoli's proof consists in showing that any positive number can be exceeded by partial sums of the harmonic series, thus (purposely) avoiding any reference to infinity (in line with classical Greek mathematics, where the "method of exhaustion" was the fundamental tool to avoid appealing to infinity). For further details, see [30] and the references therein.

2. Theorem 3.1 was obtained in [265], and the author named it a "Master Theorem of Series", as it allows to explicitly obtain closed forms for several number series, including the ones presented above and many others involving partial sums of p-series with $p \in \mathbb{N}$ (see also [B-81, Ch. 4] of the same author, containing several other related examples). A slight generalization of Theorem 3.1 is given in [29].

3. We also mention the work in [151], where significant efforts were taken to obtain sharp estimates for the remainders of the four series,

$$\sum_{n=1}^{\infty} \frac{1}{n^{s+1}}, \quad \sum_{n=1}^{\infty} \frac{(-1)^n}{n^{s+1}}, \quad \sum_{n=1}^{\infty} \frac{1}{(2n+1)^{s+1}}, \quad \text{and} \quad \sum_{n=1}^{\infty} \frac{(-1)^n}{(2n+1)^{s+1}}, \qquad s \in \mathbb{N},$$

related to the series $\sum_{n=1}^{\infty} \frac{1}{n^p}$. The authors exploit the assumption $s \in \mathbb{N}$ to derive integral representations for the relevant remainders, which in turn leads to rather sharp estimates.

4. It is proved in [38] that the numbers

$$k_m = \min\{n \in \mathbb{N} : H_n \geq m\}, \qquad m \in \mathbb{N},$$

are closely related to both the Euler number e and the Euler constant

$$\gamma = \lim_{n \to \infty} \left(\sum_{k=1}^{n} \frac{1}{k} - \log n \right) \approx 0.5772.$$

In fact, in [38] the authors were able to find the exact expression for the numbers k_m depending on m, e, and γ.

Theorem 3.32 *If $m \geq 3$ is a natural number and $e^{m-\gamma} = M + \delta$, where $M \in \mathbb{N}$ and $0 < \delta < 1$, then*

$$k_m = \begin{cases} M, & \textit{if } \delta < \frac{1}{2} - \frac{1}{10n} \quad \textit{for some } n \in \mathbb{N}, \\ M+1, & \textit{if } \delta > \frac{1}{2} + \frac{1}{n} \quad \textit{for some } n \in \mathbb{N}. \end{cases}$$

In other words, one has $k_m = \lfloor e^{m-\gamma} + \frac{1}{2} \rfloor$ unless $e^{m-\gamma} + \frac{1}{2}$ is close to a natural number.

5. In relation to the numbers k_m, in the paper [212] a similar problem is studied for the divergent p-series ($p < 1$). In particular, the author gives closed formulae for the integer parts of partial sums of the divergent p-series.

6. Theorem 3.32 is an improvement of an earlier result due to Comtet [64]. Comtet's result was further refined by DeTemple and Wang [76], and later by Zemyan [283], among others. We refer to the more recent paper by Bil and Laue [36] and the references therein for a full account of the history of determining k_m.

7. The sequence $\pi_a(n)$ defined in Subsection 3.1.1, which gives rise to the "infinitely iterated" harmonic series

$$\sum_{n=1}^{\infty} \frac{1}{n \cdot \pi_a(n)}, \tag{3.32}$$

was introduced by B. Christianson in [59], in the same paper where Theorem 3.5 was stated. The author proved such a result only for the case $a = 2$ (i.e., only a subcase of the convergence part), without the need of passing to integrals, as mentioned in (i) of Remark 3.6. In addition to proving such a result, the author also develops a relatively simple method for approximating the value of series (3.32) (for $a = 2$) with any desired degree of accuracy.

8. In Subsection 3.1.2, it is mentioned that the sum of the terms of the harmonic series that do not contain a "9" in their decimal representation is slightly less than 23. A method for computing this sum was developed by R. Baillie in 1979 [25], and the same author, together with T. Schmelzer, developed a high-accuracy method for summing harmonic series not containing the terms whose indices belong to certain subsets of the natural numbers [233]. For example, it is shown that if

$$\mathcal{N} = \{n \in \mathbb{N} : n \text{ has no even numbers in its decimal representation}\},$$

then

$$\sum_{n \in \mathcal{N}} \frac{1}{n} \approx 3.17$$

(in fact the authors obtained several additional decimal digits of accuracy), and if

$$\mathcal{M} = \{m \in \mathbb{N} : m \text{ does not contain the string 314 in its decimal representation}\},$$

then

$$\sum_{m \in \mathcal{M}} \frac{1}{m} \approx 2299.83$$

(again the authors obtained in [233] the approximation of this sum with much higher accuracy).

9. Another surprising fact concerning the harmonic series with deleted terms is related to its sum where only the digits of the indices are in a sense restricted [12]. More precisely, for an integer $b \geq 2$, we define $s_b(n)$ as the sum of the digits of n in base b.

Theorem 3.33 *For any integer $b \geq 2$, there holds*

$$\lim_{k \to \infty} \sum_{\substack{n=1 \\ s_b(n)=k}}^{\infty} \frac{1}{n} = \frac{2 \log b}{b-1}.$$

For a proof and further statements in this context, see [12]. See also the related (earlier) paper [102].

10. Boas [B-47, Ch. 3] also investigated related problems of harmonic series with deleted terms in 1979, in parallel to Baillie's work [25]. In particular, he obtained the following refinement of Theorem 3.7 for the particular choice of the digit $X =$ "0", which consists not on removing terms containing prescribed digits but rather on multiplying them by some weight.

Theorem 3.34 *Let $\{c_k\}$ be a nonnegative sequence. Define $\{\varepsilon_n\}$ as*

$$\varepsilon_n = \begin{cases} 1, & \text{if } n \text{ does not contain the digit "0" in base 10}, \\ c_k, & \text{if } n \text{ contains } k \text{ times the digit "0" in base 10}, \ k > 0. \end{cases}$$

Then, the series $\sum \frac{\varepsilon_n}{n}$ converges if and only if $\sum c_k$ converges.

Note that the harmonic series corresponds to $c_k \equiv 1$, while the harmonic series without all terms with "0" in their representation corresponds to $c_k \equiv 0$.

See also [270, 271] for related results on sums of the harmonic series with deleted terms.

11. The author of [110] developed an alternative approach for proving Theorem 3.8.

12. In [216] the reader may find further results on general series with deleted terms, as well as some results in relation to harmonic series.

13. Bonar and Khoury [B-11] showed that every positive real number is a selective sum of the harmonic series (in relation with this fact, see Theorem 3.14). In that same book, Theorem 3.9 is stated and proved, which we reproduce here. We note that the finite sums in Theorem 3.9 reproducing a positive rational number r are often referred to as *Egyptian fractions*; as such a finite sum of fractions with numerator equal to 1 was the notation for rational numbers used in ancient Egypt (for instance, $1/2+1/3+1/15$ is an Egyptian fraction).

14. A rigorous proof of the equivalence that Euler found (with lack of required rigor, perhaps heuristically in the view of modern mathematics), namely

$$\sum_{\substack{p \text{ prime} \\ p \leq n}} \frac{1}{p} \asymp \log \log n,$$

was actually proved by Franz Mertens in 1874; see [185].

15. The reader may find various proofs of the prime number theorem (Theorem 3.10) by different methods in [70, 111, 203]. More precisely, it was proved independently by Hadamard in [111], and by de la Valleé Poussin in a series of papers [70]. All those publications appeared in the same year (1896) and are based on the study of the Riemann zeta function $\zeta(s)$. In this respect, let us also mention the works of P. Chebyshev [B-50]. Although Hadamard and de la Valleé Poussin proved the result in very different ways, the starting point for both of their approaches was establishing that $\zeta(s)$ does not vanish on the line $\operatorname{Re} s = 1$. For a recent study of connections between the theory of number series and the Riemann zeta function, see [197]. In 1980, D. J. Newman published a simple proof of the prime number theorem based on analytic arguments [203].

Section 3.2 (A test resembling l'Hôpital's rule)

16. The idea and main results on which Section 3.2 is based appeared in [21].

17. Rolle's theorem (used in the proof of Theorem 3.11) reads as follows.

Theorem 3.35 (Rolle's theorem) *Let $f \colon [a,b] \to \mathbb{R}$ be continuously differentiable on (a,b) and such that $f(a) = f(b)$. Then there exists $c \in (a,b)$ such that $f'(c) = 0$.*

Section 3.3 (Selective sums of positive convergent series)

18. For the definition of selective sums, see [B-11, pp. 79–81].

19. The main results concerning selective sums of convergent positive series in Section 3.3, as well as the examples we showed, appeared in [105].

20. As pointed out earlier, the set of selective sums obtained in Theorem 3.17 has a structure that resembles that of the classical Cantor set, although it does not need to have measure zero.

21. It is also interesting to note that one may have an intermediate situation, where condition (3.19) does not hold for finitely many $n \in \mathbb{N}$, as for the series $\sum \frac{1}{n^p}$ with $p > 1$ ((3.19) is true only if $n > N = N(p)$). In this case one has a mixed structure of the set of selective sums, namely the union of 2^n intervals of the form

$$\left[\sum_{k=1}^{N} c_k a_k, \sum_{k=1}^{N} c_k a_k + \sum_{k=N+1}^{\infty} a_k\right].$$

This was proved in [143]. For several other examples and interesting remarks around Theorems 3.16 and 3.17, we refer to [105].

22. The problem of determining all the possible sums of (3.15) appeared in [B-67, Part 1, Ch. III, Problem 130].

Section 3.4 ("Change of variables" for sums)

23. Theorem 3.18 was proved in [46] for monotone sequences. Theorem 3.19 was also proved in [46]. The author used those results in order to study the smoothness of functions of several variables in Besov spaces when restricted to a smaller number of variables.

Section 3.5 (When does a transformed series inherit the convergence?)

24. Theorem 3.22 was obtained in [182], as an attempt to derive a comparison test valid for series with real-valued terms, rather than nonnegative terms as in the classical comparison test (Theorem 2.18). Hardy [B-42] had already concluded that no general comparison test was possible for conditionally convergent series. In fact, a simple generalization of the classical comparison test for series with real-valued terms (which we state as Theorem 2.19) is also mentioned as a starting point in [182], and although it is a valid test for many types of series for which other classical tests do not apply, it is not as powerful as other classical tests may be.

This was an impact for the idea to split the general term a_n in Theorem 3.22 in a sum of simpler quantities for which the comparison test can be applied. In fact, as its proof illustrates, it is based on comparing the terms of the series $\sum f(a_n)$ with those of the series $\sum a_n$ and $\sum (a_n)^{2k}$. For instance, Theorem 3.22 may be used to determine the convergence or divergence of the series $\sum b_n$, where

$$b_n = \log\left(1 + \frac{(-1)^n}{n^\gamma}\right), \qquad \gamma > 0.$$

The first attempt, based on classical tests, would be using Leibniz's test (Theorem 2.66). However, it is useful only to determine the convergence of the given series for $\gamma \geq 1$, since otherwise $|b_n|$ is not decreasing (a required assumption in Leibniz's test; this can easily be seen using the Taylor expansion $\log(1+x) = x - \frac{x^2}{2} + o(x^2)$). Nevertheless, applying Theorem 3.22, one can show that the series $\sum b_n$ converges if and only if $\gamma \geq \frac{1}{2}$.

Section 3.6 (An integral-ratio test)

25. Theorem 3.24 was obtained by Brink in [50]. Some years before, Brink had obtained in [49] a simpler version of that theorem, which stated that under the assumptions of Theorem 3.24, the series $\sum a_k$ converges if and only if

$$\int_m^\infty e^{\int_m^x \log r(t)\,dt}\,dx < \infty.$$

Using Theorem 3.24, one can see that this claim follows from the inequalities

$$1 - \frac{1}{r(t)} \leq \log r(t) \leq r(t) - 1, \qquad r(t) > 0,$$

and so Theorem 3.24 is a refinement of the original result obtained in [49]. A further generalization of Theorem 3.24 is given in the paper [226].

Section 3.7 (A divergence test for positive series)

26. The original version of Theorem 3.26 first appeared in [219] by Post, and reads as follows.

Theorem 3.36 *Let* $f\colon \mathbb{N} \to \mathbb{N}$ *be a function satisfying*

$$f(n+1) - f(n) \geq n+1, \qquad n \in \mathbb{N}, \tag{3.33}$$

and let $\{a_n\}$ *be a positive sequence satisfying*

$$a_n \leq a_{n+1} + a_{f(n)}, \qquad n \in \mathbb{N}. \tag{3.34}$$

Then the series $\sum a_n$ *diverges.*

For the proof of Theorem 3.36, Post used a combinatorial argument that cannot be used to prove the generalizations of this result presented in Section 3.7. The problem arises at the indices n such that $\lambda_n < 1$ in Theorem 3.26 (which need not exist if $\lambda_n \equiv 1$, but indeed may exist in a more general situation).

27. Other results toward Theorem 3.36 were obtained before it, typically for specific choices of f. In 1964, Graham and Jewett proved that the conclusion of Theorem 3.36 holds for $f(n) = n(n+1)$, and also roughly proved that the result is valid for any function f increasing "as fast as" $\frac{n^2}{2}$ using an apparently rather complicated method (which caused this work not to be published). Later Erdős pointed out that if $f(n) \leq cn^2$ with $0 < c < \frac{1}{2}$, then, in general, condition (3.34) alone does not imply the divergence of $\sum a_n$ (see Example 3.29, (iii)). In 1965, Newman obtained the conclusion for $f(n) = n^2$ (all these facts are mentioned in [219] without reference).

28. It is also noted in [219] that if condition (3.33) is replaced by the slightly weaker

$$f(n+1) - f(n) \geq n - C, \qquad n \geq N,$$

where N and $C > 0$ are fixed, the conclusion is still true.

29. Theorems 3.26 and 3.30 were obtained in [44] by D. Borwein and Meir. In that same paper, several further observations are made, and questions of interest around these results are posed.

Part II

Double Series

This part contains an extension of the classical theory of single number series to dimension two (Chapters 4 and 5), as well as its development by means of generalizations of the monotonicity-type conditions from Chapter 1 (this is done in Chapter 6).

Although within the scope of the book this part has merely an expository character, being nothing more but a particular case of the general theory given in Part III, it is still convenient to present it explicitly, for a number of reasons. First, two-dimensional series, being a "basic" example of multiple series, already illustrate some of the essential differences from the one-dimensional case. Second, this part supplements Part III in a way that allows an easier reading of the latter, requiring the assimilation of several new notations. Being tedious in arbitrary dimension, these notations often have simple and transparent counterparts in dimension two.

4

Rudiments of double sequences and series

For the sake of completeness, in this chapter we present some of the existing two-dimensional results on number series. In fact, most of the known results have been given in literature just in dimension two. In general, we follow [B-21, Ch. 2], since it is the most complete among the sources on the subject. One more reason is that just this source is much less accessible. At certain points, we supplement the presentation with results from other sources, or even with new results. The main point is that certain facts on double sequences are needed only for the study of the behavior of the partial sums of double series. For the sake of convenience, these partial sums and sequences are mostly denoted in a similar manner.

4.1 Double sequences

In general, the results on convergence or divergence of double series follow from the results on convergence or divergence of double sequences (of their partial sums). To adjust these results to the following study of double series, we will denote double sequences by $\{a_{k\ell}\}_{k,\ell\ge 1}$. It can be written as the infinite matrix

$$\begin{matrix} a_{11} & a_{12} & \cdots & a_{1\ell} & \cdots \\ a_{21} & a_{22} & \cdots & a_{2\ell} & \cdots \\ \vdots & \vdots & \ddots & \vdots & \ddots \\ a_{k1} & a_{k2} & \cdots & a_{k\ell} & \cdots \\ \vdots & \vdots & \ddots & \vdots & \ddots \end{matrix}$$

We will usually deal with sequences whose indices begin at 1, that is, by $\{a_{k\ell}\}$ we mean just $\{a_{k\ell}\}_{k,\ell\ge 1}$, unless otherwise stated. We may also write commas to separate indices, for instance, $a_{k,\ell+1}$ instead of $a_{k\ell+1}$, to improve readability. Abusing of language, we may often refer to the terms $a_{k\ell}$ of the sequence $\{a_{k\ell}\}$ just as a "sequence", always keeping in mind that by this we mean "the sequence whose general term is $a_{k\ell}$".

Definition 4.1 Let $\{a_{k\ell}\} \subset \mathbb{C}$ be a double sequence. We say that $A \in \mathbb{C}$ is the limit of $\{a_{k\ell}\}$ if for each $\varepsilon > 0$ there exists a natural number N such that

$$|a_{k\ell} - A| < \varepsilon \quad \text{provided } k, \ell > N,$$

written as

$$\lim_{k,\ell\to\infty} a_{k\ell} = A.$$

Equivalently, it is said that the given double sequence converges to the element A. This type of convergence is called Pringsheim convergence. Just this type of convergence, which is not the only possible one, will play an important role when studying the convergence of double series.

Definition 4.2 A real-valued double sequence $\{a_{k\ell}\}$ is divergent to $+\infty$ if for every $A > 0$ there exists a natural number N such that

$$a_{k\ell} > A \quad \text{provided } k, \ell > N,$$

written as

$$\lim_{k,\ell\to\infty} a_{k\ell} = +\infty.$$

Analogously, a real-valued double sequence $\{a_{k\ell}\}$ is divergent to $-\infty$ if for every $A > 0$ there exists a natural number N such that

$$a_{k\ell} < -A \quad \text{provided } k, \ell > N,$$

and we write

$$\lim_{k,\ell\to\infty} a_{k\ell} = -\infty.$$

Double sequences can be bounded or unbounded; there are also intermediate situations. Let us proceed to the precise definitions.

Definition 4.3 We say that a double sequence $\{a_{k\ell}\} \subset \mathbb{C}$ is bounded if there is a number $B > 0$ such that $|a_{k\ell}| \leq B$ for all $k, \ell \geq 1$.

Definition 4.4 A double sequence with real-valued terms $\{a_{k\ell}\}$ is bounded from above if there is a real number B such that $a_{k\ell} \leq B$ for all $k, \ell \geq 1$. Analogously, a double sequence with real-valued terms $\{a_{k\ell}\}$ is bounded from below if there is a real number b such that $a_{k\ell} \geq b$ for all $k, \ell \geq 1$.

Obviously, if a double sequence with real-valued terms is bounded both from above and from below, then it is bounded, and if a double sequence with real-valued terms is bounded, then it is bounded both from above and from below.

Definition 4.5 A double sequence with real-valued terms $\{a_{k\ell}\}$ is quasi-bounded from above (below) if there is a natural number N such that the double sequence $\{a_{k\ell}\}_{k,\ell\geq N}$ is bounded from above (below).

Definition 4.6 A double sequence with real-valued terms $\{a_{k\ell}\}$ is quasi-bounded if there is a natural number N such that the double sequence $\{a_{k\ell}\}_{k,\ell\geq N}$ is bounded.

Remark 4.7 Contrary to the property of every convergent one-dimensional sequence to be bounded, a convergent double sequence is, generally speaking, not bounded. Indeed, let

$$a_{k\ell} = \begin{cases} \ell, & \text{if } k = 1 \text{ and } \ell = 1, 2, \dots, \\ 0, & \text{if } k > 1 \text{ and } \ell = 1, 2, \dots \end{cases}$$

It is evident that this double sequence converges to zero but is not bounded.

Further, it is easy to observe that every convergent double sequence is quasi-bounded.

Like in the one-dimensional case, an important class of double sequences is that of monotone sequences. We present only one possible generalization of the notion of monotonicity, however, maybe the most natural one.

Definition 4.8 A double sequence $\{a_{k\ell}\}$ with real-valued terms is monotone decreasing (respectively increasing) if

$$a_{k\ell} \leq a_{mn} \quad (\text{respectively } a_{k\ell} \geq a_{mn}) \quad \text{for} \quad k \geq m,\ \ell \geq n.$$

We will call such a sequence just decreasing (respectively increasing). We may also refer to monotone decreasing sequences just as “monotone”.

We note that this definition is simply the coordinate-wise monotonicity. By introducing the notation for two-dimensional differences

$$\Delta^{(1,0)} a_{k\ell} := a_{k\ell} - a_{k+1,\ell},$$
$$\Delta^{(0,1)} a_{k\ell} := a_{k\ell} - a_{k,\ell+1},$$

we observe that $\{a_{k\ell}\}$ is decreasing (respectively increasing) if

$$\Delta^{(1,0)} a_{k\ell} \geq 0 \text{ and } \Delta^{(0,1)} a_{k\ell} \geq 0$$
$$(\text{respectively } \Delta^{(1,0)} a_{k\ell} \leq 0 \text{ and } \Delta^{(0,1)} a_{k\ell} \leq 0),$$

for all k, ℓ.

One may consider other definitions of monotonicity, as the so-called Hardy monotone sequences. To this end, we introduce the iterated differences with respect to both indices,

$$\Delta^{(1,1)} a_{k\ell} := \Delta^{(0,1)}(\Delta^{(1,0)} a_{k\ell}) = \Delta^{(1,0)}(\Delta^{(0,1)} a_{k\ell})$$
$$= a_{k\ell} - a_{k+1,\ell} - a_{k,\ell+1} + a_{k+1,\ell+1}.$$

Definition 4.9 A double sequence $\{a_{k\ell}\}$ with real-valued terms is Hardy monotone if $\Delta^{(1,1)} a_{k\ell} \geq 0$ for all $k, \ell \geq 1$.

Hardy monotone sequences play an important role in harmonic analysis, as the differences $\Delta^{(1,1)}a_{k\ell}$ appear naturally when applying Abel's transformation once in each variable, which reads as

$$\begin{aligned}\sum_{k=m_1}^{n_1}\sum_{\ell=m_2}^{n_2} a_{k\ell}b_{k\ell} &= \sum_{k=m_1}^{n_1-1}\sum_{\ell=m_2}^{n_2-1} \Delta^{(1,1)}a_{k\ell}\sum_{\mu=m_1}^{k}\sum_{\nu=m_2}^{\ell} b_{\mu\nu}\\ &+ \sum_{k=m_1}^{n_1-1} \Delta^{(1,0)}a_{k,n_2}\sum_{\mu=m_1}^{k}\sum_{\nu=m_2}^{n_2} b_{\mu\nu}\\ &+ \sum_{\ell=m_2}^{n_2-1} \Delta^{(0,1)}a_{n_1,\ell}\sum_{\mu=m_1}^{n_1}\sum_{\nu=m_2}^{\ell} b_{\mu\nu} + a_{n_1,n_2}\sum_{\mu=m_1}^{n_1}\sum_{\nu=m_2}^{n_2} b_{\mu\nu}.\end{aligned}$$

Note that neither (decreasing or increasing) monotonicity nor Hardy monotonicity imply each other. Indeed, let us consider the sequence with general term

$$a_{k\ell} = \frac{1}{\min\{k,\ell\}},$$

which is clearly monotone decreasing. However, for any $k \geq 1$, we have $\Delta^{(1,1)}a_{kk} = \frac{1}{k+1} - \frac{1}{k} < 0$; as for the converse, the sequence $a_{k\ell} = k + \ell$ is not monotone decreasing, although for any $k, \ell \geq 1$ one has $\Delta^{(1,1)}a_{k\ell} = 0$.

On the other hand, the sequence given by

$$a_{k\ell} = \max\{k,\ell\}$$

is clearly monotone increasing, but $\Delta^{(1,1)}a_{kk} = -1 < 0$. The sequence given by

$$a_{k\ell} = \begin{cases} 2, & \text{if } k = \ell = 1,\\ 1, & \text{if } \min\{k,\ell\} = 1,\ \max\{k,\ell\} > 1,\\ 0, & \text{otherwise,}\end{cases}$$

is not monotone increasing, but $\Delta^{(1,1)}a_{k\ell} = 0$ for every $k, \ell \in \mathbb{N}$, and thus it is Hardy monotone.

In Chapter 6 we consider generalizations of monotone and Hardy monotone double sequences, namely general monotonicity (Definition 6.11). Of course, their corresponding multidimensional generalizations will also be considered in Chapter 9.

Theorem 4.10 *If a double sequence $\{a_{k\ell}\}$ is increasing (decreasing), then for it to be convergent to a finite limit it is necessary and sufficient that it be bounded from above (below).*

Theorem 4.10 follows from Theorem 7.12 where the higher dimensional case is considered.

Remark 4.11 If an increasing double sequence $\{a_{k\ell}\}$ is not bounded from above, then

$$\lim_{k,\ell\to\infty} a_{k\ell} = +\infty.$$

If a decreasing double sequence $\{a_{k\ell}\}$ is not bounded from below, then

$$\lim_{k,\ell\to\infty} a_{k\ell} = -\infty.$$

For double sequences, the Cauchy criterion holds true in the following form.

Theorem 4.12 *If a double sequence $\{a_{k\ell}\}$ converges, then for any $\varepsilon > 0$, there exists a natural number N such that for $k, \ell \geq N$ and $\mu, \nu \geq 0$, the inequality*

$$|a_{k+\mu,\ell+\nu} - a_{k\ell}| < \varepsilon$$

holds. Conversely, if this condition holds, then the double sequence $\{a_{k\ell}\}$ converges.

For a multidimensional version of Theorem 4.12 and its proof, we refer to Theorem 7.14 below.

Apart from double limits, an iteration of the limits in each index separately is frequently considered. Such limits are called iterated limits and can be treated as follows. Let

$$\lim_{\ell\to\infty} a_{k\ell} = A_k$$

exist for every k. The question of the existence of the limit of the sequence $\{A_k\}_{k\geq 1}$ appears:

$$\lim_{k\to\infty} A_k = \lim_{k\to\infty}\Big(\lim_{\ell\to\infty} a_{k\ell}\Big).$$

If this limit exists, it is one of the iterated limits. The other one can be obtained if the limits are taken in the reverse order:

$$\lim_{\ell\to\infty}\Big(\lim_{k\to\infty} a_{k\ell}\Big).$$

We remark that the iterated limits should not be equal, generally speaking. Indeed, let

$$a_{k\ell} = \frac{k-\ell}{k+\ell}, \qquad k,\ell = 1,2,\ldots \tag{4.1}$$

Obviously,

$$\lim_{\ell\to\infty} a_{k\ell} = -1, \qquad \lim_{k\to\infty} a_{k\ell} = 1,$$

and consequently

$$\lim_{k\to\infty}\Big(\lim_{\ell\to\infty} a_{k\ell}\Big) = -1, \qquad \lim_{\ell\to\infty}\Big(\lim_{k\to\infty} a_{k\ell}\Big) = 1.$$

Thus, both iterated limits do exist but do not coincide.

The following statement gives sufficient conditions for both iterated limits to be equal.

Theorem 4.13 *Let a double sequence $\{a_{k\ell}\}$ converge to $A \in \mathbb{C}$. If for every k the limit $\lim_{\ell\to\infty} a_{k\ell} = A_k$ exists, then the iterated limit*

$$\lim_{k\to\infty}\left(\lim_{\ell\to\infty} a_{k\ell}\right)$$

exists and is equal to A. Correspondingly, if $\lim_{k\to\infty} a_{k\ell} = B_\ell$ exists for every ℓ, then the iterated limit

$$\lim_{\ell\to\infty}\left(\lim_{k\to\infty} a_{k\ell}\right)$$

exists and is equal to A.

The proof of Theorem 4.13 is carried out in arbitrary dimension in Theorem 7.16. Here we just observe that the double limit does not exist in (4.1). This can be proved by means of homogeneity as will be for (4.2) below.

We remark that the existence of the limit of a double sequence does not imply, generally speaking, the existence of the iterated limits. Indeed, let

$$a_{k\ell} = \frac{1}{k}\cos \ell\pi + \frac{1}{\ell}\cos k\pi, \qquad k,\ell = 1,2,\ldots$$

Since

$$|a_{k\ell}| \le \frac{1}{k} + \frac{1}{\ell},$$

we obtain

$$\lim_{k,\ell\to\infty} a_{k\ell} = 0.$$

However, $\lim_{k\to\infty}\left(\lim_{\ell\to\infty} a_{k\ell}\right)$ and $\lim_{\ell\to\infty}\left(\lim_{k\to\infty} a_{k\ell}\right)$ do not exist (in fact, not even the inner limits exist).

We also note that, in general, the existence of both iterated limits and their equality does not yield the existence of the double limit. Indeed, let

$$a_{k\ell} = \frac{k\ell}{k^2+\ell^2}, \qquad k,\ell = 1,2,\ldots \tag{4.2}$$

It is clear that

$$\lim_{k\to\infty}\left(\lim_{\ell\to\infty} a_{k\ell}\right) = \lim_{\ell\to\infty}\left(\lim_{k\to\infty} a_{k\ell}\right) = 0.$$

However, $\lim_{k,\ell\to\infty} a_{k\ell}$ does not exist. For this, it suffices to consider the limits for various subsequences of the type $k = \alpha\ell$, with $\alpha \in \mathbb{N}$. Each leads to the limit $\frac{\alpha}{1+\alpha^2}$, which means that the double limit does not exist.

Theorem 4.14 *If a double sequence $\{a_{k\ell}\}$ is increasing (respectively decreasing), then all the three limits $\lim_{k\to\infty}\left(\lim_{\ell\to\infty} a_{k\ell}\right)$, $\lim_{\ell\to\infty}\left(\lim_{k\to\infty} a_{k\ell}\right)$, and $\lim_{k,\ell\to\infty} a_{k\ell}$ exist and are equal to one another, even if they are $+\infty$ (respectively $-\infty$).*

For a proof in several dimensions, see Theorem 7.18.

4.2 Convergence of double series: definitions and basic notions

Given a double sequence of complex numbers $\{a_{k\ell}\}$, we define by S_{KL} the (K,L)th partial sum of $\{a_{k\ell}\}$:

$$S_{KL} = \sum_{k=1}^{K}\sum_{\ell=1}^{L} a_{k\ell}.$$

The double series with general term $a_{k\ell}$,

$$\sum_{k,\ell} a_{k\ell} = \sum_{k,\ell=1}^{\infty} a_{k\ell}, \tag{4.3}$$

is defined, if no otherwise is explicitly meant, as

$$\lim_{K,L\to\infty} S_{KL},$$

where the limit may be finite, infinite, or it may not exist. Such a type of convergence (if it takes place) is traditionally called *Pringsheim's convergence*, or convergence in Pringsheim's sense. The names *rectangular convergence* or *unrestricted convergence* are also widespread. As we will see in the following section, there are other types of convergence.

Note that in (4.3) we avoid writing the double series as an iteration of one-dimensional series. The reason is that such iterations not always yield the same limits, as we will show later in this chapter.

Sums with general term $a_{k\ell}$, where the lower and upper limits of summation are different from 1 and ∞, respectively, may also be used. For instance, whenever the lower and upper limits of summation are, say, M and N respectively, we may just write the sum as $\sum_{k,\ell=M}^{N} a_{k\ell}$ for the sake of simplicity. For infinite sums with different starting indices, say, K and L respectively, we will write

$$\sum_{\substack{k=K\\ \ell=L}}^{\infty} a_{k\ell}.$$

A double series $\sum_{k,\ell} a_{k\ell}$ is said to be convergent to $A \in \mathbb{C}$ in the sense of Pringsheim (or simply convergent) if

$$\lim_{K,L\to\infty} S_{KL} = A.$$

In this case A is called the sum of the double series $\sum_{k,\ell} a_{k\ell}$, and we write

$$A = \sum_{k,\ell} a_{k\ell}.$$

If a double series does not converge in the sense of Pringsheim, then we call it a *divergent* double series. However, it may be convergent in a different sense.

A standard necessary condition for the convergence of (4.3) is

$$\lim_{K,L\to\infty} a_{KL} = 0.$$

Indeed, this follows from

$$a_{KL} = S_{KL} - S_{K,L-1} - S_{K-1,L} + S_{K-1,L-1}$$

and the definition of convergence of the series. It is important to mention that the terms $a_{k\ell}$ of a convergent series may not be bounded, as shown by

$$a_{k\ell} = \begin{cases} (-1)^{\ell} k, & \text{if } \ell \in \{1,2\} \text{ and } k \geq 3, \\ (-1)^{k} \ell, & \text{if } k \in \{1,2\} \text{ and } \ell \geq 3, \\ 0, & \text{otherwise.} \end{cases} \tag{4.4}$$

It is clear that the partial sums S_{KL} converge to 0 although its terms are unbounded.

It is easy to prove the following assertions.

1. *Linearity.* Given two double series with terms $\{a_{k\ell}\}$ and $\{b_{k\ell}\}$ that converge to A and B, respectively, then the double series with terms $\{a_{k\ell}+b_{k\ell}\}$ is also convergent to $A+B$. For $\lambda \in \mathbb{C}$, the double series with terms $\{\lambda a_{k\ell}\}$ converges to λA.
2. If $\{a_{k\ell}\}$ and $\{a'_{k\ell}\}$ are two double series such that $a_{k\ell} = a'_{k\ell}$ for all k and ℓ except a finite number of pairs (k,ℓ), then either both of them converge or both diverge.

Indeed, the first assertion immediately follows from the definitions, while for the second one we have that the double series with terms $\{a_{k\ell} - a'_{k\ell}\}$ converges, since its terms all are zeros except a finite number of them (it can also be proved using Cauchy's criterion, see Theorem 4.12).

Let us now introduce the notion of iterated series. To this end, we consider the infinite matrix

$$\begin{matrix} a_{11} & a_{12} & \cdots & a_{1\ell} & \cdots \\ a_{21} & a_{22} & \cdots & a_{2\ell} & \cdots \\ \vdots & \vdots & \ddots & \vdots & \ddots \\ a_{k1} & a_{k2} & \cdots & a_{k\ell} & \cdots \\ \vdots & \vdots & \ddots & \vdots & \ddots \end{matrix}$$

Summing every row of this matrix separately, we obtain the infinite sequence of series

$$\sum_{\ell=1}^{\infty} a_{k\ell}, \qquad k = 1, 2, \ldots$$

Summing this sequence, we get

$$\sum_{k=1}^{\infty} \sum_{\ell=1}^{\infty} a_{k\ell}. \tag{4.5}$$

This symbol is called the *iterated series*, which is said to converge if for every K there exists $\lim_{L\to\infty} S_{KL}$, and besides that the limit

$$\lim_{K\to\infty}\left(\lim_{L\to\infty} S_{KL}\right)$$

exists. Such an iterated limit is called the sum of the iterated series (4.5). Similarly, summing first the columns rather than the rows, we can construct the second possible iterated series

$$\sum_{\ell=1}^{\infty}\sum_{k=1}^{\infty} a_{k\ell}, \tag{4.6}$$

whose sum is defined in a similar way.

A double series $\sum_{k,\ell} a_{k\ell}$ is convergent in rows, if for each $K \geq 1$ the limit $\lim_{L\to\infty} S_{KL}$ exists. Similarly, the double series $\sum_{k,\ell} a_{k\ell}$ is convergent in columns, if for each $L \geq 1$ the limit $\lim_{K\to\infty} S_{KL}$ exists.

Theorem 4.15 *Let a double series $\sum_{k,\ell} a_{k\ell}$ converge to $A \in \mathbb{C}$.*

If the series

$$\sum_{k=1}^{\infty} a_{k\ell}$$

converges for every ℓ (converges in columns), then the iterated series

$$\sum_{\ell=1}^{\infty}\sum_{k=1}^{\infty} a_{k\ell}$$

converges to A.

Correspondingly, if the series

$$\sum_{\ell=1}^{\infty} a_{k\ell}$$

converges for every k (converges in rows), then the iterated series

$$\sum_{k=1}^{\infty}\sum_{\ell=1}^{\infty} a_{k\ell}$$

converges to A.

This result follows immediately from Theorem 4.13.

Remark 4.16 Generally speaking, the convergence of a double series $\sum_{k,\ell} a_{k\ell}$ does not imply its convergence in rows or in columns, and the convergence of the iterated series (4.5) and (4.6) does not imply the convergence of the double series $\sum_{k,\ell} a_{k\ell}$.

Indeed, the double series with general term (4.4) converges to 0, although the row and column series $\sum_{\ell=1}^{\infty} a_{1\ell}$ and $\sum_{k=1}^{\infty} a_{k1}$ diverge.

On the other hand, the double series with general term

$$a_{k\ell} = \begin{cases} -1, & \text{if } k = \ell \geq 1, \\ 1/2, & \text{if } \ell = k+1 \text{ or } \ell = k-1, \\ 0, & \text{otherwise}, \end{cases} \tag{4.7}$$

which when represented as an infinite matrix looks like

$$\begin{matrix} -1 & 1/2 & 0 & 0 & 0 & \cdots \\ 1/2 & -1 & 1/2 & 0 & 0 & \cdots \\ 0 & 1/2 & -1 & 1/2 & 0 & \cdots \\ 0 & 0 & 1/2 & -1 & 1/2 & \cdots \\ 0 & 0 & 0 & 1/2 & -1 & \cdots \\ \vdots & \vdots & \vdots & \vdots & \vdots & \ddots \end{matrix}$$

does not converge (since $\lim_{k\to\infty} a_{kk} = -1 \neq 0$), although all the row and column series converge. Note that in this case the sequence of partial sums does not have a limit at infinity, since

$$S_{KK} = \sum_{k=1}^{K}\sum_{\ell=1}^{K} a_{k\ell} = -1, \qquad S_{K,K+1} = -\frac{1}{2}.$$

Finally, consider the series with general term

$$a_{k\ell} = \begin{cases} 1, & \text{if } k = \ell = 1, \\ 0, & \text{if } \ell = 1,\ k \geq 2, \\ -2^{-\ell+1}, & \text{if } 1 \leq k \leq 2^{\ell-2},\ \ell \geq 2, \\ 2^{-\ell+1}, & \text{if } 2^{\ell-2} < k \leq 2^{\ell-1},\ \ell \geq 2, \\ 0, & \text{otherwise}. \end{cases} \tag{4.8}$$

The terms of such a sequence are, in matrix form,

$$\begin{matrix} 1 & -1/2 & -1/4 & -1/8 & \cdots \\ 0 & 1/2 & -1/4 & -1/8 & \cdots \\ 0 & 0 & 1/4 & -1/8 & \cdots \\ 0 & 0 & 1/4 & -1/8 & \cdots \\ 0 & 0 & 0 & 1/8 & \cdots \\ \vdots & \vdots & \vdots & \vdots & \ddots \end{matrix}$$

The corresponding series does not converge, despite having all column and row series converging to 0 (except the column series with $\ell = 1$, which converges to 1), since

$$\sum_{k=1}^{\infty}\sum_{\ell=1}^{\infty} a_{k\ell} = 0 \neq 1 = \sum_{\ell=1}^{\infty}\sum_{k=1}^{\infty} a_{k\ell}.$$

We must now define a more restrictive type of convergence of double series, namely *regular convergence*. In terms of this concept, double series possess certain desirable properties not satisfied in general for Pringsheim convergence.

Definition 4.17 A double series $\sum_{k,\ell} a_{k\ell}$ with complex-valued terms is regularly convergent (or converges in the regular sense) if for each $\varepsilon > 0$ there exists a natural number N such that

$$\left| \sum_{k=K_0}^{K} \sum_{\ell=L_0}^{L} a_{k\ell} \right| < \varepsilon,$$

provided $K \geq K_0$, $L \geq L_0$, and $K_0 + L_0 > N$, or, equivalently, there exists a natural number N' such that

$$\left| \sum_{k=K_0}^{K} \sum_{\ell=L_0}^{L} a_{k\ell} \right| < \varepsilon,$$

provided $K \geq K_0$, $L \geq L_0$, and $\max\{K_0, L_0\} > N'$.

We now give some basic results concerning regular convergence.

Remark 4.18 (i) In dimension one, convergence and regular convergence of series reduce to the same phenomenon.

(ii) If a double series converges in the regular sense, then it converges in the sense of Pringsheim (i.e., its partial sums form a convergent double sequence), which readily follows from the definitions. So, if A is the sum of such a series, we say that the series converges in the regular sense to A.

(iii) In particular, the terms $a_{k\ell}$ of a regularly convergent series vanish as $k+\ell \to \infty$.

(iv) Convergence of a double series in the sense of Pringsheim does not imply regular convergence (see, for instance, Remark 4.16, or also (4.4)).

Regular convergence can be characterized as follows.

Theorem 4.19 *A double series $\sum_{k,\ell} a_{k\ell}$ converges in the regular sense to A if and only if the series $\sum_{\ell=1}^{\infty} a_{k\ell}$ and $\sum_{k=1}^{\infty} a_{k\ell}$ converge for all $k, \ell \geq 1$, respectively, and the double series $\sum_{k,\ell} a_{k\ell}$ converges to A.*

Proof Assume that $\sum_{k,\ell} a_{k\ell}$ converges to A in the regular sense. The result in one direction then readily follows from the definition of regular convergence (see also Remark 4.18).

For the converse, the convergence of the series $\sum_{k,\ell} a_{k\ell}$ implies, by the Cauchy criterion (Theorem 4.12), that given $\varepsilon > 0$ we can find $N \geq 0$ such that

$$\left| \sum_{k=K_0}^{K} \sum_{\ell=L_0}^{L} a_{k\ell} \right| < \varepsilon, \qquad \text{provided that } K \geq K_0 \geq N,\ L \geq L_0 \geq N. \tag{4.9}$$

Furthermore, the convergence of $\sum_{k=1}^{\infty} a_{k\ell}$ for every $\ell \geq 0$ implies that there exists $N' \geq N$ such that

$$\left|\sum_{k=K_0}^{K}\sum_{\ell=L_0}^{L} a_{k\ell}\right| < \varepsilon, \qquad \text{if } K \geq K_0 \geq N',\ N > L \geq L_0. \tag{4.10}$$

Let us now estimate

$$\left|\sum_{k=K_0}^{K}\sum_{\ell=L_0}^{L} a_{k\ell}\right|$$

in the case where $K \geq K_0 \to \infty$ and $L \geq L_0$ (the case $K \geq K_0$ and $L \geq L_0 \to \infty$ is treated similarly). By (4.9) and (4.10),

$$\left|\sum_{k=K_0}^{K}\sum_{\ell=L_0}^{L} a_{k\ell}\right| = \left|\sum_{k=K_0}^{K}\sum_{\ell=L_0}^{N-1} a_{k\ell} + \sum_{k=K_0}^{K}\sum_{\ell=N}^{L} a_{k\ell}\right| < 2\varepsilon,$$

as desired. □

A useful property of regularly convergent double series is that both iterated series also converge to the same sum. In other words, Fubini's theorem holds for regularly convergent double series.

Theorem 4.20 *Let $\sum_{k,\ell} a_{k\ell}$ be a regularly convergent double series. Then*

$$\sum_{k,\ell=1}^{\infty} a_{k\ell} = \lim_{K,L\to\infty} S_{KL} = \sum_{k=1}^{\infty}\sum_{\ell=1}^{\infty} a_{k\ell} = \sum_{\ell=1}^{\infty}\sum_{k=1}^{\infty} a_{k\ell}.$$

Note that this result immediately follows from Theorems 4.13 and 4.19.

4.3 Convergence of double integrals: definitions and basic notions

A parallel theory of Pringsheim and regular convergence is developed for integrals as well. We now state the integral counterparts of the above definitions and results in this section without proofs.

Given a locally integrable function $f\colon [1,\infty)^2 \to \mathbb{C}$, we define the improper double integral

$$I := \int_1^{\infty}\int_1^{\infty} f(x,y)\,dx\,dy = \lim_{s,t\to\infty}\int_1^{t}\int_1^{s} f(x,y)\,dx\,dy, \tag{4.11}$$

where the limit may be finite, infinite, or may not exist. It is worth emphasizing that such integrals are understood as two-dimensional ones, instead of as an iteration of one-dimensional integrals (for such an iteration we will use parentheses in the inner integral, as in (4.12) below). Obviously, the lower limits of integration in (4.11) may be replaced with 0 or any other positive real number. Without loss of generality, we prefer to take them to be 1. If

$$\lim_{s,t\to\infty}\int_1^t\int_1^s f(x,y)\,dx\,dy = A\in\mathbb{C},$$

we say that the double integral I *converges to* A *in the sense of Pringsheim*. Otherwise we say that the double integral I is *divergent*. The Cauchy criterion for convergence of integrals reads as follows:

The double integral I *converges in the sense of Pringsheim if and only if for every* $\varepsilon > 0$ *there exists* $z > 0$ *such that*

$$\left|\int_{x_1}^{x_2}\int_{y_1}^{y_2} f(x,y)\,dx\,dy\right| < \varepsilon, \qquad \text{for all } x_1, y_1 > z, \qquad x_2 > x_1,\ y_2 > y_1.$$

It is important to emphasize that, contrary to the case of series, the condition

$$f(x,y)\to 0 \qquad \text{as } x,y\to\infty$$

is not necessary for the convergence of the double integral I. Indeed, this fact is delivered by the following simple example:

$$f(x,y) = \begin{cases} 1, & \text{if } (x,y)\in[2^k, 2^k+2^{-k/2})\times[2^k,2^k+2^{-k/2}), \quad k\in\mathbb{N},\\ 0, & \text{otherwise,}\end{cases}$$

where

$$\int_1^\infty\int_1^\infty f(x,y)\,dx\,dy = \sum_{k=1}^\infty \frac{1}{2^k} = 1.$$

A more sophisticated example illustrating this is the so-called *Fresnel integral*. The sine and cosine Fresnel integrals are, respectively,

$$\int_0^\infty \sin x^2\,dx \quad\text{and}\quad \int_0^\infty \cos x^2\,dx,$$

and both of them converge to $\frac{\sqrt{2\pi}}{4}$. More generally, if $a > 1$, both integrals

$$\int_0^\infty \sin x^a\,dx \quad\text{and}\quad \int_0^\infty \cos x^a\,dx$$

converge. Using these integrals it is easy to produce examples of convergent double integrals where the integrand function does not vanish at infinity, as for instance $f(x,y) = \sin x^2 \sin y^2$.

As in the case of series, it is important to study iterated (double) integrals as well. An iterated double integral is defined as

$$\lim_{s\to\infty}\int_1^s\left(\lim_{t\to\infty}\int_1^t f(x,y)\,dy\right)dx, \quad\text{or}\quad \lim_{t\to\infty}\int_1^t\left(\lim_{s\to\infty}\int_1^s f(x,y)\,dx\right)dy,$$

where the inner limit is first taken with respect to a fixed t (or s, respectively). In order to distinguish between the two possible iterated double integrals we will write

$$\int_1^\infty \left(\int_1^\infty f(x,y)\,dx \right) dy, \qquad \text{or} \qquad \int_1^\infty \left(\int_1^\infty f(x,y)\,dy \right) dx \tag{4.12}$$

(note that the double integral is written without parentheses).

The following result, which is a counterpart of Theorem 4.15 for integrals, gives sufficient conditions for the iterated integrals to converge to the same limit as the double integral does.

Theorem 4.21 *Let the double integral I converge to $A \in \mathbb{C}$.*

If the integral

$$\int_1^\infty f(x,y)\,dx$$

converges for all $y > 1$, then the iterated double integral

$$\int_1^\infty \left(\int_1^\infty f(x,y)\,dx \right) dy$$

converges to A.

Correspondingly, if the integral

$$\int_1^\infty f(x,y)\,dy$$

converges for all $x > 1$, then the iterated double integral

$$\int_1^\infty \left(\int_1^\infty f(x,y)\,dy \right) dx$$

converges to A.

Based on the above examples for series, one may easily produce examples of double integrals that violate one or more of the conclusions of Theorem 4.21 in various ways; for instance:

1. The iterated integrals do not have the same limit.
2. The iterated integrals do not converge.
3. The double integral diverges, although both iterated integrals converge, and so on.

To this end, it suffices to consider the function

$$f(k,\ell) = a_{k\ell}, \qquad x \in [k, k+1) \times [\ell, \ell+1), \qquad k, \ell \in \mathbb{N},$$

for a given double sequence $\{a_{k\ell}\}$.

We now define *regular convergence* of double integrals.

Definition 4.22 Let $f\colon (1,\infty)^2 \to \mathbb{C}$. We say that the double integral

$$\int_1^\infty \int_1^\infty f(x,y)\,dx\,dy$$

is regularly convergent (or converges in the regular sense) if for each $\varepsilon > 0$ there exists a positive real number z such that

$$\left| \int_{x_1}^{x_2} \int_{y_1}^{y_2} f(x,y)\, dx\, dy \right| < \varepsilon,$$

provided $x_2 > x_1$, $y_2 > y_1$ and $x_1 + y_1 > z$, or, equivalently, there exists a positive real number z' such that

$$\left| \int_{x_1}^{x_2} \int_{y_1}^{y_2} f(x,y)\, dx\, dy \right| < \varepsilon,$$

provided $x_2 > x_1$, $y_2 > y_1$ and $\max\{x_1, y_1\} > z'$.

It readily follows from the definition that if a double integral converges in the regular sense, then it converges in the sense of Pringsheim, and its value is just the limit in the Pringsheim sense.

Remark 4.23 It is interesting to note that, unlike the case of sequences, the regular convergence of the double integral I does not imply

$$f(x,y) \to 0, \qquad \text{as } \max\{x,y\} \to \infty.$$

This phenomenon does not allow the characterization of regular convergence of integrals analogous to that of Theorem 4.19 for series. Indeed, consider

$$f(x,y) = (-1)^{i+j} 2^n, \qquad \text{for } (x,y) \in R_{ij}^{(n)} := \left[\frac{i-1}{2^n}, \frac{i}{2^n}\right) \times \left[\frac{j-1}{2^n}, \frac{j}{2^n}\right),$$

where $n \in \mathbb{N}$, and either

$$i = 1, \ldots, n2^n, \quad \text{and} \quad j = n2^n + 1, \ldots, (n+1)2^n,$$

or

$$i = n2^n + 1, \ldots, (n+1)2^n, \quad \text{and} \quad j = 1, \ldots, n2^n.$$

Let us prove that $\int_{\mathbb{R}_+^2} f(x,y)\, dx\, dy$ converges regularly. Any rectangle

$$R = [x_1, x_2] \times [y_1, y_2]$$

is contained in the union of finitely many rectangles of the type $R_{ij}^{(n)}$. We now split the integral $\int_R f(x,y)\, dx\, dy$ into five integrals as follows: four of them correspond to integrals in the corners

$$\int_{R \cap R_{i_\alpha j_\alpha}^{(n_\alpha)}} f(x,y)\, dx\, dy, \qquad \alpha = 1, \ldots, 4.$$

The last integral corresponds to the remaining part. If we further subdivide the last one in integrals over rectangles of the form $R \cap R_{ij}^{(n)}$, for most of them there is another integral with the same absolute value but opposite sign, and thus they cancel one another. The only integrals possibly remaining after this cancellation are precisely those over rectangles surrounding each of the corners (there are at most three of them for each of the corners). Hence, in total, $\int_R f(x,y)\,dx\,dy$ is the sum of at most sixteen nonzero integrals over rectangles of the form $R \cap R_{ij}^{(n)}$, four corresponding to the rectangles in the corners, plus twelve corresponding to the rectangles surrounding the corners. Since each of them satisfies the estimate

$$\left| \int_{R\cap R_{ij}^{(n)}} f(x,y)\,dx\,dy \right| \le 2^{-2\max\{\lfloor x_1\rfloor,\lfloor y_1\rfloor\}} \cdot 2^{\max\{\lfloor x_1\rfloor,\lfloor y_1\rfloor\}} = 2^{-\max\{\lfloor x_1\rfloor,\lfloor y_1\rfloor\}},$$

we conclude that

$$\left| \int_R f(x,y)\,dx\,dy \right| \le 16 \cdot 2^{-\max\{\lfloor x_1\rfloor,\lfloor y_1\rfloor\}} \to 0, \qquad \text{as } \max\{\lfloor x_1\rfloor, \lfloor y_1\rfloor\} \to \infty,$$

that is, $\int_{\mathbb{R}_2^+} f(x,y)\,dx\,dy$ converges regularly.

On the other hand, for every $y \ge 0$ the integral $\int_0^\infty f(x,y)\,dx$ diverges, since for $m = \lfloor y\rfloor, \lfloor y\rfloor + 1, \ldots$, we have

$$\left| \int_m^{m+2^{-m}} f(x,y)\,dx \right| = 2^m \cdot 2^{-m} = 1, \qquad \text{as } m \to \infty.$$

Analogously, for every $x \ge 0$ the integral $\int_0^\infty f(x,y)\,dy$ diverges.

Although a counterpart of Theorem 4.19 does not hold for integrals, the analog of Fubini's theorem for regular convergence (i.e., a converse of Theorem 4.20) does.

Theorem 4.24 *Let $f\colon (1,\infty)^2 \to \mathbb{C}$. If the double integral I converges in the regular sense, then each of its iterated double integrals converges to the same limit, namely*

$$\int_1^\infty \int_1^\infty f(x,y)\,dx\,dy = \int_1^\infty \left(\int_1^\infty f(x,y)\,dx \right) dy$$
$$= \int_1^\infty \left(\int_1^\infty f(x,y)\,dy \right) dx.$$

4.4 Different types of convergence of double series

In addition to the two types of convergence, Pringsheim's and regular convergence, we have studied in detail above, one may consider more general methods of summation (or, in other words, different ways of ordering or defining partial sums). For introduction, note that when we calculate the limit of the partial sums (in Pringsheim's sense), we are actually *summing by rectangles*. In general, we define the sum to be

$$\sum_{k,\ell=1}^{\infty} a_{k\ell} = \lim_{m\to\infty} \sum_{(k,\ell)\in\Gamma_m} a_{k\ell},$$

where

$$\Gamma_m = \{(k,\ell) \in \mathbb{N}^2 : 1 \le k \le K_m,\ 1 \le \ell \le L_m\},$$

and K_m and L_m tend to infinity as $m \to \infty$ (in particular, if $K_m = L_m$ for all m, then we *sum by squares*). By taking different expanding sets Γ_m (such that their union covers the whole $\mathbb{N}^2$), we can define more general methods of summation, called *summation by curves*.

Definition 4.25 Let $\{a_{k\ell}\}$ be a double sequence and let

$$\Gamma_1 \subset \Gamma_2 \subset \Gamma_3 \subset \cdots \tag{4.13}$$

be a sequence of subsets of $\mathbb{N}^2$ such that $\bigcup_{m\in\mathbb{N}} \Gamma_m = \mathbb{N}^2$. We say that (4.3) converges to A along the sequence of curves $\{\Gamma_m\}_{m\in\mathbb{N}}$ if

$$\lim_{m\to\infty} \sum_{(k,\ell)\in\Gamma_m} a_{k\ell} = A.$$

Again, it should be noted that by definition *a double series converges to A in the sense of Pringsheim if and only if it converges to A along any sequence of rectangles.*

Important methods of summation different from that by rectangles are *summation* by triangles, where the sequence of curves (4.13) is given by

$$\Gamma_m = T_m = \{(k,\ell) \in \mathbb{N}^2 : k + \ell \leq m\}, \qquad m \geq 2,$$

or *summation by circles* (also called *spherical summation, or circular in dimension two*), where the curves (4.13) are given by

$$\Gamma_m = \left\{(k,\ell) \in \mathbb{N}^2 : \sqrt{k^2 + \ell^2} \leq m\right\}.$$

The most natural question when facing all these different methods of summation is whether the convergence of a given series is independent of the summation ordering. With what we have seen so far, one should be readily able to answer this question: in general, the convergence of a series does depend on the method of summation. In example (4.7) above, where

$$a_{k\ell} = \begin{cases} -1, & \text{if } k = \ell \geq 1, \\ 1/2, & \text{if } \ell = k+1 \text{ or } \ell = k-1, \min\{k,\ell\} > 1, \\ 0, & \text{otherwise,} \end{cases}$$

we see that when summing by squares, one gets a limit of the partial sums different from the one obtained by summing over the rectangles

$$\{(k,\ell) : 1 \leq k \leq m,\ 1 \leq \ell \leq n\}, \qquad m \neq n.$$

More precisely,

$$\sum_{k,\ell=1}^{m} a_{k\ell} = -1, \qquad \text{for } m \geq 1,$$

and

$$\sum_{k=1}^{m}\sum_{\ell=1}^{n} a_{k\ell} = -\frac{1}{2}, \qquad \text{for } m \geq 1 \text{ and } n \neq m.$$

If we consider summation by the triangles T_m, the situation is by no means better. Indeed, in this case, one has, for $m \geq 1$,

$$\sum_{(k,\ell)\in T_m} a_{k\ell} = \begin{cases} 0, & \text{if } m \text{ is even,} \\ -1, & \text{if } m \text{ is odd,} \end{cases}$$

and hence the triangular sums cannot converge.

We have already discussed various types of summing double series or, in other words, of ordering the terms of double series as various partial sums. Except Pringsheim's convergence, spherical and square ones are considered as the most important. Let us add some counterexamples concerning their interrelations, always assuming $k, \ell \in \mathbb{N}$.

First we consider the double sequence with general term

$$a_{k\ell} = \begin{cases} 2, & \text{if } k = \ell, \\ -1, & \text{if } k = \ell + 1 \text{ or } \ell = k + 1, \\ 0, & \text{otherwise.} \end{cases}$$

Being square convergent, the double series $\sum a_{k\ell}$ converges neither in Pringsheim's sense nor spherically. Indeed,

$$S_{KK} = \sum_{k,\ell=1}^{K} a_{k\ell} = 2,$$

while

$$S_{K,K+1} = \sum_{k=1}^{K}\sum_{\ell=1}^{K+1} a_{k\ell} = 1.$$

Then there are at least two types of spherical partial sums different from one another:

$$\sum_{k^2+\ell^2\leq 2K^2} a_{k\ell} = \sum_{k,\ell=1}^{K} a_{k\ell} = 2,$$

and

$$\sum_{k^2+\ell^2\leq (K+1)^2+K^2} a_{k\ell} = \sum_{k,\ell=1}^{K} a_{k\ell} + a_{K,K+1} + a_{K+1,K} = 0.$$

Further, we consider the double series $\sum a_{k\ell}$ with general term as follows. Let $\{n_i\} \subset \mathbb{N}$ be a sequence of indices such that $5n_{i+1} > 7n_i$ for all $i \in \mathbb{N}$ and define

$$a_{k\ell} = \begin{cases} 2, & \text{if } k = \ell = 5n_i,\ i \in \mathbb{N}, \\ -1, & \text{if } k = n_i,\ \ell = 7n_i \text{ or } k = 7n_i,\ \ell = n_i,\ i \in \mathbb{N}, \\ 0, & \text{otherwise.} \end{cases}$$

The corresponding double series is spherically convergent, but neither square convergent nor in Pringsheim's sense. Indeed, since $2 \cdot (5n_i)^2 = n_i^2 + (7n_i)^2$, we have, for any $i \in \mathbb{N}$,

$$\sum_{k^2+\ell^2 \le R^2} a_{k\ell} = \sum_{50n_i^2 \le R^2} (a_{n_i,7n_i} + a_{7n_i,n_i} + a_{5n_i,5n_i}) = 0,$$

while

$$\sum_{k,\ell=0}^{5n_i} a_{k\ell} = 2, \quad \text{and} \quad \sum_{k=0}^{7n_i} a_{k\ell} = 0.$$

Finally, by Corollary 4.54 below, given $\varepsilon > 0$ we can find a convergent series $\sum_k a_k$ with $a_k = O(k^{\varepsilon-1})$ such that the series $\sum_{k,\ell} a_k a_\ell$ spherically diverges while it converges in Pringsheim's sense as the product of two convergent single series.

In view of the above issues, it is natural to strive for sufficient conditions that guarantee that the convergence of a series does not depend on the sequence of curves (4.13) chosen for defining the partial sums. This is the scope of the following theorem, which will be stated without a proof (see Notes-8 in the end of this chapter for appropriate references).

Theorem 4.26 *Let $\{a_{k\ell}\}$ be a double sequence such that the double series $\sum a_{k\ell}$ converges absolutely. Then this series converges to the same limit along any sequence of curves $\{\Gamma_m\}_{m\in\mathbb{N}}$.*

This result addresses absolutely convergent series, which will appear later in Section 4.6. As an exercise, one can prove this result by means of Theorem 4.50 below. Note also the relationship of Theorem 4.26 with the one-dimensional statement of Theorem 2.48.

We now proceed to giving a sufficient condition for the spherical and square convergence to be equivalent.

Theorem 4.27 *Let $\{a_{k\ell}\}$ be a double sequence such that*

$$a_{k\ell} = o\left(\frac{1}{k^2 + \ell^2}\right) \quad \text{as} \quad k^2 + \ell^2 \to \infty. \tag{4.14}$$

If the double series $\sum a_{k\ell}$ converges over squares to A, then it converges spherically to the same sum, and conversely.

To prove this statement, we first give two auxiliary results.

Theorem 4.28 *For $R > 0$, the number $n(R)$ of points $(k, \ell) \in (\mathbb{N} \cup \{0\})^2$ such that $k^2 + \ell^2 \le R^2$ satisfies*

$$n(R) = \frac{\pi}{4} R^2 + O(R).$$

We do not give a proof of the latter, since it falls off the scope of this book. Instead we give an appropriate reference in Notes-10.

Lemma 4.29 *If* (4.14) *holds, then*

$$\sum_{\substack{k^2+\ell^2>n^2\\ k,\ell\le n}} |a_{k\ell}| \qquad \text{and} \qquad \sum_{n^2<k^2+\ell^2\le 2n^2} |a_{k\ell}|$$

tend to zero as $n \to \infty$.

Proof By Theorem 4.28, for $n \in \mathbb{N}$, the number of points $(k,\ell) \in \mathbb{N}^2$ satisfying $n^2 < k^2 + \ell^2 \le 2n^2$ is

$$\frac{\pi}{4}2n^2 - \frac{\pi}{4}n^2 + O(n) = \frac{\pi}{4}n^2 + O(n) = O(n^2).$$

Therefore, by (4.14),

$$\sum_{\substack{k^2+\ell^2>n^2\\ k,\ell\le n}} |a_{k\ell}| \le \sum_{n^2<k^2+\ell^2\le 2n^2} |a_{k\ell}| = o\left(\frac{1}{n^2}\right)O(n^2) = o(1)$$

as $n \to \infty$. □

Proof of Theorem 4.27 For every $R > 0$, there exists $n \in \mathbb{N}$ such that $n < R \le n+1$. Denote the Rth partial spherical sum by

$$S_R^\circ = \sum_{k^2+\ell^2\le R^2} a_{k\ell}.$$

Now, by (4.14) and Lemma 4.29,

$$\begin{aligned}|S_R^\circ - A| &\le |S_R^\circ - S_{nn}| + |S_{nn} - A| \le \sum_{n^2<k^2+\ell^2\le R^2} |a_{k\ell}| + |S_{nn} - A|\\ &\le \sum_{n^2<k^2+\ell^2\le 2n^2} |a_{k\ell}| + |S_{nn} - A| = o(1) + |S_{nn} - A|,\end{aligned}$$

and, conversely,

$$\begin{aligned}|S_{nn} - A| &\le |S_{nn} - S_n^\circ| + |S_n^\circ - A| \le \sum_{n^2<k^2+\ell^2\le 2n^2} |a_{k\ell}| + |S_n^\circ - A|\\ &= o(1) + |S_n^\circ - A|,\end{aligned}$$

as desired. □

The following two theorems are about the sharpness of Theorem 4.27 with respect to condition (4.14). For convenience, we will consider sequences and series with indices k,ℓ starting at 0 there. Moreover, we will use additional notation. Let K_n be the closed square

$$K_n = \left\{(k,\ell) \in (\mathbb{N}\cup\{0\})^2 : 0 \le k,\ell \le n\right\},$$

and let C_R be the (integer points in the) quarter of the circle with radius R centered at the origin lying in the first quadrant, that is,

$$C_R = \left\{(k,\ell) \in (\mathbb{N}\cup\{0\})^2 : k^2 + \ell^2 \le R^2\right\}.$$

These results will be accompanied by appropriate pictures for further clarity. In this respect, we recall the convention that we took at the beginning of this chapter: indices of double sequences are represented as if they were indices of infinite double matrices

Theorem 4.30 *There exists a double series $\sum a_{k\ell}$ with $a_{k\ell}$ satisfying the condition*

$$a_{k\ell} = O\left(\frac{1}{k^2+\ell^2}\right), \quad k+\ell > 0, \tag{4.15}$$

which is convergent over squares but spherically divergent.

Proof Define, for $n \in \mathbb{N}$ with $n \geq 2$,

$$\mu_n := \left\lfloor \frac{2^{n+1}}{\sqrt{5}} \right\rfloor.$$

Then,

$$\frac{2^n}{\sqrt{2}} < \mu_n \leq \frac{2^{n+1}}{\sqrt{5}} < 2^n.$$

Indeed, $\mu_2 = 3 > \frac{4}{\sqrt{2}}$, while for $n > 2$, we get

$$\mu_n > \frac{2^{n+1}}{\sqrt{5}} - 1 > \frac{2^{n+1}}{\sqrt{8}} = \frac{2^n}{\sqrt{2}}.$$

For an integer $m \in \left(\frac{2^n}{\sqrt{2}}, \mu_n\right]$, we consider the horizontal interval

$$I_m := \left\{(m,\nu) \in (\mathbb{N}\cup\{0\})^2 : 0 \leq \nu \leq m\right\}.$$

Since, for $0 \leq \nu \leq \left\lfloor \frac{m}{2} \right\rfloor$,

$$m^2 + \nu^2 \leq m^2 + \left(\frac{m}{2}\right)^2 < \frac{5}{4}\left(\frac{2^{n+1}}{\sqrt{5}}\right)^2 = 2^{2n},$$

the number of points in $I_m \cap C_{2^n}$ is larger than or equal to the number of points in $I_m \backslash C_{2^n}$.

Let us now construct the sequence $\{a_{k\ell}\}$. Set $G_n := K_{\mu_n} \backslash C_{2^n}$. Intuitively, G_n is a region whose boundaries are formed by a "triangular sawtooth", a part of the horizontal interval I_{μ_n}, and a part of the vertical interval

$$\left\{(\nu,\mu_n) \in (\mathbb{N}\cup\{0\})^2 : 0 \leq \nu \leq \mu_n\right\}$$

(precisely the respective subintervals lying outside C_{2^n}); see Figure 4.1.

For $(k,\ell) \in G_n$ with $k \geq \ell$, we define

$$a_{k\ell} = \frac{1}{k^2+\ell^2}.$$

We are now going to define the terms $a_{k\ell}$ in C_{2^n} by means of its values in the intersections $I_m \cap G_n$, with $m \in \left(\frac{2^n}{\sqrt{2}}, \mu_n\right]$. As already observed, the number of

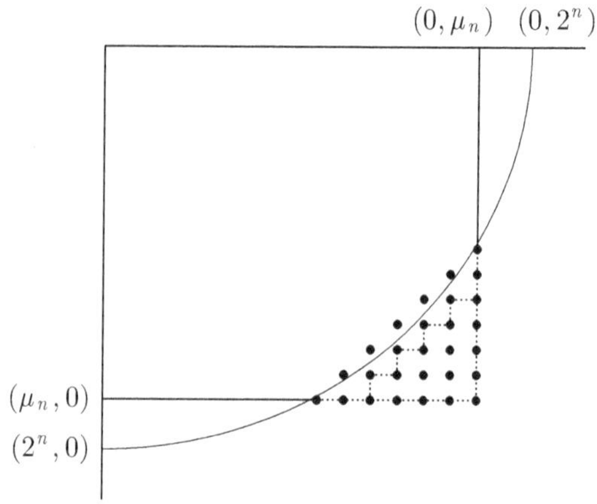

Figure 4.1 Representation of the region G_n (with the boundary highlighted as a dotted line).

points in $I_m \backslash C_{2^n} = I_m \cap G_n$ is smaller than or equal to the number of points in $I_m \cap C_{2^n} = I_m \backslash G_n$. Thus, for each $m \in \left(\frac{2^n}{\sqrt{2}}, \mu_n\right]$ and each point $(m, \ell) \in I_m \cap G_n$, we can choose another point $(m, q) \in I_m \backslash G_n$ in such a way that the correspondence between these (m, ℓ) and (m, q) be one to one. After this, we set $a_{mq} = -a_{m\ell}$. For all other points $(k, \ell) \in (\mathbb{N} \cup \{0\})^2$, we set $a_{k\ell} = 0$. Note that (4.15) holds. Indeed, if $a_{mq} \neq 0$ and $(m, q) \notin G_n$, then by definition $a_{mq} = -a_{m\ell}$ for $(m, \ell) \in G_n$ and $\ell > q$. Hence,

$$|a_{mq}| = \frac{1}{m^2 + \ell^2} \leq \frac{1}{m^2 + q^2},$$

and (4.15) follows.

From the definition of $a_{k\ell}$ it easily follows that

$$\sum_{(k,\ell) \in I_m} a_{k\ell} = 0, \qquad \text{for } m \in \mathbb{N} \cup \{0\}.$$

This yields the square convergence of the series $\sum a_{k\ell}$ to 0, since

$$S_{NN} = \sum_{m=0}^{N} \sum_{(k,\ell) \in I_m} a_{k\ell} = 0, \qquad \text{for } N \in \mathbb{N}.$$

Let us now prove that $\sum a_{k\ell}$ is spherically divergent. For any $n \in \mathbb{N}$, $n \geq 2$, it is clear that $K_{2^{n-1}} \subset C_{\frac{2^n}{\sqrt{2}}}$. Further, none of the points $(k, \ell) \in C_{\frac{2^n}{\sqrt{2}}} \backslash K_{2^{n-1}}$ belongs to any interval of the form I_m with $m \in \left(\frac{2^n}{\sqrt{2}}, \mu_n\right]$ (thus $a_{k\ell} = 0$ for all $(k, \ell) \in C_{\frac{2^n}{\sqrt{2}}} \backslash K_{2^{n-1}}$ and any $n \in \mathbb{N}$, by definition). Indeed, if $(k, \ell) \in I_m$, then $k^2 + \ell^2 \geq m^2 > 2^{2n-1}$, and hence $(k, \ell) \notin C_{\frac{2^n}{\sqrt{2}}}$ (note also that none of the points

$(k, \ell) \in C_{\frac{2^n}{\sqrt{2}}} \backslash K_{2^{n-1}}$ can belong to an interval I_m with $m \in \left(\frac{2^{n_0}}{\sqrt{2}}, \mu_{n_0}\right]$ and $n_0 \neq n$). Thus,

$$S^{\circ}_{\frac{2^n}{\sqrt{2}}} = \sum_{(k,\ell)\in C_{\frac{2^n}{\sqrt{2}}}} a_{k\ell} = \sum_{(k,\ell)\in K_{2^{n-1}}} a_{k\ell} = \sum_{m=0}^{2^{n-1}} \sum_{(k,\ell)\in I_m} a_{k\ell} = 0.$$

Therefore, for the spherical partial sums $S^{\circ}_{\frac{2^n}{\sqrt{2}}}$, we obtain

$$\lim_{n\to\infty} S^{\circ}_{\frac{2^n}{\sqrt{2}}} = 0.$$

On the other hand, we show that the difference between the 2^nth square and circular partial sums does not tend to zero as $n \to \infty$. Firstly,

$$S_{2^n,2^n} - S^{\circ}_{2^n} = \sum_{(k,\ell)\in K_{2^n}} a_{k\ell} - \sum_{(k,\ell)\in C_{2^n}} a_{k\ell} = \sum_{(k,\ell)\in G_n} \frac{1}{k^2+\ell^2}$$
$$\geq \frac{|G_n|}{2\mu_n^2} \geq \frac{|G_n|}{2^{2n+1}},$$

where $|G_n|$ is the number of points in G_n. Secondly, by Theorem 4.28,

$$|G_n| = |K_{\mu_n}| - |C_{2^n}| \geq \mu_n^2 - \frac{\pi}{4}2^{2n} + O(2^n)$$
$$\geq \left(\frac{2^{n+1}}{\sqrt{5}} - 1\right)^2 - \frac{\pi}{4}2^{2n} + O(2^n)$$
$$= \left(\frac{4}{5} - \frac{\pi}{4}\right)2^{2n} + O(2^n) \geq 2^{2n}\beta > 0,$$

where β can be chosen to be independent of n for n large enough. Hence,

$$\lim_{n\to\infty} S^{\circ}_{2^n} \neq 0.$$

Thus, the spherical partial sums of $\sum a_{k\ell}$ do not converge. □

The following theorem gives an example of a series with opposite type of convergence to that of Theorem 4.30.

Theorem 4.31 *There exists a double series $\sum a_{k\ell}$ with $a_{k\ell}$ satisfying condition* (4.15), *which is spherically convergent but divergent over squares.*

Proof For fixed $n \in \mathbb{N}$, we set

$$J_n := K_{2^n} \backslash C_{2^{n-1}\sqrt{5}},$$

and consider the sets

$$\delta_{nm} := C_{2^{n-1}\sqrt{5}+nm} \backslash C_{2^{n-1}\sqrt{5}+n(m-1)},$$

where

$$m = 1, 2, \ldots, \left\lfloor \frac{2^n\left(\sqrt{2} - \frac{\sqrt{5}}{2}\right)}{n} \right\rfloor. \tag{4.16}$$

The sets δ_{nm} split $C_{2^n\sqrt{2}} \backslash C_{2^{n-1}\sqrt{5}}$ into $\left\lfloor \frac{2^n\left(\sqrt{2}-\frac{\sqrt{5}}{2}\right)}{n} \right\rfloor + 1$ parts. Each set δ_{nm} is a circular band lying between two quarters of circles whose radii differ by n from one another (except for the last possible index of m, in which case δ_{nm} covers a residual region). We note that the number of integer points in $\delta_{nm} \cap J_n$ is less than that of $\delta_{nm} \backslash J_n$ (see Figure 4.2).

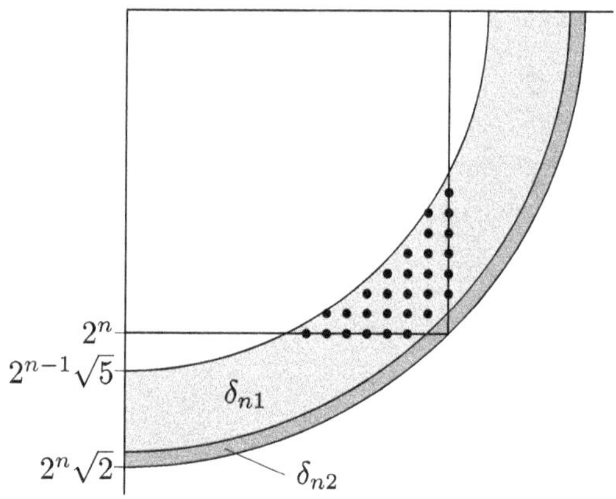

Figure 4.2 Representation of the regions δ_{nm}, with the points in $\delta_{nm} \cap J_n$ highlighted (the exact scale of the picture is for $n = 4$).

We define

$$a_{k\ell} = \frac{1}{k^2 + \ell^2}, \qquad \text{for } (k, \ell) \in \delta_{nm} \cap J_n,$$

for the range of m as in (4.16). We postpone showing that the terms $a_{k\ell}$ satisfy (4.15) until the end and instead turn to study the convergence character of $\sum a_{k\ell}$ both over squares and spherically.

In the same way as in the proof of Theorem 4.30, we take in $\delta_{nm} \backslash J_n$ as many points as we have in $\delta_{nm} \cap J_n$. Establishing one-to-one correspondence between the points $(k, \ell) \in \delta_{nm} \cap J_n$ and the points $(p, q) \in \delta_{nm} \backslash J_n$, we set $a_{pq} = -a_{k\ell}$. For all other points of the plane, we set $a_{k\ell} = 0$. By this, we have defined the values for a certain number of points $(k, \ell) \in \bigcup_m \delta_{nm}$.

Let us check that the series $\sum a_{k\ell}$ is spherically convergent. We first observe that by construction we have

$$\sum_{(k,\ell)\in\delta_{nm}} a_{k\ell} = 0$$

for the range of m as in (4.16). Thus, to prove that the series converges over circles, it suffices to prove that

$$\sum_{(k,\ell)\in\delta_{nm}} |a_{k\ell}| \to 0,$$

as $n \to \infty$, uniformly in m in the range given in (4.16). Since $(k,\ell) \in J_n$ satisfies $(k,\ell) \notin C_{2^{n-1}\sqrt{5}}$, we have

$$k^2 + \ell^2 > 5 \cdot 2^{2(n-1)}.$$

Therefore,

$$\sum_{(k,\ell)\in\delta_{nm}} |a_{k\ell}| = 2 \sum_{(k,\ell)\in\delta_{nm}\cap J_n} \frac{1}{k^2+\ell^2} < \frac{2}{5\cdot 2^{2(n-1)}} |\delta_{nm}|,$$

where $|\delta_{nm}|$ stands for the number of integer points in δ_{nm}. We have, by Theorem 4.28,

$$\begin{aligned}|\delta_{nm}| &= \frac{\pi}{4}\left(2^{n-1}\sqrt{5}+nm\right)^2 - \frac{\pi}{4}\left(2^{n-1}\sqrt{5}+n(m-1)\right)^2 + O(2^n)\\ &= \frac{\pi}{4}n\left(2^n\sqrt{5}+(2m-1)n\right) + O(2^n) = O\left(n2^n\right).\end{aligned}$$

Hence,

$$\begin{aligned}\sum_{(k,\ell)\in\delta_{nm}} |a_{k\ell}| &\le \frac{2}{5\cdot 2^{2(n-1)}}\left(\frac{\pi}{4}n\left(2^n\sqrt{5}+(2m-1)n\right)+O(2^n)\right)\\ &= O\left(\frac{n}{2^n}\right).\end{aligned}$$

Therefore, $\sum_{(k,\ell)\in\delta_{nm}} |a_{k\ell}| \to 0$ as $n \to \infty$ uniformly in m in the range given in (4.16). Thus, the double series $\sum_{k\ell} a_{k\ell}$ converges spherically to 0.

We now turn to the proof of the divergence of the series over squares. On the one hand, since for n large enough

$$C_{2^{n-1}\sqrt{2}} \subset K_{\left\lfloor \frac{2^n}{\sqrt{2}} \right\rfloor + 1} \subset C_{2^{n-1}\sqrt{5}},$$

we have

$$S_{\left\lfloor \frac{2^n}{\sqrt{2}} \right\rfloor+1, \left\lfloor \frac{2^n}{\sqrt{2}} \right\rfloor+1} = 0.$$

On the other hand, since $J_n \subset K_{2^n}$, we have $k^2+\ell^2 \le 2^{2n+1}$ for every $(k,\ell) \in J_n$. Moreover, all the terms $a_{k\ell}$ with $(k,\ell) \in J_n$ are nonnegative. Therefore, by Theorem 4.28,

$$\begin{aligned}S_{2^n,2^n} &= \sum_{(k,\ell)\in J_n} a_{k\ell} = \sum_{(k,\ell)\in J_n} \frac{1}{k^2+\ell^2} \ge \frac{1}{2^{2n+1}}|J_n|\\ &\ge \frac{1}{2^{2n+1}}\left(|K_{2^n}| - \left|C_{2^{n-1}\sqrt{5}}\right|\right)\\ &= \frac{1}{2^{2n+1}}\left(2^{2n} - \frac{\pi}{4}\cdot 5\cdot 2^{2n-2} + O(2^n)\right) \ge \beta > 0,\end{aligned}$$

where β can be chosen to be independent of n for n large enough.

It remains to show that the terms $a_{k\ell}$ satisfy (4.15). For the terms $a_{k\ell}$ with $(k,\ell) \in J_n$, condition (4.15) is satisfied. If $a_{pq} \neq 0$ and $(p,q) \in \delta_{nm}\backslash J_n$, then for the corresponding $(k,\ell) \in J_n \cap \delta_{nm}$, since $(k,\ell) \notin C_{2^{n-1}\sqrt{5}+n(m-1)}$, we have

$$k^2 + \ell^2 \geq \left(2^{n-1}\sqrt{5} + n(m-1)\right)^2.$$

Hence, for n large enough,

$$\begin{aligned}|a_{pq}| = a_{k\ell} = \frac{1}{k^2+\ell^2} &\leq \frac{1}{\left(2^{n-1}\sqrt{5}+n(m-1)\right)^2} \\ &\leq \frac{2}{\left(2^{n-1}\sqrt{5}+nm\right)^2} \leq \frac{2}{p^2+q^2},\end{aligned}$$

where $(p,q) \in C_{2^{n-1}\sqrt{5}+nm}$. By this, condition (4.15) is satisfied, which completes the proof. □

It turns out, however, that Theorem 4.27 may be strengthened in the following sense. Condition (4.14) and the convergence of the double series $\sum a_{k\ell}$ over circles or squares imply its convergence over rectangles.

Theorem 4.32 *If the double series $\sum a_{k\ell}$ converges over squares to A and the terms $a_{k\ell}$ satisfy condition* (4.14), *then the given series converges over rectangles to A.*

Proof For $K, L \in \mathbb{N} \cup \{0\}$, we put $N = \min(K, L)$. For definiteness let $N = K$ and $K < L$. Then

$$S_{KL} = S_{NN} + \sum_{k=0}^{K} \sum_{\ell=K+1}^{L} a_{k\ell}.$$

We have

$$\begin{aligned}\left|\sum_{k=0}^{K}\sum_{\ell=K+1}^{L} a_{k\ell}\right| &\leq \sum_{k=0}^{K}\sum_{\ell=K+1}^{L} |a_{k\ell}| \leq o\left(\sum_{k=0}^{K}\sum_{\ell=K+1}^{\infty} \frac{1}{k^2+\ell^2}\right) \\ &= o\left(\sum_{k=0}^{K}\sum_{s=0}^{\infty}\sum_{\ell=2^sK+1}^{2^{s+1}K} \frac{1}{k^2+\ell^2}\right) = o\left(\sum_{k=0}^{K}\sum_{s=0}^{\infty} \frac{K2^s}{k^2+K^2 2^{2s}}\right) \\ &= o\left(\sum_{k=0}^{K}\sum_{s=0}^{\infty} \frac{1}{K2^s}\right) = o(1).\end{aligned}$$

This yields

$$S_{KL} = S_{NN} + o(1),$$

as required. □

Thus, from Theorems 4.27 and 4.32 it follows that when condition (4.14) is satisfied, convergence over rectangles, squares, and circles are equivalent.

Remark 4.33 Theorem 4.32 cannot be strengthened replacing condition (4.14) by the weaker (4.15). Indeed, if we put

$$a_{k\ell} = \begin{cases} \frac{1}{k^2+\ell^2} & \text{for } k < \ell, \\ -\frac{1}{k^2+\ell^2} & \text{for } k > \ell, \\ 0 & \text{for } k = \ell, \end{cases}$$

then the double series $\sum a_{k\ell}$ converges over squares to 0 but does not converge over rectangles.

We now give sufficient conditions for double series of the form $\sum_{k,\ell} a_k b_\ell$ to converge spherically.

Theorem 4.34 *If the series $\sum_k a_k$ and $\sum_\ell b_\ell$ converge to A and B respectively, and the sequences $\{ka_k\}$, $\{\ell b_\ell\}$ are bounded (i.e., $a_k = O(\frac{1}{k})$ and $b_\ell = O(\frac{1}{\ell})$), then the double series $\sum_{k,\ell} a_k b_\ell$ converges spherically to AB.*

Remark 4.35 If we replace the conditions

$$a_k = O\Big(\frac{1}{k}\Big) \quad \text{and} \quad b_k = O\Big(\frac{1}{k}\Big)$$

in Theorem 4.34 by the weaker conditions

$$\sum_{k=n}^{2n} |a_k| = O(1), \qquad \sum_{k=n}^{2n} |b_k| = O(1),$$

respectively, then Theorem 4.34 would include absolutely convergent series and series with lacunary indices.

Proof of Theorem 4.34 For $R > 0$ large enough, we have

$$S_R^\circ = S_{\lfloor \frac{R}{\sqrt{2}} \rfloor, \lfloor \frac{R}{\sqrt{2}} \rfloor} + \sum_{\substack{k^2+\ell^2 \le R^2 \\ k > \lfloor \frac{R}{\sqrt{2}} \rfloor}} a_k b_\ell + \sum_{\substack{k^2+\ell^2 \le R^2 \\ \ell > \lfloor \frac{R}{\sqrt{2}} \rfloor}} a_k b_\ell. \tag{4.17}$$

From the assumptions $\sum_k a_k = A$ and $\sum_\ell b_\ell = B$ it follows that

$$S_{\lfloor \frac{R}{\sqrt{2}} \rfloor, \lfloor \frac{R}{\sqrt{2}} \rfloor} \to AB \quad \text{as } R \to \infty.$$

Let us now check that the remaining two terms on the right-hand side of (4.17) vanish as $R \to \infty$. We only show in detail that the first sum vanishes as $R \to \infty$, since the other one is treated analogously. We can write, for noninteger R (the case of integer R only produces a slight variation on the indices of summation),

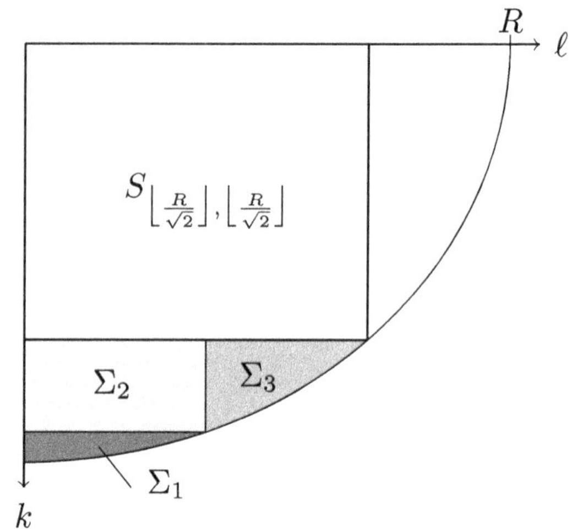

Figure 4.3 Sketch of the regions containing the indices of summation in the sums Σ_j in (4.18), as well as the indices in the square partial sum $S_{\lfloor \frac{R}{\sqrt{2}} \rfloor, \lfloor \frac{R}{\sqrt{2}} \rfloor}$.

$$\begin{aligned}\sum_{\substack{k^2+\ell^2 \le R^2 \\ k > \lfloor \frac{R}{\sqrt{2}} \rfloor}} a_k b_\ell &= \sum_{\substack{k^2+\ell^2 \le R^2 \\ k \ge \lfloor R-1 \rfloor}} a_k b_\ell + \sum_{\lfloor \frac{R}{\sqrt{2}} \rfloor < k < \lfloor R-1 \rfloor} \sum_{1 \le \ell \le \sqrt{R^2 - \lfloor R-1 \rfloor^2}} a_k b_\ell \\ &\quad + \sum_{\lfloor \frac{R}{\sqrt{2}} \rfloor < k < \lfloor R-1 \rfloor} \sum_{\sqrt{R^2-\lfloor R-1 \rfloor^2} < \ell \le \sqrt{R^2-k^2}} a_k b_\ell \\ &=: \Sigma_1 + \Sigma_2 + \Sigma_3. \end{aligned} \tag{4.18}$$

See Figure 4.3 for a sketch of the regions of the indices of summation in the sums Σ_j, $j = 1, 2, 3$. Since $a_k = o(1)$ and $\sum_\ell b_\ell$ converges, it is clear that $\Sigma_1 \to 0$ as $R \to \infty$. Secondly, by Cauchy's criterion for convergence of series,

$$\sum_{\lfloor \frac{R}{\sqrt{2}} \rfloor < k < \lfloor R-1 \rfloor} a_k \to 0 \qquad \text{as } R \to \infty,$$

since $\sum_k a_k$ converges. This, combined with the convergence of $\sum_\ell b_\ell$, implies that $\Sigma_2 \to 0$ as $R \to \infty$. Finally, by Cauchy's criterion and the convergence of $\sum_\ell b_\ell$, we have

$$\sum_{\sqrt{R^2-\lfloor R-1 \rfloor^2} < \ell \le \sqrt{R^2-k^2}} b_\ell = o(1) \qquad \text{as } R \to \infty,$$

uniformly in $\lfloor \frac{R}{\sqrt{2}} \rfloor < k < \lfloor R-1 \rfloor$. Further, since by hypotheses $ka_k = O(1)$, we derive that

$$\begin{aligned}|\Sigma_3| &\lesssim \left(\max_{\lfloor \frac{R}{\sqrt{2}} \rfloor < k < \lfloor R-1 \rfloor} k|a_k| \right) \cdot \left(\max_{\lfloor \frac{R}{\sqrt{2}} \rfloor < k < \lfloor R-1 \rfloor} \left| \sum_{\sqrt{R^2-\lfloor R-1 \rfloor^2} < \ell \le \sqrt{R^2-k^2}} b_\ell \right| \right) \\ &= o(1),\end{aligned}$$

as $R \to \infty$, which concludes the proof. □

The following method of summing double series is somewhat aside. For the needs of analysis, it was introduced for both positive and negative indices. The partial sums are taken over the sets $U \in \mathbb{Z}^2$ having the following properties:

1. U is symmetric with respect to both coordinate axes.
2. If $(k,\ell) \in U$, then

$$\big[\min\{0,k\},\max\{0,k\}\big] \times \big[\min\{0,\ell\},\max\{0,\ell\}\big] \cap \mathbb{Z}^2 \subseteq U.$$

In this case we say $U \in A^{\natural}$.

Thus, the following definitions are introduced.

Definition 4.36 Let $U \in \mathbb{Z}^2$ be bounded. Then $U \in A^{\flat}$ if:

(i) From $(k,\ell) \in U$ it follows that $(-k,\ell) \in U$ and $(k,-\ell) \in U$;
(ii) $U \in A^{\natural}$.

Definition 4.37 We say that the series $\sum_{(k,\ell)\in\mathbb{Z}^2} a_{k\ell}$ is u-convergent to a number $A \in \mathbb{C}$ if for $\varepsilon > 0$ there exists N such that

$$\Big| \sum_{(k,-\ell)\in U} a_{k\ell} - A \Big| < \varepsilon,$$

for every set $U \in A^{\flat}$ that contains the disk $\{(k,\ell) \in \mathbb{Z}^2 : k^2 + \ell^2 \le N^2\}$.

It is clear that if a double series u-converges, then it converges to the same quantity in the sense of Pringsheim, over spheres, and so on. See Notes-12 for more comments.

We conclude this section with mentioning the so-called hyperbolic crosses. For a long time in Approximation Theory and Fourier Analysis, expansions with ordering of the terms by the hyperbolic crosses have played an important role. Of course, such sums become number series at each point. Let us define the traditional hyperbolic summation for double number series. Let hyperbolic partial sums be defined as

$$\mathcal{S}_N = \sum_{\substack{k\ell \le N,\\ k,\ell > 0}} a_{k\ell}.$$

We call the series hyperbolically convergent if a finite limit of $\mathcal{S}_N$ as $N \to \infty$ exists. Observe that we sum the series with respect to integer points of N-dilations of the basic hyperbolic cross $st \le 1$, $s,t > 0$, excluding the points with at least one zero coordinate, that is, excluding the coordinate axes.

4.5 Positive double series

A double series with positive (respectively nonnegative) terms $\{a_{k\ell}\}$ is shortly called a *positive double series* (respectively *nonnegative double series*).

Theorem 4.38 *For a double series $\sum_{k,\ell} a_{k\ell}$ with nonnegative terms to be convergent, it is necessary and sufficient that its sequence of partial sums $\{S_{KL}\}$ be bounded.*

This result immediately follows from Theorem 4.10.

Theorem 4.39 *A double series $\sum_{k,\ell} a_{k\ell}$ with nonnegative terms and both iterated ones either converge to the same limit or diverge simultaneously.*

This result is a direct consequence of Theorem 4.14.

In particular, Theorems 4.19 and 4.39 imply the following:

Corollary 4.40 *If a nonnegative double series converges, then it converges in the regular sense.*

More generally, we now show that all types of convergence of double series with nonnegative terms are equivalent.

Theorem 4.41 *If a double series with nonnegative terms $\{a_{k\ell}\}$ and a one-dimensional series with terms $\{b_n\}$ have all the terms the same (that is, $\{a_{k\ell}\} = \{b_n\}$), then the convergence of one of them implies the convergence of the other, and their sums are equal.*

Proof Let the double series $\sum_{k,\ell} a_{k\ell}$ converge. Denoting its sum by A, we set

$$B_n = b_1 + b_2 + \cdots + b_n,$$

where n is an arbitrary natural number. Let us choose natural numbers K and L such that the partial sum S_{KL} of $\sum_{k,\ell} a_{k\ell}$ contains all the elements $b_1, b_2, \ldots, b_n$. Obviously,

$$B_n \le S_{KL} \le A.$$

It follows from this that the series $\sum b_n$ converges. Denoting its sum by B, we have

$$B \le A.$$

Further, for arbitrary natural numbers K and L, we choose an arbitrary natural number ν so that all the elements of the partial sum S_{KL} of the double series $\sum_{k,\ell} a_{k\ell}$ are among the elements $b_1, b_2, \ldots, b_\nu$. Hence,

$$S_{KL} \le B_\nu \le B.$$

Therefore,

$$A \le B,$$

which yields $A = B$ and concludes the proof. □

The multidimensional analog of Theorem 4.41 is Theorem 7.34, which we prove in a different manner.

For any convergent/divergent positive double series, one may find another double series that converges/diverges, having terms asymptotically larger/smaller than the original series. In particular, this implies that there is no "limiting" double series with which one can compare any other positive double series to determine its convergence or divergence.

Theorem 4.42 *For every divergent double series with positive terms $\{a_{k\ell}\}$, there exists a decreasing double sequence with positive terms $\{b_{k\ell}\}$ such that*

$$\lim_{k+\ell\to\infty} b_{k\ell} = 0,$$

and

$$\sum_{k,\ell=1}^{\infty} a_{k\ell}b_{k\ell} = \infty.$$

Theorem 4.43 *For every convergent double series with positive terms $\{a_{k\ell}\}$, there exists an increasing double sequence with positive terms $\{b_{k\ell}\}$ such that*

$$\lim_{k+\ell\to\infty} b_{k\ell} = \infty,$$

and

$$\sum_{k,\ell=1}^{\infty} a_{k\ell}b_{k\ell} < \infty.$$

Theorems 4.42 and 4.43 can be written in a slightly more general form as their multidimensional analogs are (Theorems 7.36 and 7.37, respectively). The corresponding proofs will also be included.

4.6 Absolute and conditional convergence of double series

Among all the complex-valued double sequences $\{a_{k\ell}\}$ such that $\sum_{k,\ell} a_{k\ell}$ converges, one can distinguish between those sequences for which $\sum_{k,\ell} |a_{k\ell}|$ is also convergent, and those for which $\sum_{k,\ell} |a_{k\ell}|$ is divergent. The former series represent a specific class of double series that are called *absolutely convergent series*. They possess many of the basic properties of the series with nonnegative terms.

Definition 4.44 Given a double sequence $\{a_{k\ell}\} \subset \mathbb{C}$, the series $\sum_{k,\ell} a_{k\ell}$ is said to converge absolutely if

$$\sum_{k,\ell=1}^{\infty} |a_{k\ell}| < \infty.$$

First of all, it is important to link the absolute and regular convergence.

Theorem 4.45 *An absolutely convergent double series $\sum_{k,\ell} a_{k\ell}$ converges in the regular sense. The converse is not true.*

For a proof in the multidimensional case, we refer to Theorem 7.39. This result readily yields Fubini's theorem for double series.

Corollary 4.46 *For an absolutely convergent double series, the iterated series also converge absolutely; this double series and both iterated ones converge (to the same limit).*

The proof is a consequence of Theorems 4.19, 4.39, and 4.45. In the multidimensional setting the situation is similar (cf. Corollary 7.40).

Theorem 4.47 *If, for the terms of a double series $\sum_{k,\ell} a_{k\ell}$, there holds*

$$|a_{k\ell}| \leq C b_{k\ell}$$

for all $k, \ell \geq 1$, where the double series $\sum_{k,\ell} b_{k\ell}$ is a convergent nonnegative series, then the double series $\sum_{k,\ell} a_{k\ell}$ converges absolutely.

The proof is trivial. This is actually an anticipation of the two-dimensional comparison test (cf. Theorem 5.1 below).

Theorem 4.48 *An absolutely convergent series with terms $\{a_{k\ell}\}$ converges, and*

$$\left| \sum_{k,\ell=1}^{\infty} a_{k\ell} \right| \leq \sum_{k,\ell=1}^{\infty} |a_{k\ell}|.$$

Proof Since the series $\sum_{k,\ell} |a_{k\ell}|$ converges, it follows from Theorem 4.12 that for any $\varepsilon > 0$ there exists a natural number N such that for $K \geq N, L \geq N$, and $\mu, \nu \geq 0$, we have

$$0 \leq \Sigma_{K+\mu,L+\nu} - \Sigma_{KL} < \varepsilon,$$

where

$$\Sigma_{KL} = \sum_{k=1}^{K} \sum_{\ell=1}^{L} |a_{k\ell}|.$$

It is now clear that

$$|S_{K+\mu,L+\nu} - S_{KL}| \leq \Sigma_{K+\mu,L+\nu} - \Sigma_{KL} < \varepsilon,$$

for $K \geq N, L \geq N$, and $\mu \geq 1, \nu \geq 1$. Applying again Theorem 4.12, we conclude that the series $\sum_{k,\ell} a_{k\ell}$ converges, as required.

Besides that,

$$\left| \sum_{k=1}^{K} \sum_{\ell=1}^{L} a_{k\ell} \right| \leq \sum_{k=1}^{K} \sum_{\ell=1}^{L} |a_{k\ell}|$$

for any K and L. Passing to the limit as $K, L \to \infty$, we complete the proof of the theorem. □

It is worth mentioning that the convergence of a double series $\sum_{k,\ell} a_{k\ell}$ does not imply, generally speaking, the convergence of the series $\sum_{k,\ell} |a_{k\ell}|$. A convergent series $\sum a_{k\ell}$ such that $\sum |a_{k\ell}|$ diverges is *conditionally convergent.*

Theorem 4.49 *If a double series $\sum_{k,\ell} a_{k\ell}$ with real-valued terms is conditionally convergent, then there exists a double sequence $\{\delta_{k\ell}\}$ such that*

$$\lim_{k+\ell\to\infty} \delta_{k\ell} = 0$$

and

$$\sum_{k,\ell=1}^{\infty} a_{k\ell}\delta_{k\ell} = +\infty.$$

This result is a simple consequence of Theorem 4.42 (for a detailed proof, see Theorem 7.43).

Theorem 4.50 *If a double series with terms $\{a_{k\ell}\}$ and a single series with terms $\{b_n\}$ have all the terms the same (that is, $\{a_{k\ell}\} = \{b_n\}$), then the absolute convergence of one of them implies the absolute convergence of the other, and their sums are equal.*

For a proof, see Theorem 7.44.

Finally we give sufficient conditions for the convergence to the same sum of both a two-dimensional series and the same series with rearranged terms.

Theorem 4.51 *If a double series with terms $\{a_{k\ell}\}$ is absolutely convergent, then the double series*

$$\sum_{k,\ell=1}^{\infty} a'_{k\ell},$$

with the same terms but rearranged arbitrarily, converges absolutely and has the same sum.

For a multidimensional analog, see Theorem 7.45. Note that Theorem 4.51 simply follows by combining the one-dimensional result from Theorem 2.48 and Theorem 4.50.

Remark 4.52 In view of Theorem 4.51, it is natural to ask whether the same conclusion holds under the weaker assumption that the series $\sum_{k,\ell} a_{k\ell}$ converges in the regular sense, rather than absolutely. This is not the case, as one can easily see through the example

$$a_{k\ell} = \begin{cases} \frac{(-1)^k}{k}, & \text{if } k = \ell, \\ 0, & \text{otherwise}, \end{cases}$$

and Theorem 2.51.

4.7 Notes

Section 4.1 (Double sequences)

1. Double Hardy monotone sequences (Definition 4.9) were already studied in 1903 by Krause [154].

Section 4.2 (Convergence of double series: definitions and basic notions)

2. It is worth warning that one may find different names for the familiar types of convergence. For example, Pringsheim's convergence is sometimes called *unrestricted rectangular convergence*, while the regular one is called *restricted rectangular convergence*.

3. The sequence (4.8) was extracted from [B-56, §48]. Other examples and further discussion (also for the case of double integrals) may be found in the same reference.

4. Regular convergence (Definiton 4.17) was first introduced by Hardy in [121] (see also [190]).

5. Fubini's theorem for regularly convergent double series (Theorem 4.20) was proved in [190]. In that paper, the author also gave the counterexample in Remark 4.23, showing that a characterization of regular convergence for integrals analogous to Theorem 4.19 is not possible.

Section 4.3 (Convergence of double integrals: definitions and basic notions)

6. For further properties of Fresnel integrals, see, e.g., [B-3, Ch. 7].

7. Fubini's theorem for regularly convergent double integrals (Theorem 4.24) was obtained in [193].

Section 4.4 (Different types of convergence of double series)

8. For further discussion on summation by curves (cf. Definition 4.25), as well as for a proof of Theorem 4.26, see [B-60, §6.5]).

9. Theorems 4.27 and 4.30 were proved by Bakhbukh in [26]. Surprisingly, such results have not gotten any attention for the fifty years after their publication.

10. For a proof of Theorem 4.28, see [B-17, p. 55].

11. The sufficient conditions for the spherical convergence of the double series whose terms are products of two one-dimensional sequences in Theorem 4.34 were given by Ambroladze in [13]. The following statement concerning the sharpness of Theorem 4.34 with respect to the boundedness of the sequences $\{ka_k\}$ and $\{\ell b_\ell\}$ was also given in [13]:

Theorem 4.53 *For any nonnegative sequence $\{c_k\}$ with $\lim_{k\to\infty} c_k = \infty$, there exists a convergent series $\sum_k a_k$ satisfying $ka_k \le c_k$ $(k \in \mathbb{N})$ such that the double series $\sum_{k,\ell} a_k a_\ell$ diverges spherically.*

The proof of this statement resembles those of Theorems 4.30 and 4.31, although it is more sophisticated. As a simple consequence of Theorem 4.53, we obtain the following:

Corollary 4.54 *For any $\varepsilon > 0$, there exists a convergent (one-dimensional) series $\sum_k a_k$ with $a_k = O(k^{\varepsilon-1})$ such that the series $\sum_{k,\ell} a_k a_\ell$ diverges spherically.*

12. Definition 4.37 was introduced by Arutyunyan, see [15] and [16], in connection with the problems of representation of functions by multiple trigonometric series, and developed

by Dyachenko, see [86, 87, 88]. It was shown in [86] that the notion of u-convergence is useful in the study of the Fourier series of functions of several variables that have bounded variation in the sense of Hardy.

These definitions are not difficult to generalize to the higher-dimensional case, and such a generalization is studied in [88].

13. As mentioned before, hyperbolic crosses proved to be of crucial importance in various problems of Approximation Theory and Fourier Analysis. They were introduced by K. I. Babenko in [23]. They turned out to be natural in approximation on classes of functions with bounded mixed derivative, where the symbol of the corresponding differential operator is hyperbolic. Among many researches dealing with hyperbolic crosses, we mention the contribution of Temlyakov, see, e.g., [B-28] and [253].

14. In [140], two types of convergence stronger than the regular convergence but weaker than the absolute one were considered for double series by M. Kadets: perfect and unconditional convergence.

Definition 4.55 A double series $\sum_{k,\ell} a_{k\ell}$ is said to be *perfectly convergent* if for any $\alpha_k, \beta_\ell = \pm 1$ $(k, \ell \in \mathbb{N})$ the double series $\sum_{k,\ell} a_{k\ell}\alpha_k\beta_\ell$ converges regularly.

Definition 4.56 A double series $\sum_{k,\ell} a_{k\ell}$ is said to be *unconditionally convergent* if for any permutations $\pi, \sigma : \mathbb{N} \to \mathbb{N}$ the double series $\sum_{k,\ell} a_{\pi(k)\sigma(\ell)}$ converges regularly.

It is interesting that the two notions coincide: a double series is perfectly convergent if and only if it is unconditionally convergent ([140, Theorem 2]). On the other hand, there exists a double series that converges perfectly but not absolutely ([140, Theorem 3]). Furthermore, if one replaces regular convergence by Pringsheim's convergence in Definitions 4.55 and 4.56, then they are not equivalent. A counterexample is given by the double series $\sum_{k,\ell} a_{k\ell}$ with

$$a_{k\ell} = \begin{cases} (-1)^{\ell+k-1}, & \text{if } k \in \{1,2\} \text{ and } \ell \in \mathbb{N}, \\ 0, & \text{otherwise.} \end{cases}$$

For $\alpha_1 = 1$, $\alpha_k = -1$ $(k > 1)$, and $\beta_\ell = 1$, the double series $\sum_{k,\ell} a_{k\ell}\alpha_k\beta_\ell$ diverges, while for any permutations $\pi, \sigma : \mathbb{N} \to \mathbb{N}$, the double series $\sum_{k,\ell} a_{\pi(k)\sigma(\ell)}$ converges in the sense of Pringsheim, since for any $m, n \in \mathbb{N}$ with $m \geq \max\{\pi(1), \pi(2)\}$, we have

$$\sum_{k=1}^{m}\sum_{\ell=1}^{n} a_{\pi(k)\sigma(\ell)} = \sum_{\ell=0}^{n}((-1)^{\sigma(\ell)} - (-1)^{\sigma(\ell)}) = 0.$$

See also [141] for the study of such convergence types in Banach spaces.

15. In [B-52] and [B-53] more detailed study of these types of convergence is undertaken by M. Kadets and V. Kadets, but only in dimension one.

16. In [237], Sheffer proposed a new definition of convergence of double series, the so-called σ-convergence (independently considered by Amerio in [14]), which is "closer" to a multidimensional analog of one-dimensional convergence than Pringsheim's convergence is, but appears to be stronger than the regular one.

17. Before defining σ-convergence we define a σ-region. A σ-region is the union of a finite number of rectangles with the sides parallel to the coordinate axes and containing $(1, 1)$. The lower boundary of a σ-region may be thought of as a "staircase" rising to the right, as in Figure 4.4.

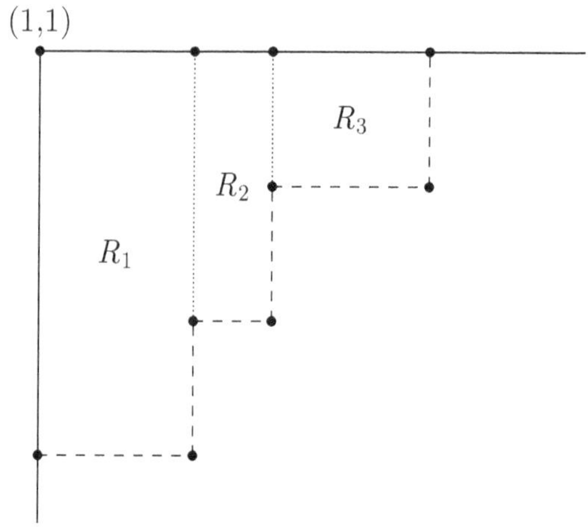

Figure 4.4 σ-region corresponding to the union of the rectangles $R_j, j = 1, 2, 3$.

Definition 4.57 A double series $\sum_{k,\ell} a_{k\ell}$ is said to be σ-convergent to a number A if for every $\varepsilon > 0$ there exist indices p and q such that

$$\left| \sum_{(k,\ell)\in R} a_{k\ell} - A \right| < \varepsilon,$$

for every σ-region R containing (p, q).

Wilansky [274] compared σ-convergence with regular convergence and showed that the former is strictly stronger than the latter, that is, every series that is σ-convergent is also regularly convergent to the same sum, but there exist regular convergent double series that are not σ-convergent.

18. In [186], Meyer gave a necessary and sufficient condition for the σ-convergence of a monotone alternating double series, namely a double series

$$\sum_{k,\ell=1}^{\infty} (-1)^{k+\ell} a_{k\ell},$$

where $\{a_{k\ell}\}$ is a positive monotone sequence (cf. Definition 4.8). Meyer showed that σ-convergence of this series is equivalent to its diagonal summability, namely to the convergence of the one-dimensional series $\sum_{n=2}^{\infty} D_n$, where

$$D_n = \sum_{k+\ell=n} (-1)^{k+\ell} a_{k\ell} = (-1)^n \sum_{k+\ell=n} a_{k\ell}$$

are the diagonal sums of the double series $\sum (-1)^{k+\ell} a_{k\ell}$.

5

Tests for convergence of double series

5.1 Tests for convergence of positive double series

Before presenting double versions of the basic tests for positive series, it is worth mentioning that these are also valid for nonnegative series in most of the cases. When possible, statements will be given for nonnegative series rather than just positive.

A simplest yet important way to figure out whether a nonnegative double series converges or diverges is to compare it with some other series about which such an information is known.

Theorem 5.1 **(Comparison test)** *Given two double series with nonnegative terms $a_{k\ell}$ and $b_{k\ell}$ such that*

$$a_{k\ell} \leq C b_{k\ell}$$

for all k, ℓ, the convergence of $\sum_{k,\ell} b_{k\ell}$ implies that of $\sum_{k,\ell} a_{k\ell}$, and the divergence of $\sum_{k,\ell} a_{k\ell}$ implies that of $\sum_{k,\ell} b_{k\ell}$.

The proof of Theorem 5.1 is trivial; nevertheless, we present its multidimensional version in Theorem 8.1 for the sake of completeness.

We continue with an extension of d'Alembert's test. Recall that such a test essentially states that if a sequence of positive terms decreases faster than a geometric progression of ratio less than 1, its underlying series converges.

Theorem 5.2 *Let $\{a_{k\ell}\}$ be a positive double sequence. If there exists a number $q < 1$ such that*

$$\frac{a_{k+1,\ell}}{a_{k\ell}} \leq q$$

for any fixed ℓ ($k = 1, 2, \ldots$), and

$$\frac{a_{1,\ell+1}}{a_{1\ell}} \leq q$$

for $\ell = 1, 2, \ldots$, then the double series $\sum_{k,\ell} a_{k\ell}$ converges. If for certain $\nu \geq 1$, there holds

$$\frac{a_{k+1,\nu}}{a_{k\nu}} \geq 1$$

for $k = 1, 2, \ldots$, then the series $\sum_{k,\ell} a_{k\ell}$ diverges.

Theorem 5.2 is actually a particular case of the following:

Theorem 5.3 *If two double series with positive terms $a_{k\ell}$ and $b_{k\ell}$ are such that*

$$\frac{a_{k+1,\ell}}{a_{k\ell}} \leq \frac{b_{k+1,\ell}}{b_{k\ell}} \tag{5.1}$$

for any fixed ℓ ($k = 1, 2, \ldots$), and

$$\frac{a_{1,\ell+1}}{a_{1\ell}} \leq \frac{b_{1,\ell+1}}{b_{1\ell}} \tag{5.2}$$

for $\ell = 1, 2, \ldots$, then the convergence of

$$\sum_{k,\ell=1}^{\infty} b_{k\ell} \tag{5.3}$$

implies that of

$$\sum_{k,\ell=1}^{\infty} a_{k\ell}, \tag{5.4}$$

and the divergence of (5.4) *implies that of* (5.3).

Indeed, we observe that Theorem 5.2 is just Theorem 5.3 with $b_{k\ell} = q^{k+\ell}$ (note that the series $\sum_{k,\ell=1}^{\infty} q^{k+\ell-2}$ converges and to $(1-q)^{-2}$, for $q < 1$). The statement concerning divergence is also a particular case. A version of Theorem 5.3 is stated and proved in full generality below (Theorem 8.3). Nevertheless, we include the proof in two dimensions to emphasize that such a result is a consequence of the most basic test, the comparison test (Theorem 5.1).

Proof of Theorem 5.3 Condition (5.1) yields

$$a_{k\ell} \leq \frac{a_{1\ell}}{b_{1\ell}} b_{k\ell}.$$

Further, it follows from (5.2) that

$$\frac{a_{1\ell}}{b_{1\ell}} \leq \frac{a_{11}}{b_{11}}.$$

Therefore,

$$a_{k\ell} \leq \frac{a_{11}}{b_{11}} b_{k\ell}$$

for all $k, \ell = 1, 2, \ldots$, and the conclusion follows from Theorem 5.1. □

Remark 5.4 The assumptions of Theorem 5.2 can be replaced with

$$\frac{a_{k,\ell+1}}{a_{k\ell}} \leq q, \quad \frac{a_{k+1,1}}{a_{k1}} \leq q, \quad \text{and} \quad \frac{a_{\nu,\ell+1}}{a_{\nu\ell}} \geq 1,$$

respectively. Similarly, the assumptions of Theorem 5.3 can be replaced with

$$\frac{a_{k,\ell+1}}{a_{k\ell}} \leq \frac{b_{k,\ell+1}}{b_{k\ell}} \quad \text{and} \quad \frac{a_{k+1,1}}{a_{k1}} \leq \frac{b_{k+1,1}}{b_{k1}},$$

respectively. In fact, this is contemplated in the statement of their multidimensional analogs, Theorems 8.2 and 8.3.

The following theorem is a generalization of Cauchy's kth root test. Again, it is a consequence of the comparison test (its multidimensional analog is Theorem 8.4).

Theorem 5.5 *Given a double series with nonnegative terms $a_{k\ell}$, if there exists a number $q < 1$ and a natural number N such that for $k + \ell > N$,*

$$a_{k\ell}^{\frac{1}{k+\ell}} \leq q,$$

then the double series $\sum_{k,\ell} a_{k\ell}$ converges. On the other hand, if there exist arbitrary large values of $k + \ell$ for which

$$a_{k\ell}^{\frac{1}{k+\ell}} \geq 1,$$

then the series $\sum_{k,\ell} a_{k\ell}$ diverges.

Theorem 5.5 may also be obtained from the following more general statement (cf. Theorem 8.5 for a proof in arbitrary dimension).

Theorem 5.6 *Given a double series with nonnegative terms $\{a_{k\ell}\}$, define*

$$L = \limsup_{k+\ell\to\infty} a_{k\ell}^{\frac{1}{k+\ell}}.$$

(i) *If $L < 1$, then $\sum_{k,\ell} a_{k\ell}$ converges.*
(ii) *If $L > 1$, then $\sum_{k,\ell} a_{k\ell}$ diverges.*
(iii) *If $L = 1$, the test is inconclusive.*

5.2 Tests for convergence involving bounded variation in two dimensions

In our overview of one-dimensional number series, we started with series whose terms are not necessarily positive. We think that there is certain logic in this, though the order of consideration in all the textbooks is just the opposite. However, our order shows additionally why there is a need in the study of positive series. The convergence of the former is less surprising than that of the latter, since the sums of positive numbers always accumulate, while the terms with arbitrary signs may interfere. Therefore, we will keep the chosen order in the multivariate presentations. We restrict ourselves to the double versions of the above considered one-dimensional tests. Since finding them collected together is impossible, the possibility to see them all even in the standard form ought to be useful.

Though the assumptions are posed on the series

$$\sum_{k,\ell=1}^{\infty} a_{k\ell}$$

and

$$\sum_{k,\ell=1}^{\infty} b_{k\ell},$$

the assertions will address the series

$$\sum_{k,\ell=1}^{\infty} a_{k\ell} b_{k\ell}.$$

What indeed lacks for the presentation, which one can quickly spot from comparison with dimension one, is the notion of bounded variation for multiple sequences. However, before this we ought to extend the multidimensional notion of monotonicity for sequences. Recall that in Chapter 4 we have defined a notion of monotonicity in dimension two by

$$a_{k\ell} \geq a_{mn} \quad (a_{k\ell} \leq a_{mn}) \quad \text{for} \quad k \geq m,\ \ell \geq n.$$

We now need a different type of it.

Definition 5.7 A double sequence $\{a_{k\ell}\}$ with real-valued terms is called (1,0)-increasing ((1,0)-decreasing) if

$$a_{k\ell} \geq a_{m\ell} \quad (a_{k\ell} \leq a_{m\ell}) \quad \text{for} \quad k \geq m, \quad \ell \in \mathbb{N},$$

and (0,1)-increasing ((0,1)-decreasing) if

$$a_{k\ell} \geq a_{kn} \quad (a_{k\ell} \leq a_{kn}) \quad \text{for} \quad \ell \geq n, \quad k \in \mathbb{N}.$$

It is called (1,1)-increasing ((1,1)-decreasing) if

$$a_{k\ell} - a_{m\ell} - a_{kn} + a_{mn} \geq 0 \quad (a_{k\ell} - a_{m\ell} - a_{kn} + a_{mn} \leq 0),$$

for $k \geq m,\ \ell \geq n$.

To reproduce the one-dimensional results, we have to introduce the concept of bounded variation in two dimensions. We are mainly interested in the so-called bounded Hardy variation. However, we first define the *Vitali-type variation* of $a = \{a_{k\ell}\}$ as

$$Vv(a) := \sum_{k,\ell=1}^{\infty} |\Delta^{(1,1)} a_{k\ell}|.$$

If $Vv(a)$ is finite we say that $\{a_{k\ell}\}$ is of bounded Vitali variation.

The sequence $\{a_{k\ell}\}$ is said to be of *bounded Hardy variation* if it is of Vitali bounded variation, and additionally $\{a_{k\ell}\}$ is of bounded variation when restricted to any of the two indices.

It is worth noting that the necessary and sufficient condition for $\{a_{k\ell}\}$ to be of *bounded Hardy variation* is that this sequence can be represented as the difference of two sequences each of which is simultaneously (1,0)-decreasing, (0,1)-decreasing, and (1,1)-decreasing (for the statement in several dimensions and a fully detailed proof, see Theorem 8.10).

One more feature of the generalizations of the above tests is that they are impossible without regular convergence. More precisely, each time one has to specify what kind of convergence is presumed. The proofs mainly lie in the application of Abel's transformation (iteratively in each variable).

The two-dimensional Dedekind's test reads as follows.

Theorem 5.8 *If $\{b_{k\ell}\}$ is of bounded Hardy variation and*

$$b_{k\ell} \to 0 \qquad \text{as } k+\ell \to 0,$$

and the partial sums S_{KL} of the series $\sum_{k,\ell} a_{k\ell}$ are uniformly bounded, then the double series $\sum_{k,\ell} a_{k\ell} b_{k\ell}$ converges regularly.

The following is an extension of Dirichlet's test in two dimensions.

Theorem 5.9 *If the partial sums S_{KL} of a double series $\sum_{k,\ell} a_{k\ell}$ are bounded, and if the sequence $\{b_{k\ell}\}$ is (1,0)-decreasing, (0,1)-decreasing, and (1,1)-decreasing, and moreover*

$$b_{k\ell} \to 0 \qquad \text{as } k+\ell \to \infty,$$

then the double series $\sum_{k,\ell} a_{k\ell} b_{k\ell}$ converges regularly.

Remark 5.10 The main point of interest in the above statement (and, correspondingly, also in Theorem 5.8 as well as in Theorem 5.11 below) is that in the conclusion one may obtain the stronger regular convergence of the series $\sum_{k,\ell} a_{k\ell} b_{k\ell}$ rather than its (Pringsheim) convergence (in fact, the corresponding multidimensional results that are presented in Section 8.2 fit this stronger conclusion).

An important and widely known consequence of Theorem 5.9 is the celebrated Leibniz's test for alternating series. One of the possible extensions of such a test may be as follows.

Corollary 5.11 *If the sequence $\{b_{k\ell}\}$ is (1,0)-decreasing, (0,1)-decreasing, and (1,1)-decreasing, and*

$$b_{k\ell} \to 0 \qquad \text{as } k+\ell \to \infty,$$

then the double series

$$\sum_{k,\ell=1}^{\infty} (-1)^{k+\ell} b_{k\ell}$$

converges regularly.

For example, we can derive the convergence of the following generalization of the classical series $\sum_k \frac{(-1)^k}{k}$:

$$\sum_{k,\ell=1}^{\infty} \frac{(-1)^{k+\ell}}{k+\ell}.$$

The du Bois-Reymond test can be generalized as follows.

Theorem 5.12 *If $\{b_{k\ell}\}$ is of bounded Hardy variation, and the double series $\sum_{k,\ell} a_{k\ell}$ converges regularly, then the double series $\sum_{k,\ell} a_{k\ell}b_{k\ell}$ converges regularly.*

It seems that also no double extension of Abel's test exists in the literature; however, it readily follows from Theorem 5.12.

Theorem 5.13 *If the double series $\sum_{k,\ell} a_{k\ell}$ converges regularly, then the double series $\sum_{k,\ell} a_{k\ell}b_{k\ell}$ also converges regularly, provided that the sequence $\{b_{k\ell}\}$ is bounded, (1,0)-decreasing, (0,1)-decreasing, and (1,1)-decreasing.*

5.3 Convergence of double series with monotone decreasing terms

In this section we give two-dimensional versions of three classical tests for series with decreasing terms, along with their proofs: Abel–Olivier's test, Cauchy's condensation test, and Maclaurin–Cauchy's integral test. The corresponding improvements of these tests and their generalizations to larger classes of sequences will be discussed in detail in Chapter 6.

We start with Abel–Olivier's test, which guarantees certain asymptotic decay for terms of convergent series.

Theorem 5.14 *Let $\{a_{k\ell}\}$ be a nonnegative decreasing double sequence. If the double series*

$$\sum_{k,\ell=1}^{\infty} a_{k\ell}$$

converges, then

$$k\ell a_{k\ell} \to 0 \qquad \text{as } k+\ell \to \infty.$$

Proof Since $\{a_{k\ell}\}$ is nonnegative and $\sum a_{k\ell}$ converges, we have, by the Cauchy criterion for convergence of sequences (applied to the partial sums of the series),

$$mna_{2m,2n} \le \sum_{k=m+1}^{2m} \sum_{\ell=n+1}^{2n} a_{k\ell} \le \sum_{k=1}^{2m}\sum_{\ell=1}^{2n} a_{k\ell} - \sum_{k=1}^{m}\sum_{\ell=1}^{n} a_{k\ell} \to 0 \quad \text{as } m+n \to \infty,$$

where in the first inequality we use the fact that $\{a_{k\ell}\}$ is decreasing. □

One of the most important tools in the theory of number series is the Cauchy condensation test. For double series, it reads as follows.

Theorem 5.15 *Let $\{a_{k\ell}\}$ be a nonnegative decreasing double sequence. Then the double series*

$$\sum_{k,\ell=1}^{\infty} a_{k\ell}$$

converges if and only if the double series

$$\sum_{m,n=0}^{\infty} 2^{m+n} a_{2^m,2^n}$$

does as well.

Proof Denoting the partial sums of the series in question by

$$S_{KL} = \sum_{k=1}^{K}\sum_{\ell=1}^{L} a_{k\ell} \quad \text{and} \quad T_{MN} = \sum_{m=0}^{M-1}\sum_{n=0}^{N-1} 2^{m+n} a_{2^m,2^n},$$

respectively, we compare them keeping in mind that, by Theorem 4.38, the convergence of a series follows from the boundedness of the corresponding partial sums.

For $K < 2^M$ and $L < 2^N$, we have

$$S_{KL} \le \sum_{m=0}^{M-1}\sum_{n=0}^{N-1}\left(\sum_{k=2^m}^{2^{m+1}-1}\sum_{\ell=2^n}^{2^{n+1}-1} a_{k\ell}\right) \le \sum_{m=0}^{M-1}\sum_{n=0}^{N-1} 2^{m+n} a_{2^m,2^n} = T_{MN}. \tag{5.5}$$

On the other hand, for $K > 2^M$ and $L > 2^N$, we have

$$\begin{aligned}
3S_{KL} &\ge \sum_{m=0}^{M-1}\sum_{n=0}^{N-1}\left(\sum_{k=2^m}^{2^{m+1}-1}\sum_{\ell=2^n}^{2^{n+1}-1} a_{k\ell}\right) \\
&\quad + \frac{1}{2}\sum_{m=0}^{M-1}\left(\sum_{k=2^m}^{2^{m+1}-1} a_{k1}\right) + \frac{1}{2}\sum_{n=0}^{N-1}\left(\sum_{\ell=2^n}^{2^{n+1}-1} a_{1\ell}\right) + a_{11} \\
&\ge \frac{1}{4}\sum_{m=0}^{M-1}\sum_{n=0}^{N-1} 2^{m+n+2} a_{2^{m+1},2^{n+1}} \\
&\quad + \frac{1}{4}\sum_{m=0}^{M-1} 2^{m+1} a_{2^{m+1},1} + \frac{1}{4}\sum_{n=0}^{N-1} 2^{n+1} a_{1,2^{n+1}} + a_{11} \\
&= \frac{1}{4}\sum_{m=1}^{M}\sum_{n=1}^{N} 2^{m+n} a_{2^m,2^n} + \frac{1}{4}\sum_{m=1}^{M} 2^m a_{2^m,1} + \frac{1}{4}\sum_{n=1}^{N} 2^n a_{1,2^n} + a_{11} \\
&\ge \frac{1}{4} T_{MN}.
\end{aligned}$$

Therefore,

$$12 S_{KL} \ge T_{MN}. \tag{5.6}$$

In virtue of (5.5) and (5.6), the sequences $\{S_{KL}\}$ and $\{T_{MN}\}$ are either both bounded or both unbounded, which proves the theorem. □

As in one dimension, the Maclaurin–Cauchy integral test is one of the most important and helpful tools. It is important to remark that driven by the one-dimensional version, one may tend to think that in two dimensions this result will ensure the equiconvergence of both the double series and the double integral simultaneously. However, this is wrong, since one must also involve the one-dimensional integrals. The example

$$f(x,y) = \begin{cases} \frac{1}{y}, & \text{if } x = 1, \\ e^{-(x+y)}, & \text{if } x > 1, \end{cases}$$

illustrates the situation where $\int_1^\infty \int_1^\infty f(x,y)\,dx\,dy$ converges but $\sum_{k,\ell=1}^\infty f(k,\ell)$ diverges.

The precise statement is as follows.

Theorem 5.16 *If $f(x,y)$ is a nonnegative function decreasing with respect to each of the variables in the domain $[a,\infty)\times[b,\infty)$, where a and b are real numbers, then the series*

$$\sum_{k,\ell=0}^{\infty} f(a+k, b+\ell)$$

converges if and only if all three integrals

$$\int_a^\infty f(x,b)\,dx, \tag{5.7}$$

$$\int_b^\infty f(a,y)\,dy, \tag{5.8}$$

and

$$\int_a^\infty \int_b^\infty f(x,y)\,dy\,dx \tag{5.9}$$

converge.

Proof Since $f(x,y)$ is a decreasing function with respect to each of the variables x and y, we have

$$f(a+k,b+\ell) \le f(x,y) \le f(a+k-1, b+\ell-1),$$

provided $a+k-1 \le x \le a+k$ and $b+\ell-1 \le y \le b+\ell$. Hence,

$$f(a+k,b+\ell) \le \int_{a+k-1}^{a+k} \int_{b+\ell-1}^{b+\ell} f(x,y)\,dy\,dx \le f(a+k-1,b+\ell-1).$$

Summing up in $k = 1, 2, \ldots, \mu$ and $\ell = 1, 2, \ldots, \nu$, we obtain

$$\sum_{k=1}^{\mu}\sum_{\ell=1}^{\nu} f(a+k, b+\ell) \le \int_a^{a+\mu}\int_b^{b+\nu} f(x,y)\,dy\,dx$$
$$\le \sum_{k=0}^{\mu-1}\sum_{\ell=0}^{\nu-1} f(a+k, b+\ell). \tag{5.10}$$

Further,

$$f(a, b+\ell) \le f(a, y) \le f(a, b+\ell-1),$$

provided $b+\ell-1 \le y \le b+\ell$. Therefore,

$$f(a, b+\ell) \le \int_{b+\ell-1}^{b+\ell} f(a,y)\,dy \le f(a, b+\ell-1),$$

and

$$f(a+k, b) \le \int_{a+k-1}^{a+k} f(x,b)\,dx \le f(a+k-1, b).$$

Letting $k = 1, 2, \ldots, \mu$ and $\ell = 1, 2, \ldots, \nu$, and summing the last inequalities, we obtain, correspondingly,

$$\sum_{\ell=1}^{\nu} f(a, b+\ell) \le \int_b^{b+\nu} f(a,y)\,dy \le \sum_{\ell=0}^{\nu-1} f(a, b+\ell), \tag{5.11}$$

and

$$\sum_{k=1}^{\mu} f(a+k, b) \le \int_a^{a+\mu} f(x,b)\,dx \le \sum_{k=0}^{\mu-1} f(a+k, b). \tag{5.12}$$

Since $f(x,y)$ is a nonnegative function, the integrals in (5.10), (5.11), and (5.12) increase with μ and ν. Hence the integrals (5.7), (5.8), and (5.9) converge. By this, the series

$$\sum_{k=0}^{\infty} f(a+k, b), \quad \sum_{\ell=0}^{\infty} f(a, b+\ell), \quad \text{and} \quad \sum_{k,\ell=1}^{\infty} f(a+k, b+\ell)$$

converge, and therefore, the double series

$$\sum_{k,\ell=0}^{\infty} f(a+k, b+\ell)$$

converges as well.

Conversely, if the last series converges, it follows from the second inequality in (5.10) that the double integral (5.9) converges. Similarly, the convergence of the integrals (5.7) and (5.8) follows from the second inequalities in (5.11) and (5.12), respectively. □

Example 5.17 (i) The double series

$$\sum_{k,\ell=1}^{\infty} \frac{1}{k^{\alpha}\ell^{\beta}}$$

converges for $\alpha > 1$ and $\beta > 1$, and diverges if $\alpha \leq 1$ or $\beta \leq 1$.

(ii) The double series

$$\sum_{k,\ell=1}^{\infty} \frac{1}{(k+\ell)^{\alpha}}$$

converges for $\alpha > 2$ and diverges for $\alpha \leq 2$.

To prove this, let us consider the double integral

$$I = \int_1^{\infty}\int_1^{\infty} \frac{dx\,dy}{(x+y)^{\alpha}}$$

and study its convergence. We denote by D_b the domain bounded by the lines $x = 1$, $y = 1$, and $x + y = b$, where $b > 2$. For $\alpha > 2$, we have

$$\begin{aligned} \int\int_{D_b} \frac{dx\,dy}{(x+y)^{\alpha}} &= \int_1^{b-1} dx \int_1^{b-x} \frac{dy}{(x+y)^{\alpha}} \\ &= \frac{b^{1-\alpha}(b-2)}{1-\alpha} - \frac{b^{2-\alpha} - 2^{2-\alpha}}{(1-\alpha)(2-\alpha)}. \end{aligned}$$

It is clear thereby that

$$I = \frac{2^{2-\alpha}}{(1-\alpha)(2-\alpha)},$$

that is, the integral converges. Further, the integrals

$$\int_1^{\infty} \frac{dx}{(1+x)^{\alpha}} \quad \text{and} \quad \int_1^{\infty} \frac{dy}{(1+y)^{\alpha}}$$

also converge. Therefore, by Theorem 5.16, the double series

$$\sum_{k,\ell=1}^{\infty} \frac{1}{(k+\ell)^{\alpha}}$$

converges.

If $\alpha \leq 2$, we have

$$\begin{aligned} \int\int_{D_b} \frac{dx\,dy}{(x+y)^{\alpha}} &\geq \int_1^{b-1} dx \int_1^{b-x} \frac{dy}{(x+y)^{2}} = \int_1^{b-1} \left(\frac{1}{x+1} - \frac{1}{b}\right) dx \\ &= \log\frac{b}{2} - \frac{b-2}{b}, \end{aligned}$$

which tends to infinity as $b \to \infty$. In conclusion, the considered series converges for $\alpha > 2$ and diverges for $\alpha \leq 2$.

(iii) The double series

$$\sum_{k,\ell=1}^{\infty} \frac{1}{(Ak^2 + 2Bk\ell + C\ell^2)^{\alpha}}, \qquad \alpha > 0,$$

with $AC - B^2 > 0$, where A, B, and C are real numbers, and $A > 0$ (which implies $C > 0$), converges for $\alpha > 1$ and diverges for $\alpha \le 1$.

Denoting $M = \max\{A, |B|, C\}$ we have

$$(Ak^2 + 2Bk\ell + C\ell^2)^{\alpha} \le M^{\alpha}(k+\ell)^{2\alpha},$$

and, equivalently,

$$\frac{1}{(Ak^2 + 2Bk\ell + C\ell^2)^{\alpha}} \ge \frac{M^{-\alpha}}{(k+\ell)^{2\alpha}}.$$

Therefore, it follows from the previous example that the series

$$\sum_{k,\ell=1}^{\infty} \frac{M^{-\alpha}}{(k+\ell)^{2\alpha}}$$

diverges for $\alpha \le 1$. On the other hand,

$$Ak^2 + 2Bk\ell + C\ell^2 = \frac{(Ak + B\ell)^2 + (AC - B^2)\ell^2}{A} \ge \frac{AC - B^2}{A}\,\ell^2.$$

Thereby,

$$\frac{1}{(Ak^2 + 2Bk\ell + C\ell^2)^{\alpha}} \le \frac{A^{\alpha}}{(AC - B^2)^{\alpha}} \frac{1}{\ell^{2\alpha}},$$

and, analogously,

$$\frac{1}{(Ak^2 + 2Bk\ell + C\ell^2)^{\alpha}} \le \frac{C^{\alpha}}{(AC - B^2)^{\alpha}} \frac{1}{k^{2\alpha}}.$$

Multiplying both sides of the last two inequalities, we get

$$\frac{1}{(Ak^2 + 2Bk\ell + C\ell^2)^{2\alpha}} \le \left(\frac{AC}{(AC - B^2)^2}\right)^{\alpha} \frac{1}{(k\ell)^{2\alpha}},$$

and, therefore,

$$\frac{1}{(Ak^2 + 2Bk\ell + C\ell^2)^{\alpha}} \le \left(\frac{\sqrt{AC}}{AC - B^2}\right)^{\alpha} \frac{1}{k^{\alpha}\ell^{\alpha}}.$$

Since, for $\alpha > 1$, the double series

$$\sum_{k,\ell=1}^{\infty} \frac{1}{k^{\alpha}\ell^{\alpha}}$$

converges, the given double series converges for $\alpha > 1$, by Theorem 5.1.

5.4 Notes

Section 5.2 (Tests for convergence involving bounded variation in two dimensions)

1. In [B-84, Ch. 1, §1, Th. 1.3], a two-dimensional generalization of Dedekind's test (Theorem 5.8; see Theorem 2.62 for the one-dimensional case) is proved (the author mistakenly attributes it to Dirichlet).

2. Theorem 5.9 may be found in Bromwich's book [B-16, §37].

3. The discussion in Remark 5.10 was already noted by Hardy [122].

4. Corollary 5.11 was obtained in a slightly weaker form in [53]. In particular, regular convergence is not derived there, although the author shows that the two iterated series converge to the same value (which follows from regular convergence; cf. Theorem 4.20).

5. The du Bois-Reymond test was generalized to the two-dimensional case in [B-84, Ch. 1, §1, Th. 1.2].

6. In the proof of Theorem 2.61, it is mentioned that the sequence space ℓ^1 is the β-dual space of c_0, and we relate it to the *BVS* property of the involved sequence $\{b_n\}$. For recent developments in this direction in the two-dimensional case, we refer to [279].

6

Convergence tests involving generalizations of monotonicity

In dimension one, we have seen that in many convergence tests one can relax the assumption of monotonicity. The situation is similar in the multivariate setting. Two-dimensional generalizations allow one to feel better both the resemblance to and difference from the one-dimensional case.

6.1 Double series and generalizations of monotonicity

We are going to focus mainly on two different classes of sequences, namely weak and general monotonicity. Both extend the class of monotone sequences in two dimensions.

6.1.1 Weak monotonicity

We begin with *weak monotonicity*, which extends the corresponding one-dimensional concept introduced in Chapter 1 (see Definition 1.29).

Definition 6.1 A nonnegative double sequence $\{a_{k\ell}\}$ is said to be a *weak monotone sequence*, written $\{a_{k\ell}\} \in WMS$, if, for any $m, n \in \mathbb{N}$,

$$a_{k\ell} \le C a_{mn}, \qquad \text{for } m \le k \le 2m,\ n \le \ell \le 2n,$$

where $C > 0$ is a constant independent of m, n.

It is clear that monotone sequences are also weak monotone. However, the class WMS is much larger, since it contains some increasing sequences (for instance, $a_{k\ell} = k\ell$). In fact, the class WMS contains all polynomials in the variables k and ℓ.

One can also define the WMS property independently in each index.

Definition 6.2 Let $\{a_{k\ell}\}$ be a nonnegative sequence. For $j = 1, 2$, $\{a_{k\ell}\}$ is said to be weak monotone in the jth variable, written $\{a_{k\ell}\} \in WMS_j$, if

$$\begin{aligned} a_{kn} &\le C a_{mn}, \qquad \text{for } m \le k \le 2m \text{ and any } n \in \mathbb{N},\ \text{if } j = 1,\\ a_{m\ell} &\le C a_{mn}, \qquad \text{for } n \le \ell \le 2n \text{ and any } m \in \mathbb{N},\ \text{if } j = 2, \end{aligned}$$

where $C > 0$ is a constant independent of m, n.

Remark 6.3 It is easy to see that $\{a_{k\ell}\} \in WMS$ if and only if $\{a_{k\ell}\} \in WMS_j$ for $j = 1, 2$ (cf. Remark 9.3 for the details in arbitrary dimension).

The following properties for *WMS* follow immediately from the definition (for the multidimensional case, see Proposition 9.5).

Proposition 6.4 *Let $\{a_{k\ell}\}$ and $\{b_{k\ell}\}$ be double WMS.*

(i) *If $\{a_{k\ell}\} \in WMS$ with constant $C \le 1$, then $\{a_{k\ell}\}$ is decreasing.*
(ii) *If there exist $k, \ell \in \mathbb{N}$ such that $a_{k\ell} = 0$, then $a_{mn} = 0$ for all $m \ge k$ and $n \ge \ell$.*
(iii) *For every $m_1 \le m_2$ and $n_1 \le n_2 \in (\mathbb{N} \cup \{0\})$, and every $2^{m_j} \le k_j \le 2^{m_j+1}$ and $2^{n_j} \le \ell_j \le 2^{n_j+1}$ with $k_j \le \ell_j$ ($j = 1, 2$), we have*

$$a_{\ell_1,\ell_2} \le C a_{2^{n_1},2^{n_2}} \le \cdots \le C^{\max_j\{n_j - m_j\}} a_{2^{m_1+1},2^{m_2+1}} \le C^{\max_j\{n_j-m_j\}+1} a_{k_1,k_2}.$$

(iv) *The double sequences $\{a_{k\ell} + b_{k\ell}\}$ and $\{a_{k\ell} b_{k\ell}\}$ are WMS.*
(v) *The double sequences $\{\min(a_{k\ell}, b_{k\ell})\}$ and $\{\max(a_{k\ell}, b_{k\ell})\}$ are WMS.*

We now present some examples of two-dimensional *WMS*.

Example 6.5 (i) If the terms of a double sequence $\{a_{k\ell}\}$ are of the form

$$a_{k\ell} = b_k c_\ell,$$

where $\{b_k\}$ and $\{c_\ell\}$ are one-dimensional positive sequences, then $\{a_{k\ell}\} \in WMS$ if and only if the sequences $\{b_k\}$ and $\{c_\ell\}$ are weak monotone in the one-dimensional sense.

(ii) Let $P, Q \ne 0$ be polynomials of two variables with nonnegative coefficients and $\{b_k\}, \{c_\ell\} \in WMS$. The sequence

$$a_{k\ell} = P\big(b_k, c_\ell\big)$$

is *WMS*. If, moreover, the sequences $\{b_k\}$ and $\{c_\ell\}$ are increasing and $\{d_{k\ell}\}$ is a nonnegative double sequence such that $0 < \alpha \le d_{k\ell} \le \beta < \infty$, then

$$a_{k\ell} = \frac{P\big(b_k, c_\ell\big)}{Q\big(b_k, c_\ell\big)} d_{k\ell}$$

is *WMS*. This is not true in general if the sequences $\{b_k\}$ and $\{c_\ell\}$ are not increasing.

(iii) Let $b > 1$, and let φ_1 and φ_2 be strictly increasing positive functions defined on $\mathbb{R}_+$. Then

$$a_{k\ell} = b^{\varphi_1(k)+\varphi_2(\ell)} \in WMS \quad \text{if and only if} \quad \varphi_j(2x) - \varphi_j(x) = O(1)$$

as $x \to \infty$, for $j = 1, 2$.

These examples will be discussed in more detail in Section 9.1.

It is evident from Example 6.5 that weak monotone sequences may increase to infinity. Actually, if $\{a_{k\ell}\} \in WMS$, it cannot grow to infinity faster than a polynomial (depending on the sequence).

Proposition 6.6 *Let $\{a_{k\ell}\}$ be a double WMS with constant $C > 0$. Then the double sequence $\left\{(k\ell)^{-\log_2 C} a_{k\ell}\right\}$ is bounded.*

For a proof in arbitrary dimension, see Proposition 9.7.

We also introduce quasi-monotone sequences. We will not focus on the study of those, but rather present their relation with *WMS*.

Definition 6.7 Let $\alpha > 0$. We say that $\{a_{k\ell}\}$ is α-quasi-monotone (written $\{a_{k\ell}\} \in QMS(\alpha)$) if the sequence $\{(k\ell)^{-\alpha} a_{k\ell}\}$ is monotone (decreasing). The whole class of quasi-monotone sequences (*QMS*) is defined as

$$QMS = \bigcup_{\alpha>0} QMS(\alpha).$$

The following is an important property relating *QMS* and *WMS*.

Theorem 6.8 *Let $\{a_{k\ell}\} \in WMS$ with constant $C \geq 1$. Then there exists a double sequence $\{b_{k\ell}\} \in QMS(\log_2 C)$ such that*

$$a_{k\ell} \leq b_{k\ell} \leq 4C^3 a_{k\ell} \qquad \text{for every } (k, \ell) \in \mathbb{N}^2.$$

The proof of this theorem is based on the restriction in the growth of $\{a_{k\ell}\}$ given in Proposition 6.6. For a detailed proof of Theorem 6.8 in the multidimensional setting, see Theorem 9.9.

We also define the concept of weak monotonicity for functions of two variables.

Definition 6.9 A nonnegative function defined on $(0, \infty)^2$ is said to be *weak monotone*, written $f \in WM$, if, for any $x, y \in (0, \infty)$,

$$f(s, t) \leq Cf(x, y), \qquad \text{for } x \leq s \leq 2x,\ y \leq t \leq 2y,$$

where $C > 0$ is a constant independent of x, y.

Remark 6.10 If we put $f(s, t) = a_{k\ell}$ for $(s, t) \in (k, k+1] \times (\ell, \ell+1]$, then $f \in WM$ if and only if $\{a_{k\ell}\} \in WMS$.

6.1.2 General monotone sequences

We present yet another generalization of monotonicity that proved to be convenient in harmonic analysis, yielding extended solutions to problems that were originally studied for just monotone sequences. Such a concept is called *general monotonicity*. Recall the notation introduced in Section 5.2:

$$\Delta^{(1,0)} a_{k\ell} = a_{k\ell} - a_{k+1,\ell}, \quad \Delta^{(0,1)} a_{k\ell} = a_{k\ell} - a_{k,\ell+1},$$
$$\Delta^{(1,1)} a_{k\ell} = \Delta^{(1,0)}(\Delta^{(0,1)} a_{k\ell}) = \Delta^{(0,1)}(\Delta^{(1,0)} a_{k\ell})$$
$$= a_{k\ell} - a_{k+1,\ell} - a_{k,\ell+1} + a_{k+1,\ell+1}.$$

Definition 6.11 Let $\{a_{k\ell}\}$ be a complex-valued double sequence. We say that $\{a_{k\ell}\}$ is a *general monotone sequence*, written $\{a_{k\ell}\} \in GMS$, if there is a constant $C > 0$ such that

$$\sum_{k=m}^{2m-1}\sum_{\ell=n}^{2n-1}|\Delta^{(1,1)}a_{k\ell}|\le C|a_{mn}|, \tag{6.1}$$

for all $m,n\in\mathbb{N}$.

Remark 6.12 As in one dimension, although *GMS* is defined for complex-valued sequences, we imply that any $\{a_{k\ell}\}\in GMS$ is nonnegative when comparing such a class with any other class of nonnegative sequences, as, for instance, *QMS* or *WMS*.

Remark 6.13 In dimension one, if $\{a_k\}\in GMS$ and $a_k=0$ for some $k\in\mathbb{N}$, then $a_n=0$ for every $n\ge k$. This is not true for two-dimensional *GMS*. A counterexample is delivered by $a_{k\ell}=k+\ell-2$. It is clear that $\Delta^{(1,1)}a_{k\ell}=0$ for all $k,\ell\in\mathbb{N}$, and therefore trivially $\{a_{k\ell}\}\in GMS$. However, $a_{11}=0$ and $a_{k\ell}>0$ for all k,ℓ with $\max\{k,\ell\}>1$.

We continue with some examples of *GMS*.

Example 6.14 (i) Let $\{a_{k\ell}\}\subset\mathbb{C}$ be of the form $a_{k\ell}=b_kc_\ell$ for some one-dimensional complex-valued sequences $\{b_k\}$ and $\{c_\ell\}$. If $\{b_k\}$ and $\{c_\ell\}$ satisfy the one-dimensional general monotonicity condition, that is,

$$\sum_{k=n}^{2n-1}|b_k-b_{k+1}|\le C_1|b_n|,\qquad\text{and}\qquad\sum_{\ell=m}^{2m-1}|c_\ell-c_{\ell+1}|\le C_2|c_m|,$$

then trivially $\{a_{k\ell}\}\in GMS$, since in this case

$$\Delta^{(1,1)}a_{k\ell}=(b_k-b_{k+1})(c_\ell-c_{\ell+1}).$$

(ii) If $a_{k\ell}\to 0$ as $k+\ell\to\infty$, $\Delta^{(1,1)}a_{k\ell}\ge 0$, and the double series

$$\sum_{k,\ell=1}^{\infty}\Delta^{(1,1)}a_{k\ell} \tag{6.2}$$

converges, then $\{a_{k\ell}\}\in GMS$. Indeed, since

$$\begin{aligned}\sum_{k=m}^{M}\sum_{\ell=n}^{N}|\Delta^{(1,1)}a_{k\ell}|&=\sum_{k=m}^{M}\sum_{\ell=n}^{N}\Delta^{(1,1)}a_{k\ell}\\&=a_{mn}-a_{m,N+1}-a_{M+1,n}+a_{M+1,N+1},\end{aligned} \tag{6.3}$$

and this expression grows with M,N (because $\Delta^{(1,1)}a_{k\ell}\ge 0$), we obtain, by using $a_{k\ell}\to 0$ as $k+\ell\to\infty$,

$$\sum_{k=m}^{2m-1}\sum_{\ell=n}^{2n-1}\Delta^{(1,1)}a_{k\ell}\le\sum_{k=m}^{\infty}\sum_{\ell=n}^{\infty}\Delta^{(1,1)}a_{k\ell}=a_{mn},$$

and hence $\{a_{k\ell}\}\in GMS$.

(iii) In (ii), the conclusion holds if one replaces the assumption on the convergence of the series (6.2) by the condition $\{|a_{k\ell}|\} \in WMS$. Indeed, in this case, by (6.3) (with $M = 2m-1$ and $N = 2n-1$),

$$\sum_{k=m}^{2m-1}\sum_{\ell=n}^{2m-1} |\Delta^{(1,1)}a_{k\ell}| \le |a_{mn}| + |a_{m,2n}| + |a_{2m,n}| + |a_{2m,2n}|$$
$$\le (3C+1)|a_{mn}|,$$

where C is the WMS constant of $\{a_{k\ell}\}$.

(iv) For $M \in \mathbb{N}$, let $a_{k\ell} = (k+\ell)^M$ (if $M = 0$ this sequence is constant, and thus trivially GMS, while for $M = 1$, we have $\Delta^{(1,1)}a_{k\ell} = 0$, and hence $\{a_{k\ell}\} \in GMS$ as well). For $M \ge 2$, using the binomial relation

$$(a+b)^M = \sum_{N=0}^{M} \binom{M}{N} a^N b^{M-N}, \qquad a, b \in \mathbb{C},$$

we have

$$\begin{aligned}
|\Delta^{(1,1)}a_{k\ell}| &= |(k+\ell)^M - 2(k+\ell+1)^M + (k+\ell+2)^M| \\
&= \left|(k+\ell)^M - 2\sum_{N=0}^{M}\binom{M}{N}(k+\ell)^N + \sum_{N=0}^{M}\binom{M}{N}(k+\ell)^N 2^{M-N}\right| \\
&= \left|\sum_{N=0}^{M-1}\binom{M}{N}(k+\ell)^N 2^{M-N} - 2\sum_{N=0}^{M-1}\binom{M}{N}(k+\ell)^N\right| \\
&= 2\sum_{N=0}^{M-2}\binom{M}{N}(k+\ell)^N(2^{M-N-1}-1) \asymp (k+\ell)^{M-2}.
\end{aligned}$$

Therefore,

$$\begin{aligned}
\sum_{k=m}^{2m-1}\sum_{\ell=n}^{2n-1} |\Delta^{(1,1)}a_{k\ell}| &\asymp \sum_{k=m}^{2m-1}\sum_{\ell=n}^{2n-1} (k+\ell)^{M-2} \asymp mn(m+n)^{M-2} \\
&\lesssim \max\{m,n\}^M \asymp (m+n)^M = a_{mn},
\end{aligned}$$

and so $\left\{a_{k\ell} = (k+\ell)^M\right\} \in GMS$ for any $M \in \mathbb{N}$. A more general version of this example will be given in the multidimensional setting, where we will also consider $M \in \mathbb{R}$ (cf. Example 9.15).

(v) The product of two double GMS is not necessarily GMS. To see this, consider

$$b_{k\ell} = \begin{cases} 1, & \text{if } k \text{ is even,} \\ 2, & \text{if } k \text{ is odd,} \end{cases} \qquad c_{k\ell} = \begin{cases} 1, & \text{if } \ell \text{ is even,} \\ 2, & \text{if } \ell \text{ is odd.} \end{cases}$$

It is clear that for any $k,\ell \in \mathbb{N}$, $\Delta^{(1,1)}b_{k\ell} = \Delta^{(1,1)}c_{k\ell} = 0$, and hence $\{b_{k\ell}\}, \{c_{k\ell}\} \in GMS$. Furthermore, if we define $a_{k\ell} = b_{k\ell}c_{k\ell}$, then

$$a_{k\ell} = \begin{cases} 1, & \text{if } k \text{ and } \ell \text{ are even,} \\ 2, & \text{if } k+\ell \text{ is odd,} \\ 4, & \text{if } k \text{ and } \ell \text{ are odd.} \end{cases}$$

It follows easily that $|\Delta^{(1,1)}a_{k\ell}| = 1$ for every $k,\ell \in \mathbb{N}$, and therefore the *GMS* condition in this case reads as

$$mn = \sum_{k=m}^{2m-1}\sum_{\ell=n}^{2n-1} |\Delta^{(1,1)}a_{k\ell}| \le Ca_{mn} \le 4C,$$

which clearly does not hold.

We now define, similarly to the one-dimensional case, the two-dimensional analog of the class *RBVS*. A double sequence $\{a_{k\ell}\}$ is said to be with rest of bounded variation, written $\{a_{k\ell}\} \in RBVS$, if there exists a constant $C > 0$ such that

$$\sum_{k=m}^{\infty}\sum_{\ell=n}^{\infty} |\Delta^{(1,1)}a_{k\ell}| \le Ca_{mn}, \qquad m,n \in \mathbb{N}.$$

The *RBVS* condition is rather restrictive, and it is clear that $RBVS \subset GMS$, where the inclusion is strict. This can be shown by using the one-dimensional counterexample $\{c_k\}$ given in Example 1.11, and defining

$$a_{k\ell} = c_k c_\ell,$$

with $\Delta^{(1,1)}a_{k\ell} = \Delta c_k \Delta c_\ell$.

We will not focus on the study of *RBVS*, although, for the sake of completeness, and due to its close relation with *GMS*, it is convenient to at least introduce it and mention such a relation.

Remark 6.15 The sequence $a_{k\ell} = k+\ell$ in Example 6.14 (iv) satisfies $\Delta^{(1,1)}a_{k\ell} = 0$ for every k,ℓ, although it tends to infinity as $k+\ell \to \infty$. This illustrates that the *GMS* condition alone may sometimes not be enough to capture the variational properties of a sequence $\{a_{k\ell}\}$ (this phenomenon is also observed in the sequences $\{b_{k\ell}\}$ and $\{c_{k\ell}\}$ in Example 6.14 (v)). In some cases, the variational properties may be represented by the one-dimensional differences rather than by two-dimensional ones (for instance, those in (6.4) below).

Remark 6.16 In dimension one, the classes *GMS* and *WMS* are related by $GMS \subsetneq WMS$ (cf. Remark 1.34). In dimension two this is not the case (i.e., $GMS \not\subset WMS$). Indeed, let

$$a_{k\ell} = \begin{cases} \frac{1}{m^2 n}, & \text{if } 2^m \le k < 2^{m+1},\ \ell = 2^n,\ m,n \in \mathbb{N}, \\ \frac{1}{mn}, & \text{if } 2^m \le k < 2^{m+1} \text{ and } 2^n < \ell < 2^{n+1},\ m,n \in \mathbb{N}. \end{cases}$$

Obviously, this sequence is not *WMS*. To check the belonging of the sequence to *GMS*, one should calculate and estimate the differences

$$a_{k\ell} - a_{k,\ell+1} - a_{k+1,\ell} + a_{k+1,\ell+1},$$

for $2^m \le k < 2^{m+1}$ and $2^n \le \ell < 2^{n+1}$. All of them are zero except the two, where $k = 2^{m+1} - 1$, $\ell = 2^n$ and $k = 2^{m+1} - 1$, $\ell = 2^{n+1} - 1$. For the first one, the absolute value of the difference in question is positive and satisfies

$$\frac{1}{m^2 n} - \frac{1}{(m+1)^2 n} - \frac{1}{mn} + \frac{1}{(m+1)n} \le \frac{2}{m^2 n}.$$

The second difference is estimated analogously; therefore, the sum of the differences is dominated by $a_{2^m,2^n}$. The other cases are the shifted versions of the described one; therefore, the sequence is *GMS*.

Remark 6.17 In relation with Remark 6.16, we also have *WMS* $\not\subset$ *GMS*, and therefore these two classes are not comparable. Indeed, consider the sequence $\{a_{k\ell}\}$ given by

$$a_{k,1} = \begin{cases} 2 \cdot 4^{-m}, & \text{if } 2^m \le k < 2^{m+1} \text{ and } k \text{ is even,} \\ 4^{-m}, & \text{if } 2^m \le k < 2^{m+1} \text{ and } k \text{ is odd,} \end{cases} \qquad m \in \mathbb{N},$$

and $a_{k\ell} = 0$ for $\ell > 1$. It is clear that $\{a_{k\ell}\} \in$ *WMS*. On the other hand, we have, for $m \in \mathbb{N}$,

$$\sum_{k=2^m}^{2^{m+1}-1} |\Delta^{(1,1)} a_{k,1}| \asymp 2^m 4^{-m} = 2^{m-1} a_{2^m,1},$$

or, in other words, the *GMS* condition does not hold for any constant $C > 0$. The extension of this example to arbitrary dimension is given in Remark 9.17.

Remark 6.18 In general, the condition $\{a_{k\ell}\} \in$ *GMS* does not imply that the corresponding one-dimensional inequalities

$$\sum_{k=m}^{2m-1} |\Delta^{(1,0)} a_{kn}| \le C|a_{mn}|, \qquad \text{and} \qquad \sum_{\ell=n}^{2n-1} |\Delta^{(0,1)} a_{m\ell}| \le C|a_{mn}|, \tag{6.4}$$

with $m, n \in \mathbb{N}$, hold. Indeed, for the sequence $\{a_{k\ell}\} \in$ *GMS* in Remark 6.16, the one-dimensional inequality

$$\sum_{\ell=n}^{2n-1} |\Delta^{(0,1)} a_{m\ell}| \le C a_{mn}$$

is not satisfied for any absolute constant C, since, for any $m, n \in \mathbb{N}$,

$$\frac{1}{mn} \asymp \frac{1}{n}\left(\frac{1}{m} - \frac{1}{m^2}\right) = |\Delta^{(0,1)} a_{2^m,2^n}| \le \sum_{\ell=2^n}^{2^{n+1}-1} |\Delta^{(0,1)} a_{2^m,\ell}| \not\lesssim a_{2^m,2^n} = \frac{1}{m^2 n}.$$

Remarks 6.15 and 6.18 motivate the definition of the following more restrictive class of general monotone sequences.

Definition 6.19 Let $\{a_{k\ell}\}$ be a complex-valued double sequence. We say that $\{a_{k\ell}\}$ is a *Hardy-general monotone sequence*, written $\{a_{k\ell}\} \in HGMS$, if $\{a_{k\ell}\} \in GMS$, and moreover (6.4) holds.

Note that the condition defining the class *GMS* relates to the Vitali-type bounded variation of a sequence (that is, only iterated differences are considered), while the conditions that define the class *HGMS* are related to the Hardy-type bounded variation concept, which, in addition to the Vitali-type variation of a sequence, involves also both of the one-dimensional differences, namely $\Delta^{(1,0)}a_{k\ell}$ and $\Delta^{(0,1)}a_{k\ell}$ (where ℓ is fixed in the first case, and k is fixed in the second). Thus, it would be appropriate to refer to *GMS* as *Vitali-general monotone sequences*, although for simplicity we will not adopt this nomenclature.

A criterion similar to Proposition 1.9 holds for *HGMS* in two dimensions.

Proposition 6.20 *A double sequence $\{a_{k\ell}\} \subset \mathbb{C}$ belongs to the class HGMS if and only if $\{|a_{k\ell}|\} \in WMS$, for every $M > m$,*

$$\sum_{k=m}^{M} |\Delta^{(1,0)}a_{k\ell}| \le C\left(|a_{m\ell}| + \sum_{k=m+1}^{M} \frac{|a_{k\ell}|}{k}\right), \qquad \ell \in \mathbb{N}, \tag{6.5}$$

for every $N > n$,

$$\sum_{\ell=n}^{N} |\Delta^{(0,1)}a_{k\ell}| \le C\left(|a_{kn}| + \sum_{\ell=n+1}^{N} \frac{|a_{k\ell}|}{\ell}\right), \qquad k \in \mathbb{N}, \tag{6.6}$$

and for every $M > m$ and every $N > n$,

$$\begin{aligned}\sum_{k=m}^{M}\sum_{\ell=n}^{N} |\Delta^{(1,1)}a_{k\ell}| \\ \le C\left(|a_{mn}| + \sum_{k=m+1}^{M}\sum_{\ell=n+1}^{N} \frac{|a_{k\ell}|}{k\ell} + \sum_{k=m+1}^{M} \frac{|a_{kn}|}{k} + \sum_{\ell=n+1}^{N} \frac{|a_{m\ell}|}{\ell}\right).\end{aligned} \tag{6.7}$$

Proof Since $\{a_{k\ell}\} \in HGMS$, by definition we have $\{|a_{k\ell_0}|\}, \{|a_{k_0\ell}|\} \in GMS$ as one-dimensional sequences, for fixed ℓ_0 and k_0, respectively. Thus, by Proposition 1.9 we obtain $\{|a_{k\ell_0}|\}, \{|a_{k_0\ell}|\} \in WMS$, for any fixed ℓ_0 and k_0, respectively (note that this is equivalent to the fact that $\{|a_{k\ell}|\} \in WMS$ as a double sequence; see Remark 9.3 for details in the multidimensional setting), and moreover (6.5) and (6.6) hold. It then remains to establish (6.7). For $m < M$ and $n < N$, let $\mu, \nu \ge 0$ be such that $2^{\mu}m \le M < 2^{\mu+1}m$ and $2^{\nu}n \le N < 2^{\nu+1}n$. Then

$$\sum_{k=m}^{M}\sum_{\ell=n}^{N}|\Delta^{(1,1)}a_{k\ell}| \le \sum_{s=0}^{\mu}\sum_{k=2^s m}^{2^{s+1}m-1}\sum_{t=0}^{\nu}\sum_{\ell=2^t n}^{2^{t+1}n-1}|\Delta^{(1,1)}a_{k\ell}|$$
$$\le C\sum_{s=0}^{\mu}\sum_{t=0}^{\nu}|a_{2^s m,2^t n}| \le C\Bigg(|a_{mn}|+\sum_{s=1}^{\mu}\sum_{t=1}^{\nu}|a_{2^s m,2^t n}|\sum_{k=2^s m}^{2^{s+1}m-1}\sum_{\ell=2^t n}^{2^{t+1}n-1}\frac{1}{k\ell}$$
$$+\sum_{s=1}^{\mu}|a_{2^s m,n}|\sum_{k=2^s m}^{2^{s+1}m-1}\frac{1}{k}+\sum_{t=1}^{\nu}|a_{m,2^t n}|\sum_{\ell=2^t n}^{2^{t+1}n-1}\frac{1}{\ell}\Bigg)$$
$$\le C\Bigg(|a_{mn}|+\sum_{k=m+1}^{M}\sum_{\ell=n+1}^{N}\frac{|a_{k\ell}|}{k\ell}+\sum_{k=m+1}^{M}\frac{|a_{kn}|}{k}+\sum_{\ell=n+1}^{N}\frac{|a_{m\ell}|}{\ell}\Bigg).$$

Note that in the above inequalities we have repeatedly used $\{|a_{k\ell}|\} \in WMS$. We also emphasize that the different occurrences of C do not refer to the same value, in general. Let us now prove the converse. By Proposition 1.9, it follows from (6.5), (6.6), and $\{|a_{k\ell}|\} \in WMS$ that $\{a_{k\ell_0}\}, \{a_{k_0\ell}\} \in GMS$ as one-dimensional complex-valued sequences, that is, both inequalities in (6.4) hold. Finally, using the *WMS* condition, we derive from (6.7) that

$$\sum_{k=m}^{2m-1}\sum_{k=n}^{2n-1}|\Delta^{(1,1)}a_{k\ell}|$$
$$\le C\Bigg(|a_{mn}|+\sum_{k=m+1}^{2m-1}\sum_{\ell=n+1}^{2n-1}\frac{|a_{k\ell}|}{k\ell}+\sum_{k=m+1}^{2m-1}\frac{|a_{kn}|}{k}+\sum_{\ell=n+1}^{2n-1}\frac{|a_{m\ell}|}{\ell}\Bigg)$$
$$\le C\Bigg(|a_{mn}|+|a_{mn}|\sum_{k=m+1}^{2m-1}\sum_{\ell=n+1}^{2n-1}\frac{1}{k\ell}+|a_{mn}|\sum_{k=m+1}^{2m-1}\frac{1}{k}+|a_{mn}|\sum_{\ell=n+1}^{2n-1}\frac{1}{\ell}\Bigg)$$
$$\le C|a_{mn}|,$$

as desired. □

Note that unlike the fact that the classes *GMS* and *WMS* are not comparable (cf. Remark 9.16), one has $HGMS \subset WMS$.

Similar to the one-dimensional case, we introduce the class $\overline{GMS}$ (cf. (1.15)). Such a class is defined by replacing the term on the right-hand side of (6.1) by the averages of $\{a_{k\ell}\}$.

Definition 6.21 Let $\{a_{k\ell}\}$ be a complex-valued double sequence. We say that $\{a_{k\ell}\} \in \overline{GMS}$ if there exist constants $c > 1$ and $C > 0$ such that

$$\sum_{k=m}^{2m-1}\sum_{\ell=n}^{2n-1}|\Delta^{(1,1)}a_{k\ell}| \le C\sum_{k=\lfloor m/c\rfloor}^{\lfloor cm\rfloor}\sum_{\ell=\lfloor n/c\rfloor}^{\lfloor cn\rfloor}\frac{|a_{k\ell}|}{k\ell},$$

for all $m, n \in \mathbb{N}$.

Remark 6.22 We have $GMS \subsetneq \overline{GMS}$. Indeed, to prove such an inclusion, let us restrict ourselves to the case $c = 2$ (for the other cases calculations are similar) and write

$$mn \sum_{k=m}^{2m-1} \sum_{\ell=n}^{2n-1} |\Delta^{(1,1)} a_{k\ell}| \lesssim \sum_{M=\lfloor m/2 \rfloor}^{2m-1} \sum_{N=\lfloor n/2 \rfloor}^{2n-1} \sum_{k=M}^{2M-1} \sum_{\ell=N}^{2N-1} |\Delta^{(1,1)} a_{k\ell}|$$
$$\leq C \sum_{M=\lfloor m/2 \rfloor}^{2m-1} \sum_{N=\lfloor n/2 \rfloor}^{2n-1} |a_{MN}|,$$

where in the last equality we applied the GMS condition. Dividing both sides by mn and using

$$\frac{1}{mn} \sum_{M=\lfloor m/2 \rfloor}^{2m-1} \sum_{N=\lfloor n/2 \rfloor}^{2n-1} |a_{MN}| \asymp \sum_{M=\lfloor m/2 \rfloor}^{2m-1} \sum_{N=\lfloor n/2 \rfloor}^{2n-1} \frac{|a_{MN}|}{MN}$$

proves the inclusion. In order to show its properness, it suffices to consider a one-dimensional sequence $\{b_k\} \in \overline{GMS} \backslash GMS$ (constructed in Example 1.13) and define

$$a_{k\ell} = b_k b_\ell.$$

Then, $\{a_{k\ell}\} \in \overline{GMS} \backslash GMS$ as a double sequence.

We aim to define a counterpart of GMS for functions of two variables. To this end, some discussion on the concept of bounded variation in two dimensions is necessary.

6.1.3 Functions of bounded variation

Let $f\colon \mathbb{R}^2 \to \mathbb{C}$ and let $\{x_n\}, \{y_n\} \subset \mathbb{R}$ be two strictly increasing sequences. We define the differences

$$\begin{aligned}
\Delta^{(1,0)} f(x_i, y) &:= f(x_i, y) - f(x_{i+1}, y), \\
\Delta^{(0,1)} f(x, y_j) &:= f(x, y_j) - f(x, y_{j+1}), \\
\Delta^{(1,1)} f(x_i, y_j) &:= \Delta^{(0,1)}(\Delta^{(1,0)} f(x_i, y_j)) = \Delta^{(1,0)}(\Delta^{(0,1)} f(x_i, y_j)) \\
&= f(x_i, y_j) - f(x_{i+1}, y_j) - f(x_i, y_{j+1}) + f(x_{i+1}, y_{j+1}).
\end{aligned}$$

Given a compact rectangle $R := [a, b] \times [c, d] \subset \mathbb{R}^2$, we denote by $\mathcal{P}(R)$ the set of all partitions of R, namely

$$\mathcal{P}(R) = \mathcal{P}([a, b]) \times \mathcal{P}([c, d]),$$

with $\mathcal{P}([a, b])$ as in Definition 1.1. The *Vitali variation of* $f\colon R \to \mathbb{C}$ *over* R is defined as

$$VV_R(f) := \sup \sum_{i=0}^{m-1} \sum_{j=0}^{n-1} |\Delta^{(1,1)} f(x_i, y_j)|,$$

where the supremum is taken over all products of two finite sequences $\{x_k\}_{k=0}^{m} \times \{y_k\}_{k=0}^{n} \subset \mathcal{P}(R)$.

If $VV_R(f)$ is finite we say that f is of bounded Vitali variation on R, and the quantity $VV_R(f)$ is also referred to as the Riemann–Stieltjes integral

$$\int_a^b \int_c^d |d^{(1,1)} f(s,t)|.$$

As in one dimension, we remark that the double Riemann–Stieltjes integral is defined slightly differently than above, although the two definitions coincide.

Likewise, we define the one-dimensional Riemann–Stieltjes integrals

$$\int_a^b |d^{(1,0)} f(s,y)| = \sup_{\{x_k\}_{k=0}^{m} \subset \mathcal{P}([a,b])} \sum_{j=0}^{m-1} |\Delta^{(1,0)} f(x_j, y)|, \qquad y \in [c,d],$$

$$\int_c^d |d^{(0,1)} f(x,t)| = \sup_{\{y_k\}_{k=0}^{n} \subset \mathcal{P}([c,d])} \sum_{k=0}^{n-1} |\Delta^{(0,1)} f(x, y_k)|, \qquad x \in [a,b].$$

We emphasize that the last two (one-dimensional) Riemann–Stieltjes integrals are functions of y and x, respectively.

The Vitali variation of f is defined over noncompact rectangles R' as

$$VV_{R'}(f) = \sup_{\substack{R \subset R' \\ R \text{ compact}}} VV_R(f).$$

It is said that f is of *bounded Hardy variation* on a rectangle R if f is of bounded Vitali variation on R, and additionally f is of bounded variation when restricted to any of the two variables.

Hardy variation neutralizes the following unpleasant property of Vitali variation: if a function f has bounded Vitali variation, the function $f(x,y)+f_1(x)+f_2(y)$ continues to possess this property even for very bad (say, not measurable) marginal functions of one variable $f_1(x)$ and $f_2(y)$.

Definition 6.23 A function f defined on a rectangle $R \subset \mathbb{R}^2$ (which may be noncompact) is locally of bounded Hardy variation on R if it is of bounded Hardy variation on every compact rectangle $R' \subset R$.

Finally, we call a function f defined on $(0,\infty)^2$ *admissible* if it is locally of bounded Hardy variation on $(0,\infty)^2$ and $f(x,y) \to 0$ as $x+y \to \infty$. Further remarks and definitions about variation in several dimensions are given in Section 8.2.

6.1.4 General monotone functions

We are now in a position to define the concept of general monotonicity in two dimensions.

Definition 6.24 A complex-valued admissible function is said to be general monotone, written $f \in GM$, if there exists a constant $C > 0$ such that for all $x, y > 0$,

$$VV_{[x,2x]\times[y,2y]}(f) = \int_x^{2x}\int_y^{2y} |d^{(1,1)}f(s,t)| \le C|f(x,y)|.$$

As usual, it is convenient to define the class GM (or any other related class) for complex-valued functions, but we will restrict ourselves to the nonnegative case when comparing the GM class with classes of nonnegative functions, as QM or WM.

The counterpart of $\overline{GMS}$ for functions is defined as follows.

Definition 6.25 Let f be a complex-valued admissible function. We say that $f \in \overline{GM}$ if there exist constants $c > 1$ and $C > 0$ such that

$$\int_x^{2x}\int_y^{2y} |d^{(1,1)}f(s,t)| \le C\int_{x/c}^{cx}\int_{y/c}^{cy} \frac{|f(s,t)|}{st}\,dt\,ds,$$

for all $x, y > 0$.

Remark 6.26 Let $f \in GM$ (respectively $f \in \overline{GM}$). If we define, for $k, \ell \in \mathbb{N}$, $a_{k\ell} = f(k,\ell)$, then the sequence $\{a_{k\ell}\}$ belongs to the class GMS (respectively $\overline{GMS}$). Indeed, since

$$|\Delta^{(1,1)}a_{k\ell}| = \left|\int_k^{k+1}\int_\ell^{\ell+1} d^{(1,1)}f(s,t)\right| \le \int_k^{k+1}\int_\ell^{\ell+1} |d^{(1,1)}f(s,t)|,$$

the claim follows by summing up in $m \le k \le 2m-1$ and $\ell \le n \le 2\ell - 1$ and applying the GM (respectively $\overline{GM}$) condition.

Although we will not contemplate it in the sequel, we include the definition of the counterpart of $HGMS$ in the context of functions for the sake of completeness.

Definition 6.27 An admissible complex-valued function f is said to be *Hardy-general monotone*, written $f \in HGM$, if $f \in GM$, and, additionally, there exists a constant $C > 0$ such that for all $x, y > 0$,

$$\int_x^{2x} |d^{(1,0)}f(s,y)| \le C|f(x,y)| \quad \text{and} \quad \int_y^{2y} |d^{(0,1)}f(x,t)| \le C|f(x,y)|.$$

Remark 6.28 A similar criterion to Proposition 6.20 may be obtained for HGM functions.

6.2 Convergence tests for double series with monotone-type coefficients

Here we present different generalizations of convergence tests known either for two-dimensional decreasing sequences (see above) or for one-dimensional weak monotone sequences (cf. Chapter 2). We will not include the proofs of most of these results, as they are given in Part III for arbitrary dimension, along with further discussion.

6.2.1 Abel–Olivier's test

Recall that this test was originally treated as a necessary condition for a one-dimensional series with decreasing coefficients $\{a_n\}$ to converge, namely $na_n \to 0$ (cf. Theorem 2.98), and that by using Abel's transformation one can easily arrive at an extended Abel–Olivier's test and its generalization, incorporating weighted series (cf. Theorems 2.99 and 2.100, respectively).

In two and higher dimensions, the Abel–Olivier test is significantly more involved than in the one-dimensional case. The two-dimensional extended version (without weights, i.e., the two-dimensional converse of Theorem 2.99) reads as follows.

Theorem 6.29 *Let $\{a_{k\ell}\} \in WMS$. Then the double series*

$$\sum_{k,\ell=1}^{\infty} a_{k\ell}$$

converges if and only if all the following conditions hold:

$$k\ell a_{k\ell} \to 0 \qquad \text{as } k+\ell \to \infty, \tag{6.8}$$

$$k\sum_{\ell=1}^{\infty} a_{k\ell} \to 0 \quad \text{and} \quad k\sum_{\ell=1}^{\infty} \ell(a_{k\ell} - a_{k,\ell+1}) \to 0 \qquad \text{as } k \to \infty,$$

$$\ell\sum_{k=1}^{\infty} a_{k\ell} \to 0 \quad \text{and} \quad \ell\sum_{k=1}^{\infty} k(a_{k\ell} - a_{k+1,\ell}) \to 0 \qquad \text{as } \ell \to \infty,$$

and the double series

$$\sum_{k,\ell=1}^{\infty} k(a_{k\ell} - a_{k+1,\ell}), \qquad \sum_{k,\ell=1}^{\infty} \ell(a_{k\ell} - a_{k,\ell+1}),$$

$$\sum_{k,\ell=1}^{\infty} k\ell(a_{k\ell} - a_{k,\ell+1} - a_{k+1,\ell} + a_{k+1,\ell+1})$$

converge.

In particular, in this case

$$\sum_{k,\ell=1}^{\infty} a_{k\ell} = \sum_{k,\ell=1}^{\infty} k(a_{k\ell} - a_{k+1,\ell}) = \sum_{k,\ell=1}^{\infty} \ell(a_{k\ell} - a_{k,\ell+1})$$
$$= \sum_{k,\ell=1}^{\infty} k\ell(a_{k\ell} - a_{k,\ell+1} - a_{k+1,\ell} + a_{k+1,\ell+1}).$$

One may also wonder whether this test is sharp or, in other words, whether we can ensure that the convergence of $\sum_{k,\ell=1}^{\infty} a_{k\ell}$ implies

$$(u_{k\ell})k\ell a_{k\ell} \to 0 \qquad \text{as } k,\ell \to \infty,$$

for some sequence $\{u_{k\ell}\}$ increasing to infinity. This question as well as related ones will be discussed in Subsection 9.2.1 in the context of multidimensional series.

For a double sequence $\{b_{k\ell}\}$, we denote

$$B_{k\ell}^{(1,0)} = \sum_{m=1}^{k} b_{m\ell}, \qquad B_{k\ell}^{(0,1)} = \sum_{n=1}^{\ell} b_{kn}, \qquad B_{k\ell}^{(1,1)} = \sum_{n=1}^{\ell}\sum_{m=1}^{k} b_{mn}.$$

The weighted generalization of Theorem 6.29 (which is the two-dimensional version of Theorem 2.100) is as follows.

Theorem 6.30 *Let $\{a_{k\ell}\} \in WMS$ and $\{b_{k\ell}\}$ be a double sequence of nonnegative numbers such that*

$$B_{k\ell}^{(1,0)} \lesssim \sum_{m=\lfloor k/2\rfloor}^{k} b_{m\ell}, \qquad B_{k\ell}^{(0,1)} \lesssim \sum_{n=\lfloor \ell/2\rfloor}^{\ell} b_{kn}, \tag{6.9}$$

for all $k, \ell \in \mathbb{N}$. Then

$$\sum_{k,\ell=1}^{\infty} a_{k\ell} b_{k\ell}$$

converges if and only if

$$B_{k\ell}^{(1,1)} a_{k\ell} \to 0 \qquad \text{as } k+\ell \to \infty,$$

$$\sum_{k=1}^{\infty} B_{k\ell}^{(0,1)} a_{k\ell} \to 0 \quad \text{and} \quad \sum_{k=1}^{\infty} B_{k\ell}^{(1,1)} \Delta^{(1,0)} a_{k\ell} \to 0, \quad \text{as } \ell \to \infty,$$

$$\sum_{\ell=1}^{\infty} B_{k\ell}^{(1,0)} a_{k\ell} \to 0 \quad \text{and} \quad \sum_{\ell=1}^{\infty} B_{k\ell}^{(1,1)} \Delta^{(0,1)} a_{k\ell} \to 0, \quad \text{as } k \to \infty,$$

and all the series

$$\sum_{k,\ell=1}^{\infty} B_{k\ell}^{(1,0)} \Delta^{(1,0)} a_{k\ell}, \quad \sum_{k,\ell=1}^{\infty} B_{k\ell}^{(0,1)} \Delta^{(0,1)} a_{k\ell}, \quad \sum_{k,\ell=1}^{\infty} B_{k\ell}^{(1,1)} \Delta^{(1,1)} a_{k\ell}$$

converge. In particular, in this case all these series converge to the same sum $\sum_{k,\ell=1}^{\infty} a_{k\ell} b_{k\ell}$.

For a proof in arbitrary dimension, we refer to Theorem 9.36.

Finally, we may drop the assumptions (6.9) on the sequence $\{b_{k\ell}\}$ in Theorem 6.30 if we assume that $\{a_{k\ell}\}$ is decreasing in each variable (for a proof of such an assertion in arbitrary dimension, see Theorem 9.37).

Theorem 6.31 *Let $\{a_{k\ell}\}$ be a nonnegative double sequence decreasing in each variable, such that $a_{k\ell} \to 0$ as $k+\ell \to \infty$. Then, for any nonnegative double sequence $\{b_{k\ell}\}$, the conclusion of Theorem 6.30 holds.*

We will show in Subsection 6.2.2 below that in relation to Abel–Olivier's test, one cannot guarantee a faster decay of the sequence $\{a_{k\ell}\}$ than (6.8). On the other hand, it is interesting to figure out under which conditions on $\{a_{k\ell}\}$ the decay rate in (6.8) can be improved. In particular, we are interested in obtaining a condition of the form

$$\sqrt{k^2+\ell^2}a_{k\ell} \to 0, \qquad \text{as } k+\ell \to \infty$$

(i.e., replacing the weight $k\ell$ in condition (6.8) by the radial weight $\sqrt{k^2+\ell^2}$). This is possible if we consider sequences in an appropriate class, where the *WMS* condition is satisfied radially. For convenience, to this end, we deal with functions rather than sequences.

Definition 6.32 Let $f\colon (0,\infty)^2 \to \mathbb{R}_+$. We say that f is *radially weak monotone*, written $f \in WM_{\mathrm{Rad}}$, if for any $x, y > 0$,

$$f(s,t) \le Cf(x,y) \quad \text{if } x^2+y^2 \le s^2+t^2 \le 2(x^2+y^2), \tag{6.10}$$

where $C > 0$ is a constant independent of x, y.

Remark 6.33 We have $WM_{\mathrm{Rad}} \subsetneq WM$. The inclusion is clear, while the fact that it is proper can be proved by means of a simple counterexample, such as

$$f(x,y) = \begin{cases} 1, & \text{if } x \le 1, \\ 0, & \text{otherwise.} \end{cases}$$

An interesting property of radially weak monotone functions is that they are equivalent to radial functions whose radial part is weak monotone in the one-dimensional sense.

Lemma 6.34 *A necessary and sufficient condition for $f \in WM_{\mathrm{Rad}}$ is that there exists a radial function $g(x,y) = g_0\big(\sqrt{x^2+y^2}\big)$, $g_0 \in WM$, such that, for some $0 < C_1 < C_2$,*

$$C_1 f(x,y) \le g(x,y) \le C_2 f(x,y),$$

for all $x, y > 0$.

The classical Abel–Olivier test for $f \in WM_{\mathrm{Rad}}$ provides faster decay of f at infinity than (6.8) (or rather than its counterpart for functions, which is $xyf(x,y) \to 0$ as $x+y \to \infty$).

Proposition 6.35 *Let $f \in WM_{\mathrm{Rad}}$. If the double integral*

$$\int_0^\infty \int_0^\infty f(x,y)\,dx\,dy$$

converges, then $(x^2+y^2)f(x,y) \to 0$ as $x+y \to \infty$.

The proof is rather straightforward by passing to polar coordinates and using Lemma 6.34 along with the one-dimensional Abel–Olivier test for *WM* functions (Theorem 2.99). For a detailed proof, we refer to the multidimensional Proposition 9.41.

The following result relates the decay properties of the terms of divergent series with those of the harmonic series.

Theorem 6.36 *Let $\{a_{k\ell}\}$ be a double WMS and assume that the series $\sum_{k,\ell=1}^{\infty} a_{k\ell}$ is divergent. Then the series*

$$\sum_{k,\ell=1}^{\infty} \min\left\{a_{k\ell}, \frac{1}{k\ell}\right\}$$

is also divergent.

A proof in arbitrary dimension is given in Theorem 9.42.

Remark 6.37 The proof of Theorem 6.36 (or rather its multidimensional analog, Theorem 9.42) uses the fact that for $\{a_{k\ell}\} \in WMS$, the sequence with general term

$$b_{k\ell} = \min\left\{a_{k\ell}, \frac{1}{k\ell}\right\}$$

is also a *WMS*, which follows from (v) of Proposition 6.4. Thus, Abel–Olivier's test may be applied to the sequence $\{b_{k\ell}\}$, which satisfies $k\ell b_{k\ell} \not\to 0$ as $k+\ell \to \infty$ and yields the divergence of the series $\sum_{k,\ell} b_{k\ell}$. However, such a proof would not work if we assume that $\{a_{k\ell}\}$ is decreasing and aim to use the classical Abel–Olivier test, since in this case the sequence $\{b_{k\ell}\}$ is not necessarily decreasing.

6.2.2 Optimality of Abel–Olivier's test

Here we study the optimality of condition (6.8) in Abel–Olivier's test (Theorem 6.29). In particular, we are interested in the question whether there exists a double sequence $\{u_{k\ell}\}$ with $\limsup_{k+\ell\to\infty} u_{k\ell} = \infty$ such that

$$u_{k\ell} k\ell a_{k\ell} \to 0 \quad \text{as } k+\ell \to \infty, \quad \text{or as } k, \ell \to \infty, \tag{6.11}$$

provided that the series $\sum_{k,\ell} a_{k\ell}$ converges. This is not the case, even if $\{a_{k\ell}\}$ is decreasing.

Proposition 6.38 *Let $\{u_{k\ell}\}$ be an increasing double sequence and assume that $\limsup_{k\to\infty} u_{k,\ell_0} = \infty$ for some fixed $\ell_0 \in \mathbb{N}$. Then there exists a positive decreasing double sequence $\{a_{k\ell}\}$ such that $\sum_{k,\ell} a_{k\ell}$ converges, and for any $\ell \ge \ell_0$,*

$$u_{k\ell} k\ell a_{k\ell} \not\to 0 \quad \textit{as } k \to \infty.$$

Likewise, if $\limsup_{\ell\to\infty} u_{k_0,\ell} = \infty$ for some fixed $k_0 \in \mathbb{N}$, there exists a positive decreasing double sequence $\{a_{k\ell}\}$ such that $\sum_{k,\ell} a_{k\ell}$ converges, and for any $k \ge k_0$,

$$u_{k\ell} k\ell a_{k\ell} \not\to 0 \quad \textit{as } \ell \to \infty.$$

Finally, if $\limsup_{k,\ell\to\infty} u_{k\ell} = \infty$ then there exists a positive decreasing double sequence $\{a_{k\ell}\}$ such that $\sum_{k,\ell} a_{k\ell}$ converges, while

$$u_{k\ell} k\ell a_{k\ell} \not\to 0 \quad \textit{as } k, \ell \to \infty.$$

In particular, the above conditions imply that if $\limsup_{k+\ell\to\infty} u_{k\ell} = \infty$, *there exists a positive decreasing double sequence* $\{a_{k\ell}\}$ *such that* $\sum_{k,\ell} a_{k\ell}$ *converges and*

$$u_{k\ell} k\ell a_{k\ell} \not\to 0 \quad \textit{as } k+\ell \to \infty.$$

For a proof in the multidimensional setting, we refer to Proposition 9.44.

Remark 6.39 Using Proposition 6.38, we can construct a decreasing positive sequence $\{a_{k\ell}\}$ such that $\sum_{k,\ell} a_{k\ell}$ converges, but

$$(k^2+\ell^2)a_{k\ell} \not\to 0 \quad \text{as } k+\ell\to\infty, \qquad \text{nor as } k,\ell\to\infty.$$

Indeed, for

$$u_{k\ell} = \frac{k^2+\ell^2}{k\ell},$$

we have

$$\limsup_{k+\ell\to\infty} u_{k\ell} = \infty \qquad \text{and} \qquad \limsup_{k,\ell\to\infty} u_{k\ell} = \infty,$$

and Proposition 6.38 applies. Moreover, since

$$k\ell \le k^2+\ell^2,$$

the Abel–Olivier test for WM_{Rad} (Proposition 6.35) yields a faster decay of $\{a_{k\ell}\}$ than the usual (6.8) in the classical Abel–Olivier test.

Remark 6.40 Similarly, the Abel–Olivier test for WM_{Rad} is also optimal. That is, for every positive function φ such that $\varphi(x,y) \to \infty$ as $x+y\to\infty$, there exists $f \in WM_{\text{Rad}}$ such that the double integral

$$\int_0^\infty \int_0^\infty f(x,y)\,dx\,dy$$

converges, but $\varphi(x,y)(x^2+y^2)f(x,y) \not\to 0$ as $x+y\to\infty$. For further details in arbitrary dimension, see Proposition 9.46.

In relation to (6.11), given $\{a_{k\ell}\} \in WMS$ such that the sequence with condensed indices $\{2^{k+\ell}a_{2^k,2^\ell}\}$ belongs to the class WMS_j (for either $j=1$ or $j=2$; see Definition 6.2), we can obtain a decay of the type (6.8) with an additional logarithmic term. More precisely, if $\{2^{k+\ell}a_{2^k,2^\ell}\} \in WMS_1$ (the case WMS_2 is symmetric) and $\sum_{k,\ell} a_{k\ell}$ converges, then

$$(\log k)k\ell a_{k\ell} \to 0 \quad \text{as } k+\ell\to\infty. \tag{6.12}$$

To this end, we first present an Abel–Olivier test for WMS_j.

Lemma 6.41 *Let* $\{a_{k\ell}\}$ *be a nonnegative double sequence and assume that* $\{a_{k\ell}\} \in WMS_1$. *If the double series* $\sum_{k,\ell} a_{k\ell}$ *converges, then*

$$k a_{k\ell} \to 0 \quad \textit{as} \quad k+\ell\to\infty.$$

We omit the proof, since it is rather trivial. With Lemma 6.41 in hand, one may derive a decay of the form (6.12) for the appropriate classes of sequences.

Proposition 6.42 *Let $\{a_{k\ell}\} \in WMS$ be such that $\{2^{k+\ell} a_{2^k,2^\ell}\} \in WMS_1$ and assume that the double series $\sum_{k,\ell} a_{k\ell}$ converges. Then,* (6.12) *holds.*

The proof is based on the Cauchy condensation test for double *WMS* (see Corollary 6.50 below). For a detailed proof of Proposition 6.42 in the multidimensional setting, see Proposition 9.48.

6.2.3 Applications of Orlicz-type classes

Let us give two-dimensional counterparts of the results in Subsection 2.4.7. Recall that these are consequences of Abel–Olivier's test (in particular, of condition (6.8)). The following statements are useful for applications in the theory of discrete and continuous Orlicz and Zygmund spaces, respectively. For appropriate references, see Notes-89 in Chapter 2. We will only state them without proofs; for details, see Subsection 9.2.3.

Theorem 6.43 *Let $\{a_{k\ell}\}$ be a nonnegative double sequence and let $\varphi\colon [0,\infty) \to [0,\infty)$ be nondecreasing such that*

$$\varphi(x^2) \le C\varphi(x), \qquad x > 1. \tag{6.13}$$

(i) *If the series*

$$\sum_{k,\ell=1}^{\infty} \frac{a_{k\ell}}{(k\ell)^2} \varphi(a_{k\ell}) \tag{6.14}$$

converges, then the series

$$\sum_{k,\ell=1}^{\infty} \frac{a_{k\ell}}{(k\ell)^2} \varphi(k\ell) \tag{6.15}$$

does as well.

(ii) *If there exists $C > 0$ such that*

$$a_{mn} \le \frac{C}{mn} \sum_{k=\lfloor m/2 \rfloor}^{m} \sum_{\ell=\lfloor n/2 \rfloor}^{n} a_{k\ell}$$

for all $m, n \in \mathbb{N}$, then the convergence of (6.15) *implies the convergence of* (6.14).

In particular, if $\{a_{k\ell}\} \in WMS$, then the series (6.14) *and* (6.15) *converge or diverge simultaneously.*

For a proof, see the multidimensional Theorem 9.49. An example of a function φ satisfying (6.13) is

$$\varphi(x) = \log^{\beta}(1+x), \qquad \beta \ge 0.$$

The counterpart of Theorem 6.43 for functions is as follows.

Theorem 6.44 *Let $f\colon (0,1]^2 \to [0,\infty)$ and let $\varphi\colon [0,\infty) \to [0,\infty)$ be nondecreasing such that*

$$\varphi(x^2) \le C\varphi(x), \qquad x > 1.$$

(i) *If the integral*

$$\int_0^1 \int_0^1 f(x,y)\varphi(f(x,y))\,dx\,dy \tag{6.16}$$

converges, then the integral

$$\int_0^1 \int_0^1 f(x,y)\varphi\Big(\frac{1}{xy}\Big)\,dx\,dy \tag{6.17}$$

does as well.

(ii) *If there exists $C > 0$ such that the estimate*

$$f(x,y) \le \frac{C}{xy}\int_{x/2}^{x}\int_{y/2}^{y} f(s,t)\,dt\,ds$$

holds for all $x, y \in (0,1]$, then the convergence of (6.17) *implies the convergence of* (6.16).

In particular, if $f \in WM$, both integrals (6.16) *and* (6.17) *converge or diverge simultaneously.*

For a proof, we refer to the corresponding multidimensional statement, Theorem 9.50.

Remark 6.45 The one-dimensional version of Theorem 6.43 (respectively Theorem 6.44), Theorem 2.115 (respectively Theorem 2.116), holds for sequences in the class $WMS \cup \overline{GMS}$ (respectively functions of the class $WM \cup \overline{GM}$). Indeed, this is because in one dimension there holds

$$a_n \le \frac{C}{n}\sum_{k=\lfloor n/2\rfloor}^{n} a_k, \qquad \{a_k\} \in WMS \cup \overline{GMS},$$

$$f(x) \le \frac{C}{x}\int_{x/2}^{x} f(t)\,dt, \qquad f \in WM \cup \overline{GM}.$$

These estimates are not true in general in two (or arbitrary) dimensions for GMS nor GM, although they still hold in the weak monotone case, that is,

$$a_{mn} \le \frac{C}{mn}\sum_{k=\lfloor m/2\rfloor}^{m}\sum_{\ell=\lfloor n/2\rfloor}^{n} a_{k\ell}, \qquad \{a_{k\ell}\} \in WMS, \tag{6.18}$$

$$f(x,y) \le \frac{C}{xy}\int_{x/2}^{x}\int_{y/2}^{y} f(s,t)\,dt\,ds \qquad f \in WM.$$

Let us show that (6.18) does not hold, in general, for $\{a_{k\ell}\} \in GMS$ (which is a subclass of $\overline{GMS}$; see Remark 6.22).

Let

$$a_{k\ell} = \begin{cases} \frac{1}{mn}, & \text{if } 2^{m-1} \le k < 2^m,\ \ell = 2^{n-1},\ m,n \in \mathbb{N}, \\ \frac{1}{m^2 n}, & \text{if } 2^{m-1} \le k < 2^m \text{ and } 2^{n-1} < \ell < 2^n,\ m,n \in \mathbb{N}. \end{cases}$$

Notice that this example is very similar to that of Remark 6.16. In order to prove that $\{a_{k\ell}\} \in GMS$, we note that if $m \in \mathbb{N}$ and $n \geq 2$, $2^{m-1} \leq i < 2^m$, and $2^{n-1} \leq j < 2^n$, then

$$\begin{aligned}
&\sum_{k=i}^{2i-1}\sum_{\ell=j}^{2j-1} |a_{k\ell} - a_{k+1,\ell} - a_{k,\ell+1} + a_{k+1,\ell+1}| \\
&\leq \sum_{k=2^{m-1}}^{2^{m+1}-3}\sum_{\ell=2^{n-1}}^{2^{n+1}-3} |a_{k\ell} - a_{k+1,\ell} - a_{k,\ell+1} + a_{k+1,\ell+1}|.
\end{aligned}$$

The last sum has exactly three nonzero terms for each $m \in \mathbb{N}$ and $n \geq 2$, namely those corresponding to the pairs of indices

$$(k,\ell) = (2^m-1, 2^{n-1}),\ (2^m-1, 2^n-1),\ \text{and}\ (2^m-1, 2^n)$$

(for $n = 1$ this sum has two nonzero terms, since in this case $2^{n-1} = 2^n - 1$, and the first two pairs of indices are just the same). Therefore, for any $m, n \in \mathbb{N}$ we have

$$\begin{aligned}
&\sum_{k=i}^{2i-1}\sum_{\ell=j}^{2j-1} |a_{k\ell} - a_{k+1,\ell} - a_{k,\ell+1} + a_{k+1,\ell+1}| \\
&\leq |a_{2^m-1,2^{n-1}} - a_{2^m,2^{n-1}} - a_{2^m-1,2^{n-1}+1} + a_{2^m,2^{n-1}+1}| \\
&\quad + |a_{2^m-1,2^n-1} - a_{2^m,2^n-1} - a_{2^m-1,2^n} + a_{2^m,2^n}| \\
&\quad + |a_{2^m-1,2^n} - a_{2^m,2^n} - a_{2^m-1,2^n+1} + a_{2^m,2^n+1}| \\
&\leq C\frac{1}{m^2 n} \leq Ca_{ij}.
\end{aligned}$$

We now show that an estimate of the form (6.18) is not possible. Indeed, for $m, n \in \mathbb{N}$,

$$\begin{aligned}
\sum_{k=2^{m-1}}^{2^m}\sum_{\ell=2^{n-1}}^{2^n} a_{k\ell} &= \sum_{k=2^{m-1}}^{2^m}\sum_{\ell=2^{n-1}+1}^{2^n-1} a_{k\ell} + \sum_{k=2^{m-1}}^{2^m} a_{k,2^{n-1}} + \sum_{k=2^{m-1}}^{2^m} a_{k,2^n} \\
&\leq \sum_{k=2^{m-1}}^{2^m}\sum_{\ell=2^{n-1}+1}^{2^n-1} \frac{1}{m^2 n} + \sum_{k=2^{m-1}}^{2^m} \frac{1}{mn} + \sum_{k=2^{m-1}}^{2^m} \frac{1}{m(n+1)} \\
&\leq \frac{2^m 2^{n-1}}{m^2 n} + \frac{2^{m+1}}{mn}.
\end{aligned}$$

Thus, if estimate (6.18) were true, one would have, for $k = 2^m$ and $\ell = 2^n$,

$$a_{2^m,2^n} = \frac{1}{(m+1)(n+1)} \leq C\left(\frac{1}{2m^2 n} + \frac{1}{2^{n-1}mn}\right),$$

for all m, n. However, it is clear that this inequality holds for no constant C as $m + n \to \infty$.

6.2.4 Variations of Dvoretzky's test

We now present a generalization of Dvoretzky's type test (Theorem 2.112) in the two-dimensional setting. To this end, we recall the notation

$$\overleftarrow{\Delta} u_k := u_{k+1} - u_k,$$

which is the "backward" difference (so that $\overleftarrow{\Delta} u_k = -\Delta u_k$).

We first state a generalized Dvoretzky type test in two dimensions and then present a two-dimensional analog of the original statement.

Theorem 6.46 *Let $\{a_{k\ell}\}, \{b_{k\ell}\} \in WMS$. Assume that the series $\sum_{k,\ell} a_{k\ell}$ and $\sum_{k,\ell} b_{k\ell}$ are convergent and divergent, respectively, and that the one-dimensional series*

$$\sum_{\ell=1}^{\infty} b_{k\ell} \qquad \text{and} \qquad \sum_{k=1}^{\infty} b_{k\ell}$$

are convergent for all fixed k and ℓ, respectively. Then, for any increasing sequences of natural numbers $\{u_k\}$ and $\{v_\ell\}$ satisfying $\overleftarrow{\Delta} u_{k+1} = O(\overleftarrow{\Delta} u_k)$ and $\overleftarrow{\Delta} v_{\ell+1} = O(\overleftarrow{\Delta} v_\ell)$, there exist infinitely many pairs (k, ℓ) such that

$$a_{mn} < b_{mn}, \tag{6.19}$$

for all $u_k \le m \le u_{k+1}$ and $v_\ell \le n \le v_{\ell+1}$.

For a proof in arbitrary dimension, see Theorem 9.53.

We mention that the reason for calling Theorem 6.46 a "generalized" Dvoretzky type test is not only that the monotonicity assumption on $\{a_{k\ell}\}$ and $\{b_{k\ell}\}$ is replaced by *WMS*. Indeed, the original result by Dvoretzky was stated for specific choices of the sequences $\{u_k\}$ and $\{v_\ell\}$ (see Theorem 2.112 and the corresponding references in Notes-88 of Chapter 2). An analog of the original test (extended to *WMS*) reads as follows.

Corollary 6.47 *Let $\{a_{k\ell}\}, \{b_{k\ell}\} \in WMS$. Assume that the series $\sum_{k,\ell} a_{k\ell}$ and $\sum_{k,\ell} b_{k\ell}$ are convergent and divergent, respectively, and that the one-dimensional series*

$$\sum_{\ell=1}^{\infty} b_{k\ell} \qquad \text{and} \qquad \sum_{k=1}^{\infty} b_{k\ell}$$

are convergent for all fixed k and ℓ, respectively. Then, for any $M, N > 1$, there exist infinitely many pairs (k, ℓ) such that

$$a_{mn} < b_{mn},$$

for every $M^k \le m \le M^{k+1}$ and $N^\ell \le n \le N^{\ell+1}$.

Remark 6.48 The conclusion of Theorem 6.46 does not hold if we do not assume that the series $\sum_{\ell=1}^{\infty} b_{k\ell}$ and $\sum_{k=1}^{\infty} b_{k\ell}$ converge for all k and ℓ, respectively. Indeed, consider the example

$$a_{k\ell} = 2^{-k-\ell}, \qquad b_{k\ell} = \begin{cases} 1, & \text{if } k = 1, \\ 2^{-k-\ell}, & \text{if } k > 1, \end{cases}$$

so that $\sum_{k,\ell} a_{k\ell}$ converges, $\sum_{k,\ell} b_{k\ell}$ diverges, and moreover $\sum_{\ell=1}^{\infty} b_{1\ell}$ also diverges (and $\{a_{k\ell}\}, \{b_{k\ell}\} \in WMS$). Clearly, (6.19) is not valid. However, in this case one may obtain a weaker assertion, which we state in full generality in the multidimensional setting (Theorem 9.53).

6.2.5 Cauchy's condensation test and its generalizations

We begin with a generalization of Cauchy's classical condensation test (see Theorem 5.15): if $\{a_{k\ell}\}$ is a nonnegative decreasing sequence, then

$$\sum_{k,\ell=1}^{\infty} a_{k\ell} \quad \text{converges if and only if} \quad \sum_{k,\ell=0}^{\infty} 2^{k+\ell} a_{2^k,2^\ell} \quad \text{does.}$$

As in the one-dimensional case (Theorem 2.91), we generalize such a statement in two different ways. First, we replace monotonicity by the *WMS*, and secondly, we consider more general condensing indices in place of 2^k and 2^ℓ.

Recall that an increasing sequence $\{u_n\}$ (which we assume to take values in $\mathbb{N} \cup \{0\}$) is said to be in the class Λ if there is $r \in \mathbb{N}$ such that

$$\frac{u_{k+r}}{u_k} \geq q > 1,$$

for all $k \geq 1$ or, equivalently, if

$$\sum_{j=1}^{n} u_j \leq C u_n,$$

for all $n \in \mathbb{N}$.

Theorem 6.49 *Let $\{a_{k\ell}\} \in WMS$. Let $\{u_k\}, \{v_\ell\} \in \Lambda$, and assume that $u_{k+1} = O(u_k)$ and $v_{\ell+1} = O(v_\ell)$. Then, the following nine series,*

$$\sum_{\substack{k=u_1 \\ \ell=v_1}}^{\infty} a_{k\ell}, \quad \sum_{\substack{k=1 \\ \ell=v_1}}^{\infty} u_k a_{u_k,\ell}, \quad \sum_{\substack{k=u_1 \\ \ell=1}}^{\infty} v_\ell a_{k,v_\ell}, \quad \sum_{k,\ell=1}^{\infty} u_k v_\ell a_{u_k,v_\ell},$$

$$\sum_{\substack{k=1 \\ \ell=v_1}}^{\infty} u_k |a_{u_k,\ell} - a_{u_{k+1},\ell}|, \quad \sum_{\substack{k=u_1 \\ \ell=1}}^{\infty} v_\ell |a_{k,v_\ell} - a_{k,v_{\ell+1}}|,$$

$$\sum_{k,\ell=1}^{\infty} u_k v_\ell |a_{u_k,v_\ell} - a_{u_{k+1},v_\ell}|, \qquad \sum_{k,\ell=1}^{\infty} u_k v_\ell |a_{u_k,v_\ell} - a_{u_k,v_{\ell+1}}|,$$

and

$$\sum_{k,\ell=1}^{\infty} u_k v_\ell |a_{u_k,v_\ell} - a_{u_{k+1},v_\ell} - a_{u_k,v_{\ell+1}} + a_{u_{k+1},v_{\ell+1}}|,$$

converge or diverge simultaneously.

For the proof, see Theorem 9.55. Some remarks are in order. First, as already mentioned, we deal with series having condensed indices (i.e., those with indices u_k and/or v_ℓ). The original Cauchy condensation test (Theorem 5.15) corresponds to $u_k = v_k = 2^k$. Second, in addition to the double series with condensed indices, series with differences are also considered in Theorem 6.49.

For convenience, we state a simplified version of the Cauchy condensation test for *WMS* (with $u_k = v_k = 2^k$, and where less series are considered).

Corollary 6.50 *Let $\{a_{k\ell}\} \in WMS$. Then the three series*

$$\sum_{k,\ell=1}^{\infty} a_{k\ell}, \quad \sum_{k,\ell=0}^{\infty} 2^{k+\ell} a_{2^k,2^\ell}, \quad \textit{and} \quad \sum_{k,\ell=0}^{\infty} 2^{k+\ell} |\Delta^{(1,1)} a_{2^k,2^\ell}|$$

converge or diverge simultaneously.

6.2.6 $\ell^p - \ell^q$ inequalities and Stechkin's lemma

We now present two-dimensional versions of Theorems 2.117 and 2.118, related with the $\ell^q \subset \ell^p$ inequality (2.77).

Theorem 6.51 *Let $\{a_{k\ell}\} \in WMS$ and $0 < q < p < \infty$. Then the inequality*

$$\Big(\sum_{k,\ell=1}^{\infty} a_{k\ell}^p\Big)^{\frac{1}{p}} \le 3^{\frac{2}{q}} C^2 \Big(\sum_{k,\ell=1}^{\infty} a_{k\ell}^q (k\ell)^{\frac{q}{p}-1}\Big)^{\frac{1}{q}}$$

holds, where C is the WMS constant of $\{a_{k\ell}\}$. In particular, the convergence of

$$\sum_{k,\ell=1}^{\infty} a_{k\ell}^q (k\ell)^{\frac{q}{p}-1}$$

implies that of

$$\sum_{k,\ell=1}^{\infty} a_{k\ell}^p.$$

For $p = \infty$ and $0 < q < \infty$, we have

$$\sup_{k,\ell} a_{k\ell} \le 2^{\frac{2}{q}} C \Big(\sum_{k,\ell=1}^{\infty} \frac{a_{k\ell}^q}{k\ell}\Big)^{\frac{1}{q}}.$$

The detailed proof is given in the multidimensional case (see Theorem 9.57).

The analog of Theorem 6.51 for functions is as follows (cf. Theorem 9.58 for the statement and the proof in the multidimensional case).

Theorem 6.52 *Let $f \in WM$ and $0 < q < p < \infty$. Then the inequality*

$$\left(\int_{(0,\infty)^2} f(x,y)^p\,dx\,dy\right)^{\frac{1}{p}} \le 2^{\frac{2}{q}} C^2 \left(\int_{(0,\infty)^2} f(x,y)^q (xy)^{\frac{q}{p}-1}\,dx\,dy\right)^{\frac{1}{q}}$$

holds, where C is the WM constant of f. In particular, the convergence of the integral

$$\int_{(0,\infty)^2} f(x,y)^q (xy)^{\frac{q}{p}-1}\,dx\,dy$$

implies that of

$$\int_{(0,\infty)^2} f(x,y)^p\,dx\,dy.$$

For $p = \infty$ and $0 < q < \infty$, we have

$$\sup_{(x,y)\in(0,\infty)^2} f(x,y) \le 2^{\frac{2}{q}} C \left(\int_{(0,\infty)^2} \frac{f(x,y)^q}{xy}\,dx\,dy\right)^{\frac{1}{q}}.$$

In particular, the convergence of the last integral implies the boundedness of f.

We can also obtain a similar statement for functions in the WM_{Rad} class.

Theorem 6.53 *Let $f \in WM_{\mathrm{Rad}}$ and $0 < q < p < \infty$. Then there exists a constant $C > 0$ (depending on the constant in condition (6.10)) such that*

$$\left(\int_{(0,\infty)^2} f(x,y)^p\,dx\,dy\right)^{\frac{1}{q}} \le C \left(\int_{(0,\infty)^2} f(x,y)^q (x^2+y^2)^{(\frac{q}{p}-1)}\,dx\,dy\right)^{\frac{1}{q}}.$$

In particular, the convergence of the integral

$$\int_{(0,\infty)^2} f(x,y)^q (x^2+y^2)^{(\frac{q}{p}-1)}\,dx\,dy$$

implies that of

$$\int_{(0,\infty)^2} f(x,y)^p\,dx\,dy.$$

For $p = \infty$ and $0 < q < \infty$, we have

$$\sup_{x,y\in(0,\infty)} f(x,y) \le C \left(\int_{(0,\infty)^2} \frac{f(x,y)^q}{x^2+y^2}\,dx\,dy\right)^{\frac{1}{q}},$$

for some constant C (depending on the constant in condition (6.10)). In particular, the convergence of the last integral implies the boundedness of f.

For a proof in arbitrary dimension, see Theorem 9.59.

To conclude, we give an analog of the one-dimensional Stechkin's lemma (Theorem 2.119) for sequences in the class $QMS(\alpha)$ (see Definition 6.7).

Theorem 6.54 *Let $\alpha \geq 0$ and $\{a_{k\ell}\} \in QMS(\alpha)$. For $0 < q < p < \infty$ and any natural numbers $m \leq M$ and $n \leq N$, we have*

$$\left(\sum_{k=m}^{M}\sum_{\ell=n}^{N}(k\ell)^{\alpha(p-q)}a_{k\ell}^{p}\right)^{\frac{1}{p}} \leq C(mn)^{-\alpha(p-q)+\frac{1}{p}-\frac{1}{q}}\left(\sum_{k=1}^{M}\sum_{\ell=1}^{N}(k\ell)^{\alpha(p-q)}a_{k\ell}^{q}\right)^{\frac{1}{q}},$$

where

$$C = \sup_{n}\left(n^{-\alpha p-1}\sum_{k=1}^{n}k^{\alpha p}\right)^{-\frac{2}{q}}$$

(in particular, if $\alpha = 0$, we have $C = 1$). For the case $p = \infty$ and $0 < q < \infty$, we have

$$(mn)^{-\alpha}a_{mn} = \max_{\substack{m\leq k\leq M\\ n\leq \ell\leq N}}(k\ell)^{-\alpha}a_{k\ell} \leq (mn)^{-\frac{1}{q}}\left(\sum_{k=1}^{M}\sum_{\ell=1}^{N}(k\ell)^{-\alpha q}a_{k\ell}^{q}\right)^{\frac{1}{q}}.$$

The idea of the proof is the same as in the one-dimensional case. For a proof in the multivariate setting, see Theorem 9.60.

As a corollary, we have the classical Stechkin lemma in two dimensions.

Corollary 6.55 *Let $\{a_{k\ell}\}$ be a nonnegative decreasing sequence. For $0<q<p<\infty$ and any natural numbers $m \leq M$ and $n \leq N$, we have*

$$\left(\sum_{k=m}^{M}\sum_{\ell=n}^{N}a_{k\ell}^{p}\right)^{\frac{1}{p}} \leq (mn)^{\frac{1}{p}-\frac{1}{q}}\left(\sum_{k=1}^{M}\sum_{\ell=1}^{N}a_{k\ell}^{q}\right)^{\frac{1}{q}}.$$

For the case $p = \infty$ and $0 < q < \infty$, we have

$$a_{mn} = \max_{\substack{m\leq k\leq M\\ n\leq \ell\leq N}}a_{k\ell} \leq (mn)^{-\frac{1}{q}}\left(\sum_{k=1}^{M}\sum_{\ell=1}^{N}a_{k\ell}^{q}\right)^{\frac{1}{q}}.$$

Finally, we also have a counterpart of Theorem 2.121. For the statement and proof in arbitrary dimension, we refer to Theorem 9.62. In order to simplify the presentation of the next result, we introduce the following notation. For $m, n \in \mathbb{N}$, define the rectangle

$$R_{mn} = \{(k,\ell) \in \mathbb{N}^2 : k \leq m,\ \ell \leq n\}.$$

Theorem 6.56 *Let $0 < q < p < \infty$ and let $\{a_{k\ell}\}$ be a nonnegative double sequence. Assume that for any $m, n \in \mathbb{N}$, the estimate*

$$a_{mn} \leq A\left(\frac{1}{mn}\sum_{k=1}^{m}\sum_{\ell=1}^{n}a_{k\ell}^{\alpha}\right)^{\frac{1}{\alpha}} + B\left(\sum_{(k,\ell)\in\mathbb{N}^2\setminus R_{mn}}\frac{a_{k\ell}^{\beta}}{k\ell}\right)^{\frac{1}{\beta}} \tag{6.20}$$

holds with some $0 < \alpha, \beta \le q$, *and constants* $A, B \ge 0$. *Then there exists a constant* C *depending on* A, B, p, q, *and* β, *such that for every* $m, n \in \mathbb{N}$,

$$\left(\sum_{k=m}^{\infty}\sum_{\ell=n}^{\infty} a_{k\ell}^{p}\right)^{\frac{1}{p}} \le C(mn)^{\frac{1}{p}-\frac{1}{q}}\left(\sum_{k,\ell=1}^{\infty} a_{k\ell}^{q}\right)^{\frac{1}{q}}.$$

In particular, if (6.20) *holds with* $B = 0$, *we have*

$$\left(\sum_{k=m}^{\infty}\sum_{\ell=n}^{\infty} a_{k\ell}^{p}\right)^{\frac{1}{p}} \le A(mn)^{\frac{1}{p}-\frac{1}{q}}\left(\sum_{k,\ell=1}^{\infty} a_{k\ell}^{q}\right)^{\frac{1}{q}}.$$

For the case $p = \infty$ *and* $0 < q < \infty$,

$$\sup_{\substack{k\ge m\\ \ell\ge n}} a_{k\ell} \le C(mn)^{-\frac{1}{q}}\left(\sum_{k,\ell=1}^{\infty} a_{k\ell}^{q}\right)^{\frac{1}{q}}.$$

We emphasize that sequences decreasing in each index satisfy (6.20) with $A = 1$, $B = 0$, and arbitrary α. In this case, Corollary 6.55 implies Theorem 6.56.

On the other hand, if $\{a_{k\ell}\} \in WMS$, then (6.20) holds with $B = 0$, arbitrary $\alpha > 0$, and $A = 2^{\frac{2}{\alpha}}C$, where C is the WMS constant of $\{a_{k\ell}\}$ (which follows easily from the definition; cf. Remark 9.63 for the details).

6.2.7 Maclaurin–Cauchy integral test

Another important test for monotone series is the Maclaurin–Cauchy integral test (Theorem 5.16), which is also valid for WM functions.

Theorem 6.57 *Let* $f \in WM$ *and* $m, n \in \mathbb{N}$. *Then the double series*

$$\sum_{\substack{k=m\\ \ell=n}}^{\infty} f(k,\ell)$$

converges if and only if all the integrals

$$\int_m^{\infty} f(x,n)\,dx, \qquad \int_n^{\infty} f(m,y)\,dy, \qquad \int_n^{\infty}\int_m^{\infty} f(x,y)\,dx\,dy$$

converge.

The proof of Theorem 6.57 (in a more general formulation) is contained in Subsection 9.2.7 (cf. Theorem 9.64; the statement of Theorem 6.57 in arbitrary dimension is given in Corollary 9.65 in the multidimensional setting). The fact that such a test is still valid under the condition $f \in WM$ makes it a powerful tool, since the class of WM functions is much larger than the class of monotone functions and gives rise to a number of interesting cases that can easily be studied in two or more dimensions (cf. Examples 6.61 below).

Remark 6.58 An interesting feature of the Maclaurin–Cauchy integral test in two dimensions (which is not revealed in dimension one) is that it involves not only the double integral

$$\int_n^\infty \int_m^\infty f(x,y)\,dx\,dy$$

but also the one-dimensional integrals

$$\int_m^\infty f(x,n)\,dx \quad \text{and} \quad \int_n^\infty f(m,y)\,dy.$$

More precisely, the convergence of the series $\sum_{k,\ell} f(k,\ell)$ implies the convergence of all three integrals, but the convergence of the double integral alone does not imply the convergence of the double series. To see this, consider the example

$$f(x,y) = \begin{cases} e^{-x-y}, & \text{if } x,y > 1, \\ \dfrac{1}{xy}, & \text{if } x = 1 \text{ or } y = 1. \end{cases}$$

It is clear that

$$\int_1^\infty \int_1^\infty f(x,y)\,dx\,dy = \int_1^\infty \int_1^\infty e^{-x-y}\,dx\,dy < \infty,$$

but

$$\sum_{k,\ell=1}^\infty f(k,\ell) \geq \sum_{k=1}^\infty f(k,1) = \sum_{k=1}^\infty \frac{1}{k} = \infty.$$

We emphasize that both one-dimensional integrals

$$\int_1^\infty f(x,1)\,dx = \int_1^\infty \frac{1}{x}\,dx \quad \text{and} \quad \int_1^\infty f(1,y)\,dy = \int_1^\infty \frac{1}{y}\,dy$$

are also divergent.

The Maclaurin–Cauchy integral test also holds for the class WM_{Rad}.

Theorem 6.59 *Let $f \in WM_{\text{Rad}}$. Then the series*

$$\sum_{k,\ell=1}^\infty f(k,\ell)$$

and the integral

$$\int_1^\infty \int_1^\infty f(x,y)\,dx\,dy$$

converge or diverge simultaneously.

For a proof in the multidimensional setting, see Theorem 9.68.

Remark 6.60 In contrast with the Maclaurin–Cauchy integral test for *WM* functions (Theorem 6.57), in Theorem 6.59 it is not necessary to assume the convergence of the one-dimensional integrals in order to derive the convergence of the corresponding series. In other words, if any of the one-dimensional integrals

$$\int_1^\infty f(1,y)\,dy \qquad \text{or} \qquad \int_1^\infty f(x,1)\,dx$$

diverges, then the corresponding double integral also diverges (and so does the double series). Indeed, for $f \in WM_{\mathrm{Rad}}$, by Lemma 6.34 there exists a function $g_0 \in WM$ such that $f(x,y) \asymp g_0\big(\sqrt{x^2+y^2}\big)$. Thus, setting $E = \{(r,\theta) : r \ge 2,\ \theta \in [\frac{\pi}{3}, \frac{\pi}{6}]\}$ in polar coordinates, we clearly have $E \subset [1,\infty)^2$, and thus

$$\begin{aligned}
\int_1^\infty \int_1^\infty f(x,y)\,dx\,dy &\ge \int\!\!\int_E f(x,y)\,dx\,dy \\
&\ge \int_{\frac{\pi}{3}}^{\frac{\pi}{6}} \left(\int_2^\infty f(r\cos\theta, r\sin\theta) r\,dr \right) d\theta \\
&\ge \int_{\frac{\pi}{3}}^{\frac{\pi}{6}} \left(\int_2^\infty f(r\cos\theta, r\sin\theta)\,dr \right) d\theta \\
&\asymp \int_{\frac{\pi}{3}}^{\frac{\pi}{6}} \left(\int_2^\infty g_0(r)\,dr \right) d\theta \\
&\asymp \int_2^\infty g_0(r)\,dr = \infty,
\end{aligned}$$

provided that either $\int_1^\infty f(1,y)\,dy = \infty$ or $\int_1^\infty f(x,1)\,dx = \infty$ (in fact, both are equivalent to $\int_1^\infty g_0(r)\,dr$, i.e., if one diverges, so does the other). This proves the claim.

Let us now present some interesting examples that can be studied by means of the Maclaurin–Cauchy integral test for *WM* functions. We only give the convergence/divergence conditions for each example without proofs, since their extensions are studied in full detail in the multidimensional setting (see Example 9.71).

Example 6.61 (i) The series

$$\sum_{k,\ell=1}^\infty \frac{1}{(k+\ell)^\gamma} \tag{6.21}$$

converges if and only if $\gamma > 2$.

(ii) The integral

$$\int_2^\infty \int_2^\infty \frac{1}{xy \log^\gamma(xy)}\,dx\,dy, \qquad \gamma > 0,$$

converges if and only if $\gamma > 2$.

(iii) The series

$$\sum_{k=1}^{\infty}\sum_{\ell=1}^{\infty}\frac{(k^{\alpha_1}+\ell^{\alpha_2})^{\gamma}}{k^{\beta_1}\ell^{\beta_2}}, \quad \alpha_j,\beta_j,\gamma>0, \quad j=1,2,$$

converges if and only if

$$\alpha_j\gamma-\beta_j<-1, \quad \text{for} \quad j=1,2.$$

(iv) The series

$$\sum_{k,\ell=1}^{\infty}\frac{k^{\alpha_1}\ell^{\alpha_2}}{(k^{\beta_1}+\ell^{\beta_2})^{\gamma}},$$

with $\alpha_j>-1$, $\beta_j>0$, and $\gamma>0$, converges if and only if

$$\frac{\alpha_1+1}{\beta_1}+\frac{\alpha_2+1}{\beta_2}<\gamma.$$

This series is a generalization of (6.21), which corresponds to the case $\alpha_j=0$ and $\beta_j=1$, for $j=1,2$.

(v) If $\alpha_1,\alpha_2\geq 0$, the series

$$\sum_{k,\ell=1}^{\infty}\frac{1}{(k^2+\ell^2)\log^{\alpha_1}(k+1)\log^{\alpha_2}(\ell+1)}$$

converges if and only if $\max\{\alpha_1,\alpha_2\}>1$.

6.2.8 Equiconvergence results involving general monotonicity

It is often convenient to consider classes of sequences different from *WMS*. The *GMS* condition, which characterizes quantitative properties related to bounded variation (in contrast with *WMS*, defined only in terms of magnitudes), provides a desirable alternative in some problems.

Here we present equiconvergence results for *GMS* that are in general not true for *WMS*.

We start with a statement that relates series having *GMS* terms with the corresponding series involving differences.

Theorem 6.62 *Let $\{a_{k\ell}\}\in GMS$ be vanishing as $k+\ell\to\infty$ and $m,n\in\mathbb{N}$. Then the four series,*

$$\sum_{k=m}^{\infty}\sum_{\ell=n}^{\infty}|a_{k\ell}|,$$

$$\sum_{k=m}^{\infty}\sum_{\ell=n}^{\infty}k|\Delta^{(1,0)}a_{k\ell}|, \quad \sum_{k=m}^{\infty}\sum_{\ell=n}^{\infty}\ell|\Delta^{(0,1)}a_{k\ell}|, \quad \textit{and} \quad \sum_{k=m}^{\infty}\sum_{\ell=n}^{\infty}k\ell|\Delta^{(1,1)}a_{k\ell}|,$$

converge or diverge simultaneously. What is more, the relation $\asymp$ holds between any pair of such series.

For a proof in the multidimensional setting, see Theorem 9.74. It is interesting to note that the conclusion of Theorem 6.62 does not hold for *WMS*. More precisely, there exists a double (positive) sequence $\{a_{k\ell}\} \in WMS \backslash GMS$ for which

$$\sum_{k,\ell=1}^{\infty} a_{k\ell} < \infty \qquad \text{and} \qquad \sum_{k,\ell=1}^{\infty} k\ell |\Delta^{(1,1)} a_{k\ell}| = \infty.$$

Indeed, consider the example $\{a_{k\ell}\} \in WMS \backslash GMS$ given in Remark 6.17, where

$$a_{k,1} = \begin{cases} 2 \cdot 4^{-m}, & \text{if } 2^m \le k < 2^{m+1} \text{ and } k \text{ is even}, \\ 4^{-m}, & \text{if } 2^m \le k < 2^{m+1} \text{ and } k \text{ is odd}, \end{cases} \qquad m \in \mathbb{N},$$

and $a_{k\ell} = 0$ for $\ell > 1$. For $m \in \mathbb{N}$, we have

$$\sum_{k=2^m}^{2^m-1} |\Delta^{(1,1)} a_{k,1}| \asymp 2^{-m}.$$

Therefore,

$$\sum_{k,\ell=1}^{\infty} k\ell |\Delta^{(1,1)} a_{k\ell}| = \sum_{k=1}^{\infty} k |\Delta^{(1,1)} a_{k,1}| = \sum_{m=0}^{\infty} \sum_{k=2^m}^{2^{m+1}-1} k |\Delta^{(1,1)} a_{k,1}|$$
$$\asymp \sum_{m=0}^{\infty} 1 = \infty.$$

On the other hand, we have

$$\sum_{k,\ell=1}^{\infty} a_{k\ell} = \sum_{k=1}^{\infty} a_{k,1} = \sum_{m=0}^{\infty} \sum_{k=2^m}^{2^{m+1}-1} a_{k,1} \le 2 \sum_{m=0}^{\infty} 2^m 4^{-m} < \infty.$$

An analog of Theorem 6.62 for *WM* functions reads as follows.

Theorem 6.63 *Let $f \in GM$ be such that $|f| \in WM$. Assume that for some $m, n \in \mathbb{R}_+$, the integrals*

$$\int_{2m}^{\infty} |f(x,n)| \, dx \quad \text{and} \quad \int_{2n}^{\infty} |f(m,y)| \, dy$$

are convergent. Then the four integrals,

$$\int_m^{\infty} \int_n^{\infty} |f(x,y)| \, dy \, dx, \qquad \int_m^{\infty} \int_n^{\infty} st |d^{(1,1)} f(s,t)|,$$
$$\int_m^{\infty} \int_n^{\infty} t |d^{(0,1)} f(x,t)| \, dx, \qquad \int_m^{\infty} \int_n^{\infty} s |d^{(1,0)} f(s,y)| \, dy,$$

converge or diverge simultaneously.

For a proof in arbitrary dimension, see Theorem 9.75 (where such a result is stated even in broader generality).

Remark 6.64 Theorem 6.63 contains more hypotheses than its analog for sequences (Theorem 6.62). In fact, the case of functions is somewhat more delicate (see the proof of Theorem 9.75). More precisely, the hypotheses on the convergence of the integrals

$$\int_{2m}^{\infty} |f(x,n)|\,dx \quad \text{and} \quad \int_{2n}^{\infty} |f(m,y)|\,dy$$

cannot be omitted. Indeed, consider the nonnegative function

$$f(x,y) = \begin{cases} 1, & 0 \le x, y \le 1, \\ 1/x, & 1 < x,\ 0 \le y \le 1, \\ 1/y, & 0 \le x \le 1,\ 1 < y, \\ 0, & 1 < x, y. \end{cases}$$

This function is GM, and being monotone in each variable, $f \in WM$. On the one hand,

$$\int_{1}^{\infty}\int_{1}^{\infty} f(x,y)\,dx\,dy = 0,$$

and on the other hand,

$$\begin{aligned}\int_{1}^{N}\int_{1}^{2} st|d^{(1,1)}f(s,t)| &= \sum_{k=1}^{N-1}\int_{k}^{k+1}\int_{1}^{2} st|d^{(1,1)}f(s,t)| \\ &\ge \sum_{k=1}^{N-1} k\int_{k}^{k+1}\int_{1}^{2} |d^{(1,1)}f(s,t)| \\ &= \sum_{k=1}^{N-1} k\left(\frac{1}{k} - \frac{1}{k+1}\right) \asymp \sum_{k=1}^{N} \frac{1}{k}.\end{aligned}$$

Letting $N \to \infty$, we find that the integral

$$\int_{1}^{\infty}\int_{1}^{\infty} st|d^{(1,1)}f(s,t)|$$

diverges.

Theorems 6.62 and 6.63 hold for a larger class than GMS and GM, respectively. If $\beta := \{\beta_{k\ell}\}$ is a nonnegative double sequence depending on $a := \{a_{k\ell}\} \subset \mathbb{C}$, we say that the couple (a, β) determines a general monotone sequence a with majorant β (denoted $(a, \beta) \in \mathcal{GMS}$), if

$$\sum_{k=m}^{2m-1}\sum_{\ell=n}^{2n-1} |\Delta^{(1,1)} a_{k\ell}| \le C\beta_{mn},$$

for all $m, n \in \mathbb{N}$. The corresponding classes of functions are defined similarly. If f is a complex-valued function defined on $(0,\infty)^2$ and β is a nonnegative function of two variables, we say that the couple (f, β) determines a general monotone function f with majorant β, written $(f, \beta) \in \mathcal{GM}$, if

$$\int_x^{2x} \int_y^{2y} |d^{(1,1)} f(s,t)| \leq C\beta(x,y),$$

for all $x, y > 0$.

The generalizations of Theorems 6.62 and 6.63 to the classes $\mathcal{GMS}$ and $\mathcal{GM}$, respectively, read as follows.

Proposition 6.65 *Let $(a, \beta) \in \mathcal{GMS}$ be such that*

$$\sum_{k=m}^{\infty} \sum_{\ell=n}^{\infty} \beta_{k\ell} \leq C \sum_{k=m}^{\infty} \sum_{\ell=n}^{\infty} |a_{k\ell}|,$$

for some constant $C > 0$. Then the conclusion of Theorem 6.62 holds.

Proposition 6.66 *Let $(f, \beta) \in \mathcal{GM}$ be such that*

$$\int_m^{\infty} \int_n^{\infty} \beta(x,y)\, dy\, dx \leq C \int_m^{\infty} \int_n^{\infty} |f(x,y)|\, dy\, dx,$$

for some constant $C > 0$. Then the conclusion of Theorem 6.63 holds.

For the proofs of Propositions 6.65 and 6.66, we refer to their multidimensional analogs, Propositions 9.77 and 9.78, respectively.

An example of a double sequence $\{\beta_{k\ell}\}$ for which the hypothesis of Proposition 6.65 holds is

$$\beta_{k\ell} = \sum_{k=\lfloor m/c \rfloor}^{\lfloor cm \rfloor} \sum_{\ell=\lfloor n/c \rfloor}^{\lfloor cn \rfloor} \frac{|a_{k\ell}|}{k\ell}, \qquad c > 1.$$

The class given by the pairs $(a, \beta) \in \mathcal{GMS}$ with such a choice of $\beta = \{\beta_{k\ell}\}$ corresponds to the class $\overline{GMS}$ (see Definition 6.21). In this case, there holds

$$\sum_{k,\ell=1}^{\infty} \beta_{k\ell} \leq C \sum_{k,\ell=1}^{\infty} |a_{k\ell}|. \tag{6.22}$$

In the case of functions, the situation is similar. For

$$\beta(x,y) = \int_{x/c}^{cx} \int_{y/c}^{cy} \frac{|f(s,t)|}{st}\, dt\, ds, \qquad c > 1,$$

the class given by the pairs $(f, \beta) \in \mathcal{GM}$ with such a β corresponds to the class $\overline{GM}$ (cf. Definition 6.25). In this case,

$$\int_0^{\infty} \int_0^{\infty} \beta(x,y)\, dx\, dy \leq C \int_0^{\infty} \int_0^{\infty} |f(x,y)|\, dx\, dy.$$

We now give a version of Theorem 2.127 in two dimensions. An issue regarding such a statement is that it requires the additional hypothesis $\{a_{k\ell}\} \in WMS$. First, we recall the definition of the backward differences. Given a (one-dimensional) sequence $\{u_k\}$, we define $\overleftarrow{\Delta} u_k := u_{k+1} - u_k = -\Delta u_k$.

Theorem 6.67 *Let $\{a_{k\ell}\} \in GMS$ be such that $\{|a_{k\ell}|\} \in WMS$. Let $\{u_k\}$ and $\{v_\ell\}$ be strictly increasing sequences of natural numbers such that*

$$\overleftarrow{\Delta} u_{k+1} = O(\overleftarrow{\Delta} u_k), \qquad \overleftarrow{\Delta} v_{\ell+1} = O(\overleftarrow{\Delta} v_\ell).$$

Then, all the double series,

$$\sum_{\substack{k=u_1\\ \ell=v_1}}^{\infty} |a_{k\ell}|, \quad \sum_{k,\ell=1}^{\infty} u_k v_\ell |a_{u_k,v_\ell} - a_{u_{k+1},v_\ell} - a_{u_k,v_{\ell+1}} + a_{u_{k+1},v_{\ell+1}}|,$$

$$\sum_{k,\ell=1}^{\infty} \overleftarrow{\Delta} u_k v_\ell |a_{u_k,v_\ell} - a_{u_k,v_{\ell+1}}|, \quad \sum_{k,\ell=1}^{\infty} u_k \overleftarrow{\Delta} v_\ell |a_{u_k,v_\ell} - a_{u_{k+1},v_\ell}|,$$

$$\sum_{k,\ell=1}^{\infty} \overleftarrow{\Delta} u_k \overleftarrow{\Delta} v_\ell |a_{u_k,v_\ell}|,$$

converge or diverge simultaneously.

Remark 6.68 By choosing $u_k = k$ and $v_\ell = \ell$ in Theorem 6.67, we recover the conclusion of Theorem 6.62. However, in Theorem 6.67, we require the extra assumption that $\{|a_{k\ell}|\} \in WMS$ (in particular, this is the case if $\{a_{k\ell}\} \in HGMS$). It is also worth mentioning that Theorem 6.67 remains true for $(a, \beta) \in \mathcal{GMS}$ satisfying (6.22). For further details, see the proof of its multidimensional analog, Theorem 9.80.

The conclusion of Theorem 6.67 is not true if either of the conditions

$$\overleftarrow{\Delta} u_{k+1} = O(\overleftarrow{\Delta} u_k) \quad \text{or} \quad \overleftarrow{\Delta} v_{\ell+1} = O(\overleftarrow{\Delta} v_\ell)$$

is not satisfied. Indeed, consider the decreasing sequences given by $a_k = \frac{1}{k^2}$ and $b_\ell = \frac{1}{\ell \log \ell}$. The double sequence defined by $c_{k\ell} = a_k b_\ell$ belongs to $GMS \cap WMS$. On the one hand,

$$\sum_{k=1}^{\infty}\sum_{\ell=2}^{\infty} c_{k\ell} = \sum_{k=1}^{\infty} \frac{1}{k^2} \sum_{\ell=2}^{\infty} \frac{1}{\ell \log \ell} = \infty.$$

On the other hand, if $u_k = k$ and $v_\ell = 2^{\ell!}$, then $\Delta v_{\ell+1} \neq O(\Delta v_\ell)$, and

$$\sum_{k=1}^{\infty}\sum_{\ell=1}^{\infty} u_k v_\ell |c_{u_k,v_\ell} - c_{u_{k+1},v_\ell} - c_{u_k,v_{\ell+1}} + c_{u_{k+1},v_{\ell+1}}|$$
$$= \sum_{k=1}^{\infty}\sum_{\ell=1}^{\infty} u_k v_\ell |(a_{u_k} - a_{u_{k+1}})(b_{v_\ell} - b_{v_{\ell+1}})|$$

$$= \sum_{k=1}^{\infty} k\left(\frac{1}{k^2} - \frac{1}{(k+1)^2}\right)\sum_{\ell=1}^{\infty} 2^{\ell!}\left(\frac{1}{2^{\ell!}\log 2^{\ell!}} - \frac{1}{2^{(\ell+1)!}\log 2^{(\ell+1)!}}\right)$$

$$\lesssim \sum_{k=1}^{\infty} \frac{1}{k^2} \sum_{\ell=1}^{\infty} \frac{1}{\ell!\log 2} < \infty.$$

6.3 Notes

Section 6.1 (Double series and generalizations of monotonicity)

1. The idea on how to construct larger classes of (Hardy)-general monotone sequences was first given in two dimensions in [91]. Those may be defined as follows.

Definition 6.69 Let $\{a_{k\ell}\}$ be a complex-valued double sequence, and let $\{\beta_{k\ell}^{(0,1)}\}$, $\{\beta_{k\ell}^{(1,0)}\}$, and $\{\beta_{k\ell}^{(1,1)}\}$ be nonnegative sequences depending on $\{a_{k\ell}\}$. We say that $\{a_{k\ell}\}$ is a $(\beta^{(0,1)}, \beta^{(1,0)}, \beta^{(1,1)})$ *Hardy-general monotone sequence*, written as

$$\{a_{k\ell}\} \in HGMS\big(\beta^{(0,1)}, \beta^{(1,0)}, \beta^{(1,1)}\big),$$

if there is a constant $C > 0$ such that

$$\sum_{k=m}^{2m-1} |\Delta^{(1,0)} a_{kn}| \le C\beta_{mn}^{(0,1)},$$

$$\sum_{\ell=n}^{2n-1} |\Delta^{(0,1)} a_{m\ell}| \le C\beta_{mn}^{(1,0)},$$

$$\sum_{k=m}^{2m-1}\sum_{\ell=n}^{2n-1} |\Delta^{(1,1)} a_{k\ell}| \le C\beta_{mn}^{(1,1)},$$

for all $m, n \in \mathbb{N}$.

One can readily see that in order to construct larger classes than *HGMS* (Definition 6.19), where $\beta_{mn}^{\eta} = a_{mn}$, $\eta = (1,0), (0,1), (1,1)$, it suffices to find appropriate choices of $\{\beta_{mn}^{\eta}\}$, as, for instance,

$$\beta_{mn}^{(0,1)} = \frac{1}{m}\sum_{k=\lfloor m/\lambda\rfloor}^{\lfloor \lambda m\rfloor} |a_{kn}|, \quad \beta_{mn}^{(1,0)} = \frac{1}{n}\sum_{\ell=\lfloor n/\lambda\rfloor}^{\lfloor \lambda n\rfloor} |a_{m\ell}|,$$

$$\beta_{mn}^{(1,1)} = \frac{1}{mn}\sum_{k=\lfloor m/\lambda\rfloor}^{\lfloor \lambda m\rfloor}\sum_{\ell=\lfloor n/\lambda\rfloor}^{\lfloor \lambda n\rfloor} |a_{k\ell}|,$$

where $\lambda > 1$ is an arbitrary fixed parameter. It is clear that if a sequence $\{a_{mn}\}$ satisfies (6.1) and (6.4), then also $\{a_{mn}\} \in HGMS\big(\beta^{(0,1)}, \beta^{(1,0)}, \beta^{(1,1)}\big)$, and the converse is not true. Indeed, this can easily be seen by considering any positive sequence $\{b_{mn}\}$ that satisfies (6.1) and (6.4) and setting one term to be zero. More precisely, defining

$$a_{k\ell} = \begin{cases} b_{k\ell}, & \text{if } (k,\ell) \ne (10,10), \\ 0, & \text{if } (k,\ell) = (10,10), \end{cases}$$

we find that $\{a_{k\ell}\}$ satisfies neither (6.1) nor (6.4), but

$$\{a_{k\ell}\} \in HGMS(\beta^{(0,1)}, \beta^{(1,0)}, \beta^{(1,1)}).$$

2. The idea of using sequences belonging to some general monotone class of the type $HGMS(\beta^{(0,1)}, \beta^{(1,0)}, \beta^{(1,1)})$ is pivotal in the study of trigonometric series, and it has attracted many authors in the recent past (although the classes GMS and $\overline{GMS}$ not involving one-dimensional differences have also proved useful; see, e.g., [93]). The study of those classes is too specialized and falls out of the scope of this book. For the interested reader, we refer to the recent research papers [84, 91, 92, 146, 148, 150, 168, 281] for a deeper study of the $HGMS(\beta^{(0,1)}, \beta^{(1,0)}, \beta^{(1,1)})$ classes and their applications.

3. It is important to stress that the Vitali variation is not the only type of variation that has been introduced. Different other variations may be obtained upon considering other types of differences, which give rise to different notions of the concept of bounded variation. Several definitions are given in [61].

Section 6.2 (Convergence tests for double series with monotone-type coefficients)

4. The following statement may be seen as a counterexample for the Cauchy condensation test when monotonicity is not assumed.

Theorem 6.70 *Let $0 < p < q \leq \infty$. There exists a nonnegative double sequence $\{a_{k\ell}\}$ such that*

$$\left(\sum_{\ell=1}^{\infty}\left(\sum_{k=1}^{\infty} a_{k\ell}^{p}\right)^{\frac{q}{p}}\right)^{\frac{1}{q}} < \infty, \qquad \text{or } \sup_{\ell\geq 1}\left(\sum_{k=1}^{\infty} a_{k\ell}^{p}\right)^{\frac{1}{p}} < \infty \quad \text{if } q = \infty,$$

for which

$$\sup_{k\geq 0} 2^{k/p} a_{k,\lfloor 2^k x\rfloor} = \infty, \qquad \text{for all } x \in [1,2].$$

This result was obtained in [46] in order to build functions that serve as counterexamples in relation to Besov spaces. Its proof is based on the existence of a sequence $\{b_n\}$ such that

$$\sup_{k\geq 0}\left(\frac{1}{2^k}\sum_{n=2^k}^{2^{k+1}-1} b_n\right) \leq 1,$$

and

$$\sup_{k\geq 0} b_{\lfloor 2^k x\rfloor} = \infty, \qquad \text{for all } x \in [1,2].$$

Part III

Multiple Series

In this part we develop the theory of number series in arbitrary dimension, one of the primary goals of this book. As a prelude, we start by introducing a rather extensive list of notations that allow us to write the matter as concisely as possible. The reader who is interested in seeing the results written in a more classical manner may consult Part II.

After this prelude, the rest is a direct extension of Part II to the multivariate case. By this, the basic theory of multiple number series (along with some of the most remarkable convergence tests) is given in Chapters 7 and 8. In Chapter 9, we explore novel convergence tests involving multidimensional versions of the monotonicity-type conditions introduced in Chapter 1.

7

Rudiments of multiple sequences and series

From now on, we are concerned with multidimensional sequences and series. Unlike Chapters 4–6, we work with series in any dimension, not only in dimension two. Many of the multidimensional results that we will present in this part have never appeared in such a general form (a few of them have, and some others have appeared only in dimension two). As our goal is to give a comprehensive description of the theory of convergence of multidimensional number series, we start with the basic theory of multiple sequences, just as done in Chapter 4.

It is worth mentioning that one of the main difficulties in the context of multidimensional sequences is the notation. Here we attempt to adopt a notation that is both accurate and as simple as possible.

7.1 Notation

For $d \in \mathbb{N}$, a d-dimensional complex-valued sequence $a = \{a_{\mathbf{k}}\}$ (or a d-tuple sequence) is a function defined on $\mathbb{N}^d$, which takes values on $\mathbb{C}$,

$$\begin{aligned} a\colon \mathbb{N}^d &\to \mathbb{C} \\ \mathbf{k} &\mapsto a_{\mathbf{k}} = a_{k_1,\dots,k_d}. \end{aligned}$$

For convenience, we sometimes consider the domain of a to be $(\mathbb{N} \cup \{0\})^d$, or a subset of it, not being necessarily $\mathbb{N}^d$.

For $\mathbf{k} = (k_1, \dots, k_d)$ and $\mathbf{n} = (n_1, \dots, n_d)$, we use the notations

$$\mathbf{n}^{\mathbf{k}} := (n_1^{k_1}, \dots, n_d^{k_d}), \quad |\mathbf{k}| := k_1 + \cdots + k_d, \quad \|\mathbf{k}\|^2 := k_1^2 + \cdots + k_d^2.$$

For $c \in \mathbb{R}$, $c\mathbf{n} = (cn_1, \dots, cn_d)$, and for $c \in \mathbb{R} \cup \{\infty\}$, we set $\mathbf{c} = (c, \dots, c)$. In particular, $\mathbf{0} = (0, \dots, 0)$ and $\mathbf{1} = (1, \dots, 1)$. The product $\mathbf{kn}$ is understood coordinate-wise, that is, $\mathbf{kn} = (k_1 n_1, \dots, k_d n_d)$. Similarly, inequalities such as $\mathbf{x} \le \mathbf{y}$ or $\mathbf{x} < \mathbf{y}$ with $\mathbf{x}, \mathbf{y} \in \mathbb{R}^d$ are also meant coordinate-wise.

For $\mathbf{x} \in \mathbb{R}^d$, we define the d-tuple floor function of $\mathbf{x}$, written $\lfloor \mathbf{x} \rfloor$, by just applying the floor function to each of the components of $\mathbf{x}$.

By $\mathbf{k} \to \boldsymbol{\infty}$ we mean that $k_1, \ldots, k_d \to \infty$, or, equivalently,

$$\min_{j=1,\ldots,d} k_j \to \infty,$$

while $|\mathbf{k}| \to \infty$ stands for $k_1 + \cdots + k_d \to \infty$, or, equivalently,

$$\max_{j=1,\ldots,d} k_j \to \infty.$$

For $i = 1, \ldots, d$, denote by $\mathbf{e}_i$ the d-tuple having 1 on the ith place and 0's otherwise, namely

$$(\mathbf{e}_i)_j := \begin{cases} 1, & \text{if } i = j, \\ 0, & \text{if } i \neq j, \end{cases} \qquad 1 \leq j \leq d.$$

The d-dimensional vectors $\eta = (\eta_1, \ldots, \eta_d)$, $\zeta = (\zeta_1, \ldots, \zeta_d)$, and $\theta = (\theta_1, \ldots, \theta_d)$ will always be vectors with the entries either 0 or 1 only. Correspondingly, $|\eta| = \eta_1 + \cdots + \eta_d$. Clearly, the vectors $\mathbf{e}_j$, $\mathbf{0}$, and $\mathbf{1}$ are among these η, ζ, or θ.

Given $\mathbf{t} \in \mathbb{R}^d$ and $\mathbf{0} \leq \eta \leq \mathbf{1}$ with $\eta \neq \mathbf{0}$, we denote by $\mathbf{t}_\eta$ the $|\eta|$-dimensional vector where only the coordinates of $\mathbf{t}$ such that $\eta_i = 1$ are involved. For example, $\mathbf{t}_{\mathbf{e}_j} = (x_j)$. Respectively, the notation $\mathbb{N}^\eta$ means that only the indices for which $\eta_j = 1$ are involved (i.e., for every $\mathbf{n} \in \mathbb{N}^d$, we have $\mathbf{n}_\eta \in \mathbb{N}^\eta$). Under this definition, $\mathbb{N}^{\mathbf{1}} = \mathbb{N}^d$. This leads to the following notation, which is useful to express the index $\mathbf{u}$ in $a_{\mathbf{u}}$ by means of two or more different d-dimensional vectors of indices. For $\mathbf{0} \leq \eta \leq \mathbf{1}$ and $\mathbf{k}, \mathbf{n} \in \mathbb{N}^d$, we denote, in order to distinguish between each of the two parts,

$$a_{\mathbf{n}_\eta, \mathbf{k}_{\mathbf{1}-\eta}} := a_{u_1,\ldots,u_d} = a_{\mathbf{u}},$$

where

$$u_j = \begin{cases} n_j, & \text{if } \eta_j = 1, \\ k_j, & \text{if } \eta_j = 0. \end{cases}$$

For instance, if $d = 4$ and $\eta = (0, 1, 1, 0)$, we have $\mathbf{k}_{\mathbf{1}-\eta} = (k_1, k_4)$, $\mathbf{n}_\eta = (n_2, n_3)$, and $\mathbf{k}_{\mathbf{1}-\eta}, \mathbf{n}_\eta = (k_1, n_2, n_3, k_4)$. Thus,

$$a_{u_1,\ldots,u_4} = a_{k_1,n_2,n_3,k_4}.$$

That is, we represent the given indices in a special way, separating two different groups of them for future treatment. If $\eta = \mathbf{0}$ or $\eta = \mathbf{1}$, we just take $a_{\mathbf{k}}$ or $a_{\mathbf{n}}$, respectively. Sometimes we need to split the indices into three or more groups to treat them separately. To this end, for $\mathbf{0} \leq \zeta \leq \mathbf{1} - \eta$ and $\mathbf{m} \in \mathbb{N}^d$, we set

$$a_{\mathbf{m}_\zeta, \mathbf{n}_\eta, \mathbf{k}_{\mathbf{1}-\eta-\zeta}} = a_{u_1,\ldots,u_d},$$

where

$$u_j = \begin{cases} m_j, & \text{if } \zeta_j = 1, \\ n_j, & \text{if } \eta_j = 1, \\ k_j, & \text{if } \eta_j = \zeta_j = 0. \end{cases}$$

If $\zeta = \mathbf{0}$ or $\zeta = \mathbf{1} - \eta$, we take $a_{\mathbf{n}_\eta,\mathbf{k}_{\mathbf{1}-\eta}}$ or $a_{\mathbf{m}_\zeta,\mathbf{n}_\eta}$, respectively. In a similar way, it is possible to extend this notation to the case where $\mathbf{u}$ is given by several different vectors (groups) of indices. However, for our purpose, the above definitions are enough.

The (rectangular) partial sums are simply denoted, for any $\mathbf{n} \in \mathbb{N}^d$, as

$$S_\mathbf{n} := \sum_{\mathbf{k}=\mathbf{1}}^{\mathbf{n}} a_\mathbf{k} = \sum_{k_1=1}^{n_1} \cdots \sum_{k_d=1}^{n_d} a_\mathbf{k},$$

where we have split the last sum into d iterated one-dimensional sums. This is possible because each of the indices of $\mathbf{n}$ is finite (this is not true in the limit $\mathbf{n} \to \infty$, as we will discuss in detail in Section 7.2 below).

In some situations, we may need a special notation for certain kind of partial sums. For instance, this is the case if we want to sum not over all indices but over a strict subset of them. If $\mathbf{0} \leq \eta \leq \mathbf{1}$ with $\eta \neq \mathbf{0}$ and $\mathbf{m} \leq \mathbf{n} \in \mathbb{N}^d$, we define

$$\sum_{\mathbf{k}_\eta=\mathbf{m}_\eta}^{\mathbf{n}_\eta} a_\mathbf{k} := \sum_{j:\eta_j=1} \sum_{k_j=m_j}^{n_j} a_\mathbf{k},$$

where the indices $\mathbf{k}_{\mathbf{1}-\eta}$ are fixed.

Further notation regarding sums will be introduced when needed.

The following is a useful decomposition of the partial sums $S_\mathbf{n}$ (or more generally, arbitrary rectangular sums) of any given d-tuple series into sums over 2^d smaller rectangles. For $\boldsymbol{\ell}, \mathbf{m}, \mathbf{n} \in \mathbb{N}^d$ such that $\boldsymbol{\ell} \leq \mathbf{m} \leq \mathbf{n}$, we have

$$\sum_{\mathbf{k}=\boldsymbol{\ell}}^{\mathbf{n}} a_\mathbf{k} = \sum_{\mathbf{0}\leq\theta\leq\mathbf{1}} \sum_{\mathbf{k}_\theta=\boldsymbol{\ell}_\theta}^{(\mathbf{m}-\mathbf{1})_\theta} \sum_{\mathbf{k}_{\mathbf{1}-\theta}=\mathbf{m}_{\mathbf{1}-\theta}}^{\mathbf{n}_{\mathbf{1}-\theta}} a_\mathbf{k}, \tag{7.1}$$

where, in the case of $m_j = \ell_j$ for some $1 \leq j \leq d$, the convention

$$\sum_{\mathbf{k}_\theta=\boldsymbol{\ell}_\theta}^{(\mathbf{m}-\mathbf{1})_\theta} \sum_{\mathbf{k}_{\mathbf{1}-\theta}=\mathbf{m}_{\mathbf{1}-\theta}}^{\mathbf{n}_{\mathbf{1}-\theta}} a_\mathbf{k} = 0$$

is taken for any θ such that $\theta_j = 1$ (i.e., in this case the sum over θ in (7.1) would not produce 2^d summands, but less). To illustrate this, we consider the case $d = 2$, $\boldsymbol{\ell} = (1,1)$, $\mathbf{m} = (3,1)$, and $\mathbf{n} = (5,5)$. We have

$$\sum_{s=1}^{5}\sum_{t=1}^{5} a_{st} = \sum_{s=1}^{2}\sum_{t=1}^{5} a_{st} + \sum_{s=3}^{5}\sum_{t=1}^{5} a_{st}.$$

On the other hand, for $\mathbf{m} = (2,3)$ (and the same values of $\boldsymbol{\ell}$, $\mathbf{n}$), the decomposition corresponding to (7.1) reads as

$$\sum_{s=1}^{5}\sum_{t=1}^{5} a_{st} = \sum_{t=1}^{2} a_{1t} + \sum_{t=3}^{5} a_{1t} + \sum_{s=2}^{5}\sum_{t=1}^{2} a_{st} + \sum_{s=2}^{5}\sum_{t=3}^{5} a_{st}.$$

We also introduce notation in relation to successive differences. Let

$$\Delta^{\mathbf{e}_j} a_{\mathbf{k}} := a_{\mathbf{k}} - a_{\mathbf{k}+\mathbf{e}_j}$$

be the difference with respect to the jth coordinate, and

$$\Delta^{\eta} a_{\mathbf{k}} := \Bigg(\prod_{j:\eta_j=1} \Delta^{\mathbf{e}_j} \Bigg) a_{\mathbf{k}}$$

be the "ηth" difference. For example, let in dimension three $\eta = (0, 1, 1)$. Then

$$\Delta^{\eta} a_{\mathbf{k}} = a_{\mathbf{k}} - a_{k_1,k_2+1,k_3} - a_{k_1,k_2,k_3+1} + a_{k_1,k_2+1,k_3+1}.$$

Consequently, $\Delta^{\mathbf{1}} a_{\mathbf{k}}$ denotes the iterated difference with respect to all the variables. We also follow the convention $\Delta^{\mathbf{0}} a_{\mathbf{k}} = a_{\mathbf{k}}$. If $\{a_k\}$ is a one-dimensional sequence, we just write $\Delta a_k = a_k - a_{k+1}$.

The following explicit formula for $\Delta^{\eta} a_{\mathbf{k}}$ will be useful.

Proposition 7.1 *For $\mathbf{0} \le \eta \le \mathbf{1}$ with $\eta \ne \mathbf{0}$ and any d-tuple sequence $\{a_{\mathbf{k}}\}$, we have*

$$\Delta^{\eta} a_{\mathbf{k}} = a_{\mathbf{k}} + \sum_{\zeta \le \eta, \zeta \ne \mathbf{0}} (-1)^{|\zeta|} a_{\mathbf{k}+\zeta} = a_{\mathbf{k}} + \sum_{\zeta \le \eta, \zeta \ne 0} (-1)^{|\zeta|} a_{\mathbf{k}_{\mathbf{1}-\zeta}, (\mathbf{k}+\mathbf{1})_{\zeta}}. \tag{7.2}$$

Proof We prove the first equality in (7.2). The second one is clear, as it is to the effect that the indices are just rewritten in a different way. We proceed by induction on $|\eta|$, the number of nonzero components of η. If $|\eta| = 1$ we have $\eta = \mathbf{e}_j$ for some $j \in \{1, \ldots, d\}$, and by definition

$$\Delta^{\mathbf{e}_j} a_{\mathbf{k}} = a_{\mathbf{k}} - a_{\mathbf{k}+\mathbf{e}_j},$$

so that (7.2) holds.

Suppose now that $1 \le \ell < d$ and that (7.2) holds for all η with $|\eta| \le \ell$. We show that in this case (7.2) holds for every θ satisfying $|\theta| \le \ell + 1$ in place of η. By induction hypothesis, it is enough to prove (7.2) for every θ with $|\theta| = \ell + 1$. We pick η with $|\eta| = \ell$, $j \in \{1, \ldots, d\}$ such that $\eta_j = 0$, and define $\theta = \eta + \mathbf{e}_j$. By induction hypothesis and the definitions of $\Delta^{\mathbf{e}_j} a_{\mathbf{k}}$ and $\Delta^{\theta} a_{\mathbf{k}}$, we have

$$\begin{aligned}
\Delta^{\theta} a_{\mathbf{k}} = \Delta^{\eta}\big(\Delta^{\mathbf{e}_j} a_{\mathbf{k}}\big) &= \Delta^{\mathbf{e}_j} a_{\mathbf{k}} + \sum_{\zeta \le \eta, \zeta \ne \mathbf{0}} (-1)^{|\zeta|} \Delta^{\mathbf{e}_j} a_{\mathbf{k}+\zeta} \\
&= a_{\mathbf{k}} - a_{\mathbf{k}+\mathbf{e}_j} + \sum_{\zeta \le \eta, \zeta \ne \mathbf{0}} (-1)^{|\zeta|} \big(a_{\mathbf{k}+\zeta} - a_{\mathbf{k}+\zeta+\mathbf{e}_j}\big) \\
&= a_{\mathbf{k}} - a_{\mathbf{k}+\mathbf{e}_j} + \sum_{\zeta \le \eta, \zeta \ne \mathbf{0}} (-1)^{|\zeta|} a_{\mathbf{k}+\zeta} + \sum_{\zeta \le \eta, \zeta \ne \mathbf{0}} (-1)^{|\zeta|+1} a_{\mathbf{k}+\zeta+\mathbf{e}_j}.
\end{aligned}$$

It remains to observe that we can rewrite the last expression as

$$a_{\mathbf{k}} + \sum_{\substack{\zeta \le \theta, \zeta \ne \mathbf{0} \\ \zeta_j = 0}} (-1)^{|\zeta|} a_{\mathbf{k}+\zeta} + \sum_{\substack{\zeta \le \theta, \zeta \ne \mathbf{0} \\ \zeta_j = 1}} (-1)^{|\zeta|} a_{\mathbf{k}+\zeta} = a_{\mathbf{k}} + \sum_{\zeta \le \theta, \zeta \ne \mathbf{0}} (-1)^{|\zeta|} a_{\mathbf{k}+\zeta},$$

as desired. □

Remark 7.2 Note that identity (7.2) can be rewritten as

$$\Delta^{\eta} a_{\mathbf{k}} = \sum_{\mathbf{0} \le \zeta \le \eta} (-1)^{|\zeta|} a_{\mathbf{k}+\zeta} = \sum_{\mathbf{0} \le \zeta \le \eta} (-1)^{|\zeta|} a_{\mathbf{k}_{1-\zeta},(\mathbf{k}+\mathbf{1})_{\zeta}}.$$

Another useful formula is the multidimensional telescopic sum, which incorporates not only one-dimensional differences but higher-dimensional ones as well.

Proposition 7.3 *For $\mathbf{0} \le \eta \le \mathbf{1}$ with $\eta \ne \mathbf{0}$, any d-tuple sequence $\{a_{\mathbf{k}}\}$, any $\mathbf{1} \le \mathbf{n} \le \mathbf{m} \in \mathbb{N}^d$, and any fixed $\mathbf{k}_{1-\eta} \in \mathbb{N}^{1-\eta}$, we have*

$$\sum_{\mathbf{k}_{\eta}=\mathbf{n}_{\eta}}^{\mathbf{m}_{\eta}} \Delta^{\eta} a_{\mathbf{k}} = a_{\mathbf{k}_{1-\eta},\mathbf{n}_{\eta}} + \sum_{\zeta \le \eta,\, \zeta \ne \mathbf{0}} (-1)^{|\zeta|} a_{\mathbf{k}_{1-\eta},\mathbf{n}_{\eta-\zeta},(\mathbf{m}+\mathbf{1})_{\zeta}}. \tag{7.3}$$

Remark 7.4 Note that the term $a_{\mathbf{k}_{1-\eta},\mathbf{n}_{\eta-\zeta},(\mathbf{m}+\mathbf{1})_{\zeta}}$ appearing in the sum on the right-hand side of (7.3) can equivalently be rewritten as $a_{\mathbf{k}_{1-\eta},\mathbf{n}_{\eta-\zeta},(\mathbf{m}+\eta)_{\zeta}}$, since $\zeta \le \eta$.

Proof of Proposition 7.3 We proceed by induction on $|\mathbf{m}_{\eta} - \mathbf{n}_{\eta}|$. Note that the case $\mathbf{m}_{\eta} = \mathbf{n}_{\eta}$ is trivial, as the sum on the left-hand side of (7.3) contains only one term, and in this case it follows from (7.2) that

$$\begin{aligned} \Delta^{\eta} a_{\mathbf{k}_{1-\eta},\mathbf{n}_{\eta}} &= a_{\mathbf{k}_{1-\eta},\mathbf{n}_{\eta}} + \sum_{\zeta \le \eta,\, \zeta \ne \mathbf{0}} (-1)^{|\zeta|} a_{\mathbf{k}_{1-\eta},\mathbf{n}_{\eta-\zeta},(\mathbf{n}+\mathbf{1})_{\zeta}} \\ &= a_{\mathbf{k}_{1-\eta},\mathbf{n}_{\eta}} + \sum_{\zeta \le \eta,\, \zeta \ne \mathbf{0}} (-1)^{|\zeta|} a_{\mathbf{k}_{1-\eta},\mathbf{n}_{\eta-\zeta},(\mathbf{m}+\mathbf{1})_{\zeta}}, \end{aligned} \tag{7.4}$$

where the last equality follows from $\mathbf{m}_{\eta} = \mathbf{n}_{\eta}$.

We begin the induction procedure by taking $|\mathbf{m}_{\eta} - \mathbf{n}_{\eta}| = 1$. In this case $\mathbf{m}_{\eta} = \mathbf{n}_{\eta} + \mathbf{e}_j$, where j is such that $\eta_j = 1$. It follows from (7.4) that

$$\begin{aligned} \sum_{\mathbf{k}_{\eta}=\mathbf{n}_{\eta}}^{\mathbf{m}_{\eta}} \Delta^{\eta} a_{\mathbf{k}} &= \Delta^{\eta} a_{\mathbf{k}_{1-\eta},\mathbf{n}_{\eta}} + \Delta^{\eta} a_{\mathbf{k}_{1-\eta},(\mathbf{n}+\mathbf{e}_j)_{\eta}} \\ &= a_{\mathbf{k}_{1-\eta},\mathbf{n}_{\eta}} + \sum_{\zeta \le \eta,\, \zeta \ne \mathbf{0}} (-1)^{|\zeta|} a_{\mathbf{k}_{1-\eta},\mathbf{n}_{\eta-\zeta},(\mathbf{n}+\mathbf{1})_{\zeta}} \\ &\quad + a_{\mathbf{k}_{1-\eta},(\mathbf{n}+\mathbf{e}_j)_{\eta}} + \sum_{\zeta \le \eta,\, \zeta \ne \mathbf{0}} (-1)^{|\zeta|} a_{\mathbf{k}_{1-\eta},(\mathbf{n}+\mathbf{e}_j)_{\eta-\zeta},(\mathbf{n}+\mathbf{1}+\mathbf{e}_j)_{\zeta}}. \end{aligned} \tag{7.5}$$

We rewrite the two sums on the right-hand side of (7.5) as

$$\begin{aligned} &\sum_{\zeta \le \eta,\, \zeta \ne \mathbf{0}} (-1)^{|\zeta|} a_{\mathbf{k}_{1-\eta},\mathbf{n}_{\eta-\zeta},(\mathbf{n}+\mathbf{1})_{\zeta}} \\ &= \Bigg(\sum_{\substack{\zeta \le \eta \\ \zeta_j = 1}} + \sum_{\substack{\zeta \le \eta,\, \zeta \ne \mathbf{0} \\ \zeta_j = 0}} \Bigg) (-1)^{|\zeta|} a_{\mathbf{k}_{1-\eta},\mathbf{n}_{\eta-\zeta},(\mathbf{n}+\mathbf{1})_{\zeta}} =: A_1 + A_2, \end{aligned}$$

and

$$\sum_{\zeta\le\eta,\,\zeta\neq\mathbf{0}} (-1)^{|\zeta|} a_{\mathbf{k}_{\mathbf{1}-\eta},(\mathbf{n}+\mathbf{e}_j)_{\eta-\zeta},(\mathbf{n}+\mathbf{1}+\mathbf{e}_j)_\zeta}$$

$$= \Bigg(\sum_{\substack{\theta\le\eta\\ \theta_j=1}} + \sum_{\substack{\theta\le\eta,\,\theta\neq\mathbf{0}\\ \theta_j=0}}\Bigg)(-1)^{|\theta|} a_{\mathbf{k}_{\mathbf{1}-\eta},(\mathbf{n}+\mathbf{e}_j)_{\eta-\theta},(\mathbf{n}+\mathbf{1}+\mathbf{e}_j)_\theta} =: A_3 + A_4$$

(note that we have replaced ζ by θ in the last equality, which will make the following steps clearer). First of all, we observe that the right-hand side of (7.5) can be rewritten as

$$a_{\mathbf{k}_{\mathbf{1}-\eta},\mathbf{n}_\eta} + a_{\mathbf{k}_{\mathbf{1}-\eta},(\mathbf{n}+\mathbf{e}_j)_\eta} + \sum_{i=1}^{4} A_i.$$

A careful examination of the terms in the last expression shows that in order to prove (7.3), it is enough to show that $a_{\mathbf{k}_{\mathbf{1}-\eta},(\mathbf{n}+\mathbf{e}_j)_\eta} + A_1 + A_4 = 0$, since in this case

$$\begin{aligned}
a_{\mathbf{k}_{\mathbf{1}-\eta},\mathbf{n}_\eta} &+ a_{\mathbf{k}_{\mathbf{1}-\eta},(\mathbf{n}+\mathbf{e}_j)_\eta} + \sum_{i=1}^{4} A_i = a_{\mathbf{k}_{\mathbf{1}-\eta},\mathbf{n}_\eta} + A_2 + A_3\\
&= a_{\mathbf{k}_{\mathbf{1}-\eta},\mathbf{n}_\eta} + \sum_{\substack{\zeta\le\eta,\,\zeta\neq\mathbf{0}\\ \zeta_j=0}} (-1)^{|\zeta|} a_{\mathbf{k}_{\mathbf{1}-\eta},\mathbf{n}_{\eta-\zeta},(\mathbf{n}+\mathbf{1})_\zeta}\\
&\quad + \sum_{\substack{\theta\le\eta\\ \theta_j=1}} (-1)^{|\theta|} a_{\mathbf{k}_{\mathbf{1}-\eta},(\mathbf{n}+\mathbf{e}_j)_{\eta-\theta},(\mathbf{n}+\mathbf{1}+\mathbf{e}_j)_\theta}\\
&= a_{\mathbf{k}_{\mathbf{1}-\eta},\mathbf{n}_\eta} + \sum_{\substack{\zeta\le\eta,\,\zeta\neq\mathbf{0}\\ \zeta_j=0}} (-1)^{|\zeta|} a_{\mathbf{k}_{\mathbf{1}-\eta},\mathbf{n}_{\eta-\zeta},(\mathbf{n}+\mathbf{1}+\mathbf{e}_j)_\zeta}\\
&\quad + \sum_{\substack{\theta\le\eta\\ \theta_j=1}} (-1)^{|\theta|} a_{\mathbf{k}_{\mathbf{1}-\eta},\mathbf{n}_{\eta-\theta},(\mathbf{n}+\mathbf{1}+\mathbf{e}_j)_\theta}\\
&= a_{\mathbf{k}_{\mathbf{1}-\eta},\mathbf{n}_\eta} + \sum_{\zeta\le\eta,\,\zeta\neq\mathbf{0}} (-1)^{|\zeta|} a_{\mathbf{k}_{\mathbf{1}-\eta},\mathbf{n}_{\eta-\zeta},(\mathbf{n}+\mathbf{1}+\mathbf{e}_j)_\zeta},
\end{aligned}$$

which is the desired equality.

Let us proceed to showing that $a_{\mathbf{k}_{\mathbf{1}-\eta},(\mathbf{n}+\mathbf{e}_j)_\eta} + A_1 + A_4 = 0$. On the one hand,

$$\begin{aligned}
A_4 &= \sum_{\substack{\theta\le\eta,\,\theta\neq\mathbf{0}\\ \theta_j=0}} (-1)^{|\theta|} a_{\mathbf{k}_{\mathbf{1}-\eta},(\mathbf{n}+\mathbf{e}_j)_{\eta-\theta},(\mathbf{n}+\mathbf{1}+\mathbf{e}_j)_\theta}\\
&= \sum_{\substack{\theta\le\eta,\,\theta\neq\mathbf{0}\\ \theta_j=0}} (-1)^{|\theta|} a_{\mathbf{k}_{\mathbf{1}-\eta},(\mathbf{n}+\mathbf{e}_j)_{\eta-\theta},(\mathbf{n}+\mathbf{1})_\theta}
\end{aligned}$$

$$= \sum_{\substack{\theta \le \eta, \theta \ne \mathbf{0} \\ \theta_j = 0}} (-1)^{|\theta|} a_{\mathbf{k}_{\mathbf{1}-\eta}, \mathbf{n}_{\eta-\theta-\mathbf{e}_j}, (\mathbf{n}+\mathbf{1})_{\theta+\mathbf{e}_j}}$$

$$= \sum_{\substack{\zeta \le \eta \\ \zeta_j = 1}} (-1)^{|\zeta|+1} a_{\mathbf{k}_{\mathbf{1}-\eta}, \mathbf{n}_{\eta-\zeta}, (\mathbf{n}+\mathbf{1})_{\zeta}} - a_{\mathbf{k}_{\mathbf{1}-\eta}, (\mathbf{n}+\mathbf{e}_j)_\eta},$$

where in the last equality we made the change $\zeta = \theta + \mathbf{e}_j$ (note that on the right-hand side we subtract the term $a_{\mathbf{k}_{\mathbf{1}-\eta}, (\mathbf{n}+\mathbf{e}_j)_\eta}$, since it does not appear in the sum A_4). Thus, the last expression equals $-A_1 - a_{\mathbf{k}_{\mathbf{1}-\eta}, (\mathbf{n}+\mathbf{e}_j)_\eta}$, and we conclude that

$$a_{\mathbf{k}_{\mathbf{1}-\eta}, (\mathbf{n}+\mathbf{e}_j)_\eta} + A_1 + A_4 = 0,$$

which completes the proof of the first inductive step. In order to complete the proof, we assume that (7.3) holds for any $\mathbf{m}_\eta \ge \mathbf{n}_\eta$ with $|\mathbf{m}_\eta - \mathbf{n}_\eta| = \ell \ge 1$. We have to prove that (7.3) indeed holds with $\mathbf{m}_\eta + \mathbf{e}_j$ in place of $\mathbf{m}_\eta$, where j is such that $\eta_j = 1$. To this end, we use the induction hypothesis and (7.2) to obtain

$$\sum_{\mathbf{k}_\eta = \mathbf{n}_\eta}^{\mathbf{m}_\eta + \mathbf{e}_j} \Delta^\eta a_{\mathbf{k}} = \sum_{\mathbf{k}_\eta = \mathbf{n}_\eta}^{\mathbf{m}_\eta} \Delta^\eta a_{\mathbf{k}} + \Delta^\eta a_{\mathbf{k}_{\mathbf{1}-\eta}, (\mathbf{m}+\mathbf{e}_j)_\eta}$$
$$= a_{\mathbf{k}_{\mathbf{1}-\eta}, \mathbf{n}_\eta} + \sum_{\zeta \le \eta, \zeta \ne \mathbf{0}} (-1)^{|\zeta|} a_{\mathbf{k}_{\mathbf{1}-\eta}, \mathbf{n}_{\eta-\zeta}, (\mathbf{m}+\mathbf{1})_\zeta}$$
$$+ a_{\mathbf{k}_{\mathbf{1}-\eta}, (\mathbf{m}+\mathbf{e}_j)_\eta} + \sum_{\zeta \le \eta, \zeta \ne \mathbf{0}} (-1)^{|\zeta|} a_{\mathbf{k}_{\mathbf{1}-\eta}, (\mathbf{m}+\mathbf{e}_j)_{\eta-\zeta}, (\mathbf{m}+\mathbf{1}+\mathbf{e}_j)_\zeta}.$$

The rest of the proof is omitted, as it follows exactly the same lines as that of the first inductive step, in which the two sums of the last expression are split into two parts, and then the cancellations that occur between them are found. □

We now introduce further notation. If $\{u_k^1\}, \ldots, \{u_k^d\}$ are sequences of natural numbers and $\mathbf{j}, \mathbf{k} \in \mathbb{N}^d$, we define the d-tuple $\mathbf{u_k} := (u_{k_1}^1, \ldots, u_{k_d}^d)$, and

$$a_{\mathbf{u_k}} := a_{u_{k_1}^1, \ldots, u_{k_d}^d},$$
$$a_{(\mathbf{u_k})_\eta, \mathbf{j}_{\mathbf{1}-\eta}} := a_{\eta_1 u_{k_1}^1 + (1-\eta_1) j_1, \ldots, \eta_d u_{k_d}^d + (1-\eta_d) j_d}.$$

For the sake of clarity, we emphasize that $a_{(\mathbf{u_k})_\eta, \mathbf{j}_{\mathbf{1}-\eta}}$ equals $a_{m_1, \ldots, m_d}$, with $m_\ell = u_{k_\ell}^\ell$ if $\eta_\ell = 1$, and $m_\ell = j_\ell$ if $\eta_\ell = 0$.

For the differences of the terms $a_{(\mathbf{u_k})_\eta, \mathbf{j}_{\mathbf{1}-\eta}}$, one has, given a d-tuple $\mathbf{u_k}$ and $1 \le \ell \le d$,

$$\Delta^{\mathbf{e}_\ell} a_{(\mathbf{u_k})_\eta, \mathbf{j}_{\mathbf{1}-\eta}} = \begin{cases} a_{(\mathbf{u_k})_\eta, \mathbf{j}_{\mathbf{1}-\eta}} - a_{(\mathbf{u}_{\mathbf{k}+\mathbf{e}_\ell})_\eta, \mathbf{j}_{\mathbf{1}-\eta}}, & \text{if } \eta_\ell = 1, \\ a_{(\mathbf{u_k})_\eta, \mathbf{j}_{\mathbf{1}-\eta}} - a_{(\mathbf{u_k})_\eta, (\mathbf{j}+\mathbf{e}_\ell)_{\mathbf{1}-\eta}}, & \text{otherwise.} \end{cases}$$

For the sake of clarity, let us consider an example of the iterated difference

$$\Delta^\zeta a_{(\mathbf{u_k})_\eta, \mathbf{j}_{\mathbf{1}-\eta}} = \left(\prod_{\ell : \zeta_\ell = 1} \Delta^{\mathbf{e}_\ell} \right) a_{(\mathbf{u_k})_\eta, \mathbf{j}_{\mathbf{1}-\eta}}$$

in dimension three. Let $\eta = (0,1,1)$ and $\zeta = (1,0,1)$. For increasing sequences of natural numbers $\{u_k^2\}$ and $\{u_k^3\}$, we have

$$a_{(\mathbf{u_k})_\eta, \mathbf{j_{1-\eta}}} = a_{j_1, u_{k_2}^2, u_{k_3}^3},$$

and

$$\begin{aligned}\Delta^\zeta a_{(\mathbf{u_k})_\eta, \mathbf{j_{1-\eta}}} &= \Delta^{\mathbf{e}_1}\left(\Delta^{\mathbf{e}_3} a_{j_1, u_{k_2}^2, u_{k_3}^3}\right) = \Delta^{\mathbf{e}_1}\left(a_{j_1, u_{k_2}^2, u_{k_3}^3} - a_{j_1, u_{k_2}^2, u_{k_3+1}^3}\right) \\ &= a_{j_1, u_{k_2}^2, u_{k_3}^3} - a_{j_1, u_{k_2}^2, u_{k_3+1}^3} - a_{j_1+1, u_{k_2}^2, u_{k_3}^3} + a_{j_1+1, u_{k_2}^2, u_{k_3+1}^3}.\end{aligned}$$

We also note that if $u_k^\ell = k$ for all $\ell = 1, \dots, d$, then $a_{(\mathbf{u_k})_\eta, \mathbf{j_{1-\eta}}} = a_{\mathbf{k}_\eta, \mathbf{j_{1-\eta}}}$.

Recall that an increasing sequence $\{u_k\}$ (which we assume to consist of natural numbers) is said to be lacunary if

$$\frac{u_{k+1}}{u_k} \ge q > 1.$$

Further, $\{u_k\} \in \Lambda$ if it can be split into finitely many lacunary sequences or, equivalently, if and only if there exists $r \in \mathbb{N}$ such that

$$\frac{u_{k+r}}{u_k} \ge q > 1, \quad k \in \mathbb{N}.$$

Finally, we introduce certain notation concerning multidimensional integration. By $\mathbf{x}_\eta$ and $d\mathbf{x}_\eta$ we denote the $|\eta|$-tuple consisting only of x_j such that $\eta_j = 1$ and $\prod_{j:\eta_j=1} dx_j$, respectively. Naturally, we can replace the notation $\mathbb{R}_+^d = [0,\infty)^d$ with $\mathbb{R}_+^d := [\mathbf{0},\boldsymbol{\infty})^{\mathbf{1}} = [\mathbf{0},\boldsymbol{\infty})$, while $\mathbb{R}_+^\eta := [\mathbf{0},\boldsymbol{\infty})^\eta$ means that only the variables for which $\eta_j = 1$ are involved. Similarly, $[\mathbf{a},\mathbf{b}]^{\mathbf{1}}$ is a legitimate notation for $[\mathbf{a},\mathbf{b}]$ if $[\mathbf{a},\mathbf{b}]^\eta$ is understood as

$$\prod_{j:\eta_j=1} [a_j, b_j],$$

the Cartesian product of intervals (generally we will omit the interval superindices whenever $\eta = \mathbf{1}$). We will use this notation for multiple "intervals" systematically. Replacing "[" or "]" with "(" or ")", respectively, clearly means replacing the intervals denoted with the corresponding signs product-wise. Further definitions concerning functions and integration are postponed to Section 9.1.

In certain cases, in order to formulate our results, we need to introduce additional notation.

7.2 Multiple sequences

Let us now extend to higher dimensions all the concepts given for two-dimensional sequences in Chapter 4. By $a = \{a_\mathbf{k}\}$ we understand $\{a_\mathbf{k}\}_{\mathbf{k}=\mathbf{0}}^\infty$, or also $\{a_\mathbf{k}\}_{\mathbf{k}=\mathbf{1}}^\infty$. In other words, the domain of the d-tuple sequence a will be either $(\mathbb{N} \cup \{0\})^d$ or just $\mathbb{N}^d$. If the domain is not specified, it means that either this is irrelevant, or that it is clear from the context.

Definition 7.5 Let $\{a_{\mathbf{k}}\} \subset \mathbb{C}$ be a d-tuple sequence. We say that $A \in \mathbb{C}$ is the limit of the sequence $\{a_{\mathbf{k}}\}$ if for every $\varepsilon > 0$ there exists $\mathbf{n} \in \mathbb{N}^d$ such that

$$|a_{\mathbf{k}} - A| < \varepsilon, \qquad \text{if } \mathbf{k} > \mathbf{n}.$$

In this case we write

$$\lim_{\mathbf{k}\to\infty} a_{\mathbf{k}} = A,$$

and, equivalently, we say that $\{a_{\mathbf{k}}\}$ converges to A.

The above limit is often referred to as the *Pringsheim limit* of the sequence $\{a_{\mathbf{k}}\}$. Here we let all the indices tend to infinity, but there are other ways of defining limits of sequences, and they are not always equivalent (see, for instance, (7.7) below and the subsequent discussion).

Definition 7.6 A real-valued d-tuple sequence $\{a_{\mathbf{k}}\}$ is called divergent to $+\infty$ if for every $A > 0$ there exists $\mathbf{n} \in \mathbb{N}^d$ such that

$$a_{\mathbf{k}} > A \quad \text{provided} \quad \mathbf{k} > \mathbf{n},$$

and we write

$$\lim_{\mathbf{k}\to\infty} a_{\mathbf{k}} = +\infty.$$

Analogously, a real-valued d-tuple sequence $\{a_{\mathbf{k}}\}$ is called divergent to $-\infty$ if for every $A > 0$ there exists $\mathbf{n} \in \mathbb{N}^d$ such that

$$a_{\mathbf{k}} < -A \quad \text{provided} \quad \mathbf{k} > \mathbf{n},$$

and we write

$$\lim_{\mathbf{k}\to\infty} a_{\mathbf{k}} = -\infty.$$

Likewise, we say that the sequence diverges if it does not have a limit

We also introduce the notion of boundedness for d-tuple sequences.

Definition 7.7 A d-tuple sequence $\{a_{\mathbf{k}}\} \subset \mathbb{C}$ is called bounded if there is a number $B > 0$ such that $|a_{\mathbf{k}}| \leq B$ for all $\mathbf{k} \in \mathbb{N}^d$.

Definition 7.8 A d-tuple sequence with real-valued terms $\{a_{\mathbf{k}}\}$ is called bounded from above if there is a real number B such that $a_{\mathbf{k}} \leq B$ for all $\mathbf{k} \in \mathbb{N}^d$. Analogously, a d-tuple sequence with real-valued terms $\{a_{\mathbf{k}}\}$ is called bounded from below if there is a real number b such that $a_{\mathbf{k}} \geq b$ for all $\mathbf{k} \in \mathbb{N}^d$.

A d-tuple sequence with real-valued terms is bounded both from above and from below if and only if it is bounded.

Definition 7.9 A d-tuple sequence $\{a_{\mathbf{k}}\} \subset \mathbb{C}$ is called bounded from above (below) at infinity if there exists $\mathbf{n} \in \mathbb{N}^d$ such that the sequence $\{a_{\mathbf{k}}\}_{\mathbf{k} \geq \mathbf{n}}$ is bounded from above (below).

Definition 7.10 A d-tuple sequence $\{a_{\mathbf{k}}\} \subset \mathbb{C}$ is called bounded at infinity if there exists $\mathbf{n} \in \mathbb{N}^d$ such that the d-tuple sequence $\{a_{\mathbf{k}}\}_{\mathbf{k} \geq \mathbf{n}}$ is bounded.

As in two dimensions, the fact that a d-tuple sequence $\{a_{\mathbf{k}}\}$ converges does not imply its boundedness. To see this, it suffices to extend the two-dimensional example (4.4) to any dimension by setting the rest of terms all zero.

We now aim to define one of the variants of monotone sequences in several dimensions.

Definition 7.11 A d-tuple sequence $\{a_{\mathbf{k}}\}$ with real-valued terms is called increasing (decreasing) if

$$a_{\mathbf{n}} \geq a_{\mathbf{k}} \quad (a_{\mathbf{n}} \leq a_{\mathbf{k}}) \quad \text{for} \quad \mathbf{n} \geq \mathbf{k}.$$

In both cases we will call such a sequence monotone.

Theorem 7.12 *If a real-valued d-tuple sequence* $\{a_{\mathbf{k}}\}$ *is increasing (respectively decreasing), then for it to be convergent to a finite limit, it is necessary and sufficient that it is bounded from above (respectively below). In this case,* $\lim_{\mathbf{k} \to \infty} a_{\mathbf{k}} = \sup_{\mathbf{k}}\{a_{\mathbf{k}}\}$ *(respectively* $\lim_{\mathbf{k} \to \infty} a_{\mathbf{k}} = \inf_{\mathbf{k}}\{a_{\mathbf{k}}\}$*).*

Proof It suffices to prove the case of increasing sequences. Let the d-tuple sequence $\{a_{\mathbf{k}}\}$ be bounded from above and set $A = \sup_{\mathbf{k}}\{a_{\mathbf{k}}\}$. Given $\varepsilon > 0$, we can find $\mathbf{n} \in \mathbb{N}^d$ such that

$$a_{\mathbf{n}} > A - \varepsilon.$$

Since the sequence $\{a_{\mathbf{k}}\}$ is increasing, we obtain

$$A \geq a_{\mathbf{k}} > A - \varepsilon, \qquad \text{for all } \mathbf{k} \geq \mathbf{n},$$

so that

$$\lim_{\mathbf{k} \to \infty} a_{\mathbf{k}} = A,$$

which proves the sufficiency.

We now prove the necessity part. Assume that the increasing d-tuple sequence $\{a_{\mathbf{k}}\}$ converges. Then it is bounded from above at infinity, and since the sequence $\{a_{\mathbf{k}}\}$ is increasing in each variable, we obtain $a_{\mathbf{k}} \leq A$ for all $\mathbf{k} \in \mathbb{N}^d$, or, in other words, the sequence $\{a_{\mathbf{k}}\}$ is bounded from above and $\lim_{\mathbf{k} \to \infty} a_{\mathbf{k}} = A$. □

Remark 7.13 If an increasing d-tuple sequence $\{a_{\mathbf{k}}\}$ is not bounded from above, then

$$\lim_{\mathbf{k} \to \infty} a_{\mathbf{k}} = +\infty.$$

If a decreasing d-tuple sequence $\{a_{\mathbf{k}}\}$ is not bounded from below, then

$$\lim_{\mathbf{k}\to\infty} a_{\mathbf{k}} = -\infty.$$

The multidimensional version of the Cauchy criterion is as follows.

Theorem 7.14 *If a d-tuple sequence $\{a_{\mathbf{k}}\}$ converges, then for any $\varepsilon > 0$ there exists $\mathbf{n} \in \mathbb{N}^d$ such that for $\mathbf{k} \geq \mathbf{n}$ and $\mathbf{m} \geq \mathbf{0}$, the inequality*

$$|a_{\mathbf{k}+\mathbf{m}} - a_{\mathbf{k}}| < \varepsilon \tag{7.6}$$

holds. Conversely, if this condition holds, then the d-tuple sequence $\{a_{\mathbf{k}}\}$ converges.

Proof The necessity is obvious. Let us prove the sufficiency. Given $\varepsilon > 0$, there exists $\mathbf{n} \in \mathbb{N}^d$ such that for $\mathbf{k} \geq \mathbf{n}$ and $\mathbf{m} \geq \mathbf{0}$, the inequality (7.6) holds. Consider the one-dimensional sequence $\{\alpha_r = a_{r,\ldots,r}\}_{r=0}^{\infty}$. If $r \geq \max_i n_i$, we obtain

$$|\alpha_{r+s} - \alpha_r| < \varepsilon$$

for every $s \geq 0$. It follows from the Cauchy criterion for one-dimensional sequences that the sequence $\{\alpha_r\}$ converges to a limit, say, A. Letting $\mathbf{m} \to \infty$ in (7.6) in such a way that $k_i + m_i = k_j + m_j$ for all $i, j = 1, \ldots, d$, we obtain

$$|A - a_{\mathbf{k}}| \leq \varepsilon$$

provided $\mathbf{k} \geq \mathbf{n}$, which is exactly the definition of convergence of a d-tuple sequence and thus completes the proof. □

In the multidimensional setting there are many ways of defining the limits of sequences. One can take the multiple limit (as in Definition 7.5) with respect to all variables, or, in other words, as $\mathbf{k} \to \infty$. Another option is first taking the limit with respect to a set of indices and then with respect to the rest of indices, that is, first as $\mathbf{k}_\eta \to \infty$, with fixed $\mathbf{k}_{\mathbf{1}-\eta}$ ($\mathbf{0} \leq \eta \leq \mathbf{1}$), and then as $\mathbf{k}_{\mathbf{1}-\eta} \to \infty$, or even as an iteration of d one-dimensional limits.

For a d-tuple sequence $\{a_{\mathbf{k}}\}$ and $\mathbf{0} \leq \eta \leq \mathbf{1}$, we denote

$$a_{\mathbf{k}_\eta} = \lim_{\mathbf{k}_{\mathbf{1}-\eta}\to\infty} a_{\mathbf{k}},$$

if the limit exists for all $\mathbf{k}_\eta \in (\mathbb{N} \cup \{0\})^\eta$. In this way we obtain an $|\eta|$-tuple sequence by taking the limit in the rest of indices. For any $\zeta \leq \eta$, we define

$$\lim_{\mathbf{k}_{\eta-\zeta}\to\infty} a_{\mathbf{k}_\eta} = \lim_{\mathbf{k}_{\eta-\zeta}\to\infty} \left(\lim_{\mathbf{k}_{\mathbf{1}-\eta}\to\infty} a_{\mathbf{k}} \right) \tag{7.7}$$

as the *iterated limit* of $\{a_{\mathbf{k}}\}$, first with respect to $\mathbf{1} - \eta$ and then with respect to $\eta - \zeta$. Since this last part will always be clear from the context, such a nomenclature will be systematically omitted.

Any iteration of more than two limits of the form (7.7) will also be called an iterated limit. Note that in two dimensions we could take iterated limits in two different ways only, first with respect to one of the variables and then with respect to the other. In several dimensions there are many more possibilities.

Example 7.15 Let $\mathbf{z} \in \mathbb{C}^d$ and define, in the same spirit as in (4.1),

$$a_{\mathbf{k}} = \frac{1}{|\mathbf{k}|} \sum_{j=1}^{d} k_j z_j, \qquad \mathbf{k} \geq \mathbf{1}.$$

It is clear that for any $j \in \{1, \ldots, d\}$ and any fixed $\mathbf{k}_{\mathbf{1}-\mathbf{e}_j}$,

$$\lim_{k_j \to \infty} a_{\mathbf{k}} = z_j,$$

and thus

$$\lim_{\mathbf{k}_{\mathbf{1}-\mathbf{e}_j} \to \infty} \left(\lim_{k_j \to \infty} a_{\mathbf{k}} \right) = z_j.$$

On the other hand, note that any limit of the type

$$\lim_{\mathbf{k}_\eta \to \infty} a_{\mathbf{k}}$$

may not exist if $|\eta| \geq 2$ (this can easily be proved similarly to the two-dimensional example (4.2)).

In view of Example 7.15, a relevant question is to study how iterated limits with respect to a set of indices relate to one another when they are taken in different order within this set.

Theorem 7.16 *Let $\{a_{\mathbf{k}}\}$ be a d-tuple sequence and assume that for some $\mathbf{0} \leq \eta \leq \mathbf{1}$ with $\eta \neq \mathbf{0}, \mathbf{1}$ and every fixed $\mathbf{k}_{\mathbf{1}-\eta}$, the limit*

$$\lim_{\mathbf{k}_\eta \to \infty} a_{\mathbf{k}} = a_{\mathbf{k}_{\mathbf{1}-\eta}}$$

exists. Assume moreover that for some $\eta \leq \zeta \leq \mathbf{1}$ and all $\mathbf{k}_{\mathbf{1}-\zeta}$, the limit

$$\lim_{\mathbf{k}_\zeta \to \infty} a_{\mathbf{k}} = a_{\mathbf{k}_{\mathbf{1}-\zeta}}$$

exists (if $\zeta = \mathbf{1}$ put $a_{\mathbf{k}_{\mathbf{1}-\zeta}} = A$). Then the iterated limit

$$\lim_{\mathbf{k}_{\zeta-\eta} \to \infty} \left(\lim_{\mathbf{k}_\eta \to \infty} a_{\mathbf{k}} \right)$$

exists for all $\mathbf{k}_{\mathbf{1}-\zeta}$ and is equal to $a_{\mathbf{k}_{\mathbf{1}-\zeta}}$.

Proof Given an arbitrary $\varepsilon > 0$, fix $\mathbf{k}_{\mathbf{1}-\zeta}$ and let $\mathbf{n} \in \mathbb{N}^d$ be such that

$$|a_{\mathbf{k}} - a_{\mathbf{k}_{\mathbf{1}-\zeta}}| < \varepsilon, \quad \text{provided} \quad \mathbf{k}_\zeta \geq \mathbf{n}_\zeta.$$

We now fix $\mathbf{k}_{\zeta-\eta} \geq \mathbf{n}_{\zeta-\eta}$ and pass to the limit as $\mathbf{k}_\eta \to \infty$ in the last inequality, which yields

$$|a_{\mathbf{k}_{\mathbf{1}-\eta}} - a_{\mathbf{k}_{\mathbf{1}-\zeta}}| \leq \varepsilon.$$

This is precisely

$$\lim_{\mathbf{k}_{\zeta-\eta}\to\infty}\left(\lim_{\mathbf{k}_\eta\to\infty} a_{\mathbf{k}}\right) = a_{\mathbf{k}_{1-\zeta}}. \qquad \square$$

Example 7.17 As in two dimensions, it may happen that a d-tuple sequence $\{a_{\mathbf{k}}\}$ tends to a limit as $\mathbf{k} \to \infty$, but some of its iterated limits do not exist, as the example

$$a_{\mathbf{k}} = \frac{1}{|\mathbf{k}| - k_1}\cos k_1\pi + \frac{1}{k_1}\cos|\mathbf{k}|\pi, \qquad \mathbf{k} \geq \mathbf{1},$$

shows. Indeed, it is clear that

$$\lim_{\mathbf{k}\to\infty}|a_{\mathbf{k}}| \leq \lim_{\mathbf{k}\to\infty}\frac{1}{k_1} + \frac{1}{k_2 + \cdots + k_d} = 0,$$

although the limits

$$\lim_{\mathbf{k}_{1-\mathbf{e}_1}\to\infty}\left(\lim_{k_1\to\infty} a_{\mathbf{k}}\right) \quad \text{and} \quad \lim_{k_1\to\infty}\left(\lim_{\mathbf{k}_{1-\mathbf{e}_1}\to\infty} a_{\mathbf{k}}\right)$$

do not exist.

Similarly, the existence of all iterated limits does not imply the existence of the multiple limit, even if all of them are equal. To see this, consider the example

$$a_{\mathbf{k}} = \frac{k_1\cdots k_d}{k_1^d + \cdots + k_d^d}. \tag{7.8}$$

It is clear that for any $\mathbf{0} \leq \eta \leq \mathbf{1}$, with $\eta \neq \mathbf{0}, \mathbf{1}$,

$$a_{\mathbf{k}} \leq \left(\max_{j:\eta_j=1} k_j\right)^{|\eta|-d} \prod_{j:\eta_j=0} k_j,$$

which implies $\lim_{\mathbf{k}_\eta\to\infty} a_{\mathbf{k}} = 0$, so that

$$\lim_{\mathbf{k}_{1-\eta}\to\infty}\left(\lim_{\mathbf{k}_\eta\to\infty} a_{\mathbf{k}}\right) = 0.$$

However, if $k_1 = k_2 = \cdots = k_{d-1}$ and $k_d = \alpha k_1$ with $\alpha \in \mathbb{N}$, then

$$a_{k_1,k_1,\ldots,\alpha k_1} = \frac{\alpha}{d - 1 + \alpha^d},$$

which means that the limit $\lim_{\mathbf{k}\to\infty} a_{\mathbf{k}}$ does not exist.

We observe that for monotone sequences, all the iterated limits coincide.

Theorem 7.18 *If a real-valued d-tuple sequence $\{a_{\mathbf{k}}\}$ is increasing (respectively decreasing), then for every $\mathbf{0} \leq \eta \leq \zeta \leq \mathbf{1}$ with $\eta \neq \mathbf{0}, \mathbf{1}$ and every $\mathbf{k}_{1-\zeta}$, the iterated limit*

$$\lim_{\mathbf{k}_{\zeta-\eta}\to\infty}\left(\lim_{\mathbf{k}_\eta\to\infty} a_{\mathbf{k}}\right)$$

exists and is equal to the multiple limit

$$\lim_{\mathbf{k}_\zeta\to\infty} a_{\mathbf{k}},$$

even if it is $+\infty$ (respectively $-\infty$).

Remark 7.19 The assumption that $\{a_{\mathbf{k}}\}$ is monotone is essential in Theorem 7.18. Indeed, it cannot be replaced by weak monotonicity, as example (7.8) (which defines a weak monotone d-tuple sequence) shows (we refer to Section 9.1 below for the corresponding definitions of weak monotonicity).

Proof of Theorem 7.18 Assume that $\{a_{\mathbf{k}}\}$ is increasing. By Theorem 7.12, for all $\mathbf{0} \le \eta \le \zeta \le \mathbf{1}$ with $\eta \ne \mathbf{0}, \mathbf{1}$ and every $\mathbf{k}_{\mathbf{1}-\zeta}$, the limit $\lim_{\mathbf{k}_\zeta \to \infty} a_{\mathbf{k}}$ exists, either finite or infinite. Assume first that this limit is finite, that is,

$$\lim_{\mathbf{k}_\zeta \to \infty} a_{\mathbf{k}} = a_{\mathbf{k}_{\mathbf{1}-\zeta}}.$$

Clearly, for every $\mathbf{k} \in \mathbb{N}^d$, there holds $a_{\mathbf{k}} \le a_{\mathbf{k}_{\mathbf{1}-\zeta}}$, and since for every fixed $\mathbf{k}_{\mathbf{1}-\eta}$, the limit

$$\lim_{\mathbf{k}_\eta \to \infty} a_{\mathbf{k}} = a_{\mathbf{k}_{\mathbf{1}-\eta}}$$

exists, we have, by Theorem 7.12 (applied to the increasing sequence $\{a_{\mathbf{k}_{\mathbf{1}-\eta}}\}$), that the iterated limit

$$\lim_{\mathbf{k}_{\zeta-\eta} \to \infty} a_{\mathbf{k}_{\mathbf{1}-\eta}} = \lim_{\mathbf{k}_{\zeta-\eta} \to \infty} \left(\lim_{\mathbf{k}_\eta \to \infty} a_{\mathbf{k}} \right)$$

also exists and is equal to

$$\sup_{\mathbf{k}_{\zeta-\eta}} a_{\mathbf{k}_{\mathbf{1}-\eta}} = a_{\mathbf{k}_{\mathbf{1}-\zeta}}.$$

Assume now that for a fixed $\mathbf{k}_{\mathbf{1}-\eta}$, one has

$$\lim_{\mathbf{k}_\zeta \to \infty} a_{\mathbf{k}} = +\infty.$$

Then, for any $B > 0$, there is $\mathbf{n} \in \mathbb{N}^d$ such that

$$a_{\mathbf{k}} > B \quad \text{if } \mathbf{k}_\zeta \ge \mathbf{n}_\zeta,$$

and, therefore,

$$\lim_{\mathbf{k}_\eta \to \infty} a_{\mathbf{k}} > B$$

for $\mathbf{k}_{\zeta-\eta} \ge \mathbf{n}_{\zeta-\eta}$. Thus,

$$\lim_{\mathbf{k}_{\zeta-\eta} \to \infty} \left(\lim_{\mathbf{k}_\eta \to \infty} a_{\mathbf{k}} \right) > B,$$

for arbitrary $B > 0$, which implies that the iterated limit is actually $+\infty$ and completes the proof. The statement concerning decreasing sequences follows automatically by considering the d-tuple sequence $\{-a_{\mathbf{k}}\}$, with $\{a_{\mathbf{k}}\}$ increasing. □

7.3 Convergence of multiple series: definitions and basic notions

As mentioned earlier, if $\{a_{\mathbf{k}}\}$ is a nonnegative multiple sequence ($a_{\mathbf{k}} \geq 0$), all types of convergence of $\sum_{\mathbf{k}=\mathbf{1}}^{\infty} a_{\mathbf{k}}$ are equivalent. In this case we just say that the series converges. However, recall that in two dimensions we have already defined various types of convergence. Most of the series $\sum_{\mathbf{k}=\mathbf{1}}^{\infty} a_{\mathbf{k}}$ considered in this part will have nonnegative terms, although sometimes series with real-valued coefficients of any sign (or even complex-valued ones) may arise, for example, when the differences of the sequence $a_{\mathbf{k}}$ are involved. Because of this, we have to define different types of convergence of multiple series. Let us first define *Pringsheim convergence* and state some of its basic properties. In this section all sequences are assumed to be complex-valued.

Definition 7.20 We say that the multidimensional series (or d-tuple series)

$$\sum_{\mathbf{k}=\mathbf{1}}^{\infty} a_{\mathbf{k}} \tag{7.9}$$

converges in the sense of Pringsheim (or just converges) if

$$\lim_{\mathbf{n}\to\infty} S_{\mathbf{n}} = S \in \mathbb{C}.$$

In other words, convergence of multidimensional series is equivalent to the convergence of the d-tuple sequence formed by its partial sums in the sense of Definition 7.5. A d-tuple series is said to be divergent if it does not converge in the sense of Pringsheim.

Note that if $d = 1$ then the convergence in the sense of Pringsheim is equivalent to the usual convergence of one-dimensional series.

Let us now list some of the basic properties of convergent series.

Properties. Let the series (7.9) converge to A. Then

1. The sequence $a_{\mathbf{k}}$ vanishes as $\mathbf{k} \to \infty$.
2. The series $\sum_{\mathbf{k}=\mathbf{1}}^{\infty} \lambda a_{\mathbf{k}}$ converges to λA for any $\lambda \in \mathbb{C}$.
3. If the series $\sum_{\mathbf{k}=\mathbf{1}}^{\infty} b_{\mathbf{k}}$ converges to B, then $\sum_{\mathbf{k}=\mathbf{1}}^{\infty}(a_{\mathbf{k}} + b_{\mathbf{k}})$ converges to $A + B$.

Furthermore, any nonnegative multiple series is convergent if and only if its partial sums $S_{\mathbf{n}}$ are bounded (cf. Theorem 7.12).

In the one-dimensional setting, Cauchy's criterion for the convergence states that a series $\sum_{k=1}^{\infty} a_k$ converges if and only if

$$\sum_{k=m}^{n} a_k \to 0 \qquad \text{as } n \geq m \to \infty.$$

In other words, a series is convergent if and only if its partial sums $S_n = \sum_{k=1}^{n} a_k$ form a *Cauchy sequence*. In the multidimensional setting we have a similar criterion.

Theorem 7.21 *A d-tuple series* $\sum_{\mathbf{k}=\mathbf{1}}^{\infty} a_{\mathbf{k}}$ *converges in the sense of Pringsheim if and only if*

$$|S_{\mathbf{m}} - S_{\mathbf{n}}| \to 0 \qquad \textit{as} \quad \mathbf{n}, \mathbf{m} \to \boldsymbol{\infty}.$$

This result readily follows from Theorem 7.14.

Theorem 7.22 *If the terms of two d-tuple sequences* $\{a_{\mathbf{k}}\}$ *and* $\{a'_{\mathbf{k}}\}$ *are such that* $a_{\mathbf{k}} \neq a'_{\mathbf{k}}$ *only for finitely many d-tuples* $\mathbf{k} \in \mathbb{N}^d$*, then the series* $\sum_{\mathbf{k}=\mathbf{1}}^{\infty} a_{\mathbf{k}}$ *converges if and only if the series* $\sum_{\mathbf{k}=\mathbf{1}}^{\infty} a'_{\mathbf{k}}$ *does as well.*

Proof Denote by $S_{\mathbf{m}}$ and $S'_{\mathbf{m}}$ the partial sums of the series $\sum_{\mathbf{k}=\mathbf{1}}^{\infty} a_{\mathbf{k}}$ and $\sum_{\mathbf{k}=\mathbf{1}}^{\infty} a'_{\mathbf{k}}$, respectively. Note that there exists $n_0 \in \mathbb{N}$ such that $a_{\mathbf{k}} = a'_{\mathbf{k}}$ for every $\mathbf{k}$ with $|\mathbf{k}| \geq n_0$. Hence, for every $\mathbf{k}, \mathbf{m}$ with $|\mathbf{k}|, |\mathbf{m}| \geq n_0$, one has

$$|S_{\mathbf{k}} - S_{\mathbf{m}}| = |S'_{\mathbf{k}} - S'_{\mathbf{m}}|,$$

and the result follows by Theorem 7.21. □

In some cases it is useful to compute a d-tuple series as an iteration of lower-dimensional series (although in general this is not possible, as we have seen in the two-dimensional case; cf. Remark 4.16, which gives examples that easily extend to the multidimensional case). Theorem 7.16 gives simple sufficient conditions for computing the value of a d-tuple series through iterated series.

Theorem 7.23 *Let* $\sum_{\mathbf{k}=\mathbf{1}}^{\infty} a_{\mathbf{k}}$ *be a d-tuple series and assume that for some* $\mathbf{0} \leq \eta \leq \mathbf{1}$ *with* $\eta \neq \mathbf{0}, \mathbf{1}$ *and every fixed* $\mathbf{k}_{\mathbf{1}-\eta}$*, the finite limit*

$$\lim_{\mathbf{n}_\eta \to \boldsymbol{\infty}} \sum_{\mathbf{k}_\eta = \mathbf{1}}^{\mathbf{n}_\eta} a_{\mathbf{k}} = \sum_{\mathbf{k}_\eta = \mathbf{1}}^{\boldsymbol{\infty}} a_{\mathbf{k}}$$

exists. Assume moreover that for some $\eta \leq \zeta \leq \mathbf{1}$ *and all* $\mathbf{k}_{\mathbf{1}-\zeta}$*, the finite limit*

$$\lim_{\mathbf{n}_\zeta \to \boldsymbol{\infty}} \sum_{\mathbf{k}_\zeta = \mathbf{1}}^{\mathbf{n}_\zeta} a_{\mathbf{k}} = \sum_{\mathbf{k}_\zeta = \mathbf{1}}^{\boldsymbol{\infty}} a_{\mathbf{k}}$$

exists (if $\zeta = \mathbf{1}$*, put* $\sum_{\mathbf{k}_\zeta=\mathbf{1}}^{\infty} a_{\mathbf{k}} = \sum_{\mathbf{k}=\mathbf{1}}^{\infty} a_{\mathbf{k}} = S$*). Then the iterated limit*

$$\lim_{\mathbf{n}_{\zeta-\eta} \to \boldsymbol{\infty}} \left(\lim_{\mathbf{n}_\eta \to \boldsymbol{\infty}} \sum_{\mathbf{k}_\zeta = \mathbf{1}}^{\mathbf{n}_\zeta} a_{\mathbf{k}} \right) \tag{7.10}$$

exists for all $\mathbf{k}_{\mathbf{1}-\zeta}$ *and is equal to* $\sum_{\mathbf{k}_\zeta=\mathbf{1}}^{\infty} a_{\mathbf{k}}$.

The proof readily follows from Theorem 7.16 applied to the sequence of partial sums $\{S_{\mathbf{n}}\}$ in place of $\{a_{\mathbf{n}}\}$. We call any iterated limit of the form (7.10) an *iterated series*.

It is important to emphasize that the convergence of a series in the sense of Pringsheim does not imply that its coefficients $a_{\mathbf{k}}$ tend to zero as $|\mathbf{k}| \to \boldsymbol{\infty}$, nor even their boundedness. For the corresponding counterexample in two dimensions, see (4.4). There are no such phenomena if we restrict ourselves to the *regular convergence*. Essentially, regularly convergent series are those whose all ηth partial series (with $\eta \neq \mathbf{0}$) are convergent for any set of $(\mathbf{1}-\eta)$th fixed indices. We note that in example (4.4) the series $\sum_{k\geq 1} a_{k,1}$ is divergent.

Definition 7.24 We say that the series $\sum_{\mathbf{k}=\mathbf{1}}^{\infty} a_{\mathbf{k}}$ converges in the *regular sense* if

$$\sum_{\mathbf{k}=\mathbf{m}}^{\mathbf{n}} a_{\mathbf{k}} \to 0 \qquad \text{as } |\mathbf{m}| \to \infty, \mathbf{m} \leq \mathbf{n}.$$

It follows from the definition that if a series converges in the regular sense, its terms $a_{\mathbf{k}}$ vanish as $|\mathbf{k}| \to \infty$. Moreover, if a series converges in the regular sense, it also converges in the sense of Pringsheim, and the sum is $\lim_{\mathbf{n}\to\infty} S_{\mathbf{n}}$. More precisely, we can characterize regular convergence as follows.

Theorem 7.25 *Let $\sum a_{\mathbf{k}}$ be a d-tuple series. Then, the following are equivalent.*

(i) *The series $\sum a_{\mathbf{k}}$ converges regularly.*
(ii) *The series $\sum a_{\mathbf{k}}$ as well as for any fixed $\mathbf{0} \leq \eta \leq \mathbf{1}$, $\eta \neq \mathbf{0}, \mathbf{1}$, and $\mathbf{k}_{\mathbf{1}-\eta} \in \mathbb{N}^{\mathbf{1}-\eta}$, the series*

$$\sum_{\mathbf{k}_\eta=\mathbf{1}}^{\infty} a_{\mathbf{k}} \tag{7.11}$$

converges in the sense of Pringsheim.
(iii) *The series $\sum a_{\mathbf{k}}$ converges in the sense of Pringsheim and for any fixed $\mathbf{0} \leq \eta \leq \mathbf{1}$, $\eta \neq \mathbf{0}, \mathbf{1}$, and $\mathbf{k}_{\mathbf{1}-\eta} \in \mathbb{N}^{\mathbf{1}-\eta}$, the series* (7.11) *converges regularly.*

Note that in two dimensions the equivalence (ii)⇔(iii) in Theorem 7.25 is vacuous, since one-dimensional regular convergence of the lower-dimensional (one-dimensional) series is just ordinary convergence.

Proof of Theorem 7.25 The fact that (i) implies (ii) simply follows from the definition of regular convergence and Theorem 7.21, while (i) trivially implies (iii) by definition. Also, (iii) clearly implies (ii).

Let us now prove that (iii) implies (i). First, since the d-tuple series $\sum a_{\mathbf{k}}$ converges in the sense of Pringsheim, for every $\varepsilon > 0$ there exists $N^{(1)} \in \mathbb{N}$ such that if $\min n_j \geq N^{(1)}$, then

$$\left|\sum_{\mathbf{k}=\mathbf{n}}^{\mathbf{m}} a_{\mathbf{k}}\right| < \varepsilon, \qquad \mathbf{m} \geq \mathbf{n}. \tag{7.12}$$

Secondly, since for every $1 \leq j \leq d$ and every fixed $k_j \in \mathbb{N}$ the $(d-1)$-tuple series $\sum_{\mathbf{k}_{\mathbf{1}-\mathbf{e}_j}=\mathbf{1}}^{\infty} a_{\mathbf{k}}$ converges in the regular sense, for every $\varepsilon > 0$ and $M_j \in \mathbb{N}$, there exists $N_j^{(2)} \in \mathbb{N}$ such that if $\max_{\ell\neq j} n_\ell^{(0)} \geq N_j^{(2)}$ and $m_j^{(0)} \leq m_j^{(1)} \leq M_j$, then

$$\left|\sum_{k_j=m_j^{(0)}}^{m_j^{(1)}} \sum_{\mathbf{k}_{\mathbf{1}-\mathbf{e}_j}=\mathbf{n}_{\mathbf{1}-\mathbf{e}_j}^{(0)}}^{\mathbf{n}_{\mathbf{1}-\mathbf{e}_j}^{(1)}} a_{\mathbf{k}}\right| < \varepsilon, \qquad \mathbf{n}_{\mathbf{1}-\mathbf{e}_j}^{(1)} \geq \mathbf{n}_{\mathbf{1}-\mathbf{e}_j}^{(0)}. \tag{7.13}$$

Fix $\varepsilon > 0$ and let $N^{(1)}$ be given by condition (7.12). We now apply condition (7.13) with such an $\varepsilon > 0$ and $M_j = N^{(1)}$ for all j, thus obtaining $N_1^{(2)}, N_2^{(2)}, \ldots, N_d^{(2)}$. Set $N^{(2)} := \max_j N_j^{(2)}$. Let now $\mathbf{n}^{(0)} \in \mathbb{N}^d$ be such that $\max_j n_j^{(0)} \geq \max\{N^{(1)}, N^{(2)}\}$ and ℓ be such that $n_\ell^{(0)} = \max_j n_j^{(0)}$. Now, for $\mathbf{n}^{(1)} \geq \mathbf{n}^{(0)}$ we can write, by (7.1),

$$\begin{aligned}
\sum_{\mathbf{k}=\mathbf{n}^{(0)}}^{\mathbf{n}^{(1)}} a_{\mathbf{k}} &= \sum_{k_\ell=n_\ell^{(0)}}^{n_\ell^{(1)}} \sum_{\mathbf{k}_{\mathbf{1}-\mathbf{e}_\ell}=\mathbf{n}_{\mathbf{1}-\mathbf{e}_\ell}^{(0)}}^{\mathbf{n}_{\mathbf{1}-\mathbf{e}_\ell}^{(1)}} a_{\mathbf{k}} \\
&= \sum_{k_\ell=n_\ell^{(0)}}^{n_\ell^{(1)}} \sum_{\mathbf{0}\leq\theta\leq\mathbf{1}-\mathbf{e}_\ell} \sum_{\mathbf{k}_\theta=\mathbf{n}_\theta^{(0)}}^{(\mathbf{N}^{(1)}-\mathbf{1})_\theta} \sum_{\mathbf{k}_{\mathbf{1}-\theta-\mathbf{e}_\ell}=\mathbf{N}_{\mathbf{1}-\theta-\mathbf{e}_\ell}^{(1)}}^{\mathbf{n}_{\mathbf{1}-\theta-\mathbf{e}_\ell}^{(1)}} a_{\mathbf{k}} \\
&= \sum_{k_\ell=n_\ell^{(0)}}^{n_\ell^{(1)}} \sum_{\mathbf{k}_{\mathbf{1}-\mathbf{e}_\ell}=\mathbf{N}_{\mathbf{1}-\mathbf{e}_\ell}^{(1)}}^{\mathbf{n}_{\mathbf{1}-\mathbf{e}_\ell}^{(1)}} a_{\mathbf{k}} \\
&\quad + \sum_{\substack{\mathbf{0}\leq\theta\leq\mathbf{1}-\mathbf{e}_\ell \\ \theta\neq\mathbf{0}}} \sum_{k_\ell=n_\ell^{(0)}}^{n_\ell^{(1)}} \sum_{\mathbf{k}_\theta=\mathbf{n}_\theta^{(0)}}^{(\mathbf{N}^{(1)}-\mathbf{1})_\theta} \sum_{\mathbf{k}_{\mathbf{1}-\theta-\mathbf{e}_\ell}=\mathbf{N}_{\mathbf{1}-\theta-\mathbf{e}_\ell}^{(1)}}^{\mathbf{n}_{\mathbf{1}-\theta-\mathbf{e}_\ell}^{(1)}} a_{\mathbf{k}}, \qquad (7.14)
\end{aligned}$$

where $\mathbf{N}^{(1)} = (N^{(1)}, N^{(1)}, \ldots, N^{(1)})$. On the one hand, since $n_\ell^{(0)} \geq N^{(1)}$, condition (7.12) yields

$$\left| \sum_{k_\ell=n_\ell^{(0)}}^{n_\ell^{(1)}} \sum_{\mathbf{k}_{\mathbf{1}-\mathbf{e}_\ell}=\mathbf{N}_{\mathbf{1}-\mathbf{e}_\ell}^{(1)}}^{\mathbf{n}_{\mathbf{1}-\mathbf{e}_\ell}^{(1)}} a_{\mathbf{k}} \right| < \varepsilon,$$

while on the other hand, since $n_\ell^{(0)} = \max \mathbf{n}^{(0)} \geq N^{(2)}$, condition (7.13) implies that for every $\mathbf{0} \leq \theta \leq \mathbf{1} - \mathbf{e}_\ell, \theta \neq \mathbf{0}$,

$$\left| \sum_{k_\ell=n_\ell^{(0)}}^{n_\ell^{(1)}} \sum_{\mathbf{k}_\theta=\mathbf{n}_\theta^{(0)}}^{(\mathbf{N}^{(1)}-\mathbf{1})_\theta} \sum_{\mathbf{k}_{\mathbf{1}-\theta-\mathbf{e}_\ell}=\mathbf{N}_{\mathbf{1}-\theta-\mathbf{e}_\ell}^{(1)}}^{\mathbf{n}_{\mathbf{1}-\theta-\mathbf{e}_\ell}^{(1)}} a_{\mathbf{k}} \right| < \varepsilon.$$

Combining these estimates with (7.14), we finally get

$$\left| \sum_{\mathbf{k}=\mathbf{n}^{(0)}}^{\mathbf{n}^{(1)}} a_{\mathbf{k}} \right| < \varepsilon + (2^{d-1} - 1)\varepsilon = 2^{d-1}\varepsilon,$$

which proves the regular convergence of $\sum a_{\mathbf{k}}$.

Finally, it remains to prove that (ii) implies (iii). We argue by induction on dimension d, with $d = 2$ being the base inductive step. In this case, the regular and Pringsheim convergence of the one-dimensional series $\sum_{k=1}^{\infty} a_{k\ell}$ and $\sum_{\ell=1}^{\infty} a_{k\ell}$ for any fixed ℓ and k, respectively, is equivalent. Therefore, for $d = 2$, (ii) and (iii) are essentially the same statement.

Let us now assume that the result is true for $(d-1)$-tuple series, and let us prove the conclusion for d-tuple series. By induction hypothesis, for every $\mathbf{0} \le \eta \le \mathbf{1}$ satisfying $\eta \neq \mathbf{0}, \mathbf{1}$ and $|\eta| \le d-2$, and any fixed $\mathbf{k}_{\mathbf{1}-\eta} \in \mathbb{N}^{\mathbf{1}-\eta}$, the series (7.11) converges regularly. Since by assumption any $(d-1)$-tuple series

$$\sum_{\mathbf{k}_{\eta+\mathbf{e}_j}=\mathbf{1}}^{\infty} a_{\mathbf{k}}, \tag{7.15}$$

converges in the sense of Pringsheim, where $1 \le j \le d$ is such that $\eta_j = 0$ and $\mathbf{k}_{\mathbf{1}-\eta-\mathbf{e}_j}$ is fixed, we derive that (7.15) converges regularly, by the "(iii) implies (i)" part of the statement, valid in any dimension. Thus, we conclude that for any fixed $\mathbf{0} \le \eta \le \mathbf{1}$ satisfying $|\eta| \le d-1$ (i.e., $\eta \neq \mathbf{1}$) and $\mathbf{k}_{\mathbf{1}-\eta} \in \mathbb{N}^{\mathbf{1}-\eta}$, the series (7.11) converges regularly. □

Remark 7.26 Note that if a series $\sum_{\mathbf{k}=\mathbf{1}}^{\infty} a_{\mathbf{k}}$ converges in the regular sense, then for every $\mathbf{0} \le \eta \le \mathbf{1}$ with $\eta \neq \mathbf{0}$ and every fixed $\mathbf{k}_{\mathbf{1}-\eta}$, the series $\sum_{\mathbf{k}_\eta=\mathbf{1}}^{\infty} a_{\mathbf{k}}$ also converges in the regular sense. This readily follows from the definition of regular convergence and in fact corresponds to the part "(i) implies (iii)" in Theorem 7.25.

It is important to remark that the partial sums $S_{\mathbf{n}}$ correspond to rectangular sums, and the convergence of a series is defined in terms of these rectangular sums. In many cases, it is useful to compute the sum of a convergent series $\sum_{\mathbf{k}=\mathbf{1}}^{\infty} a_{\mathbf{k}}$ as an iteration of one-dimensional series. However, the convergence of the series $\sum_{\mathbf{k}=\mathbf{1}}^{\infty} a_{\mathbf{k}}$ does not imply nor follows from the convergence of the iterated series. For the corresponding two-dimensional examples, see Remark 4.16.

A useful property of regularly convergent series is that we may rearrange the order of summation and compute the sum $\sum_{\mathbf{k}=\mathbf{1}}^{\infty} a_{\mathbf{k}}$ as an iteration of one-dimensional series or, more generally, as any iterated series. In other words, a Fubini-type theorem holds for regularly convergent series. Surprisingly enough, such a strong statement comes as a simple consequence of Theorem 7.25.

Theorem 7.27 *Let $\sum_{\mathbf{k}=\mathbf{1}}^{\infty} a_{\mathbf{k}}$ be a regularly convergent series. Then, for any $\mathbf{0} \le \eta \le \zeta \le \mathbf{1}$ with $\eta \neq \mathbf{0}, \zeta$ and any fixed $\mathbf{k}_{\mathbf{1}-\zeta}$,*

$$\sum_{\mathbf{k}_\zeta=\mathbf{1}}^{\infty} a_{\mathbf{k}} = \sum_{\mathbf{k}_{\zeta-\eta}=\mathbf{1}}^{\infty} \left(\sum_{\mathbf{k}_\eta=\mathbf{1}}^{\infty} a_{\mathbf{k}} \right). \tag{7.16}$$

In particular, the series $\sum_{\mathbf{k}=\mathbf{1}}^{\infty} a_{\mathbf{k}}$ can be computed as an iteration of one-dimensional series in any order.

Proof First of all, Theorem 7.25 implies that the series $\sum_{\mathbf{k}_\zeta=\mathbf{1}}^{\infty} a_{\mathbf{k}}$ and $\sum_{\mathbf{k}_\eta=\mathbf{1}}^{\infty} a_{\mathbf{k}}$ converge in the sense of Pringsheim for any fixed $\mathbf{k}_{\mathbf{1}-\zeta}$ and $\mathbf{k}_{\mathbf{1}-\eta}$, respectively (in fact, they converge regularly; cf. Remark 7.26). Denoting

$$A_{\mathbf{k}_{\mathbf{1}-\zeta}} = \sum_{\mathbf{k}_\zeta=\mathbf{1}}^{\infty} a_{\mathbf{k}} \quad \text{and} \quad A_{\mathbf{k}_{\mathbf{1}-\eta}} = \sum_{\mathbf{k}_\eta=\mathbf{1}}^{\infty} a_{\mathbf{k}},$$

our goal is to prove

$$A_{\mathbf{k}_{\mathbf{1}-\zeta}} = \sum_{\mathbf{k}_{\zeta-\eta}=\mathbf{1}}^{\infty} A_{\mathbf{k}_{\mathbf{1}-\eta}}, \tag{7.17}$$

for any $\mathbf{k}_{\mathbf{1}-\zeta}$. Since the series on the left-hand side of (7.16) converges, for every $\varepsilon > 0$ there exists $\mathbf{n} \in \mathbb{N}^d$ such that for $\mathbf{m}_\zeta \geq \mathbf{n}_\zeta$ one has

$$\left| A_{\mathbf{k}_{\mathbf{1}-\zeta}} - \sum_{\mathbf{k}_\zeta=\mathbf{1}}^{\mathbf{m}_\zeta} a_{\mathbf{k}} \right| = \left| A_{\mathbf{k}_{\mathbf{1}-\zeta}} - \sum_{\mathbf{k}_{\zeta-\eta}=\mathbf{1}}^{\mathbf{m}_{\zeta-\eta}} \left(\sum_{\mathbf{k}_\eta=\mathbf{1}}^{\mathbf{m}_\eta} a_{\mathbf{k}} \right) \right| < \varepsilon.$$

We then arrive at (7.17) by simply letting first $\mathbf{m}_\eta \to \infty$ and then $\mathbf{m}_{\zeta-\eta} \to \infty$ in the last inequality. □

We note that since any convergent series with nonnegative terms is regularly convergent (see Corollary 7.35 below), Theorem 7.27 is an improved version of the classical Fubini's theorem, which states that convergent series with nonnegative terms may be computed by means of iterated series (in fact, this is true in the case of absolutely convergent series, which are discussed in more detail in Section 7.6 below).

7.4 Convergence of multiple integrals: definitions and basic notions

Let us also introduce relevant notions of convergence for integrals. We will also state an integral counterpart of Theorem 7.27 without proof.

Given $\mathbf{a} \in \mathbb{R}^d_+$ and a locally integrable function $f\colon [\mathbf{a}, \infty) \to \mathbb{C}$, we define the improper multiple integral

$$\int_{[\mathbf{a},\infty)} f(\mathbf{t})\, d\mathbf{t} = \lim_{\mathbf{b}\to\infty} \int_{[\mathbf{a},\mathbf{b}]} f(\mathbf{t})\, d\mathbf{t}, \tag{7.18}$$

where the limit may be finite, infinite, or may not exist. In what follows we will always assume f to be locally integrable. The function f may also be defined in $(\mathbf{a}, \infty)$, rather than in $[\mathbf{a}, \infty)$ (unless otherwise stated, we will consider either $\mathbf{a} = \mathbf{0}$ or $\mathbf{a} = \mathbf{1}$). If

$$\int_{[\mathbf{a},\infty)} f(\mathbf{t})\, d\mathbf{t} = A \in \mathbb{C},$$

we say that the multiple integral $\int_{[\mathbf{a},\infty)} f(\mathbf{t})\, d\mathbf{t}$ *converges to A in the sense of Pringsheim*, or *diverges* otherwise. For $\mathbf{0} \leq \eta \leq \mathbf{1}$ and $\mathbf{x}_{\mathbf{1}-\eta} \in [\mathbf{a}, \infty)^{\mathbf{1}-\eta}$, we define the $|\eta|$-dimensional integral

$$\int_{[\mathbf{a},\infty)^\eta} f(\mathbf{t}_\eta, \mathbf{x}_{\mathbf{1}-\eta})\, d\mathbf{t}_\eta = \lim_{\mathbf{b}_\eta\to\infty} \int_{[\mathbf{a},\mathbf{b}]^\eta} f(\mathbf{t}_\eta, \mathbf{x}_{\mathbf{1}-\eta})\, d\mathbf{t}_\eta,$$

which is a function of the variables indexed by $\mathbf{1} - \eta$ defined at the points $\mathbf{x}_{\mathbf{1}-\eta} \in [\mathbf{a}, \infty)^{\mathbf{1}-\eta}$ where the integral converges.

The multidimensional Cauchy criterion for convergence states that *the multiple integral* $\int_{[\mathbf{a},\infty)} f(\mathbf{t})\,d\mathbf{t}$ *converges in the sense of Pringsheim if and only if for every* $\varepsilon > 0$ *there exists* $\mathbf{z} > \mathbf{0}$ *such that*

$$\left| \int_{\mathbf{x}}^{\mathbf{y}} f(\mathbf{t})\,d\mathbf{t} \right| < \varepsilon, \qquad \text{for all } \mathbf{y} > \mathbf{x} > \mathbf{z}.$$

Recall that in the context of integrals, the condition

$$f(\mathbf{t}) \to 0 \qquad \text{as } \mathbf{t} \to \infty$$

is not necessary for the convergence of the multiple integral (7.18), unlike in the case of series (see p. 209 for a detailed discussion and examples in two dimensions, which easily extend to the multiple case).

The counterpart of Theorem 7.23, for computing a multidimensional integral as an iteration of two lower-dimensional integrals, reads as follows.

Theorem 7.28 *Let* $f\colon (\mathbf{1},\infty) \to \mathbb{C}$ *and assume that for some* $\mathbf{0} \le \eta \le \mathbf{1}$ *with* $\eta \ne \mathbf{0}, \mathbf{1}$ *and every fixed* $\mathbf{x}_{\mathbf{1}-\eta} \in (\mathbf{1},\infty)^{\mathbf{1}-\eta}$, *the finite limit*

$$\lim_{\mathbf{b}_\eta \to \infty} \int_{(\mathbf{1},\mathbf{b})^\eta} f(\mathbf{t}_\eta, \mathbf{x}_{\mathbf{1}-\eta})\,d\mathbf{t}_\eta = \int_{(\mathbf{1},\infty)^\eta} f(\mathbf{t}_\eta, \mathbf{x}_{\mathbf{1}-\eta})\,d\mathbf{t}_\eta$$

exists. Assume moreover that for some $\eta \le \zeta \le \mathbf{1}$ *and every* $\mathbf{x}_{\mathbf{1}-\zeta} \in (\mathbf{1},\infty)^{\mathbf{1}-\zeta}$, *the finite limit*

$$\lim_{\mathbf{b}_\zeta \to \infty} \int_{(\mathbf{1},\mathbf{b})^\zeta} f(\mathbf{t}_\zeta, \mathbf{x}_{\mathbf{1}-\zeta})\,d\mathbf{t}_\zeta = \int_{(\mathbf{1},\infty)^\zeta} f(\mathbf{t}_\zeta, \mathbf{x}_{\mathbf{1}-\zeta})\,d\mathbf{t}_\zeta$$

exists, where if $\zeta = \mathbf{1}$, *we set*

$$\int_{(\mathbf{1},\infty)^\zeta} f(\mathbf{t}_\zeta, \mathbf{x}_{\mathbf{1}-\zeta})\,d\mathbf{t}_\zeta = \int_{(\mathbf{1},\infty)} f(\mathbf{t})\,d\mathbf{t}.$$

Then the iterated limit

$$\lim_{\mathbf{b}_{\zeta-\eta} \to \infty} \left(\lim_{\mathbf{b}_\eta \to \infty} \int_{(\mathbf{1},\mathbf{b})^\zeta} f(\mathbf{t}_\zeta, \mathbf{x}_{\mathbf{1}-\zeta})\,d\mathbf{t}_\zeta \right) \tag{7.19}$$

exists for all $\mathbf{x}_{\mathbf{1}-\zeta} \in (\mathbf{1},\infty)^{\mathbf{1}-\zeta}$ *and is equal to* $\int_{(\mathbf{1},\infty)^\zeta} f(\mathbf{t}_\zeta, \mathbf{x}_{\mathbf{1}-\zeta})\,d\mathbf{t}_\zeta$.

We do not give a proof of this theorem. We only mention that it readily follows from a counterpart of Theorem 7.16 on iterated limits for functions, just as Theorem 7.23 in the case of series. The iterated limit (7.19) is called an *iterated integral.*

We now define regular convergence of multiple integrals and state the corresponding Fubini-type theorem.

Definition 7.29 Let $f\colon (\mathbf{1},\infty) \to \mathbb{C}$ be locally integrable. We say that the (d-dimensional) multiple integral

$$\int_{(\mathbf{1},\infty)} f(\mathbf{t})\,d\mathbf{t}$$

is regularly convergent (or converges in the regular sense) if for each $\varepsilon > 0$ there exists a positive real number z such that

$$\left| \int_{\mathbf{x}}^{\mathbf{y}} f(\mathbf{t})\, d\mathbf{t} \right| < \varepsilon, \tag{7.20}$$

provided $\mathbf{y} > \mathbf{x}$ and $|\mathbf{x}| > z$, or, equivalently, there exists a positive real number z' such that (7.20) holds whenever $\mathbf{y} > \mathbf{x}$ and $\max_{1 \le \ell \le d} x_\ell > z'$.

Remark 7.30 Recall that in several dimensions a criterion for regular convergence of integrals analogous to Theorem 7.25 is not possible (see the two-dimensional Remark 4.23 for a counterexample).

To conclude this section, we state Fubini's theorem for regularly convergent integrals (i.e., a counterpart of Theorem 7.27). We do not include its proof, as it is analogous to the corresponding one for series.

Theorem 7.31 *Let $f\colon (\mathbf{1}, \boldsymbol{\infty}) \to \mathbb{C}$ and assume that the integral $\int_{(\mathbf{1},\boldsymbol{\infty})} f(\mathbf{t})\, d\mathbf{t}$ converges in the regular sense. Then, for any $\mathbf{0} \le \eta \le \zeta \le \mathbf{1}$ with $\eta \ne \mathbf{0}, \zeta$ and almost every $\mathbf{x}_{\mathbf{1}-\zeta} \in (\mathbf{1}, \boldsymbol{\infty})^{\mathbf{1}-\zeta}$,*

$$\int_{(\mathbf{1},\boldsymbol{\infty})^{\zeta}} f(\mathbf{t}_\zeta, \mathbf{x}_{\mathbf{1}-\zeta})\, d\mathbf{t}_\zeta = \int_{(\mathbf{1},\boldsymbol{\infty})^{\zeta-\eta}} \left(\int_{(\mathbf{1},\boldsymbol{\infty})^{\eta}} f(\mathbf{t}_\zeta, \mathbf{x}_{\mathbf{1}-\zeta})\, d\mathbf{t}_\eta \right) d\mathbf{t}_{\zeta-\eta}.$$

In particular, the integral $\int_{(\mathbf{1},\boldsymbol{\infty})} f(\mathbf{t})\, d\mathbf{t}$ can be computed as an iteration of one-dimensional integrals in any order.

7.5 Positive multiple series

A d-tuple series with positive (respectively nonnegative) terms $\{a_{\mathbf{k}}\}$ is called a *positive series* (respectively *nonnegative series*). Such series form important classes and are closely connected with absolutely convergent series (see Section 7.6).

Theorem 7.32 *For a nonnegative d-tuple series $\sum_{\mathbf{k}=\mathbf{1}}^{\boldsymbol{\infty}} a_{\mathbf{k}}$ to be convergent, it is necessary and sufficient that its sequence of partial sums $\{S_{\mathbf{n}}\}$ is bounded.*

This result is a direct consequence of Theorem 7.12.

Theorem 7.33 *A nonnegative d-tuple series $\sum_{\mathbf{k}=\mathbf{1}}^{\boldsymbol{\infty}} a_{\mathbf{k}}$ and any of its iterated series either converge to the same limit or diverge simultaneously.*

Theorem 7.33 readily follows from Theorem 7.18, since the sequence of partial sums $\{S_{\mathbf{n}}\}$ of a nonnegative series is increasing.

As mentioned in the introductory material, all types of convergence are equivalent for nonnegative series. More precisely, we have the following.

Theorem 7.34 *If a nonnegative d-tuple series $\{a_{\mathbf{k}}\}$ and a one-dimensional series with terms $\{b_n\}$ have all the terms the same (that is, $\{a_{\mathbf{k}}\} = \{b_n\}$), then the convergence of one of them implies the convergence of the other, and their sums are equal.*

Proof By assumption, for every $\mathbf{k} \in \mathbb{N}^d$, there exists $n \in \mathbb{N}$ such that $a_{\mathbf{k}} = b_n$, and vice versa. This establishes a bijection $\sigma : \mathbb{N} \to \mathbb{N}^d$ given by

$$a_{\mathbf{k}} = a_{\sigma(n)} = b_n,$$

and thus,

$$\sum_{n=1}^{\infty} b_n = \sum_{\mathbf{k}=\mathbf{1}}^{\infty} a_{\mathbf{k}}.$$

Note that in the last identity the series were rearranged, which is justified by the nonnegativity of the involved sequences. □

As a particular case of possible types of convergence we address the regular convergence.

Corollary 7.35 *If a nonnegative d-tuple series converges, then it converges in the regular sense.*

It is natural to ask whether there exists a "limiting" series that can determine the convergence or divergence of any positive series by comparing the terms of both. We now show that such a test can by no means exist. Roughly speaking, for any convergent/divergent series, we can find a series that converges/diverges, having terms asymptotically larger/smaller than the original series.

Theorem 7.36 *Let* $\sum_{\mathbf{k}=\mathbf{1}}^{\infty} a_{\mathbf{k}}$ *be a divergent series with positive terms. Then there exists a positive sequence* $\{b_{\mathbf{k}}\}$ *decreasing in each variable, with*

$$\lim_{|\mathbf{k}|\to\infty} b_{\mathbf{k}} = 0,$$

and for any subset E of $\{\mathbf{0} \le \eta \le \mathbf{1}\}$,

$$\sum_{\mathbf{k}=\mathbf{1}}^{\infty} a_{\mathbf{k}} \left(\prod_{\eta \in E} b_{\mathbf{k}_{\eta}, \mathbf{1}_{1-\eta}} \right) = \infty. \tag{7.21}$$

In particular,

$$\sum_{\mathbf{k}=\mathbf{1}}^{\infty} a_{\mathbf{k}} b_{\mathbf{k}} = \infty.$$

Proof Set

$$S_r = \sum_{k_1,\ldots,k_d=1}^{r} a_{\mathbf{k}}.$$

Fix $0 < \alpha < 2^{-d}$ and let $b_{\mathbf{k}} = \left(S_{\left\lfloor \frac{|\mathbf{k}|}{d} \right\rfloor} \right)^{-\alpha}$. The sequence $\{b_{\mathbf{k}}\}$ is decreasing and

$$\lim_{|\mathbf{k}|\to\infty} b_{\mathbf{k}} = 0.$$

Further, since E has at most 2^d elements, for any $\mathbf{m} = (m, \dots, m) \in \mathbb{N}^d$ such that $S_m \geq 1$ (which can be found because $S_m \to \infty$ as $m \to \infty$), one has

$$\begin{aligned}\sum_{\mathbf{k}=\mathbf{1}}^{\infty} a_{\mathbf{k}}\Big(\prod_{\eta\in E} b_{\mathbf{k}_\eta,\mathbf{1}_{1-\eta}}\Big) &\geq \sum_{\mathbf{k}=\mathbf{1}}^{\mathbf{m}} a_{\mathbf{k}}\Big(\prod_{\eta\in E} b_{\mathbf{k}_\eta,\mathbf{1}_{1-\eta}}\Big)\\ &\geq \sum_{\mathbf{k}=\mathbf{1}}^{\mathbf{m}} a_{\mathbf{k}}\Big(\prod_{\eta\in E}\Big(S_{\lfloor\frac{|\mathbf{k}|}{d}\rfloor}\Big)^{-\alpha}\Big) \geq \sum_{\mathbf{k}=\mathbf{1}}^{\mathbf{m}} a_{\mathbf{k}}\Big(\prod_{\eta\in E}\Big(S_{\lfloor\frac{|\mathbf{m}|}{d}\rfloor}\Big)^{-\alpha}\Big)\\ &= S_m^{-\alpha|E|}\sum_{\mathbf{k}=\mathbf{1}}^{\mathbf{m}} a_{\mathbf{k}} \geq S_m^{-2^d\alpha}\sum_{\mathbf{k}=\mathbf{1}}^{\mathbf{m}} a_{\mathbf{k}} = S_m^{1-2^d\alpha},\end{aligned}$$

which tends to infinity as $m \to \infty$, and thus proves (7.21). □

Theorem 7.37 *Let $\sum_{\mathbf{k}=\mathbf{1}}^{\infty} a_{\mathbf{k}}$ be a positive convergent series. Then there exists a nonnegative sequence $\{b_{\mathbf{k}}\}$, increasing in each variable and satisfying $\lim_{|\mathbf{k}|\to\infty} b_{\mathbf{k}} = \infty$, such that the series*

$$\sum_{\mathbf{k}=\mathbf{1}}^{\infty} a_{\mathbf{k}} b_{\mathbf{k}}$$

converges.

Proof The convergence of the series $\sum_{\mathbf{k}=\mathbf{1}}^{\infty} a_{\mathbf{k}}$ implies that we can find an increasing sequence $\{\mathbf{m}_j\}_{j=1}^{\infty} \subset \mathbb{N}^d$ (i.e., $\mathbf{m}_j \leq \mathbf{m}_{j+1}$) for which

$$\sum_{\mathbf{k}=\mathbf{1}}^{\mathbf{m}_{j+1}} a_{\mathbf{k}} - \sum_{\mathbf{k}=\mathbf{1}}^{\mathbf{m}_j} a_{\mathbf{k}} < \frac{1}{j^3}.$$

Setting $b_{\mathbf{k}} = 0$ for $\mathbf{1} \leq \mathbf{k} \leq \mathbf{m}_1$ and

$$b_{\mathbf{k}} = j, \qquad \text{for } \mathbf{k} \in ([\mathbf{1}, \mathbf{m}_{j+1}] \backslash [\mathbf{1}, \mathbf{m}_j]) \cap \mathbb{N}^d, \qquad j \geq 1,$$

we immediately see that $\{b_{\mathbf{k}}\}$ is increasing in each variable, $b_{\mathbf{k}} \to \infty$ as $|\mathbf{k}| \to \infty$, and

$$\sum_{\mathbf{k}=\mathbf{1}}^{\infty} a_{\mathbf{k}} b_{\mathbf{k}} = \sum_{j=1}^{\infty} j\Big(\sum_{\mathbf{k}=\mathbf{1}}^{\mathbf{m}_{j+1}} a_{\mathbf{k}} - \sum_{\mathbf{k}=\mathbf{1}}^{\mathbf{m}_j} a_{\mathbf{k}}\Big) < \sum_{j=1}^{\infty} \frac{1}{j^2} < \infty.$$ □

7.6 Absolute and conditional convergence of d-tuple series

Given a complex-valued d-tuple series $\sum_{\mathbf{k}=\mathbf{1}}^{\infty} a_{\mathbf{k}}$, the series $\sum_{\mathbf{k}=\mathbf{1}}^{\infty} |a_{\mathbf{k}}|$ may be convergent or divergent.

Definition 7.38 Given a d-tuple sequence $\{a_{\mathbf{k}}\} \subset \mathbb{C}$, the series $\sum_{\mathbf{k}=\mathbf{1}}^{\infty} a_{\mathbf{k}}$ is said to converge absolutely if

$$\sum_{\mathbf{k}=\mathbf{1}}^{\infty} |a_{\mathbf{k}}| < \infty.$$

As we have seen in the one- and two-dimensional contexts (Sections 2.2 and 4.6, respectively), absolutely convergent series are closely related to nonnegative series. Here we aim to extend the results in Section 4.6 to any dimension.

First of all, let us establish the relation between regular and absolute convergence.

Theorem 7.39 *Any absolutely convergent d-tuple series $\sum_{\mathbf{k}=\mathbf{1}}^{\infty} a_{\mathbf{k}}$ converges in the regular sense. The converse is not true.*

Proof As noted in Section 7.5, any convergent series with nonnegative terms converges in the regular sense. It remains to apply the definition of regular convergence together with the inequality

$$\left|\sum_{\mathbf{k}=\mathbf{m}}^{\mathbf{n}} a_{\mathbf{k}}\right| \leq \sum_{\mathbf{k}=\mathbf{m}}^{\mathbf{n}} |a_{\mathbf{k}}|.$$

In order to prove that the converse is not true it suffices to consider a convergent one-dimensional series $\sum b_k$ that is not absolutely convergent (for instance, $b_k = \frac{(-1)^k}{k}$) and set

$$a_{\mathbf{k}} = \begin{cases} b_k, & \text{if } k_1 = k_2 = \cdots = k_d = k, \\ 0, & \text{otherwise.} \end{cases}$$

It is clear that the series $\sum_{\mathbf{k}=\mathbf{1}}^{\infty} a_{\mathbf{k}}$ converges in the regular sense, but does not converge absolutely. □

As a corollary, we can obtain Fubini's theorem for multidimensional series. It states that the sum of an absolutely convergent series may be computed as an iterated series.

Corollary 7.40 *For an absolutely convergent d-tuple series $\sum_{\mathbf{k}=\mathbf{1}}^{\infty} a_{\mathbf{k}}$, all the iterated series also converge absolutely. Moreover, for any $\mathbf{0} \leq \eta \leq \zeta \leq \mathbf{1}$ with $\eta \neq \mathbf{0}, \zeta$ and any fixed $\mathbf{k}_{\mathbf{1}-\zeta}$,*

$$\sum_{\mathbf{k}_\zeta=\mathbf{1}}^{\infty} a_{\mathbf{k}} = \sum_{\mathbf{k}_{\zeta-\eta}=\mathbf{1}}^{\infty} \left(\sum_{\mathbf{k}_\eta=\mathbf{1}}^{\infty} a_{\mathbf{k}} \right). \tag{7.22}$$

In particular, all iterated series converge to the same limit.

Proof By Theorem 7.33, the convergence of $\sum_{\mathbf{k}=\mathbf{1}}^{\infty} |a_{\mathbf{k}}|$ implies the convergence of any of its iterated series (with the modulus incorporated).

Finally, (7.22) follows just by combining Theorems 7.27 and 7.39. □

Theorem 7.41 *Assume that the terms of a d-tuple sequence $\{a_{\mathbf{k}}\}$ are such that*

$$|a_{\mathbf{k}}| \leq C b_{\mathbf{k}},$$

for all $\mathbf{k} \geq \mathbf{1}$, where $\sum_{\mathbf{k}=\mathbf{1}}^{\infty} b_{\mathbf{k}}$ is a d-tuple convergent nonnegative series. Then the d-tuple series $\sum_{\mathbf{k}=\mathbf{1}}^{\infty} a_{\mathbf{k}}$ converges absolutely.

This readily follows from the multidimensional comparison test (see Theorem 8.1 below).

Theorem 7.42 *An absolutely convergent d-tuple series with terms $\{a_{\mathbf{k}}\}$ converges and*

$$\left|\sum_{\mathbf{k}=\mathbf{1}}^{\infty} a_{\mathbf{k}}\right| \le \sum_{\mathbf{k}=\mathbf{1}}^{\infty} |a_{\mathbf{k}}|.$$

Proof Since the series $\sum_{\mathbf{k}=\mathbf{1}}^{\infty} |a_{\mathbf{k}}|$ converges, it follows from Theorem 7.14 that for any $\varepsilon > 0$ there exists $\mathbf{n} \in \mathbb{N}^d$ such that for any $\mathbf{k} \ge \mathbf{n}$ and any $\mathbf{m} \ge \mathbf{0}$ we have

$$0 \le \Sigma_{\mathbf{k}+\mathbf{m}} - \Sigma_{\mathbf{k}} < \varepsilon,$$

where

$$\Sigma_{\mathbf{k}} = \sum_{\boldsymbol{\ell}=\mathbf{1}}^{\mathbf{k}} |a_{\boldsymbol{\ell}}|.$$

Now, it is clear that

$$|S_{\mathbf{k}+\mathbf{m}} - S_{\mathbf{k}}| \le \Sigma_{\mathbf{k}+\mathbf{m}} - \Sigma_{\mathbf{k}} < \varepsilon.$$

Applying again Theorem 7.14, we conclude that the series $\sum_{\mathbf{k}=\mathbf{1}}^{\infty} a_{\mathbf{k}}$ converges, as required.

Besides that,

$$\left|\sum_{\boldsymbol{\ell}=\mathbf{1}}^{\mathbf{k}} a_{\boldsymbol{\ell}}\right| \le \sum_{\boldsymbol{\ell}=\mathbf{1}}^{\mathbf{k}} |a_{\boldsymbol{\ell}}|,$$

for any $\mathbf{k} \ge \mathbf{1}$. Passing to the limit as $\mathbf{k} \to \infty$, we complete the proof of the theorem. □

Given a convergent d-tuple series $\sum_{\mathbf{k}=\mathbf{1}}^{\infty} a_{\mathbf{k}}$, it may happen that $\sum_{\mathbf{k}=\mathbf{1}}^{\infty} |a_{\mathbf{k}}|$ diverges. If this is the case, we call the series $\sum_{\mathbf{k}=\mathbf{1}}^{\infty} a_{\mathbf{k}}$ *conditionally convergent*. The following result illustrates the difference between conditional and absolute convergence.

Theorem 7.43 *Let a d-tuple series with real-valued terms $\sum_{\mathbf{k}=\mathbf{1}}^{\infty} a_{\mathbf{k}}$ be conditionally convergent. Then there exists a d-tuple sequence $\{\delta_{\mathbf{k}}\}$ such that*

$$\lim_{|\mathbf{k}|\to\infty} \delta_{\mathbf{k}} = 0$$

and

$$\sum_{\mathbf{k}=\mathbf{1}}^{\infty} a_{\mathbf{k}}\delta_{\mathbf{k}} = \infty.$$

Proof Since the series $\sum_{\mathbf{k}=\mathbf{1}}^{\infty} |a_{\mathbf{k}}|$ diverges, by Theorem 7.36, there exists a decreasing d-tuple sequence with positive terms $\{b_{\mathbf{k}}\}$ such that

$$\lim_{|\mathbf{k}|\to\infty} b_{\mathbf{k}} = 0$$

and

$$\sum_{\mathbf{k}=\mathbf{1}}^{\infty} |a_{\mathbf{k}}| b_{\mathbf{k}} = \infty.$$

Setting

$$\delta_{\mathbf{k}} = b_{\mathbf{k}} \operatorname{sgn} a_{\mathbf{k}},$$

we have

$$a_{\mathbf{k}} \delta_{\mathbf{k}} = |a_{\mathbf{k}}| b_{\mathbf{k}}.$$

Therefore, the assertion of the theorem holds. □

Next we have a characterization of absolutely convergent d-tuple series in terms of absolutely convergent one-dimensional series.

Theorem 7.44 *If a d-tuple series with terms $\{a_{\mathbf{k}}\}$ and a one-dimensional series with terms $\{b_n\}$ have all the terms the same (that is, $\{a_{\mathbf{k}}\} = \{b_n\}$), then the absolute convergence of one of them implies the absolute convergence of the other, and their sums are equal.*

Proof Assume that the series $\sum_{\mathbf{k}=\mathbf{1}}^{\infty} a_{\mathbf{k}}$ converges absolutely. Then, by Theorem 7.34, the series $\sum_{n=1}^{\infty} b_n$ also converges absolutely (and the converse is also true). Denoting $A = \sum_{\mathbf{k}=\mathbf{1}}^{\infty} a_{\mathbf{k}}$, let us show that we also have $A = \sum_{n=1}^{\infty} b_n$. Put

$$\Sigma_{\mathbf{k}} = \sum_{\boldsymbol{\ell}=\mathbf{1}}^{\mathbf{k}} |a_{\boldsymbol{\ell}}|, \qquad \Sigma = \sum_{\mathbf{k}=\mathbf{1}}^{\infty} |a_{\mathbf{k}}|, \quad \text{and} \quad B_n = \sum_{k=1}^{n} b_k.$$

It follows from the absolute convergence of the series $\sum_{\mathbf{k}=\mathbf{1}}^{\infty} a_{\mathbf{k}}$ that for an arbitrary $\varepsilon > 0$, there is $\mathbf{n} \in \mathbb{N}^d$ such that for any $\mathbf{k} \geq \mathbf{n}$, the inequalities

$$\Sigma - \Sigma_{\mathbf{k}} < \frac{\varepsilon}{2}$$

and

$$|A - S_{\mathbf{k}}| < \frac{\varepsilon}{2}$$

hold, where $S_{\mathbf{k}}$ is the partial sum of the series $\sum_{\mathbf{k}=\mathbf{1}}^{\infty} a_{\mathbf{k}}$. For fixed $\mathbf{k} \geq \mathbf{n}$, let us choose a natural number $N \geq \max_i n_i$ so that for $P > N$, all the elements of the partial sum $S_{\mathbf{k}}$ are contained among the elements $b_1, b_2, \ldots, b_P$. This implies

$$|B_P - S_{\mathbf{k}}| \leq \Sigma - \Sigma_{\mathbf{k}} < \frac{\varepsilon}{2}.$$

Consequently, for $P > N$, we have

$$|A - B_P| \leq |A - S_{\mathbf{k}}| + |B_P - S_{\mathbf{k}}| < \varepsilon,$$

or, in other words, $\lim_{P \to \infty} B_P = A$. □

The following problem is of importance. Suppose that the d-tuple series $\sum_{\mathbf{k}} a_{\mathbf{k}}$ converges. Rearranging its terms arbitrarily, we get a new series $\sum_{\mathbf{k}} a'_{\mathbf{k}}$. The question is whether this new series converges, and if it does, whether it converges to the sum of the original series or to a different value. Though both series have the same terms, the answer is, generally speaking, negative. However, thc following theorem suggests conditions for a positive result.

Theorem 7.45 *If a multidimensional series with terms* $\{a_{\mathbf{k}}\}$ *is absolutely convergent, then any multidimensional series with terms* $\{a'_{\mathbf{k}}\}$ *having the same terms but rearranged arbitrarily converges absolutely and has the same sum.*

Proof We take all the terms of the series $\sum_{\mathbf{k}=\mathbf{1}}^{\infty} a_{\mathbf{k}}$ and form a one-dimensional series $\sum_n b_n$. Since the series $\sum_{\mathbf{k}=\mathbf{1}}^{\infty} a_{\mathbf{k}}$ converges absolutely, the series $\sum_n b_n$ also converges absolutely, by Theorem 7.44, and to the same sum. Further, since the series $\sum_{\mathbf{k}=\mathbf{1}}^{\infty} a'_{\mathbf{k}}$ and $\sum_{\mathbf{k}=\mathbf{1}}^{\infty} a_{\mathbf{k}}$ have exactly the same elements and the latter converges absolutely, Theorem 7.44 again yields the absolute convergence, now of the series $\sum_{\mathbf{k}=\mathbf{1}}^{\infty} a'_{\mathbf{k}}$, and all of them have the same sum. □

7.7 Notes

Section 7.3 (Convergence of multiple series: definitions and basic notions)

1. In the early works concerning multidimensional series [221, 222], Pringsheim already noted that if $\{a_{\mathbf{k}}\}$ is a nonnegative multiple sequence ($a_{\mathbf{k}} \geq 0$), all types of convergence of $\sum_{\mathbf{k}=\mathbf{1}}^{\infty} a_{\mathbf{k}}$ are equivalent.

2. On the other hand, regular convergence was introduced by Hardy in [121]. The motivation for such a definition is that "convergent" series (in any sense) should possess certain desirable properties, mainly, those satisfied in one dimension. For example, the terms $a_{\mathbf{k}}$ of a regularly convergent series vanish as $|\mathbf{k}| \to \infty$ (as in one dimension), while this is not necessarily true for Pringsheim convergent series. Regular convergence was later reconsidered by Móricz in [190].

3. Theorem 7.25 was erroneously stated in [190], along with a proof of the two-dimensional case (this is another instance where the behavior of two-dimensional series does not reveal the nature of multidimensional series). The correct statement appeared later in [191], and a full proof in the three-dimensional case was given in [195]. On the other hand, Theorem 7.27 was stated in [190] (without proof), for the case of iterated one-dimensional series, with a full proof in the three-dimensional case given in [195]. Finally, Theorem 7.31 was stated in [195] and partially proved for $d = 4$. All those results were announced by the author in [194].

4. In [114], the author gives a thorough exposition of the matrix transformations that preserve certain convergence types of multiple series. These matrix transformations generalize the well-known Cesàro summability (cf. Theorem 2.6 in one dimension). The author had already obtained partial results in two dimensions [113], and other related (two-dimensional) works existed at that time; see, e.g., [144, 170, 229]. In [114], the author points out that each of those three papers contains some mistake or inaccuracy and provides a corrected answer to the corresponding questions. This is another proof of the delicacy one must have when dealing with multiple series.

8

Tests for convergence of multiple series

Here we aim to extend all the results already presented in the two-dimensional setting in Chapter 5 to the multidimensional case. Most of the proofs, examples, and further discussion will be given in full detail. The proofs not given here are either trivial or easily accessible in the literature.

Of course, the tests that come as generalizations of their one-dimensional classical counterparts deserve proper references to the original works. We remind the reader that these references are already given in Notes of Chapter 2 and so will not be included again.

8.1 Basic tests for convergence of positive multiple series

The basic convergence tests for positive series often apply to nonnegative series as well. In such cases, we will give more general statements for the latter.

We start with the comparison test, which will be systematically used to deduce other convergence tests.

Theorem 8.1 *Given two nonnegative d-tuple sequences $\{a_{\mathbf{k}}\}$ and $\{b_{\mathbf{k}}\}$ such that*

$$a_{\mathbf{k}} \le C b_{\mathbf{k}}$$

for all $\mathbf{k} \ge \mathbf{1}$, the convergence of $\sum_{\mathbf{k}=\mathbf{1}}^{\infty} b_{\mathbf{k}}$ implies that of $\sum_{\mathbf{k}=\mathbf{1}}^{\infty} a_{\mathbf{k}}$, and the divergence of $\sum_{\mathbf{k}=\mathbf{1}}^{\infty} a_{\mathbf{k}}$ implies that of $\sum_{\mathbf{k}=\mathbf{1}}^{\infty} b_{\mathbf{k}}$.

Proof Denoting by $S_{\mathbf{k}}$ and $\Sigma_{\mathbf{k}}$ the partial sums of the given series, respectively, we see that

$$S_{\mathbf{k}} \le C\Sigma_{\mathbf{k}}, \qquad \mathbf{k} \ge \mathbf{1}.$$

If the d-tuple series $\sum_{\mathbf{k}=\mathbf{1}}^{\infty} b_{\mathbf{k}}$ converges, then the d-tuple sequence $\{\Sigma_{\mathbf{k}}\}$ is bounded, which yields the boundedness of the d-tuple sequence $\{S_{\mathbf{k}}\}$ and, in turn, the convergence of $\sum_{\mathbf{k}=\mathbf{1}}^{\infty} a_{\mathbf{k}}$. Similarly, the unboundedness of $\{S_{\mathbf{k}}\}$ in the case of divergence of the d-tuple series $\sum_{\mathbf{k}=\mathbf{1}}^{\infty} a_{\mathbf{k}}$ implies the unboundedness of $\{\Sigma_{\mathbf{k}}\}$ and, in turn, the divergence of the d-tuple series $\sum_{\mathbf{k}=\mathbf{1}}^{\infty} b_{\mathbf{k}}$. □

Our following result is a d'Alembert type test, which, as in the two-dimensional case, is stated but not proved, since it will follow from a more general result (namely Theorem 8.3 below). As already mentioned, d'Alembert's test ensures the convergence of series whose terms decrease faster than a geometric progression of ratio less than 1.

Theorem 8.2 *Let $\{a_{\mathbf{k}}\}$ be a positive d-tuple sequence. If there exists a number $q < 1$ such that for every $p = 1, \ldots, d-1$, there is $\eta := \eta^{(p)}$ satisfying $|\eta| = p$, with $\eta^{(p)} = \eta^{(p-1)} + \mathbf{e}_j$, and where j is such that $\eta_j^{(p-1)} = 0$, for which*

$$\frac{a_{\mathbf{1}_\eta, (\mathbf{m}+\mathbf{1})_{\mathbf{e}_j}, \mathbf{m}_{\mathbf{1}-\eta-\mathbf{e}_j}}}{a_{\mathbf{1}_\eta, \mathbf{m}_{\mathbf{1}-\eta}}} \leq q$$

for all $m_j \geq 1$ and any fixed $\mathbf{m}_{\mathbf{1}-\eta-\mathbf{e}_j}$, then the series $\sum_{\mathbf{k}=\mathbf{1}}^{\infty} a_{\mathbf{k}}$ converges. If for some j there holds

$$\frac{a_{(\mathbf{m}+\mathbf{1})_{\mathbf{e}_j}, \mathbf{m}_{\mathbf{1}-\mathbf{e}_j}}}{a_{\mathbf{m}}} \geq 1$$

for all $m_j \geq 1$ and some fixed $\mathbf{m}_{\mathbf{1}-\mathbf{e}_j}$, then the series $\sum_{\mathbf{k}=\mathbf{1}}^{\infty} a_{\mathbf{k}}$ diverges.

This multivariate version of the classical d'Almebert test is a consequence of its following generalization.

Theorem 8.3 *Let $\{a_{\mathbf{k}}\}$ and $\{b_{\mathbf{k}}\}$ be two positive d-tuple sequences. Assume that for every $p = 1, \ldots, d-1$, there is $\eta := \eta^{(p)}$ satisfying $|\eta| = p$, with $\eta^{(p)} = \eta^{(p-1)} + \mathbf{e}_j$, and where j is such that $\eta_j^{(p-1)} = 0$ (with $\eta^{(0)} = \mathbf{0}$), for which*

$$\frac{a_{\mathbf{1}_\eta, (\mathbf{m}+\mathbf{1})_{\mathbf{e}_j}, \mathbf{m}_{\mathbf{1}-\eta-\mathbf{e}_j}}}{a_{\mathbf{1}_\eta, \mathbf{m}_{\mathbf{1}-\eta}}} \leq \frac{b_{\mathbf{1}_\eta, (\mathbf{m}+\mathbf{1})_{\mathbf{e}_j}, \mathbf{m}_{\mathbf{1}-\eta-\mathbf{e}_j}}}{b_{\mathbf{1}_\eta, \mathbf{m}_{\mathbf{1}-\eta}}}$$

for $m_j \geq 1$ and any fixed $\mathbf{m}_{\mathbf{1}-\eta-\mathbf{e}_j}$. Then the convergence of $\sum_{\mathbf{k}=\mathbf{1}}^{\infty} b_{\mathbf{k}}$ implies the convergence of $\sum_{\mathbf{k}=\mathbf{1}}^{\infty} a_{\mathbf{k}}$, and the divergence of $\sum_{\mathbf{k}=\mathbf{1}}^{\infty} a_{\mathbf{k}}$ implies the divergence of $\sum_{\mathbf{k}=\mathbf{1}}^{\infty} b_{\mathbf{k}}$.

Proof For any $\eta = \eta^{(p)}$, there is j satisfying $\eta_j = 0$ such that for every $m_j > 1$,

$$\begin{aligned}
\frac{a_{\mathbf{1}_\eta, \mathbf{m}_{\mathbf{1}-\eta}}}{a_{\mathbf{1}_{\eta+\mathbf{e}_j}, \mathbf{m}_{\mathbf{1}-\eta-\mathbf{e}_j}}} &= \frac{a_{\mathbf{1}_\eta, \mathbf{m}_{\mathbf{1}-\eta}}}{a_{\mathbf{1}_\eta, (\mathbf{m}-\mathbf{1})_{\mathbf{e}_j}, \mathbf{m}_{\mathbf{1}-\eta-\mathbf{e}_j}}} \frac{a_{\mathbf{1}_\eta, (\mathbf{m}-\mathbf{1})_{\mathbf{e}_j}, \mathbf{m}_{\mathbf{1}-\eta-\mathbf{e}_j}}}{a_{\mathbf{1}_\eta, (\mathbf{m}-\mathbf{2})_{\mathbf{e}_j}, \mathbf{m}_{\mathbf{1}-\eta-\mathbf{e}_j}}} \cdots \frac{a_{\mathbf{1}_\eta, \mathbf{2}_{\mathbf{e}_j}, \mathbf{m}_{\mathbf{1}-\eta-\mathbf{e}_j}}}{a_{\mathbf{1}_{\eta+\mathbf{e}_j}, \mathbf{m}_{\mathbf{1}-\eta-\mathbf{e}_j}}} \\
&\leq \frac{b_{\mathbf{1}_\eta, \mathbf{m}_{\mathbf{1}-\eta}}}{b_{\mathbf{1}_\eta, (\mathbf{m}-\mathbf{1})_{\mathbf{e}_j}, \mathbf{m}_{\mathbf{1}-\eta-\mathbf{e}_j}}} \frac{b_{\mathbf{1}_\eta, (\mathbf{m}-\mathbf{1})_{\mathbf{e}_j}, \mathbf{m}_{\mathbf{1}-\eta-\mathbf{e}_j}}}{b_{\mathbf{1}_\eta, (\mathbf{m}-\mathbf{2})_{\mathbf{e}_j}, \mathbf{m}_{\mathbf{1}-\eta-\mathbf{e}_j}}} \cdots \frac{b_{\mathbf{1}_\eta, \mathbf{2}_{\mathbf{e}_j}, \mathbf{m}_{\mathbf{1}-\eta-\mathbf{e}_j}}}{b_{\mathbf{1}_{\eta+\mathbf{e}_j}, \mathbf{m}_{\mathbf{1}-\eta-\mathbf{e}_j}}} \\
&= \frac{b_{\mathbf{1}_\eta, \mathbf{m}_{\mathbf{1}-\eta}}}{b_{\mathbf{1}_{\eta+\mathbf{e}_j}, \mathbf{m}_{\mathbf{1}-\eta-\mathbf{e}_j}}},
\end{aligned}$$

which implies

$$a_{\mathbf{1}_\eta, \mathbf{m}_{\mathbf{1}-\eta}} \leq \frac{b_{\mathbf{1}_\eta, \mathbf{m}_{\mathbf{1}-\eta}}}{b_{\mathbf{1}_{\eta+\mathbf{e}_j}, \mathbf{m}_{\mathbf{1}-\eta-\mathbf{e}_j}}} a_{\mathbf{1}_{\eta+\mathbf{e}_j}, \mathbf{m}_{\mathbf{1}-\eta-\mathbf{e}_j}}.$$

Taking this inequality first with $\eta = \mathbf{0}$ and applying it iteratively to the term $a_{\mathbf{1}_{\eta+\mathbf{e}_j},\mathbf{m}_{\mathbf{1}-\eta-\mathbf{e}_j}}$ with suitable $\eta = \eta^{(p)}$ and j, we eventually arrive at

$$a_{\mathbf{m}} \leq \frac{b_{\mathbf{m}}}{b_{\mathbf{1}}} a_{\mathbf{1}},$$

and the result follows by Theorem 8.1 with $C = a_{\mathbf{1}}/b_{\mathbf{1}}$. □

Note that taking $b_{\mathbf{m}} = q^{-|\mathbf{m}|}$ $(0 < q < 1)$ in Theorem 8.3 yields Theorem 8.2.
We now proceed to a multidimensional analog of the nth root test.

Theorem 8.4 *Let $\{a_{\mathbf{n}}\}$ be a d-tuple sequence with nonnegative terms. If there exist $q < 1$ and $N \in \mathbb{N}$ such that for $|\mathbf{n}| \geq N$ we have*

$$a_{\mathbf{n}}^{1/|\mathbf{n}|} \leq q,$$

then the series $\sum_{\mathbf{k}=\mathbf{1}}^{\infty} a_{\mathbf{k}}$ converges. If

$$a_{\mathbf{n}}^{1/|\mathbf{n}|} \geq 1$$

for infinitely many choices of $\mathbf{n}$, then the series $\sum_{\mathbf{k}=\mathbf{1}}^{\infty} a_{\mathbf{k}}$ diverges.

Instead of proving Theorem 8.4, we derive it as a consequence of the following more general statement, for which we include a proof.

Theorem 8.5 *Let $\{a_{\mathbf{n}}\}$ be a d-tuple sequence of nonnegative terms. Define*

$$L = \limsup_{|\mathbf{n}|\to\infty} a_{\mathbf{n}}^{1/|\mathbf{n}|}.$$

(i) *If $L < 1$, then $\sum_{\mathbf{k}=\mathbf{1}}^{\infty} a_{\mathbf{k}}$ converges.*
(ii) *If $L > 1$, then $\sum_{\mathbf{k}=\mathbf{1}}^{\infty} a_{\mathbf{k}}$ diverges.*
(iii) *If $L = 1$, the test is inconclusive.*

Proof If $L < 1$, holds, we can find $q < 1$ and $N \in \mathbb{N}$ such that, for every $\mathbf{n} \in \mathbb{N}^d$ with $|\mathbf{n}| \geq N$, we have $a_{\mathbf{n}}^{1/|\mathbf{n}|} \leq q$ or, equivalently,

$$a_{\mathbf{n}} \leq q^{|\mathbf{n}|}.$$

Since $q < 1$, the series

$$\sum_{|\mathbf{n}|<N} a_{\mathbf{n}} + \sum_{|\mathbf{n}|\geq N} q^{|\mathbf{n}|}$$

converges and so does $\sum_{\mathbf{k}=\mathbf{1}}^{\infty} a_{\mathbf{k}}$, by Theorem 8.1.

On the other hand, if $L > 1$, then $a_{\mathbf{n}}^{1/|\mathbf{n}|} \geq 1$ for infinitely many choices of $\mathbf{n}$. Hence, $a_{\mathbf{n}} \not\to 0$ as $|\mathbf{n}| \to \infty$, and thus the series $\sum_{\mathbf{k}=\mathbf{1}}^{\infty} a_{\mathbf{k}}$ diverges.

In order to show that the test is inconclusive for $L = 1$, we take the d-tuple sequence with general term

$$a_{\mathbf{n}} = \frac{1}{(n_1 \cdots n_d)^p}, \qquad p > 0.$$

This series converges if and only if $p > 1$ (as it is a product of d one-dimensional p-series), while

$$\limsup_{|\mathbf{n}|\to\infty} a_{\mathbf{n}}^{1/|\mathbf{n}|} = 1$$

for any $p > 0$. □

8.2 Tests for convergence involving d-dimensional bounded variation

We are going to establish the results for the series

$$\sum_{\mathbf{k}=\mathbf{1}}^{\infty} a_{\mathbf{k}} b_{\mathbf{k}},$$

that is, the assertions will be given just for series of this type, while the assumptions will be posed on the series

$$\sum_{\mathbf{k}=\mathbf{1}}^{\infty} a_{\mathbf{k}}$$

and

$$\sum_{\mathbf{k}=\mathbf{1}}^{\infty} b_{\mathbf{k}},$$

or on the sequences of their terms.

As in dimension two, what lacks for the presentation is the notion of bounded variation for multiple sequences. Such a notion is intimately related to the notion of monotonicity. A d-tuple sequence $\{a_{\mathbf{k}}\}$ with real-valued terms is called monotone increasing (monotone decreasing) if

$$a_{\mathbf{n}} \ge a_{\mathbf{k}} \ (a_{\mathbf{n}} \le a_{\mathbf{k}}), \quad \text{for } \mathbf{n} \ge \mathbf{k}.$$

We will typically refer to monotone decreasing sequences just as "monotone" or "decreasing".

Now, as in dimension two, but with much richer variety of options, we define a notion of monotonicity with respect to a restricted number of indices.

Definition 8.6 Let $\eta \ne \mathbf{0}$. A d-tuple sequence $\{a_{\mathbf{k}}\}$ with real-valued terms is called η-increasing (η-decreasing) if

$$a_{\mathbf{n}_\eta,\mathbf{k}_{\mathbf{1}-\eta}} \ge a_{\mathbf{k}} \quad (a_{\mathbf{n}_\eta,\mathbf{k}_{\mathbf{1}-\eta}} \le a_{\mathbf{k}}) \quad \text{for} \quad \mathbf{n}_\eta \ge \mathbf{k}_\eta$$

and any fixed $\mathbf{k}_{\mathbf{1}-\eta}$.

Although Definition 8.6 is rather general, we have included it just for showing how wide the framework may be in the context of multiple sequences. In what follows we will only be interested in η-monotonicity with $\eta = \mathbf{1}$, which was previously defined just as monotonicity.

We now define the Hardy type bounded variation for sequences in several dimensions. Recall that in Section 7.1 the difference with respect to the jth index is defined as

$$\Delta^{\mathbf{e}_j} a_{\mathbf{k}} = a_{\mathbf{k}} - a_{\mathbf{k}+\mathbf{e}_j},$$

while the "ηth" difference is defined by

$$\Delta^{\eta} a_{\mathbf{k}} = \left(\prod_{j:\eta_j=1} \Delta^{\mathbf{e}_j} \right) a_{\mathbf{k}}.$$

Note that according to the convention $\Delta^{\mathbf{0}} a_{\mathbf{k}} = a_{\mathbf{k}}$, the condition $\Delta^{\mathbf{0}} a_{\mathbf{k}} \geq 0$ is equivalent to the nonnegativity of the sequence $\{a_{\mathbf{k}}\}$.

The *Vitali type variation* of $\{a_{\mathbf{k}}\}$ is

$$Vv(a) := \sum_{\mathbf{k}=\mathbf{1}}^{\infty} |\Delta^{\mathbf{1}} a_{\mathbf{k}}|.$$

Naturally, $\Delta^{\mathbf{1}} a_{\mathbf{k}}$ denotes the iterated difference with respect to each of the coordinates. If $Vv(a)$ is finite, we say that $\{a_{\mathbf{k}}\}$ is of bounded Vitali variation.

The sequence $\{a_{\mathbf{k}}\}$ is said to be of *bounded Hardy variation* if it is of bounded Vitali variation, and additionally $\{a_{\mathbf{k}}\}$ is also of bounded Vitali variation when restricted to any group of the indices. This means that for every $\eta \neq \mathbf{0}$ and any fixed $\mathbf{k}_{\mathbf{1}-\eta} \in \mathbb{N}^{\mathbf{1}-\eta}$, the series

$$\sum_{\mathbf{k}_\eta=\mathbf{1}}^{\infty} |\Delta^{\eta} a_{\mathbf{k}}| \tag{8.1}$$

converges.

It is worth noting that the necessary and sufficient condition for $\{a_{\mathbf{k}}\}$ to be of bounded Hardy variation is that this sequence can be represented as the difference of two sequences with all differences being nonnegative (cf. Theorem 8.10 below). To prove such an assertion, we first need some basic facts.

Proposition 8.7 *Let $\{a_{\mathbf{k}}\}$ be a d-tuple sequence of bounded Hardy variation. Then, for every $\eta \geq \mathbf{0}$ with $\eta \neq \mathbf{0}, \mathbf{1}$ and every $\mathbf{k}_{\mathbf{1}-\eta} \in \mathbb{N}^{\mathbf{1}-\eta}$, the limit*

$$a_{\mathbf{k}_{\mathbf{1}-\eta}} = \lim_{\mathbf{n}_\eta \to \infty} a_{\mathbf{k}_{\mathbf{1}-\eta}, \mathbf{n}_\eta}$$

exists. Furthermore, in this case, the $|\mathbf{1}-\eta|$-tuple sequence $\left\{a_{\mathbf{k}_{\mathbf{1}-\eta}}\right\}$ is of bounded Hardy variation, and the sequence $\{a_{\mathbf{n}}\}$ converges to a finite limit as $\mathbf{n} \to \infty$.

Proof We proceed by induction on $|\eta|$. Let $|\eta| = 1$, that is, $\eta = \mathbf{e}_j$ for some $1 \le j \le d$. For any fixed $\mathbf{k}_{\mathbf{1}-\mathbf{e}_j} \in \mathbb{N}^{\mathbf{1}-\mathbf{e}_j}$, we have

$$\lim_{n_j \to \infty} a_{\mathbf{k}_{\mathbf{1}-\mathbf{e}_j}, \mathbf{n}_{\mathbf{e}_j}} = a_{\mathbf{k}_{\mathbf{1}-\mathbf{e}_j}, \mathbf{1}_{\mathbf{e}_j}} - \sum_{k_j=1}^{\infty} \Delta^{\mathbf{e}_j} a_{\mathbf{k}}.$$

Since $\{a_{\mathbf{k}}\}$ is of bounded Hardy variation, the above limit exists (i.e., the sequence $\left\{a_{\mathbf{k}_{\mathbf{1}-\mathbf{e}_j}}\right\}$ is well defined). Let us now prove that such a sequence is of bounded Hardy variation. For any $\zeta \le \mathbf{1} - \mathbf{e}_j$ and every fixed $\mathbf{k}_{\mathbf{1}-\zeta-\mathbf{e}_j}$, we have

$$|\Delta^{\zeta} a_{\mathbf{k}_{\mathbf{1}-\mathbf{e}_j}}| \le |\Delta^{\zeta} a_{\mathbf{k}_{\mathbf{1}-\mathbf{e}_j}, \mathbf{1}_{\mathbf{e}_j}}| + \sum_{k_j=1}^{\infty} |\Delta^{\zeta+\mathbf{e}_j} a_{\mathbf{k}}|.$$

Summing up in $\mathbf{k}_{\zeta}$, we get

$$\sum_{\mathbf{k}_{\zeta}=\mathbf{1}}^{\infty} |\Delta^{\zeta} a_{\mathbf{k}_{\mathbf{1}-\mathbf{e}_j}}| \le \sum_{\mathbf{k}_{\zeta}=\mathbf{1}}^{\infty} |\Delta^{\zeta} a_{\mathbf{k}_{\mathbf{1}-\mathbf{e}_j}, \mathbf{1}_{\mathbf{e}_j}}| + \sum_{\mathbf{k}_{\zeta+\mathbf{e}_j}=\mathbf{1}}^{\infty} |\Delta^{\zeta+\mathbf{e}_j} a_{\mathbf{k}}|,$$

where both series on the right side converge due to the fact that $\{a_{\mathbf{k}}\}$ is of bounded Hardy variation. This proves the first step of the inductive argument.

Suppose now that for some η such that $1 \le |\eta| < d-1$, the $|\mathbf{1} - \eta|$-tuple sequence $\left\{a_{\mathbf{k}_{\mathbf{1}-\eta}}\right\}$ is of bounded Hardy variation. Let $1 \le j \le d$ be such that $\eta_j = 0$. Since

$$\lim_{n_j \to \infty} a_{\mathbf{k}_{\mathbf{1}-\eta-\mathbf{e}_j}, \mathbf{n}_{\mathbf{e}_j}} = a_{\mathbf{k}_{\mathbf{1}-\eta-\mathbf{e}_j}, \mathbf{1}_{\mathbf{e}_j}} - \sum_{k_j=1}^{\infty} \Delta^{\mathbf{e}_j} a_{\mathbf{k}_{\mathbf{1}-\eta}}$$

and the last series converges (by induction hypothesis), the limit exists (the sequence $\left\{a_{\mathbf{k}_{\mathbf{1}-\eta-\mathbf{e}_j}}\right\}$ is well defined). In the same way as before we can show that $\left\{a_{\mathbf{k}_{\mathbf{1}-\eta-\mathbf{e}_j}}\right\}$ is of bounded Hardy variation.

Finally, in order to prove that $\{a_{\mathbf{n}}\}$ converges to a finite limit, we simply apply (7.3) with $\eta = \mathbf{1}$, $\mathbf{n} = \mathbf{1}$, and $\mathbf{m} \in \mathbb{N}^d$ to obtain

$$(-1)^d a_{\mathbf{m}+\mathbf{1}} = \sum_{\mathbf{k}=\mathbf{1}}^{\mathbf{m}} \Delta^{\mathbf{1}} a_{\mathbf{k}} - \sum_{\substack{\mathbf{0} \le \zeta \le \mathbf{1} \\ \zeta \ne \mathbf{1}}} (-1)^{|\zeta|} a_{\mathbf{1}_{\mathbf{1}-\zeta}, (\mathbf{m}+\mathbf{1})_{\zeta}}, \tag{8.2}$$

(note that here we have rearranged the terms of (7.3)). Letting $\mathbf{m} \to \infty$, we have that the series $\sum_{\mathbf{k}=\mathbf{1}}^{\infty} \Delta^{\mathbf{1}} a_{\mathbf{k}}$ converges and so do all terms in the last sum of (8.2), which yields the convergence of the sequence $\{a_{\mathbf{n}}\}$. □

The following is a simple consequence of Proposition 8.7.

Corollary 8.8 *Let $\{a_{\mathbf{k}}\}$ be a d-tuple sequence of bounded Hardy variation. Then $\{a_{\mathbf{k}}\}$ is bounded.*

We now show that nonnegative sequences all of whose differences are nonnegative are actually of bounded Hardy variation.

Proposition 8.9 *Let $\{a_{\mathbf{k}}\}$ be a nonnegative d-tuple sequence. Suppose that for every $\mathbf{0} \le \eta \le \mathbf{1}$ and every $\mathbf{k} \in \mathbb{N}^d$ there holds $\Delta^\eta a_{\mathbf{k}} \ge 0$. Then $\{a_{\mathbf{k}}\}$ is of bounded Hardy variation.*

Proof We prove by induction on $|\eta|$ that the series (8.1) converges. First, let $|\eta| = 1$, that is, $\eta = \mathbf{e}_j$ for some $1 \le j \le d$. For any fixed $\mathbf{k}_{\mathbf{1}-\mathbf{e}_j} \in \mathbb{N}^{\mathbf{1}-\mathbf{e}_j}$ and $\mathbf{m} \in \mathbb{N}^d$, we have

$$\sum_{k_j=1}^{m_j} |\Delta^{\mathbf{e}_j} a_{\mathbf{k}}| = \sum_{k_j=1}^{m_j} \Delta^{\mathbf{e}_j} a_{\mathbf{k}} = a_{\mathbf{k}_{\mathbf{1}-\mathbf{e}_j}, \mathbf{1}_{\mathbf{e}_j}} - a_{\mathbf{k}_{\mathbf{1}-\mathbf{e}_j}, (\mathbf{m}+\mathbf{1})_{\mathbf{e}_j}} \le a_{\mathbf{k}_{\mathbf{1}-\mathbf{e}_j}, \mathbf{1}_{\mathbf{e}_j}}.$$

The left-hand side is increasing in m_j and bounded from above. This yields the convergence of the series

$$\sum_{k_j=1}^{\infty} |\Delta^{\mathbf{e}_j} a_{\mathbf{k}}|,$$

which completes the proof of the first inductive step.

Assume now that the series (8.1) converges for some $\mathbf{0} \le \eta \le \mathbf{1}$ with $\eta \ne \mathbf{1}$ and every fixed $\mathbf{k}_{\mathbf{1}-\eta} \in \mathbb{N}^{\mathbf{1}-\eta}$. We show that for every $1 \le j \le d$ such that $\eta_j = 0$, the series corresponding to (8.1) with $\eta + \mathbf{e}_j$ in place of η and any fixed $\mathbf{k}_{\mathbf{1}-\eta-\mathbf{e}_j} \in \mathbb{N}^{\mathbf{1}-\eta-\mathbf{e}_j}$ converges. We have, for $\mathbf{m} \in \mathbb{N}^d$,

$$\begin{aligned}
\sum_{\mathbf{k}_{\eta+\mathbf{e}_j}=\mathbf{1}}^{\mathbf{m}_{\eta+\mathbf{e}_j}} |\Delta^{\eta+\mathbf{e}_j} a_{\mathbf{k}}| &= \sum_{\mathbf{k}_{\eta+\mathbf{e}_j}=\mathbf{1}}^{\mathbf{m}_{\eta+\mathbf{e}_j}} \Delta^{\eta+\mathbf{e}_j} a_{\mathbf{k}} = \sum_{\mathbf{k}_\eta=\mathbf{1}}^{\mathbf{m}_\eta} \sum_{k_j=1}^{m_j} \Delta^{\mathbf{e}_j} (\Delta^\eta a_{\mathbf{k}}) \\
&= \sum_{\mathbf{k}_\eta=\mathbf{1}}^{\mathbf{m}_\eta} \Delta^\eta a_{\mathbf{k}_{\mathbf{1}-\mathbf{e}_j}, \mathbf{1}_{\mathbf{e}_j}} - \sum_{\mathbf{k}_\eta=\mathbf{1}}^{\mathbf{m}_\eta} \Delta^\eta a_{\mathbf{k}_{\mathbf{1}-\mathbf{e}_j}, (\mathbf{m}+\mathbf{1})_{\mathbf{e}_j}} \\
&\le \sum_{\mathbf{k}_\eta=\mathbf{1}}^{\infty} \Delta^\eta a_{\mathbf{k}_{\mathbf{1}-\mathbf{e}_j}, \mathbf{1}_{\mathbf{e}_j}} < \infty,
\end{aligned}$$

where the convergence of the last series is due to the induction hypothesis. Therefore, since the left-hand side is increasing in $\mathbf{m}_{\eta+\mathbf{e}_j}$ (in each variable) and bounded from above, the convergence of the desired series follows (by Theorem 7.12). □

We mention that d-tuple sequences $\{a_{\mathbf{k}}\}$ such that $\Delta^{\mathbf{1}} a_{\mathbf{k}} \ge 0$ for all $\mathbf{k}$ are called *Hardy monotone*. Thus, Theorem 8.10 characterizes sequences of bounded Hardy variation as those that can be written as a difference of sequences such that, for any $\mathbf{0} \le \eta \le \mathbf{1}$ and any fixed $\mathbf{n}_{\mathbf{1}-\eta}$, the corresponding $|\eta|$-dimensional sequences $\{b_{\mathbf{k}_\eta, \mathbf{n}_{\mathbf{1}-\eta}}\}$ and $\{c_{\mathbf{k}_\eta, \mathbf{n}_{\mathbf{1}-\eta}}\}$ are Hardy monotone.

We are now in a position to obtain the aforementioned necessary and sufficient conditions for a sequence to be of bounded Hardy variation.

Theorem 8.10 *A d-tuple sequence* $\{a_{\mathbf{k}}\} \subset \mathbb{R}$ *is of bounded Hardy variation if and only if it can be represented as*

$$a_{\mathbf{k}} = b_{\mathbf{k}} - c_{\mathbf{k}}, \tag{8.3}$$

where $\{b_{\mathbf{k}}\}$ *and* $\{c_{\mathbf{k}}\}$ *satisfy* $\Delta^{\eta} b_{\mathbf{k}}, \Delta^{\eta} c_{\mathbf{k}} \geq 0$ *for all* $\mathbf{0} \leq \eta \leq \mathbf{1}$.

Proof Let $\{a_{\mathbf{k}}\}$ be of bounded Hardy variation. We prove the desired fact by induction on dimension d. Assume that the result is true for $(d-1)$-tuple sequences (in the one-dimensional case, the bounded Hardy variation reduces to just bounded variation; further, the assertion is true in this case; see Section 2.3). We introduce two sequences

$$\delta^{+}_{\mathbf{k}} = \begin{cases} \Delta^{\mathbf{1}} a_{\mathbf{k}}, & \text{if } \Delta^{\mathbf{1}} a_{\mathbf{k}} \geq 0, \\ 0, & \text{otherwise,} \end{cases}$$

and

$$\delta^{-}_{\mathbf{k}} = \begin{cases} -\Delta^{\mathbf{1}} a_{\mathbf{k}}, & \text{if } \Delta^{\mathbf{1}} a_{\mathbf{k}} \leq 0, \\ 0, & \text{otherwise,} \end{cases}$$

and then two more sequences

$$b^{\mathbf{1}}_{\mathbf{n}} = \sum_{\mathbf{k}=\mathbf{n}}^{\infty} \delta^{+}_{\mathbf{k}}$$

and

$$c^{\mathbf{1}}_{\mathbf{n}} = \sum_{\mathbf{k}=\mathbf{n}}^{\infty} \delta^{-}_{\mathbf{k}}.$$

Note that since $\{a_{\mathbf{k}}\}$ is of bounded Hardy variation, the sequences $\{b^{\mathbf{1}}_{\mathbf{n}}\}$ and $\{c^{\mathbf{1}}_{\mathbf{n}}\}$ are well defined and vanish as $|\mathbf{n}| \to \infty$. Thus, for any $\mathbf{0} \leq \eta \leq \mathbf{1}$ (recall that $\Delta^{\mathbf{0}} d_{\mathbf{k}} = d_{\mathbf{k}}$),

$$\Delta^{\eta} b^{\mathbf{1}}_{\mathbf{n}} = \sum_{\mathbf{k}_{\mathbf{1}-\eta}=\mathbf{n}_{\mathbf{1}-\eta}}^{\infty} \delta^{+}_{\mathbf{k}_{\mathbf{1}-\eta},\mathbf{n}_{\eta}} \geq 0 \quad \text{and} \quad \Delta^{\eta} c^{\mathbf{1}}_{\mathbf{n}} = \sum_{\mathbf{k}_{\mathbf{1}-\eta}=\mathbf{n}_{\mathbf{1}-\eta}}^{\infty} \delta^{-}_{\mathbf{k}_{\mathbf{1}-\eta},\mathbf{n}_{\eta}} \geq 0.$$

Further, applying (7.3) with $\eta = \mathbf{1}$ and $\mathbf{m} \to \infty$, we get

$$b^{\mathbf{1}}_{\mathbf{n}} - c^{\mathbf{1}}_{\mathbf{n}} = \sum_{\mathbf{k}=\mathbf{n}}^{\infty} \Delta^{\mathbf{1}} a_{\mathbf{k}} = a_{\mathbf{n}} + \sum_{\substack{\mathbf{0} \leq \zeta \leq \mathbf{1} \\ \zeta \neq \mathbf{0}}} (-1)^{|\zeta|} \lim_{\mathbf{m}_{\zeta} \to \infty} a_{\mathbf{n}_{\mathbf{1}-\zeta},(\mathbf{m}+\mathbf{1})_{\zeta}}. \tag{8.4}$$

By Proposition 8.7, all the involved limits in the last equality exist, and the corresponding sequences

$$a_{\mathbf{n}_{\mathbf{1}-\zeta}} = \lim_{\mathbf{m}_{\zeta} \to \infty} a_{\mathbf{n}_{\mathbf{1}-\zeta},(\mathbf{m}+\mathbf{1})_{\zeta}}, \qquad \zeta \neq \mathbf{1},$$

are of bounded Hardy variation (for $\zeta = \mathbf{1}$, we have $\lim_{\mathbf{k} \to \infty} a_{\mathbf{k}} = A \in \mathbb{R}$). Thus, by induction hypotheses, each of the sequences $\{a_{\mathbf{n}_{\mathbf{1}-\zeta}}\}$ may be written as a difference $a_{\mathbf{n}_{\mathbf{1}-\zeta}} = b^{\mathbf{1}-\zeta}_{\mathbf{n}_{\mathbf{1}-\zeta}} - c^{\mathbf{1}-\zeta}_{\mathbf{n}_{\mathbf{1}-\zeta}}$, for certain sequences $\{b^{\mathbf{1}-\zeta}_{\mathbf{n}_{\mathbf{1}-\zeta}}\}$ and $\{c^{\mathbf{1}-\zeta}_{\mathbf{n}_{\mathbf{1}-\zeta}}\}$ satisfying $\Delta^{\theta} b^{\mathbf{1}-\zeta}_{\mathbf{n}_{\mathbf{1}-\zeta}}, \Delta^{\theta} c^{\mathbf{1}-\zeta}_{\mathbf{n}_{\mathbf{1}-\zeta}} \geq 0$ for every $\theta \leq \mathbf{1} - \zeta$ with $\theta \neq \mathbf{0}$. Grouping all the sequences $\{b^{\mathbf{1}-\zeta}_{\mathbf{n}_{\mathbf{1}-\zeta}}\}$ and $\{c^{\mathbf{1}-\zeta}_{\mathbf{n}_{\mathbf{1}-\zeta}}\}$ (as well as limit A) on the right-hand side of (8.4), the desired sequences $\{b_{\mathbf{n}}\}$ and $\{c_{\mathbf{n}}\}$ can be defined as the sum of the former ones, according to the sign of each of them.

The sufficiency of the claimed representation (8.3) follows readily from Proposition 8.9. Indeed, we observe that for any $\mathbf{0} \leq \eta \leq \mathbf{1}$ with $\eta \neq \mathbf{0}$,

$$|\Delta^{\eta} a_{\mathbf{k}}| \leq \Delta^{\eta} b_{\mathbf{k}} + \Delta^{\eta} c_{\mathbf{k}},$$

where, in accordance with the assumptions, there is no need to apply the absolute value to the differences of $b_{\mathbf{k}}$ and $c_{\mathbf{k}}$. Thus, for any fixed $\mathbf{k}_{\mathbf{1}-\eta} \in \mathbb{N}^{\mathbf{1}-\eta}$, we have

$$\sum_{\mathbf{k}_{\eta}=\mathbf{1}}^{\infty} |\Delta^{\eta} a_{\mathbf{k}}| \leq \sum_{\mathbf{k}_{\eta}=\mathbf{1}}^{\infty} \Delta^{\eta} b_{\mathbf{k}} + \sum_{\mathbf{k}_{\eta}=\mathbf{1}}^{\infty} \Delta^{\eta} c_{\mathbf{k}}.$$

By Proposition 8.9, the sequences $\{b_{\mathbf{k}}\}$ and $\{c_{\mathbf{k}}\}$ are of bounded Hardy variation. Therefore, both series on the right-hand side converge and we conclude that

$$\sum_{\mathbf{k}_{\eta}=\mathbf{1}}^{\infty} |\Delta^{\eta} a_{\mathbf{k}}|$$

converges, or, in other words, $\{a_{\mathbf{k}}\}$ is of bounded Hardy variation. □

Theorem 8.10 will give us a way to pass from the generalizations of classical results for sequences of bounded variation to generalizations of not-less-classical results for monotone sequences. More precisely, the situation is as follows. There are two "key" tests of the former type, due to Dedekind and du Bois-Reymond, respectively, and corresponding corollaries of the latter one, due to Dirichlet and Abel.

One more feature of the generalizations of these tests is that they are impossible without regular convergence. More precisely, each time one has to be observant in what kind of convergence is presumed. The proofs mainly lie in the successive application of Abel's transformation with respect to each of the variables and estimating the obtained sums according to the given hypotheses. The multidimensional Abel's transformation reads as follows.

Lemma 8.11 *Let $\{a_{\mathbf{k}}\}, \{b_{\mathbf{k}}\}$ be d-tuple sequences. For any $\mathbf{m} \leq \mathbf{n} \in \mathbb{N}^d$, the identity*

$$\sum_{\mathbf{k}=\mathbf{m}}^{\mathbf{n}} a_{\mathbf{k}} b_{\mathbf{k}} = \sum_{\substack{\mathbf{0}\leq\eta\leq\mathbf{1}\\ \eta\neq\mathbf{0}}} \sum_{\mathbf{k}_{\eta}=\mathbf{m}_{\eta}}^{(\mathbf{n}-\mathbf{1})_{\eta}} \Delta^{\eta} b_{\mathbf{k}_{\eta},\mathbf{n}_{\mathbf{1}-\eta}} \sum_{\mathbf{s}_{\eta}=\mathbf{m}_{\eta}}^{\mathbf{k}_{\eta}} \sum_{\mathbf{k}_{\mathbf{1}-\eta}=\mathbf{m}_{\mathbf{1}-\eta}}^{\mathbf{n}_{\mathbf{1}-\eta}} a_{\mathbf{s}_{\eta},\mathbf{k}_{\mathbf{1}-\eta}}$$
$$+ b_{\mathbf{n}} \sum_{\mathbf{k}=\mathbf{m}}^{\mathbf{n}} a_{\mathbf{k}} \tag{8.5}$$

holds, where, if $m_j = n_j$ for some $j \in \{1,\ldots,d\}$, we take the convention

$$\sum_{\mathbf{k}_\eta=\mathbf{m}_\eta}^{(\mathbf{n}-\mathbf{1})_\eta} d_{\mathbf{k}_\eta,\mathbf{n}_{\mathbf{1}-\eta}} = 0, \qquad \text{for any } \eta \text{ with } \eta_j = 1,$$

and an arbitrary d-tuple sequence $\{d_{\mathbf{k}}\}$.

Proof We proceed by induction on the dimension d. For $d = 1$, the statement is obviously true, as it corresponds to the one-dimensional Abel's transformation (2.2) (also note that for $m = n$ we have $\sum_{k=m}^{m} a_k b_k = a_m b_m$, i.e., in this case the identity does not involve the sum containing the differences Δb_k). Assume now that the statement is true for some $d \geq 1$. Let $\{a_{\mathbf{k}}\}, \{b_{\mathbf{k}}\}$ be $(d+1)$-tuple sequences. By induction hypothesis,

$$\begin{aligned}
\sum_{\mathbf{k}=\mathbf{m}}^{\mathbf{n}} a_{\mathbf{k}} b_{\mathbf{k}} &= \sum_{k_1=m_1}^{n_1} \sum_{\mathbf{k}_{\mathbf{1}-\mathbf{e}_1}=\mathbf{m}_{\mathbf{1}-\mathbf{e}_1}}^{\mathbf{n}_{\mathbf{1}-\mathbf{e}_1}} a_{\mathbf{k}} b_{\mathbf{k}} \\
&= \sum_{k_1=m_1}^{n_1} \Bigg(\sum_{\substack{\mathbf{0}\le\eta\le\mathbf{1}-\mathbf{e}_1\\ \eta\neq\mathbf{0}}} \sum_{\mathbf{k}_\eta=\mathbf{m}_\eta}^{(\mathbf{n}-\mathbf{1})_\eta} \Delta^\eta b_{\mathbf{k}_\eta,\mathbf{n}_{\mathbf{1}-\eta}} \sum_{\mathbf{s}_\eta=\mathbf{m}_\eta}^{\mathbf{k}_\eta} \sum_{\mathbf{k}_{\mathbf{1}-\eta}=\mathbf{m}_{\mathbf{1}-\eta}}^{\mathbf{n}_{\mathbf{1}-\eta}} a_{\mathbf{s}_\eta,\mathbf{k}_{\mathbf{1}-\eta}} \\
&\quad + b_{\mathbf{n}_{\mathbf{1}-\mathbf{e}_1},\mathbf{k}_{\mathbf{e}_1}} \sum_{\mathbf{k}=\mathbf{m}_{\mathbf{1}-\mathbf{e}_1}}^{\mathbf{n}_{\mathbf{1}-\mathbf{e}_1}} a_{\mathbf{k}} \Bigg) \\
&= \sum_{\substack{\mathbf{0}\le\eta\le\mathbf{1}-\mathbf{e}_1\\ \eta\neq\mathbf{0}}} \sum_{\mathbf{k}_\eta=\mathbf{m}_\eta}^{(\mathbf{n}-\mathbf{1})_\eta} \sum_{\mathbf{s}_\eta=\mathbf{m}_\eta}^{\mathbf{k}_\eta} \sum_{\mathbf{k}_{\mathbf{1}-\eta}=\mathbf{m}_{\mathbf{1}-\eta}}^{\mathbf{n}_{\mathbf{1}-\eta}} \sum_{k_1=m_1}^{n_1} \Delta^\eta b_{\mathbf{k}_\eta,\mathbf{n}_{\mathbf{1}-\eta}} a_{\mathbf{s}_\eta,\mathbf{k}_{\mathbf{1}-\eta}} \\
&\quad + \sum_{\mathbf{k}_{\mathbf{1}-\eta}=\mathbf{m}_{\mathbf{1}-\eta}}^{\mathbf{n}_{\mathbf{1}-\eta}} \sum_{k_1=m_1}^{n_1} a_{\mathbf{k}} b_{\mathbf{n}_{\mathbf{1}-\mathbf{e}_1},\mathbf{k}_{\mathbf{e}_1}},
\end{aligned} \tag{8.6}$$

where in the last equality we have just changed the order of summation (note that if $m_1 = n_1$ the result just corresponds to the second equality in (8.6)). Assuming $m_1 < n_1$, applying the one-dimensional Abel's transformation to each of the above sums

$$\sum_{k_1=m_1}^{n_1} \Delta^\eta b_{\mathbf{k}_\eta,\mathbf{n}_{\mathbf{1}-\eta}} a_{\mathbf{s}_\eta,\mathbf{k}_{\mathbf{1}-\eta}} \qquad \text{and} \qquad \sum_{k_1=m_1}^{n_1} a_{\mathbf{k}} b_{\mathbf{n}_{\mathbf{1}-\mathbf{e}_1},\mathbf{k}_{\mathbf{e}_1}},$$

and rearranging the obtained expressions in (8.6) yield the desired identity. □

We will now see that all the convergence tests involving Hardy bounded variation are simple consequences of the multidimensional Abel's transformation (Lemma 8.11).

Let us proceed to Dedekind's test for arbitrary dimension.

Theorem 8.12 *If the sequence $\{b_{\mathbf{k}}\}$ is of bounded Hardy variation and satisfies*

$$\lim_{|\mathbf{k}|\to\infty} b_{\mathbf{k}} = 0, \tag{8.7}$$

and the partial sums $S_{\mathbf{n}}$ of the series $\sum a_{\mathbf{k}}$ are uniformly bounded, then the multiple series $\sum a_{\mathbf{k}} b_{\mathbf{k}}$ converges regularly.

Proof For each η on the right-hand side of (8.5), the sum

$$\sum_{\mathbf{k}_\eta=\mathbf{m}_\eta}^{(\mathbf{n}-\mathbf{1})_\eta} \Delta^\eta b_{\mathbf{k}_\eta,\mathbf{n}_{\mathbf{1}-\eta}} \sum_{\mathbf{s}_\eta=\mathbf{m}_\eta}^{\mathbf{k}_\eta} \sum_{\mathbf{k}_{\mathbf{1}-\eta}=\mathbf{m}_{\mathbf{1}-\eta}}^{\mathbf{n}_{\mathbf{1}-\eta}} a_{\mathbf{s}_\eta,\mathbf{k}_{\mathbf{1}-\eta}}$$

is dominated by

$$\sum_{\mathbf{k}_\eta=\mathbf{m}_\eta}^{(\mathbf{n}-\mathbf{1})_\eta} |\Delta^\eta b_{\mathbf{k}_\eta,\mathbf{n}_{\mathbf{1}-\eta}}| \left| \sum_{\mathbf{s}_\eta=\mathbf{m}_\eta}^{\mathbf{k}_\eta} \sum_{\mathbf{k}_{\mathbf{1}-\eta}=\mathbf{m}_{\mathbf{1}-\eta}}^{\mathbf{n}_{\mathbf{1}-\eta}} a_{\mathbf{s}_\eta,\mathbf{k}_{\mathbf{1}-\eta}} \right|.$$

The inner sums concerning $\{a_{\mathbf{k}}\}$ are uniformly bounded, by hypotheses, while the sums of the ηth differences of the sequence $\{b_{\mathbf{k}}\}$ vanish as $\mathbf{m} \le \mathbf{n}$ and $|\mathbf{m}| \to \infty$, by the absolute (and therefore regular) convergence of the corresponding series, since $\{b_{\mathbf{k}}\}$ is of bounded Hardy variation. Furthermore, the term $b_{\mathbf{n}} \sum_{\mathbf{k}=\mathbf{m}}^{\mathbf{n}} a_{\mathbf{k}}$ vanishes as $|\mathbf{n}| \to \infty$, by (8.7). Thus, the sums on the right-hand side of (8.5) vanish as $\mathbf{m} \le \mathbf{n}$ and $|\mathbf{m}| \to \infty$, or, in other words, $\sum a_{\mathbf{k}} b_{\mathbf{k}}$ converges in the regular sense. □

We now state an extension of Dirichlet's test (Theorem 2.64) to arbitrary dimension. It is a direct consequence of Theorem 8.12.

Theorem 8.13 *If the partial sums $S_{\mathbf{k}}$ of a multiple series $\sum a_{\mathbf{k}}$ are uniformly bounded, the multiple series $\sum a_{\mathbf{k}} b_{\mathbf{k}}$ converges regularly, provided the sequence $\{b_{\mathbf{k}}\}$ satisfies*

$$\Delta^\eta b_{\mathbf{k}} \ge 0 \quad \text{for all} \quad \eta, \quad \mathbf{0} \le \eta \le \mathbf{1}, \tag{8.8}$$

and (8.7).

Recall that any sequence satisfying (8.8) is of bounded Hardy variation (Proposition 8.9).

An important and widely known consequence of Theorem 8.13 (at least in the one-dimensional case) is the celebrated Leibniz test for alternating series. One of the possible extensions of such a test (perhaps the most standard one) may be as follows.

Theorem 8.14 *If a sequence $\{b_{\mathbf{k}}\}$ satisfies*

$$\lim_{|\mathbf{k}|\to\infty} b_{\mathbf{k}} = 0 \qquad \textit{and} \qquad \Delta^\eta b_{\mathbf{k}} \ge 0 \quad \text{for all} \quad \eta, \quad \mathbf{0} \le \eta \le \mathbf{1},$$

then the multiple series

$$\sum_{\mathbf{k}=\mathbf{1}}^{\infty} (-1)^{|\mathbf{k}|} b_{\mathbf{k}}$$

converges regularly.

Further, the du Bois-Reymond test and its corollary, Abel's test, are proved in the same manner as the preceding results. Therefore, we will present only the formulations (with a general idea of the proof) of these results.

Theorem 8.15 *If the sequence* $\{b_{\mathbf{k}}\}$ *is of bounded Hardy variation and the series* $\sum a_{\mathbf{k}}$ *converges regularly, then the series* $\sum a_{\mathbf{k}} b_{\mathbf{k}}$ *does as well.*

Proof As in the proof of Theorem 8.12, the hypotheses imply that for each η on the right-hand side of (8.5), the sum

$$\sum_{\mathbf{k}_\eta=\mathbf{m}_\eta}^{(\mathbf{n}-\mathbf{1})_\eta} \Delta^\eta b_{\mathbf{k}_\eta,\mathbf{n}_{1-\eta}} \sum_{\mathbf{s}_\eta=\mathbf{m}_\eta}^{\mathbf{k}_\eta} \sum_{\mathbf{k}_{1-\eta}=\mathbf{m}_{1-\eta}}^{\mathbf{n}_{1-\eta}} a_{\mathbf{s}_\eta,\mathbf{k}_{1-\eta}}$$

vanishes as $\mathbf{m} \le \mathbf{n}$ and $|\mathbf{m}| \to \infty$. Since $\{b_{\mathbf{k}}\}$ is of bounded Hardy variation, it is bounded, by Corollary 8.8. Thus,

$$\left| b_{\mathbf{n}} \sum_{\mathbf{k}=\mathbf{m}}^{\mathbf{n}} a_{\mathbf{k}} \right| \le \left| \sup_{\mathbf{k}} b_{\mathbf{k}} \right| \left| \sum_{\mathbf{k}=\mathbf{m}}^{\mathbf{n}} a_{\mathbf{k}} \right| \to 0,$$

if $\mathbf{m} \le \mathbf{n}$ and $|\mathbf{m}| \to \infty$, by the regular convergence of $\sum a_{\mathbf{k}}$. Thus, the left-hand side of (8.5) vanishes as $\mathbf{m} \le \mathbf{n}$ and $|\mathbf{m}| \to \infty$, or, in other words, the series $\sum a_{\mathbf{k}} b_{\mathbf{k}}$ converges regularly. □

Finally, we state the multidimensional Abel's test.

Theorem 8.16 *If the multiple series* $\sum a_{\mathbf{k}}$ *converges regularly, then the multiple series* $\sum a_{\mathbf{k}} b_{\mathbf{k}}$ *does as well, provided the sequence* $\{b_{\mathbf{k}}\} \subset \mathbb{R}$ *satisfies* (8.8) *and is bounded.*

Corresponding definitions concerning classical notions of bounded variations for functions of several variables will be given in Subsection 9.1.3.

8.3 Convergence of *d*-tuple series with decreasing terms

In this section we present extensions of the results in Section 5.3 to the multidimensional setting. These are an overview of three basic classical tests for convergence of series with decreasing terms. The proofs will be skipped, as we will present the corresponding results in a more general framework (mostly for *WMS*), along with the detailed proofs in Chapter 9.

We start this section with Abel–Olivier's test.

Theorem 8.17 *Let* $\{a_{\mathbf{k}}\}$ *be a decreasing d-tuple sequence. If the series*

$$\sum_{\mathbf{k}=\mathbf{1}}^{\infty} a_{\mathbf{k}}$$

converges, then

$$k_1 \cdots k_d a_{\mathbf{k}} \to 0 \qquad \text{as } |\mathbf{k}| \to \infty.$$

We note that this result is the most basic one among all the Abel–Olivier type tests, and that it implies numerous consequences. We study this topic in depth in Subsection 9.2.1.

Another very important test, applicable to easily derive the convergence of the p-series

$$\sum_{\mathbf{n}=\mathbf{1}}^{\infty} \frac{1}{(n_1 \cdots n_d)^p},$$

is the Cauchy condensation test. In the multidimensional setting it reads as follows (in its simplest version).

Theorem 8.18 *Let $\{a_{\mathbf{k}}\}$ be a decreasing d-tuple sequence. Then the d-tuple series*

$$\sum_{\mathbf{k}=\mathbf{1}}^{\infty} a_{\mathbf{k}}$$

converges if and only if the d-tuple series

$$\sum_{\mathbf{k}=\mathbf{0}}^{\infty} 2^{|\mathbf{k}|} a_{\mathbf{2^k}}$$

does.

We observe that this result relates the convergence of two series, namely the original one and the one with condensed indices ($\sum 2^{|\mathbf{k}|} a_{\mathbf{2^k}}$). However, one may consider intermediate series, where only certain indices are condensed, namely series of the type

$$\sum_{\mathbf{k}_\eta=\mathbf{0}}^{\infty} \sum_{\mathbf{n}_{\mathbf{1}-\eta}=\mathbf{1}}^{\infty} 2^{|\mathbf{k}_\eta|} a_{(\mathbf{2^k})_\eta, \mathbf{n}_{\mathbf{1}-\eta}},$$

for some $\mathbf{0} \leq \eta \leq \mathbf{1}$ with $\eta \neq \mathbf{0}, \mathbf{1}$ (the cases $\eta = \mathbf{0}, \mathbf{1}$ correspond to the two series appearing in Theorem 8.18). This is in fact done in Subsection 9.2.5, along with the corresponding generalization to the *WMS* case. It is interesting to note that certain series involving differences with condensed indices are considered there (see Theorem 9.55).

The last result we present in this section is the Maclaurin–Cauchy integral test. In the two-dimensional case, we have already observed the phenomenon that, given a function of two variables f decreasing in each variable, the convergence of the double integral

$$\int_1^{\infty} \int_1^{\infty} f(x, y)\, dx\, dy$$

is not sufficient for the series $\sum_{k,\ell=1} f(k,\ell)$ to converge. One needs to assume, additionally, that both one-dimensional series

$$\sum_{k=1}^{\infty} f(k,1) \qquad \text{and} \qquad \sum_{\ell=1}^{\infty} f(1,\ell)$$

converge. In the multidimensional setting the situation is similar. In this case the statement is written in terms of a decreasing function, rather than of a decreasing sequence.

Theorem 8.19 *Let $f\colon \mathbb{R}_+^d \to \mathbb{R}_+$ be a decreasing function in each variable. Then the series $\sum_{\mathbf{k}=\mathbf{1}}^{\infty} f(\mathbf{k})$ converges if and only if all the integrals*

$$\int_{(\mathbf{1},\infty)^\eta} f(\mathbf{t}_\eta, \mathbf{1}_{\mathbf{1}-\eta})\, d\mathbf{t}_\eta$$

with $\mathbf{0} \le \eta \le \mathbf{1}$, $\eta \ne \mathbf{0}$, converge. In particular, the convergence of

$$\int_{(\mathbf{1},\infty)} f(\mathbf{t})\, d\mathbf{t}$$

is necessary but not sufficient to guarantee the convergence of the series $\sum_{\mathbf{k}=\mathbf{1}}^{\infty} f(\mathbf{k})$.

Further discussion around the integral test in the *WM* setting, along with several examples, can be found in Subsection 9.2.7.

8.4 Notes

Section 8.1 (Tests for convergence of positive multiple series)

1. The history of appearance of the following results is quite interesting. It shows possible relations between the theory of number series and other areas of mathematics. In order to determine the radius of convergence of a one-variable d-tuple power series of the form

$$\sum_{\mathbf{k}=\mathbf{0}}^{\infty} a_{\mathbf{k}} x^{|\mathbf{k}|}$$

one of the main tools we can use is the $|\mathbf{n}|$-th root test (Theorem 8.4). However, if we want to study the region of convergence of a multidimensional power series of several variables, say

$$\sum_{\mathbf{k}=\mathbf{0}}^{\infty} a_{\mathbf{k}} x_1^{k_1} x_2^{k_2} \cdots x_d^{k_d}, \tag{8.9}$$

we already encounter that the $|\mathbf{n}|$-th root test has some limitations, since in general it is difficult to calculate

$$\limsup_{|\mathbf{n}|\to\infty} a_{\mathbf{n}}^{1/|\mathbf{n}|} x_1^{\frac{n_1}{|\mathbf{n}|}} x_2^{\frac{n_2}{|\mathbf{n}|}} \cdots x_d^{\frac{n_d}{|\mathbf{n}|}}.$$

We now give the statements of two different generalizations (or rather specializations) of the $|\mathbf{n}|$-th root test, which were obtained by Asadullin in [17] and [18]. Asadullin's motivation for obtaining those tests was to study the convergence of the multivariate power series (8.9). For convenience, in the following we consider the series with terms $a_{\mathbf{k}} = f(\mathbf{k})$, where $f\colon \mathbb{R}_+^d \to \mathbb{R}_+$. The first statement is as follows.

Theorem 8.20 *If for any d-tuple $\mathbf{c} = (c_1, c_2, \ldots, c_d) \in \mathbb{R}_+^d$ with $\mathbf{c} \neq \mathbf{0}$ the inequality*

$$\lim_{t\to\infty} f^{1/t}(c_1 t, \ldots, c_d t) = q(\mathbf{c}) < 1$$

holds, then the series

$$\sum_{\mathbf{k}=\mathbf{0}}^{\infty} f(\mathbf{k}) \tag{8.10}$$

converges. If for at least one d-tuple $\mathbf{c}$*, the inequality* $q(\mathbf{c}) > 1$ *holds, then the series* (8.10) *diverges.*

We observe that the sufficiency part follows immediately from Theorem 8.5. For the proof of the necessity part, we refer to [17]. It is also worth noting that Theorem 8.20 requires the existence of the involved limit, while in Theorem 8.5 it is enough to study the inferior and superior limits of the corresponding sequence.

2. The second "specialization" of the $|\mathbf{n}|$-th root test is based on the observation that if, for a positive sequence $\{a_k\}$, the limit

$$\lim_{k\to\infty} \frac{a_{k+1}}{a_k}$$

exists, then it is equal to

$$\lim_{k\to\infty} a_k^{1/k};$$

see (2.15). Taking this fact into account, Theorem 8.20 leads to the following result.

Theorem 8.21 *If for any d-tuple $\mathbf{c} = (c_1, \ldots, c_d) \in \mathbb{R}_+^d$ with $\mathbf{c} \neq \mathbf{0}$ there holds*

$$\lim_{k\to\infty} \frac{f(c_1(k+1), \ldots, c_d(k+1))}{f(c_1 k, \ldots, c_d k)} = q(\mathbf{c}) < 1,$$

then the series (8.10) *converges. If for at least one d-tuple $\mathbf{c} \in \mathbb{R}_+^d$, $\mathbf{c} \neq \mathbf{0}$, there holds $q(\mathbf{c}) > 1$ and the sequence $\{f(c_1 k, c_2 k, \ldots, c_d k)\}_{k=1}^{\infty}$ is decreasing, then the series* (8.10) *diverges.*

3. Furthermore, in [18], generalizations to several dimensions of d'Alembert's and Ermakov's tests were given (cf. Theorems 8.2 and 2.87, respectively).

4. We stress the close relation between Theorems 8.20 and 8.21 with Cauchy's and d'Alembert's tests, respectively (Theorems 8.2 and 8.4).

5. Theorems 8.20 and 8.21 are among the few examples of tests for convergence of multidimensional series known in the mid-twentieth century.

Section 8.2 (Tests for convergence involving *d*-dimensional bounded variation)

6. For an account of several basic properties of Vitali type variation (in the case of functions of two variables), see, e.g., [8, 61]. For the multivariate case, we refer to [173].

7. For a proof of the statement in Remark 9.28 and further discussion, see, e.g., [B-69, Ch. III, §59].

8. The representation of sequences of bounded Hardy variation in Theorem 8.10 already appeared in [126] for the two-dimensional case (see also [B-84, Ch. 1, §1, Th. 1.1]). The multidimensional representation in Theorem 8.10 was obtained by Leonov in [171].

9. In the paper [121] of 1904 by Hardy, Dirichlet's test (Theorem 8.13) is given in arbitrary dimension. In fact, in that paper, only Pringsheim convergence of the series is obtained. Hardy himself noted in 1905 [122] that the stronger conclusion of regular convergence of the series $\sum a_{\mathbf{k}} b_{\mathbf{k}}$ could be derived. In the subsequent paper [126] of 1917, having the same title as [121], Hardy gave an explicit proof of the two-dimensional Dirichlet test, including the conclusion on regular convergence (in fact, he proved a version of Dedekind test (Theorem 8.12) in two dimensions and then deduced Dirichlet's test). We present a multidimensional Dirichlet test is presented in a different, perhaps more convenient notation. In fact, in [121] the result was actually proved for Pringsheim convergence.

10. In [43], more sophisticated versions of Leibniz's test (Theorem 8.14) are studied in arbitrary dimension. See also [191], where the test is derived under slightly more restrictive assumptions.

9

Multiple series with relaxed monotonicity

9.1 Generalizations of monotonicity

Proceeding to d-tuple sequences for any $d \geq 2$, we first and foremost aim to define the notion of weak and general monotonicity and establish basic properties of such sequences.

9.1.1 Weak monotonicity

Definition 9.1 We say that a nonnegative d-tuple sequence $\{a_{\mathbf{k}}\}$ is *weak monotone*, written as $\{a_{\mathbf{k}}\} \in WMS$, if for any $\mathbf{n} \in \mathbb{N}^d$

$$a_{\mathbf{k}} \leq C a_{\mathbf{n}} \quad \text{with } n_j \leq k_j \leq 2n_j \quad (j = 1, \ldots, d),$$

where $C > 0$ is a constant independent of $\mathbf{n}$.

We also define the *WMS* condition independently in each variable.

Definition 9.2 Let $1 \leq j \leq d$. We say that a nonnegative d-tuple sequence $\{a_{\mathbf{k}}\}$ is weak monotone in the jth variable, written as $\{a_{\mathbf{k}}\} \in WMS_j$, if for every $\mathbf{n} \in \mathbb{N}^d$

$$a_{\mathbf{k}_{\mathbf{e}_j}, \mathbf{n}_{\mathbf{1}-\mathbf{e}_j}} \leq C a_{\mathbf{n}} \quad \text{with } n_j \leq k_j \leq 2n_j,$$

where $C > 0$ is a constant independent of $\mathbf{n}$.

Remark 9.3 Let $\{a_{\mathbf{k}}\}$ be a nonnegative d-tuple sequence. Then $\{a_{\mathbf{k}}\} \in WMS$ if and only if $\{a_{\mathbf{k}}\} \in WMS_j$ for every $1 \leq j \leq d$. Indeed, on the one hand, if $\{a_{\mathbf{k}}\} \in WMS$, it trivially satisfies the *WMS* condition in each variable separately. On the other hand, if $\{a_{\mathbf{k}}\} \in WMS_j$ for every $1 \leq j \leq d$, then for any $\mathbf{n} \in \mathbb{N}^d$

$$a_{k_1,\ldots,k_d} \leq C a_{k_1,\ldots,k_{d-1},n_d} \leq C^2 a_{k_1,\ldots,k_{d-2},n_{d-1},n_d} \leq \cdots \leq C^d a_{n_1,\ldots,n_d},$$

where $n_j \leq k_j \leq 2n_j$ for $j = 1, \ldots, d$.

Remark 9.4 The class consisting of nonnegative d-tuple sequences $\{a_{\mathbf{k}}\}$ such that for any $\mathbf{k} \in \mathbb{N}^d$ one has

$$a_{\mathbf{k}} \leq C a_{\mathbf{n}} \quad \text{whenever } n_j \leq k_j \leq \lambda n_j \quad (j = 1, \ldots, d),$$

with $\lambda > 1$ fixed, coincides with the class *WMS*.

Let us list some simple yet useful properties of *WMS*.

Proposition 9.5 *Let $\{a_{\mathbf{k}}\}$ and $\{b_{\mathbf{k}}\}$ be d-tuple WMS.*

(i) *If $\{a_{\mathbf{n}}\} \in WMS$ with constant $C \le 1$, then $\{a_{\mathbf{n}}\}$ is decreasing.*
(ii) *If there exists $\mathbf{n} \in \mathbb{N}^d$ such that $a_{\mathbf{n}} = 0$, then $a_{\mathbf{k}} = 0$ for all $\mathbf{k} \ge \mathbf{n}$.*
(iii) *For every $\mathbf{m} \le \mathbf{n} \in (\mathbb{N} \cup \{0\})^d$ and every $\mathbf{2^m} \le \mathbf{k} \le \mathbf{2^{m+1}}$ and $\mathbf{2^n} \le \boldsymbol{\ell} \le \mathbf{2^{n+1}}$ with $\mathbf{k} \le \boldsymbol{\ell}$,*

$$a_{\boldsymbol{\ell}} \le C a_{\mathbf{2^n}} \le \cdots \le C^{\max_j\{n_j - m_j\}} a_{\mathbf{2^{m+1}}} \le C^{\max_j\{n_j - m_j\}+1} a_{\mathbf{k}}.$$

(iv) *The d-tuple sequences $\{a_{\mathbf{k}} + b_{\mathbf{k}}\}$ and $\{a_{\mathbf{k}} b_{\mathbf{k}}\}$ are WMS.*
(v) *The d-tuple sequences $\{\min(a_{\mathbf{k}}, b_{\mathbf{k}})\}$ and $\{\max(a_{\mathbf{k}}, b_{\mathbf{k}})\}$ are WMS.*

Proof Items (i) and (ii) follow trivially from the *WMS* condition, while item (iii) is simply an iteration. Item (iv) also follows readily from the *WMS* condition.

Let us now prove (v). Fix a d-tuple of natural numbers $\mathbf{n}$ and note that for any $\mathbf{n} \le \mathbf{k} \le \mathbf{2n}$

$$\max(a_{\mathbf{k}}, b_{\mathbf{k}}) \le a_{\mathbf{k}} + b_{\mathbf{k}} \le C(a_{\mathbf{n}} + b_{\mathbf{n}}) \le 2C \max(a_{\mathbf{n}}, b_{\mathbf{n}}).$$

Now assume that $\min(a_{\mathbf{n}}, b_{\mathbf{n}}) = a_{\mathbf{n}}$. Then

$$\min(a_{\mathbf{k}}, b_{\mathbf{k}}) \le a_{\mathbf{k}} \le C a_{\mathbf{n}} = C \min(a_{\mathbf{n}}, b_{\mathbf{n}}).$$

If $\min(a_{\mathbf{n}}, b_{\mathbf{n}}) = b_{\mathbf{n}}$, the proof is similar. □

As the following examples show, the *WMS* class is much wider than just that of monotone sequences.

Example 9.6 (i) Assume that the terms of a d-tuple sequence are of the form

$$a_{\mathbf{n}} = a_{n_1}^{(1)} a_{n_2}^{(2)} \cdots a_{n_d}^{(d)},$$

where $\left\{a_n^{(i)}\right\}$ are one-dimensional positive sequences for $i = 1, \ldots, d$. Then, $\{a_{\mathbf{n}}\}$ is a d-tuple *WMS* if and only if the sequence $\left\{a_n^{(i)}\right\}$ is one-dimensional weak monotone for $i = 1, \ldots, d$.

(ii) Let P, $Q \ne 0$ be polynomials of d variables with nonnegative coefficients and $\left\{a_n^{(i)}\right\} \in WMS$, $i = 1, \ldots, d$. The sequence

$$a_{\mathbf{n}} = P\left(a_{n_1}^{(1)}, \ldots, a_{n_d}^{(d)}\right)$$

is *WMS*. Indeed, if $\mathbf{n} \le \mathbf{k} \le \mathbf{2n}$, then

$$P\left(a_{k_1}^{(1)}, \ldots, a_{k_d}^{(d)}\right) \le P\left(C a_{n_1}^{(1)}, \ldots, C a_{n_d}^{(d)}\right) \le C^m P\left(a_{n_1}^{(1)}, \ldots, a_{n_d}^{(d)}\right),$$

where $m = \deg P$. If, moreover, the sequences $\left\{a_n^{(i)}\right\} \in WMS$ are increasing for all $i = 1, \ldots, d$ and $c_{\mathbf{n}}$ is a nonnegative sequence such that $0 < \alpha \le c_{\mathbf{n}} \le \beta < \infty$, then

$$b_{\mathbf{n}} = \frac{P\left(a_{n_1}^{(1)}, \ldots, a_{n_d}^{(d)}\right)}{Q\left(a_{n_1}^{(1)}, \ldots, a_{n_d}^{(d)}\right)} c_{\mathbf{n}} \tag{9.1}$$

is *WMS*. Indeed, since for any $\mathbf{n} \le \mathbf{k} \le \mathbf{2n}$ there holds

$$P\big(a_{k_1}^{(1)}, \dots, a_{k_d}^{(d)}\big) \le C^m P\big(a_{n_1}^{(1)}, \dots, a_{n_d}^{(d)}\big),$$

we have

$$\frac{P\big(a_{k_1}^{(1)}, \dots, a_{k_d}^{(d)}\big)}{Q\big(a_{k_1}^{(1)}, \dots, a_{k_d}^{(d)}\big)} c_{\mathbf{k}} \le C^m \frac{P\big(a_{n_1}^{(1)}, \dots, a_{n_d}^{(d)}\big)}{Q\big(a_{n_1}^{(1)}, \dots, a_{n_d}^{(d)}\big)} c_{\mathbf{k}}$$
$$\le \frac{\beta C^m}{\alpha} \frac{P\big(a_{n_1}^{(1)}, \dots, a_{n_d}^{(d)}\big)}{Q\big(a_{n_1}^{(1)}, \dots, a_{n_d}^{(d)}\big)} c_{\mathbf{n}}.$$

In general, if the sequences $\{a_{k_i}^{(i)}\}$ are not increasing, the sequence $\{b_{\mathbf{n}}\}$ from (9.1) need not be *WMS*. This can be seen in dimension one by taking $c_k = 1$, $P(x) = x$, $Q(x) = x^2$, and $a_k = 2^{-k}$. Then $b_n = 2^n$, which is not *WMS*.

(iii) Let $b > 1$ and let $\varphi_1, \dots, \varphi_d$ be strictly increasing positive functions defined on $\mathbb{R}_+$. Then

$$a_{\mathbf{n}} = b^{\varphi_1(n_1)+\cdots+\varphi_d(n_d)} \in WMS \quad \text{if and only if} \quad \varphi_j(2x) - \varphi_j(x) = O(1)$$

as $x \to \infty$, for $j = 1, \dots, d$. We have to verify that

$$b^{\varphi_1(k_1)+\cdots+\varphi_d(k_d)} \le C b^{\varphi_1(n_1)+\cdots+\varphi_d(n_d)}$$

for every $\mathbf{n} \le \mathbf{k} \le \mathbf{2n}$. This is equivalent to

$$b^{\sum_{j=1}^d \varphi_j(k_j)-\varphi_j(n_j)} \le C, \tag{9.2}$$

which holds if and only if $\varphi_j(2x) - \varphi_j(x) = O(1)$ as $x \to \infty$, for $j = 1, \dots, d$.

A simple example of φ_j for which (9.2) holds is $\varphi_j(x) = \log x$.

In particular, if φ_j are differentiable, it follows from (9.2) and the mean value theorem that

$$b^{\varphi_1(n_1)+\cdots+\varphi_d(n_d)} \in WMS \quad \text{provided} \quad \max_{y\in[x,2x]} y\varphi_j'(y) \le C,$$

uniformly in x, for all j, and

$$b^{\varphi_1(n_1)+\cdots+\varphi_d(n_d)} \notin WMS \quad \text{if} \quad \limsup_{x\to\infty} x\varphi_j'(x) = \infty, \tag{9.3}$$

for some j. Thus, for any function $f(x)$ differentiable and increasing to infinity, if we set $\varphi_k(x) = f(x)\log x$ for some $1 \le k \le d$, then

$$b^{\varphi_1(n_1)+\cdots+\varphi_d(n_d)} \notin WMS.$$

Indeed, this follows from (9.3) and the fact that

$$x\big(f(x)\log x\big)' \ge x\frac{f(x)}{x} = f(x) \to \infty \quad \text{as } x \to \infty.$$

In view of Example 9.6, weak monotone sequences can increase to infinity. We now show that if $\{a_{\mathbf{k}}\} \in WMS$, then it cannot grow to infinity faster than a polynomial (which depends on the *WMS* constant of $\{a_{\mathbf{k}}\}$).

Proposition 9.7 *Let $\{a_{\mathbf{n}}\}$ be a d-tuple WMS with constant $C > 0$. Then the d-tuple sequence $\left\{(n_1 \cdots n_d)^{-\log_2 C} a_{\mathbf{n}}\right\}$ is bounded.*

Proof For $\mathbf{n} \in \mathbb{N}^d$, let $\mathbf{k} \in (\mathbb{N} \cup \{0\})^d$ be such that $\mathbf{n} \in [\mathbf{2^k}, \mathbf{2^{k+1}}] \cap \mathbb{N}^d$. Applying the *WMS* inequality repeatedly, we obtain

$$\begin{aligned} a_{\mathbf{n}} &\leq C a_{\mathbf{2^k}} \leq C^{k_1+1} a_{(\mathbf{2^k})_{\mathbf{1}-\mathbf{e}_1}, \mathbf{1}_{\mathbf{e}_1}} \leq C^{k_1+k_2+1} a_{(\mathbf{2^k})_{\mathbf{1}-\mathbf{e}_1-\mathbf{e}_2}, \mathbf{1}_{\mathbf{e}_1+\mathbf{e}_2}} \\ &\leq \cdots \leq C^{|\mathbf{k}|+1} a_{\mathbf{1}} = C \cdot \left(2^{|\mathbf{k}|}\right)^{\log_2 C} a_{\mathbf{1}} \asymp (n_1 \cdots n_d)^{\log_2 C} a_{\mathbf{1}}, \end{aligned}$$

as desired. □

We now define quasi-monotone d-tuple sequences. As in dimension two, we will not focus on the study of such a class of sequences, but only relate it to the class of *WMS*.

Definition 9.8 Let $\alpha > 0$. We say that a d-tuple sequence $\{a_{\mathbf{n}}\}$ is α-quasi-monotone (written as $\{a_{\mathbf{n}}\} \in QMS(\alpha)$) if the sequence $\{(n_1 \cdots n_d)^{-\alpha} a_{\mathbf{n}}\}$ is nonincreasing. We define the class of all quasi-monotone sequences (*QMS*) as

$$QMS = \bigcup_{\alpha > 0} QMS(\alpha).$$

We can use Proposition 9.7 to prove an important property of *WMS*, namely that each $\{a_{\mathbf{n}}\} \in WMS$ is equivalent to a quasi-monotone sequence, although $QMS \subsetneq WMS$ (in general, quasi-monotone sequences enjoy much better variational properties than *WMS*).

Theorem 9.9 *Let $\{a_{\mathbf{n}}\}$ be a d-tuple WMS with constant $C \geq 1$. Then there exists a d-tuple sequence $\{b_{\mathbf{n}}\} \in QMS(\log_2 C)$ such that*

$$a_{\mathbf{n}} \leq b_{\mathbf{n}} \leq 2C^{d+1} a_{\mathbf{n}} \qquad \textit{for every } \mathbf{n} \in \mathbb{N}^d.$$

Proof Let $s = \log_2 C$ and define

$$c_{\mathbf{n}} = \sup_{\mathbf{k} \geq \mathbf{n}} (k_1 \cdots k_d)^{-s} a_{\mathbf{k}}.$$

By Proposition 9.7, $\{c_{\mathbf{n}}\}$ is well defined (since $\left\{(k_1 \cdots k_d)^{-s} a_{\mathbf{k}}\right\}$ is bounded), and moreover, it is nonincreasing (by definition). Therefore, writing $b_{\mathbf{n}} = (n_1 \cdots n_d)^s c_{\mathbf{n}}$, we have that $\{b_{\mathbf{n}}\} \in QMS(s)$, and $b_{\mathbf{n}} \geq a_{\mathbf{n}}$. Now, for any $\mathbf{n} \in \mathbb{N}$, let $\mathbf{m} = \mathbf{m}(\mathbf{n}) \geq \mathbf{n}$ be the least element of the set

$$\left\{ \mathbf{m} \in \mathbb{N}^d : \mathbf{m} \geq \mathbf{n} \text{ and } (m_1 \cdots m_d)^{-s} a_{\mathbf{m}} \geq \frac{c_{\mathbf{n}}}{2} \right\}$$

(this set is nonempty, by the properties of the supremum; here by "least" we mean with respect to the lexicographical order). Finally, let $\boldsymbol{\ell}$ be such that $\mathbf{n} \in [\mathbf{2}^{\boldsymbol{\ell}}, \mathbf{2}^{\boldsymbol{\ell}+\mathbf{1}}] \cap \mathbb{N}^d$ and $\mathbf{t}$ be such that $\mathbf{m} \in [\mathbf{2}^{\mathbf{t}}, \mathbf{2}^{\mathbf{t}+\mathbf{1}}] \cap \mathbb{N}^d$. Applying the *WMS* condition, we have

$$\begin{aligned} b_{\mathbf{n}} &= (n_1 \cdots n_d)^s c_{\mathbf{n}} \le 2^{s(|\boldsymbol{\ell}|+d)+1} (m_1 \cdots m_d)^{-s} a_{\mathbf{m}} \le C 2^{s(|\boldsymbol{\ell}|+d)+1} 2^{-|\mathbf{t}|s} a_{\mathbf{2}^{\mathbf{t}}} \\ &\le 2^{s(|\boldsymbol{\ell}|+d)+1} 2^{-|\mathbf{t}|s} C^{|\mathbf{t}|-|\boldsymbol{\ell}|+1} a_{\mathbf{n}} = 2 \cdot (2^s)^d C \cdot 2^{s(|\boldsymbol{\ell}|-|\mathbf{t}|)} C^{|\mathbf{t}|-|\boldsymbol{\ell}|} a_{\mathbf{n}} \\ &= 2C^{d+1} a_{\mathbf{n}}, \end{aligned}$$

as desired. □

As in dimension two, we must define weak monotonicity for functions.

Definition 9.10 We say that a nonnegative function f defined on $(\mathbf{0}, \boldsymbol{\infty})$ is *WM* if for any $\mathbf{x} \in (\mathbf{0}, \boldsymbol{\infty})$

$$f(\mathbf{t}) \le Cf(\mathbf{x}) \quad \text{with} \quad \mathbf{x} \le \mathbf{t} \le 2\mathbf{x},$$

where $C > 0$ is a constant independent of $\mathbf{x}$.

Remark 9.11 The discrete *WMS* and continuous *WM* classes are associated with each other in the following way. Let $f(\mathbf{x}) = a_{\mathbf{n}}$ for $\mathbf{x} \in [\mathbf{n}, \mathbf{n}+\mathbf{1})$. Then

$$f \in WM \qquad \text{if and only if} \qquad \{a_{\mathbf{n}}\} \in WMS.$$

9.1.2 General monotone sequences

As we have seen in one and two dimensions, in some problems one should consider a different class of monotone-type sequences than *WMS*. In many cases, a proper replacement for weak monotonicity is general monotonicity (*GMS*). In one dimension, the *GMS* condition is more restrictive than the *WMS* condition, but in higher dimensions these conditions are incomparable in general (see Remark 6.18).

We proceed to the definition and main properties of multidimensional *GMS*. We emphasize that the constants C in the definitions below are independent of the indices $\mathbf{n}$.

Definition 9.12 We say that a complex-valued sequence $\{a_{\mathbf{k}}\}$ is *GMS* if there exists a constant $C > 0$ such that

$$\sum_{\mathbf{k}=\mathbf{n}}^{\mathbf{2n}-\mathbf{1}} |\Delta^{\mathbf{1}} a_{\mathbf{k}}| \le C|a_{\mathbf{n}}|, \tag{9.4}$$

for every $\mathbf{n} \in \mathbb{N}^d$.

Remark 9.13 For convenience, we define the class *GMS* for complex-valued sequences. However, as above, in order to compare the class *GMS* with other classes defined only for nonnegative sequences, we will implicitly restrict ourselves to the nonnegative case. The same applies for the classes related to *GMS* that will be defined subsequently.

Remark 9.14 Unlike for *WMS* or the one-dimensional *GMS* themselves, the property that if $\{a_{\mathbf{k}}\} \in GMS$ and $a_{\mathbf{n}} = 0$ for some $\mathbf{n} \in \mathbb{N}^d$, then $a_{\mathbf{m}} = 0$ for all $\mathbf{m} \geq \mathbf{n}$ does not hold in several dimensions. Indeed, a counterexample is provided by the sequence

$$a_{\mathbf{k}} = |\mathbf{k}| - d.$$

It is clear that the differences of such a sequence satisfy $\Delta^{\mathbf{1}} a_{\mathbf{k}} = 0$ for every $\mathbf{k} \in \mathbb{N}^d$ (thus, $\{a_{\mathbf{k}}\} \in GMS$ trivially). However, $a_{\mathbf{1}} = 0$ and $a_{\mathbf{k}} > 0$ for every $\mathbf{k} \neq \mathbf{1}$.

Example 9.15 (i) Let $\{a_{\mathbf{k}}\} \subset \mathbb{C}$ be of the form

$$a_{\mathbf{k}} = \prod_{j=1}^{d} a_{k_j}^{(j)} \tag{9.5}$$

for some one-dimensional sequences $\left\{a_k^{(j)}\right\} \subset \mathbb{C}$, $j = 1, \ldots d$. Then $\{a_{\mathbf{k}}\} \in GMS$ if and only if $\left\{a_k^{(j)}\right\} \in GMS$ for all j as one-dimensional sequences, that is,

$$\sum_{k=n}^{2n-1} |a_k^{(j)} - a_{k+1}^{(j)}| \leq C_j |a_n^{(j)}|, \qquad j = 1, \ldots, d.$$

Indeed, this equivalence follows from the simple observation that for sequences $\{a_{\mathbf{k}}\}$ with the structure (9.5), one has

$$\Delta^{\mathbf{1}} a_{\mathbf{k}} = \prod_{j=1}^{d} \Delta a_{k_j}^{(j)}.$$

(ii) If $a_{\mathbf{k}} \to 0$ as $|\mathbf{k}| \to \infty$, $\Delta^{\mathbf{1}} a_{\mathbf{k}} \geq 0$, and the d-tuple series

$$\sum_{\mathbf{k}=\mathbf{1}}^{\infty} \Delta^{\mathbf{1}} a_{\mathbf{k}}$$

converges, then $\{a_{\mathbf{k}}\} \in GMS$. Indeed, since $\Delta^{\mathbf{1}} a_{\mathbf{k}} \geq 0$, for any $\mathbf{m} \leq \mathbf{M} \in \mathbb{N}^d$, it follows from (7.3) that

$$\sum_{\mathbf{k}=\mathbf{m}}^{\mathbf{M}} |\Delta^{\mathbf{1}} a_{\mathbf{k}}| = \sum_{\mathbf{k}=\mathbf{m}}^{\mathbf{M}} \Delta^{\mathbf{1}} a_{\mathbf{k}} = a_{\mathbf{m}} - \sum_{\zeta \leq \mathbf{1}, \zeta \neq \mathbf{0}} (-1)^{|\zeta|} a_{\mathbf{m}_{\mathbf{1}-\zeta}, (\mathbf{M}+\mathbf{1})_\zeta}.$$

For a fixed $\mathbf{m}$, this expression increases with $\mathbf{M}$ (because $\Delta^{\mathbf{1}} a_{\mathbf{k}} \geq 0$). Hence, using $a_{\mathbf{k}} \to 0$ as $|\mathbf{k}| \to \infty$, we obtain

$$\sum_{\mathbf{k}=\mathbf{m}}^{2\mathbf{m}-\mathbf{1}} \Delta^{\mathbf{1}} a_{\mathbf{k}} \leq \sum_{\mathbf{k}=\mathbf{m}}^{\infty} \Delta^{\mathbf{1}} a_{\mathbf{k}} = a_{\mathbf{m}},$$

and thus $\{a_{\mathbf{k}}\} \in GMS$.

(iii) For $\alpha \in \mathbb{R}\backslash\{0\}$, consider the d-tuple sequence $a_{\mathbf{k}} = |\mathbf{k}|^{\alpha}$ (the case $\alpha = 0$ is excluded as it yields a constant sequence $a_{\mathbf{k}}$, which is trivially *GMS*). We now show in detail that the sequence $\{a_{\mathbf{k}}\}$ belongs to the class *GMS*. Let us first consider the case $\alpha \neq 1, 2, \ldots, d-1$. To prove this claim, it suffices to observe that for the function $f(x) = x^{\alpha}$, $x > 0$, we have

$$\Delta^{\mathbf{1}} a_{\mathbf{k}} = \sum_{j=0}^{d} \binom{d}{j} (-1)^j f(|\mathbf{k}| + j) = f^{(d)}(c), \tag{9.6}$$

for some $c \in [|\mathbf{k}|, |\mathbf{k}| + d]$ (here $f^{(d)}$ denotes the dth derivative of f). The last equality is, roughly speaking, an iterated mean value theorem (see Theorem 9.82 in Notes-3). Since $f^{(d)}(x) = Cx^{\alpha-d}$, for some constant $C \in \mathbb{R}$, we have

$$|\Delta^{\mathbf{1}} a_{\mathbf{k}}| \asymp |\mathbf{k}|^{\alpha-d},$$

and, therefore,

$$\begin{aligned}\sum_{\mathbf{k}=\mathbf{n}}^{2\mathbf{n}-\mathbf{1}} |\Delta^{\mathbf{1}} a_{\mathbf{k}}| &\asymp n_1 \cdots n_d |\Delta^{\mathbf{1}} a_{\mathbf{n}}| \\ &\leq |\Delta^{\mathbf{1}} a_{\mathbf{n}}| \max_{1 \leq j \leq d} n_j^d \asymp |\mathbf{n}|^{\alpha-d} |\mathbf{n}|^d = |\mathbf{n}|^{\alpha} = a_{\mathbf{n}}.\end{aligned}$$

On the other hand, if $\alpha = 1, 2, \ldots d-1$, then (9.6) is still valid, but in this case $f^{(d)}(c) = 0$ for all $c > 0$. Therefore,

$$\sum_{\mathbf{k}=\mathbf{n}}^{2\mathbf{n}-\mathbf{1}} |\Delta^{\mathbf{1}} a_{\mathbf{k}}| = 0 \quad \text{for all } \mathbf{n} \in \mathbb{N}^d,$$

which trivially yields $\{a_k\} \in GMS$.

(iv) The product of two multidimensional *GMS* is not necessarily *GMS*. A counterexample was constructed in two dimensions in Example 6.14 (v). To prove this claim, it suffices to consider the two sequences $\{b_{k\ell}\}$ and $\{c_{k\ell}\}$ in that two-dimensional counterexample and define

$$b_{\mathbf{k}} = \begin{cases} b_{k_1 k_2}, & \text{if } k_3 = \cdots = k_d = 1, \\ 0, & \text{otherwise,} \end{cases}$$

and

$$c_{\mathbf{k}} = \begin{cases} c_{k_1 k_2}, & \text{if } k_3 = \cdots = k_d = 1, \\ 0, & \text{otherwise.} \end{cases}$$

It can easily be checked that these sequences satisfy $\{b_{\mathbf{k}}\}, \{c_{\mathbf{k}}\} \in GMS$. In fact, we have for any $\mathbf{k} \in \mathbb{N}^d$

$$\Delta^{\mathbf{1}} b_{\mathbf{k}} = \Delta^{\mathbf{e}_1+\mathbf{e}_2} b_{\mathbf{k}} = 0,$$

since either all terms in the difference are zero (if $k_j > 1$ for some $3 \le j \le d$), or such a difference equals $\Delta^{(1,1)} b_{k_1,k_2} = 0$ (if $k_j = 1$ for all $3 \le j \le d$). An analogous argument holds for $\{c_{\mathbf{k}}\}$, while for the product $\{a_{\mathbf{k}} := b_{\mathbf{k}} c_{\mathbf{k}}\}$, one has

$$a_{k_1,k_2,1,\ldots,1} = \begin{cases} 1, & \text{if } k_1 \text{ and } k_2 \text{ are even,} \\ 2, & \text{if } k_1 + k_2 \text{ is odd,} \\ 4, & \text{if } k_1 \text{ and } k_2 \text{ are odd,} \end{cases}$$

and $a_{\mathbf{k}} = 0$ if $k_j > 1$ for some $3 \le j \le d$. Thus, $|\Delta^{\mathbf{1}} a_{k_1,k_2,1,\ldots,1}| = 1$ for all $k_1, k_2 \in \mathbb{N}$, and a similar argument as in Example 6.14 (v) shows that $\{a_{\mathbf{k}}\} \notin GMS$.

Similar to the one-dimensional case, we define the multidimensional class of *RBVS* as the class of nonnegative d-tuple sequences $\{a_{\mathbf{k}}\}$ satisfying

$$\sum_{\mathbf{k}=\mathbf{n}}^{\infty} |\Delta^{\mathbf{1}} a_{\mathbf{k}}| \le C a_{\mathbf{n}}, \qquad n \in \mathbb{N},$$

with some constant $C > 0$ independent of $\mathbf{n}$. Since the class *RBVS* is a rather restrictive one, and moreover $RBVS \subsetneq GMS$, we will not use it in any of the results we present. However, for the sake of completeness, it is worth mentioning its definition and some of its properties.

The fact that the inclusion $RBVS \subsetneq GMS$ is strict may easily be proved by means of the one-dimensional counterexample $\{c_k\}$ given in Example 1.11 and defining

$$a_{\mathbf{k}} = \prod_{j=1}^{d} c_{k_j},$$

so that $\Delta^{\mathbf{1}} a_{\mathbf{k}} = \prod_{j=1}^{d} \Delta c_{k_j}$.

Remark 9.16 Contrary to the one-dimensional case, where $GMS \subset WMS$, in higher dimensions this is not the case. A counterexample in two dimensions is given in Remark 6.16.

Remark 9.17 As a complement to Remark 9.16, we also show that $WMS \not\subset GMS$, which does yield that these two classes of sequences are incomparable. To prove this, we consider the following example: let

$$a_{\mathbf{k}} = \begin{cases} 2 \cdot 4^{-m}, & \text{if } \mathbf{k_{1-e_1}} = \mathbf{1},\ 2^m \le k_1 < 2^{m+1},\ m \in \mathbb{N}, \text{ and } k_1 \text{ is even}, \\ 4^{-m}, & \text{if } \mathbf{k_{1-e_1}} = \mathbf{1},\ 2^m \le k_1 < 2^{m+1},\ m \in \mathbb{N}, \text{ and } k_1 \text{ is odd}, \\ 0, & \text{otherwise}. \end{cases}$$

It is easy to check that $\{a_{\mathbf{k}}\} \in WMS$. On the other hand, $\{a_{\mathbf{k}}\} \notin GMS$. Indeed, for $\mathbf{m} = (m_1, 0, \dots, 0) \in (\mathbb{N} \cup \{0\})^d$,

$$\sum_{\mathbf{k}=\mathbf{2^m}}^{\mathbf{2^{m+1}-1}} |\Delta^{\mathbf{1}} a_{\mathbf{k}}| = \sum_{k_1=2^{m_1}}^{2^{m_1+1}-1} |\Delta^{\mathbf{1}} a_{\mathbf{k_{e_1}},\mathbf{1_{1-e_1}}}| \ge \sum_{k_1=2^{m_1}}^{2^{m_1+1}-2} 4^{-m_1}$$
$$\ge 2^{m_1-1} 4^{-m_1} = 2^{m_1-2} a_{\mathbf{2^m}}.$$

Remark 9.18 In general, the *GMS* condition

$$\sum_{\mathbf{k}=\mathbf{n}}^{\mathbf{2n-1}} |\Delta^{\mathbf{1}} a_{\mathbf{k}}| \le C |a_{\mathbf{n}}|$$

does not imply the corresponding inequalities for the lower-dimensional differences or, in other words, the inequalities

$$\sum_{\mathbf{k}_\eta=\mathbf{n}_\eta}^{\mathbf{2n}_\eta-\mathbf{1}} |\Delta^{\eta} a_{\mathbf{k}_\eta,\mathbf{n_{1-\eta}}}| \le C |a_{\mathbf{n}}|, \quad \text{with} \quad \eta \ne \mathbf{0}, \mathbf{1}.$$

This was shown in the two-dimensional case in Remark 6.18.

We can define more general *GMS* classes, where estimates similar to (9.4) hold, with full dimensional differences $\Delta^{\mathbf{1}} a_{\mathbf{k}}$ replaced by lower-dimensional differences $\Delta^{\eta} a_{\mathbf{k}}$, with $\mathbf{0} \le \eta \le \mathbf{1}$, $\eta \ne \mathbf{0}$. Even more generally, we can assume that a similar condition holds for arbitrarily many choices of $\mathbf{0} \le \eta \le \mathbf{1}$, $\eta \ne \mathbf{0}$.

Definition 9.19 Let $E \subset \{\eta \ne \mathbf{0} : \mathbf{0} \le \eta \le \mathbf{1}\}$. We say that a complex-valued sequence $\{a_{\mathbf{k}}\}$ is GMS_E if there exists a constant $C > 0$ such that for every $\eta \in E$ and every $\mathbf{n} \in \mathbb{N}^d$

$$\sum_{\mathbf{k}_\eta=\mathbf{n}_\eta}^{\mathbf{2n}_\eta-\mathbf{1}} |\Delta^{\eta} a_{\mathbf{k}_\eta,\mathbf{n_{1-\eta}}}| \le C |a_{\mathbf{n}}|.$$

Trivially, if $E \subset F \subset \{\eta \ne \mathbf{0} : \mathbf{0} \le \eta \le \mathbf{1}\}$ then $GMS_F \subset GMS_E$.

The classes GMS_E are rather general. In fact, we have introduced them for the sake of completeness; we will use only a reduced number of them for applications. For instance, the classes GMS_E we are most interested in are those where $E = \{\mathbf{1}\}$ (this is the usual *GMS* class in Definition 9.12), $E = \{\mathbf{e}_1, \dots, \mathbf{e}_d, \mathbf{1}\}$ (i.e., the usual *GMS* class where the corresponding estimates concerning one-dimensional differences also hold), or $E = \{\mathbf{e}_1, \dots, \mathbf{e}_d\}$, the class of d-tuple sequences satisfying the one-dimensional *GMS* condition in each variable.

The motivation to introduce the GMS_F classes comes from the fact that sometimes the "simpler" GMS condition (or $GMS_{\{\mathbf{1}\}}$ condition) does not provide enough regularity.

Remark 9.20 We note that $GMS_{\{\mathbf{e}_1,\dots,\mathbf{e}_d\}} \subsetneq WMS$. Indeed, to prove this we observe that for every $\mathbf{n} \in \mathbb{N}^d$, $j = 1,\dots,d$, and $n_j \le k_j \le 2n_j$, it follows from the $GMS_{\{\mathbf{e}_1,\dots,\mathbf{e}_d\}}$ condition that

$$a_{\mathbf{k}_{\mathbf{e}_j},\mathbf{n}_{\mathbf{1}-\mathbf{e}_j}} - a_{\mathbf{n}} \le \sum_{k_j=n_j}^{2n_j-1} |\Delta^{\mathbf{e}_j} a_{\mathbf{k}_{\mathbf{e}_j},\mathbf{n}_{\mathbf{1}-\mathbf{e}_j}}| \le C a_{\mathbf{n}},$$

and therefore $\{a_{\mathbf{k}}\} \in WMS$, by Remark 9.3. In order to show that the inclusion is proper, it suffices to consider the one-dimensional example mentioned in Remark 1.34, namely an oscillating sequence such as $\{a_n = \frac{2+\sin n}{n}\}$ (which belongs to $WMS \backslash GMS$ in one dimension), and define

$$b_{\mathbf{n}} = \prod_{j=1}^{d} a_{n_j}.$$

Then clearly $\{b_{\mathbf{k}}\} \in WMS \backslash GMS_{\{\mathbf{e}_1,\dots,\mathbf{e}_d\}}$.

Remark 9.21 As a counterpart of Remark 9.18, we now construct a d-tuple sequence satisfying the estimates

$$\sum_{\mathbf{k}_\eta=\mathbf{n}_\eta}^{2\mathbf{n}_\eta-1} |\Delta^\eta a_{\mathbf{k}_\eta,\mathbf{n}_{\mathbf{1}-\eta}}| \le C a_{\mathbf{n}},$$

for all $\mathbf{0} \le \eta \le \mathbf{1}$ with $\eta \ne \mathbf{0}, \mathbf{1}$, but not the estimate

$$\sum_{\mathbf{k}=\mathbf{n}}^{2\mathbf{n}-\mathbf{1}} |\Delta^{\mathbf{1}} a_{\mathbf{k}}| \le C a_{\mathbf{n}}.$$

Define

$$a_{\mathbf{k}} = \begin{cases} \frac{1}{2|\mathbf{k}|^2}, & \text{if } k_1 = k_2 = \cdots = k_d,\ k_1 \text{ is even}, \\ \frac{1}{|\mathbf{k}|^2}, & \text{otherwise}. \end{cases}$$

If $\mathbf{m} = (m,\dots,m)$ with $m \in \mathbb{N}$, then we have

$$\sum_{\mathbf{k}=\mathbf{m}}^{2\mathbf{m}-\mathbf{1}} |\Delta^{\mathbf{1}} a_{\mathbf{k}}| \ge \sum_{k=m}^{2m-1} |\Delta^{\mathbf{1}} a_{k,\dots,k}|.$$

Defining $b_{\mathbf{k}} = \frac{1}{|\mathbf{k}|^2}$, we see that

$$|\Delta^{\mathbf{1}} a_{k,\dots,k}| = \left| \Delta^{\mathbf{1}} b_{k,\dots,k} + \frac{\varepsilon_k}{2|\mathbf{k}|^2} \right| \ge \frac{1}{2|\mathbf{k}|^2} - |\Delta^{\mathbf{1}} b_{k,\dots,k}|,$$

where ε_k is either $+1$ or -1. Moreover, we observe that $|\Delta^1 b_{k,\ldots,k}| \asymp \frac{1}{|\mathbf{k}|^{d+2}}$ (see Theorem 9.82 in Notes-3 for details). We conclude that

$$\frac{1}{|\mathbf{m}|} \asymp \sum_{k=m}^{2m-1} \left(\frac{1}{2|\mathbf{k}|^2} - |\Delta^1 b_{k,\ldots,k}| \right)$$
$$\le \sum_{k=m}^{2m-1} |\Delta^1 a_{k,\ldots,k}| \le \sum_{\mathbf{k}=\mathbf{m}}^{\mathbf{2m-1}} |\Delta^1 a_{\mathbf{k}}| \not\lesssim a_{m,\ldots,m} \asymp \frac{1}{|\mathbf{m}|^2},$$

or, in other words, the *GMS* condition with respect to the d-dimensional differences does not hold. On the other hand, if $\eta \neq \mathbf{0}, \mathbf{1}$, we observe that if $\mathbf{k}, \mathbf{n} \in \mathbb{N}^d$, we have, by Theorem 9.82,

$$\Delta^{\eta} a_{\mathbf{k}_\eta, \mathbf{n}_{\mathbf{1}-\eta}} = \Delta^{\eta} b_{\mathbf{k}_\eta, \mathbf{n}_{\mathbf{1}-\eta}} \asymp \frac{1}{(|\mathbf{n}_{\mathbf{1}-\eta}| + |\mathbf{k}_\eta|)^{2+|\eta|}},$$

unless there exists $m \in \mathbb{N}$ such that $\mathbf{n}_\eta + \mathbf{k}_\eta + \zeta = (m, \ldots, m)$ for some $\mathbf{0} \le \zeta \le \eta$ with $\zeta \neq \mathbf{0}$ (that is, always except the cases where any of the indices of the terms in the sum on the right-hand side of (7.2) is of the form $(m, \ldots, m)$). Now, if $\mathbf{n}$ is fixed and $\mathbf{k}_\eta$ ranges between $[\mathbf{n}_\eta, 2\mathbf{n}_\eta - \mathbf{1}]$, the number of times that an index of the form $(m, \ldots, m)$ may be present in a difference $\Delta^{\eta} a_{\mathbf{k}_\eta, \mathbf{n}_{\mathbf{1}-\eta}}$ (i.e., the number of times that such an index appears in the sum on the right-hand side of (7.2)) is bounded by a constant depending only on $|\eta|$, say C. Therefore, for any $\mathbf{n} \in \mathbb{N}^d$, one has

$$\sum_{\mathbf{k}_\eta=\mathbf{n}_\eta}^{2\mathbf{n}_\eta-\mathbf{1}} |\Delta^{\eta} a_{\mathbf{k}_\eta, \mathbf{n}_{\mathbf{1}-\eta}}| \le \sum_{\mathbf{k}_\eta=\mathbf{n}_\eta}^{2\mathbf{n}_\eta-\mathbf{1}} |\Delta^{\eta} b_{\mathbf{k}_\eta, \mathbf{n}_{\mathbf{1}-\eta}}| + C\frac{1}{|\mathbf{n}|^2}$$
$$\asymp \frac{1}{|\mathbf{n}|^2} + \sum_{\mathbf{k}_\eta=\mathbf{n}_\eta}^{2\mathbf{n}_\eta-\mathbf{1}} \frac{1}{(|\mathbf{n}_{\mathbf{1}-\eta}| + |\mathbf{k}_\eta|)^{2+|\eta|}} \asymp \frac{1}{|\mathbf{n}|^2} \asymp a_{\mathbf{n}},$$

as desired.

Among all the GMS_E classes, we distinguish the one where $E = \{\eta : \mathbf{0} \le \eta \le \mathbf{1}\}$, that is, the class for which the *GMS* condition is satisfied for all η-differences. Subsequent to the two-dimensional case (Definition 6.19), we introduce the following.

Definition 9.22 The class of d-tuple sequences GMS_E with $E = \{\eta : \mathbf{0} \le \eta \le \mathbf{1}\}$ is defined as the class of *general monotone sequences of Hardy type*, and it is denoted by *HGMS*.

The class *HGMS* involves conditions on all possible types of differences, just as the Hardy bounded variation property in several dimensions. Thus, it would be reasonable to refer to the class *GMS* (which involves only the d-dimensional differences) as Vitali-general monotone sequences (just as the Vitali bounded variation in several dimensions involves only d-dimensional differences). For simplicity, we will omit this nomenclature.

For d-dimensional *HGMS* we have a criterion similar to Proposition 6.20.

Proposition 9.23 *A d-tuple complex-valued sequence $\{a_{\mathbf{k}}\}$ belongs to the class HGMS if and only if $\{|a_{\mathbf{k}}|\} \in WMS$, and for each $\mathbf{0} \leq \eta \leq \mathbf{1}$ with $\eta \neq \mathbf{0}$ and any $\mathbf{N} \geq \mathbf{n}$,*

$$\sum_{\mathbf{k}_\eta=\mathbf{n}_\eta}^{\mathbf{N}_\eta} |\Delta^\eta a_{\mathbf{k}_\eta,\mathbf{n}_{\mathbf{1}-\eta}}| \leq C\Bigg(|a_{\mathbf{n}}| + \sum_{\substack{\mathbf{0}\leq\theta\leq\eta\\ \theta\neq\mathbf{0}}} \sum_{\mathbf{k}_\theta=(\mathbf{n}+\mathbf{1})_\theta}^{\mathbf{N}_\theta} \frac{|a_{\mathbf{k}_\theta,\mathbf{n}_{\mathbf{1}-\theta}}|}{\prod_{j:\theta_j=1} k_j}\Bigg). \tag{9.7}$$

Proof Recall that by Remark 9.3 one has $\{|a_{\mathbf{k}}|\} \in WMS$ if and only if for every $1 \leq j \leq d$, $\{|a_{\mathbf{k}}|\}_{k_j=1}^{\infty} \in WMS$ as a one-dimensional sequence, for any fixed $\mathbf{k}_{\mathbf{1}-\mathbf{e}_j}$. In particular, this is equivalent to $\{|a_{\mathbf{k}}|\}_{\mathbf{k}_\eta=\mathbf{1}}^{\infty} \in WMS$ as an $|\eta|$-tuple sequence, for any fixed $\mathbf{k}_{\mathbf{1}-\eta}$.

We proceed to proving this result by induction on dimension. With Proposition 1.9 in hand, which deals with the one-dimensional case (and thus is the basis of induction), we assume that the result is true in dimension $d-1$ and derive the corresponding one in d dimensions.

We first assume that $\{a_{\mathbf{k}}\} \in HGMS$. Then $\{|a_{\mathbf{k}}|\} \in WMS$, since $HGMS \subset GMS_{\{\mathbf{e}_1,\dots,\mathbf{e}_d\}} \subset WMS$ (the first inclusion follows from the definition of the classes GMS_E, while the second is established in Remark 9.20). By induction hypothesis, (9.7) is true for all $\mathbf{0} \leq \eta \leq \mathbf{1}$ with $\eta \neq \mathbf{0}, \mathbf{1}$, and hence it remains to prove that (9.7) holds for $\eta = \mathbf{1}$. Let $\mathbf{N} \geq \mathbf{n}$ and $\boldsymbol{\ell} \in (\mathbb{N} \cup \{0\})^d$ be such that $\mathbf{2}^{\boldsymbol{\ell}}\mathbf{n} \leq \mathbf{N} < \mathbf{2}^{\boldsymbol{\ell}+\mathbf{1}}\mathbf{n}$. We have

$$\sum_{\mathbf{k}=\mathbf{n}}^{\mathbf{N}} |\Delta^{\mathbf{1}} a_{\mathbf{k}}| \leq \sum_{\mathbf{s}=\mathbf{0}}^{\boldsymbol{\ell}} \sum_{\mathbf{k}=\mathbf{2}^{\mathbf{s}}\mathbf{n}}^{\mathbf{2}^{\mathbf{s}+\mathbf{1}}\mathbf{n}-\mathbf{1}} |\Delta^{\mathbf{1}} a_{\mathbf{k}}| \leq C \sum_{\mathbf{s}=\mathbf{0}}^{\boldsymbol{\ell}} |a_{\mathbf{2}^{\mathbf{s}}\mathbf{n}}| = C \sum_{\mathbf{0}\leq\theta\leq\mathbf{1}} \sum_{\mathbf{s}_\theta=\mathbf{1}}^{\boldsymbol{\ell}_\theta} |a_{(\mathbf{2}^{\mathbf{s}}\mathbf{n})_\theta,\mathbf{n}_{\mathbf{1}-\theta}}|,$$

where the last equality is an application of (7.1). Using the *WMS* condition and the fact that $0 < c \leq \sum_{k=m}^{2m-1} \frac{1}{k} \leq C$ for every $m \in \mathbb{N}$, we estimate the last sum from above by

$$\begin{aligned} &C\Bigg(|a_{\mathbf{n}}| + \sum_{\substack{0\leq\theta\leq\mathbf{1}\\ \theta\neq 0}} \sum_{\mathbf{s}_\theta=\mathbf{1}}^{\boldsymbol{\ell}_\theta} |a_{(\mathbf{2}^{\mathbf{s}}\mathbf{n})_\theta,\mathbf{n}_{\mathbf{1}-\theta}}| \sum_{\mathbf{k}_\theta=(\mathbf{2}^{\mathbf{s}-\mathbf{1}}\mathbf{n})_\theta}^{(\mathbf{2}^{\mathbf{s}}\mathbf{n})_\theta-\mathbf{1}} \frac{1}{\prod_{j:\theta_j=1} k_j}\Bigg) \\ &\leq C\Bigg(|a_{\mathbf{n}}| + \sum_{\substack{\mathbf{0}\leq\theta\leq\mathbf{1}\\ \theta\neq\mathbf{0}}} \sum_{\mathbf{s}_\theta=\mathbf{1}}^{\boldsymbol{\ell}_\theta} \sum_{\mathbf{k}_\theta=(\mathbf{2}^{\mathbf{s}-\mathbf{1}}\mathbf{n})_\theta}^{(\mathbf{2}^{\mathbf{s}}\mathbf{n})_\theta-\mathbf{1}} \frac{|a_{\mathbf{k}_\theta,\mathbf{n}_{\mathbf{1}-\theta}}|}{\prod_{j:\theta_j=1} k_j}\Bigg) \\ &\leq C\Bigg(|a_{\mathbf{n}}| + \sum_{\substack{\mathbf{0}\leq\theta\leq\eta\\ \theta\neq\mathbf{0}}} \sum_{\mathbf{k}_\theta=(\mathbf{n}+\mathbf{1})_\theta}^{\mathbf{N}_\theta} \frac{|a_{\mathbf{k}_\theta,\mathbf{n}_{\mathbf{1}-\theta}}|}{\prod_{j:\theta_j=1} k_j}\Bigg), \end{aligned}$$

which establishes one direction of the proof. For the reverse direction, since the result is assumed to be true in $d-1$ dimensions by induction hypothesis, one has $\{a_{\mathbf{k}}\} \in GMS_E$, with $E = \{\eta : \mathbf{0} \leq \eta \leq \mathbf{1}, \eta \neq \mathbf{1}\}$. Thus, it suffices to show that $\{a_{\mathbf{k}}\} \in GMS = GMS_{\{\mathbf{1}\}}$. Applying (9.7) with $\eta = \mathbf{1}$ and $\mathbf{N} = 2\mathbf{n} - \mathbf{1}$, we obtain, by the *WMS* condition,

$$\sum_{\mathbf{k}=\mathbf{n}}^{2\mathbf{n}-1}|\Delta^{\mathbf{1}}a_{\mathbf{k}}|\le C\Bigg(|a_{\mathbf{n}}|+\sum_{\substack{\mathbf{0}\le\theta\le\mathbf{1}\\ \theta\ne\mathbf{0}}}\sum_{\mathbf{k}_\theta=(\mathbf{n}+\mathbf{1})_\theta}^{(2\mathbf{n}-\mathbf{1})_\theta}\frac{|a_{\mathbf{k}_\theta,\mathbf{n}_{\mathbf{1}-\theta}}|}{\prod_{j:\theta_j=1}k_j}\Bigg)$$

$$\le C\Bigg(|a_{\mathbf{n}}|+|a_{\mathbf{n}}|\sum_{\mathbf{k}_\theta=(\mathbf{n}+\mathbf{1})_\theta}^{(2\mathbf{n}-\mathbf{1})_\theta}\frac{1}{\prod_{j:\theta_j=1}k_j}\Bigg)\le C|a_{\mathbf{n}}|. \qquad \square$$

Remark 9.24 By Remark 9.20, the reason why one has the inclusion $HGMS\subset WMS$ (when restricted to nonnegative sequences) is that

$$HGMS\subset GMS_{\{\mathbf{e}_1,\ldots,\mathbf{e}_d\}}\subset WMS.$$

The class GMS, or any other class GMS_E, cannot satisfy $GMS_E\subset WMS$ if $\mathbf{e}_j\notin E$ for some $1\le j\le d$ (cf. Remark 9.16).

One may go a step further and define the $\overline{GMS_E}$ classes, combining the ideas of the one-dimensional $\overline{GMS}$ class (cf. Example 1.15) and those from Definition 9.19.

Definition 9.25 Let $E\subset\{\mathbf{0}\le\eta\le\mathbf{1}:\eta\ne\mathbf{0}\}$. We say that a complex-valued sequence $\{a_{\mathbf{k}}\}$ is $\overline{GMS_E}$ if there exist constants $C>0$ and $c>1$ such that for every $\eta\in E$ and every $\mathbf{n}\in\mathbb{N}^d$,

$$\sum_{\mathbf{k}_\eta=\mathbf{n}_\eta}^{2\mathbf{n}_\eta-1}|\Delta^\eta a_{\mathbf{k}_\eta,\mathbf{n}_{\mathbf{1}-\eta}}|\le C\Bigg(\prod_{j:\eta_j=1}\frac{1}{n_j}\Bigg)\sum_{\mathbf{k}_\eta=\lfloor\mathbf{n}_\eta/c\rfloor}^{\lfloor c\mathbf{n}_\eta\rfloor}|a_{\mathbf{k}_\eta,\mathbf{n}_{\mathbf{1}-\eta}}|.$$

If $E=\{\mathbf{1}\}$, then we denote $\overline{GMS_E}=\overline{GMS}$.

Remark 9.26 (i) We note that if $\{a_{\mathbf{k}}\}\in GMS_E$, then $\{a_{\mathbf{k}}\}\in\overline{GMS_E}$. Indeed, note that for any $\eta\in E$ and $\mathbf{n}\in\mathbb{N}^d$,

$$\Bigg(\prod_{j:\eta_j=1}n_j\Bigg)\sum_{\mathbf{k}_\eta=\mathbf{n}_\eta}^{2\mathbf{n}_\eta-1}|\Delta^\eta a_{\mathbf{k}_\eta,\mathbf{n}_{\mathbf{1}-\eta}}|\lesssim\sum_{\mathbf{m}_\eta=\lfloor\mathbf{n}_\eta/2\rfloor}^{2\mathbf{n}_\eta-1}\sum_{\mathbf{k}_\eta=\mathbf{m}_\eta}^{2\mathbf{m}_\eta-1}|\Delta^\eta a_{\mathbf{k}_\eta,\mathbf{n}_{\mathbf{1}-\eta}}|$$

$$\lesssim\sum_{\mathbf{m}_\eta=\lfloor\mathbf{n}_\eta/2\rfloor}^{2\mathbf{n}_\eta-1}|a_{\mathbf{m}_\eta,\mathbf{n}_{\mathbf{1}-\eta}}|,$$

and the claim follows after dividing by $\big(\prod_{j:\eta_j=1}n_j\big)$ on both sides.

(ii) On the other hand, the above inclusion is proper (i.e., $GMS_E\subsetneq\overline{GMS_E}$). To prove this, it suffices to consider a one-dimensional sequence $\{b_k\}\in\overline{GMS}\setminus GMS$ (which is constructed in Example 1.13) and define

$$a_{\mathbf{k}}=\prod_{j=1}^{d}b_{k_j}.$$

Since for every $\mathbf{0} \le \eta \le \mathbf{1}$ there holds

$$\Delta^{\eta} a_{\mathbf{k}_{\eta},\mathbf{n}_{\mathbf{1}-\eta}} = \Big(\prod_{j:\eta_j=1} \Delta b_{k_j}\Big) \cdot \Big(\prod_{j:\eta_j=0} b_{k_j}\Big),$$

the claim follows from the one-dimensional inclusion $GMS_E \subsetneq \overline{GMS}_E$.

Finally, we show a diagram relating the most relevant sequence classes introduced so far (Figure 9.1).

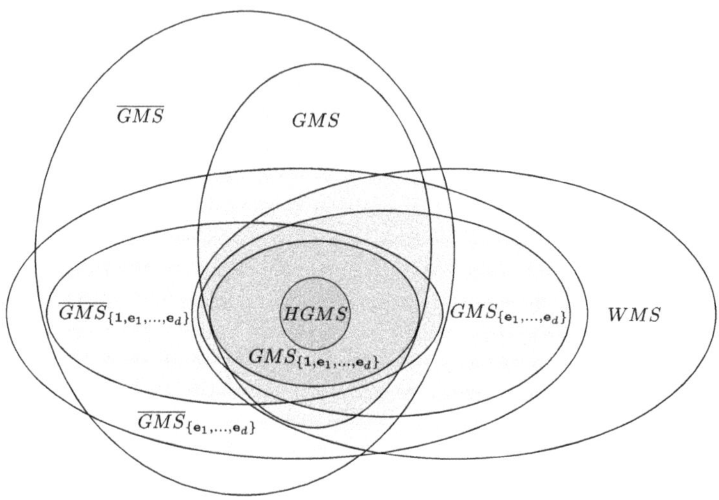

Figure 9.1 Embeddings between *GMS*-type classes and *WMS*.

As in dimension one, we can define similar counterparts of GMS, GMS_E, and $\overline{GMS}_E$ for functions.

9.1.3 Functions of bounded variation

Before defining general monotonicity in the context of functions, we proceed to giving definitions concerning bounded variation for functions of several variables. Let $f\colon \mathbb{R}^d \to \mathbb{C}$ and consider d strictly increasing (one-dimensional) sequences $\{x_k^{(1)}\}, \ldots, \{x_k^{(d)}\}$. For a d-tuple $\mathbf{k} = (k_1, \ldots, k_d) \in \mathbb{N}^d$, let

$$\mathbf{x_k} = \big(x_{k_1}^{(1)}, \ldots, x_{k_d}^{(d)}\big).$$

The map $\mathbf{k} \mapsto f(\mathbf{x_k})$ defines a d-tuple sequence (say, $\{c_{\mathbf{k}}\}$), and the η-differences of f with respect to the sequences $\{x_k^{(1)}\}, \ldots \{x_k^{(d)}\}$ are defined as

$$\Delta^{\eta} f(\mathbf{x_k}) := \Delta^{\eta} c_{\mathbf{k}}, \qquad \mathbf{0} \le \eta \le \mathbf{1},\ \eta \ne \mathbf{0}.$$

Definition 9.27 Let $f\colon \mathbb{R}^d \to \mathbb{C}$ and consider a compact rectangle $R = [\mathbf{a}, \mathbf{b}] \subset \mathbb{R}^d$. The *Vitali variation of f over R* is defined as

$$VV_R(f) = \sup_{\{\mathbf{x_k}\}_{\mathbf{k}=\mathbf{0}}^{\mathbf{n}} \subset \mathcal{P}(R)} \sum_{\mathbf{k}=\mathbf{0}}^{\mathbf{n}-\mathbf{1}} |\Delta^{\mathbf{1}} f(\mathbf{x_k})|,$$

where

$$\mathcal{P}(R) = \mathcal{P}([a_1, b_1]) \times \mathcal{P}([a_2, b_2]) \times \cdots \times \mathcal{P}([a_d, b_d])$$

is the set of all (rectangular) partitions of R. If $VV_R(f) < \infty$, we say that f is of *bounded Vitali variation* on R.

Remark 9.28 As is well known, the variation can be represented as the Stieltjes integral on the corresponding set, namely

$$VV_{[\mathbf{a},\mathbf{b}]}(f) = \int_{[\mathbf{a},\mathbf{b}]} |d^{\mathbf{1}} f(\mathbf{x})|.$$

Similarly, we can define the Vitali variation with respect to a smaller number of variables. We can define such a variation in terms of Stieltjes integrals, in view of Remark 9.28. For $\mathbf{0} \le \eta \le \mathbf{1}$ with $\eta \ne \mathbf{0}$ and fixed $\mathbf{x}_{\mathbf{1}-\eta} \in \mathbb{R}^{\mathbf{1}-\eta}$, the Vitali variation with respect to the η-differences is defined as the integral

$$\int_{[\mathbf{a},\mathbf{b}]^{\eta}} |d^{\eta} f(\mathbf{x})|.$$

We also remark that one can define the Vitali variation of a function over noncompact rectangles R' (for instance, $R' = \mathbb{R}^d_+$) by taking

$$VV_{R'}(f) = \sup_{\substack{R \subset R' \\ R \text{ compact}}} VV_R(f).$$

We observe that adding to a function f any function g of any smaller number of variables does not change this variation, that is, $VV(f) = VV(f+g)$, since for any $\mathbf{x} \in \mathbb{R}^d$ one has $\Delta^{\mathbf{1}} g(\mathbf{x}) = 0$. Such a g can be rather "bad", for example, not of bounded variation with respect to the corresponding variables, or even not measurable.

A function f is said to be of *bounded Hardy variation* if it is of bounded Vitali variation and is of bounded variation with respect to any smaller number of variables (in fact, if it is of bounded Vitali variation with respect to any number of variables not less than two and of bounded variation with respect to every single variable).

Definition 9.29 A function f defined on a rectangle $R \subset \mathbb{R}^d$ (which may be noncompact) is *locally of bounded Hardy variation* on R if it is of bounded Hardy variation on every compact rectangle $R' \subset R$.

We call a function f defined on $(\mathbf{0}, \boldsymbol{\infty})$ *admissible* if it is locally of bounded Hardy variation and vanishing at infinity, that is, $f(\mathbf{x}) \to 0$ as $|\mathbf{x}| \to \infty$.

9.1.4 General monotone functions

Definition 9.30 We say that an admissible function $f\colon (\mathbf{0},\boldsymbol{\infty}) \to \mathbb{C}$ is *general monotone*, written as GM, if there exists $C > 0$ such that for all $\mathbf{x} \in (\mathbf{0},\boldsymbol{\infty})$

$$\int_{[\mathbf{x},2\mathbf{x}]} |d^{\mathbf{1}} f(\mathbf{s})| \le C|f(\mathbf{x})|.$$

Definition 9.31 Let $E \subset \{\eta \neq \mathbf{0} : \mathbf{0} \le \eta \le \mathbf{1}\}$. We say that a complex-valued admissible function $f \in GM_E$ if there exists a constant $C > 0$ such that for every $\eta \in E$ and every $\mathbf{x} \in (\mathbf{0},\boldsymbol{\infty})$ one has

$$\int_{[\mathbf{x},2\mathbf{x}]^\eta} |d^\eta f(\mathbf{s}_\eta, \mathbf{x}_{\mathbf{1}-\eta})| \le C|f(\mathbf{x})|.$$

Note that under this definition, $GM_{\{\mathbf{1}\}} = GM$.

Definition 9.32 Let $E \subset \{\mathbf{0} \le \eta \le \mathbf{1} : \eta \neq \mathbf{0}\}$. We say that a complex-valued admissible function $f \in \overline{GM}_E$ if there exist constants $C > 0$ and $c > 1$ such that for every $\eta \in E$ and every $\mathbf{x} \in (\mathbf{0},\boldsymbol{\infty})$ one has

$$\int_{[\mathbf{x},2\mathbf{x}]^\eta} |d^\eta f(\mathbf{s}_\eta, \mathbf{x}_{\mathbf{1}-\eta})| \le C\Bigg(\prod_{j:\eta_j=1} \frac{1}{x_j}\Bigg) \int_{[c^{-1}\mathbf{x},c\mathbf{x}]^\eta} |f(\mathbf{s}_\eta, \mathbf{x}_{\mathbf{1}-\eta})|\, d\mathbf{s}_\eta.$$

If $E = \{\mathbf{1}\}$, we will write $\overline{GM}_E = \overline{GM}$.

Remark 9.33 We note that if $f \in GM_E$ or $f \in \overline{GM}_E$, and we define $a_{\mathbf{k}} = f(\mathbf{k})$, then the sequence $\{a_{\mathbf{k}}\}$ satisfies the corresponding general monotonicity condition (that is, either GMS_E or $\overline{GMS}_E$, respectively). Indeed, since for any $\mathbf{0} \le \eta \le \mathbf{1}$ with $\eta \neq 0$ and $\mathbf{k} \in \mathbb{N}^d$ one has

$$|\Delta^\eta a_{\mathbf{k}}| = \left| \int_{[\mathbf{k},\mathbf{k}+\mathbf{1}]} d^\eta f(\mathbf{u}) \right| \le \int_{[\mathbf{k},\mathbf{k}+\mathbf{1}]} |d^\eta f(\mathbf{u})|,$$

the claim follows by summing up on $\mathbf{k}$ from $\mathbf{n}$ to $2\mathbf{n} - \mathbf{1}$ and applying either the GM_E or $\overline{GM}_E$ condition.

Remark 9.34 One may define the class HGM in the same spirit as the class $HGMS$ is. In that case, a similar criterion as Proposition 9.23 could be obtained for the class HGM. For brevity, we omit such a result.

Finally, we note that if $\{a_{\mathbf{k}}\} \in \overline{GMS}_E$ ($f \in \overline{GM}_E$) with some constant $c > 1$, then the corresponding condition is fulfilled for any constant $c' > c$.

9.2 Convergence tests for multiple series with monotone-type coefficients

With the prerequisites introduced in the previous section in hand, we are now in a position to formulate and prove high-dimensional extensions of many classical tests with relaxed monotonicity.

9.2.1 Abel–Olivier's test

In most textbooks the material concerning Abel–Olivier's test is a matter of a couple of lines, or one paragraph. It turns out that this topic is much richer. This is already clear from the previous parts, but in arbitrary dimension this is completely apparent.

We start with the multidimensional version of the extended Abel–Olivier test (Theorems 2.99 and 6.29 in one and two dimensions, respectively).

Theorem 9.35 *Let $\{a_{\mathbf{k}}\}$ be a d-tuple WMS. Then*

$$\sum_{\mathbf{k}=\mathbf{1}}^{\infty} a_{\mathbf{k}}$$

converges if and only if the following three conditions hold:

$$n_1 \cdots n_d a_{\mathbf{n}} \to 0 \qquad \text{as } |\mathbf{n}| \to \infty; \tag{9.8}$$

for every $\eta \ne \mathbf{0}, \mathbf{1}$ and every $\mathbf{0} \le \zeta \le \mathbf{1} - \eta$, there holds

$$\left(\prod_{j:\eta_j=1} n_j\right) \sum_{\mathbf{k}_{\mathbf{1}-\eta}=\mathbf{1}}^{\infty} \left(\prod_{j:\zeta_j=1} k_j\right) \Delta^{\zeta} a_{\mathbf{n}_\eta, \mathbf{k}_{\mathbf{1}-\eta}} \to 0 \qquad \text{as } |\mathbf{n}_\eta| \to \infty;$$

and the series

$$\sum_{\mathbf{k}=\mathbf{1}}^{\infty} \left(\prod_{j:\eta_j=1} k_j\right) \Delta^{\eta} a_{\mathbf{k}}$$

converge (in the sense of Pringsheim) for all $\eta \ne \mathbf{0}, \mathbf{1}$. Moreover, we have for any $\mathbf{0} \le \eta \le \mathbf{1}$

$$\sum_{\mathbf{k}=\mathbf{1}}^{\infty} a_{\mathbf{k}} = \sum_{\mathbf{k}=\mathbf{1}}^{\infty} \left(\prod_{j:\eta_j=1} k_j\right) \Delta^{\eta} a_{\mathbf{k}}.$$

Instead of proving this claim, we prove its generalization, namely the extended Abel–Olivier test for series with weights (see the one- and two-dimensional Theorems 2.100 and 6.30, respectively). Theorem 9.35 will then follow as a corollary.

Before generalizing the extended Abel–Olivier test, we introduce some notation. For any $\mathbf{0} \le \eta \le \mathbf{1}$ with $\eta \ne \mathbf{0}, \mathbf{1}$ and a d-tuple sequence $\{b_{\mathbf{k}}\}$, we define

$$B_{\mathbf{k}_{\mathbf{1}-\eta}}^{\mathbf{n}_\eta} = \sum_{\mathbf{m}_\eta=\mathbf{1}}^{\mathbf{n}_\eta} b_{\mathbf{m}_\eta, \mathbf{k}_{\mathbf{1}-\eta}}.$$

If $\eta = \mathbf{1}$, we put

$$B_{\mathbf{k}_{\mathbf{1}-\eta}}^{\mathbf{n}_\eta} = B^{\mathbf{n}} = \sum_{\mathbf{m}=\mathbf{1}}^{\mathbf{n}} b_{\mathbf{m}},$$

and if $\eta = \mathbf{0}$,

$$B^{\mathbf{n}_\eta}_{\mathbf{k}_{\mathbf{1}-\eta}} = B_{\mathbf{k}_\mathbf{1}} = b_\mathbf{k}.$$

We emphasize that the definition of $B^{\mathbf{n}_\eta}_{\mathbf{k}_{\mathbf{1}-\eta}}$ has nothing to do with the definition of $A_{\mathbf{k}_\eta,\mathbf{j}_{\mathbf{1}-\eta}}$ in the beginning of the present chapter.

We are also in need of a notation for the sum of $b_\mathbf{k}$ with the upper limits of summation given by two different sets of indices. For this purpose, we define, for any $\eta \le \mathbf{1} - \zeta$ with $\zeta \ne \mathbf{0}, \mathbf{1}$,

$$B^{\mathbf{n}_\zeta,\mathbf{s}_\eta}_{\mathbf{k}_{\mathbf{1}-\zeta-\eta}} = \sum_{\mathbf{m}_\zeta=\mathbf{1}}^{\mathbf{n}_\zeta} \sum_{\mathbf{m}_\eta=\mathbf{1}}^{\mathbf{s}_\eta} b_{\mathbf{m}_{\eta+\zeta},\mathbf{k}_{\mathbf{1}-\eta-\zeta}}.$$

If $\eta + \zeta = \mathbf{1}$, we put $B^{\mathbf{n}_\zeta,\mathbf{s}_\eta}_{\mathbf{k}_{\mathbf{1}-\zeta-\eta}} = B^{\mathbf{n}_\zeta,\mathbf{s}_\eta}$. On the other hand, if $\eta = \mathbf{0}$ we arrive at

$$B^{\mathbf{n}_\zeta,\mathbf{s}_\mathbf{0}}_{\mathbf{k}_{\mathbf{1}-\zeta}} = B^{\mathbf{n}_\zeta}_{\mathbf{k}_{\mathbf{1}-\zeta}},$$

and for $\eta = \mathbf{1}$,

$$B^{\mathbf{n}_\mathbf{0},\mathbf{s}_\mathbf{1}}_{\mathbf{k}_\mathbf{0}} = B^\mathbf{s}.$$

Roughly speaking, since in all the sums defined here we use $\mathbf{1}$ as a lower limit of summation, with this notation (e.g., $B^{\mathbf{n}_\zeta,\mathbf{s}_\eta}_{\mathbf{k}_{\mathbf{1}-\zeta-\eta}}$) we indicate the upper limits of summation with the superscripts, while the subscripts indicate the indices not involved in the sums.

Theorem 9.36 *Let $\{a_\mathbf{k}\}$ be a d-tuple WMS and $\{b_\mathbf{k}\}$ be a d-tuple of nonnegative numbers such that*

$$B^{\mathbf{n}_{\mathbf{e}_j}}_{\mathbf{k}_{\mathbf{1}-\mathbf{e}_j}} \lesssim \sum_{m_j=\lfloor n_j/2 \rfloor}^{n_j} b_{\mathbf{m}_{\mathbf{e}_j},\mathbf{k}_{\mathbf{1}-\mathbf{e}_j}} \tag{9.9}$$

for any $j \in \{1, \dots, d\}$, and any $\mathbf{k}_{\mathbf{1}-\mathbf{e}_j} \in \mathbb{N}^{\mathbf{1}-\mathbf{e}_j}$ and $n_j \in \mathbb{N}$. Then

$$\sum_{\mathbf{k}=\mathbf{1}}^{\infty} a_\mathbf{k} b_\mathbf{k} \tag{9.10}$$

converges if and only if the following three conditions hold: for every $\mathbf{0} \le \eta \le \mathbf{1}$ and every $\mathbf{n}_{\mathbf{1}-\eta} \in \mathbb{N}^{\mathbf{1}-\eta}$,

$$B^{\mathbf{n}_\eta}_{\mathbf{n}_{\mathbf{1}-\eta}} a_\mathbf{n} \to 0 \qquad \text{as } |\mathbf{n}_\eta| \to \infty; \tag{9.11}$$

for every $\eta \ne \mathbf{0}, \mathbf{1}$ and every $\mathbf{0} \le \zeta \le \mathbf{1} - \eta$,

$$\sum_{\mathbf{k}_{\mathbf{1}-\eta}=\mathbf{1}}^{\infty} \Delta^\zeta a_{\mathbf{n}_\eta,\mathbf{k}_{\mathbf{1}-\eta}} B^{\mathbf{n}_\eta,\mathbf{k}_\zeta}_{\mathbf{k}_{\mathbf{1}-\eta-\zeta}} \to 0 \qquad \text{as } |\mathbf{n}_\eta| \to \infty; \tag{9.12}$$

the series

$$\sum_{\mathbf{k}=\mathbf{1}}^{\infty} \Delta^{\eta} a_{\mathbf{k}} B_{\mathbf{k}_{1-\eta}}^{\mathbf{k}_\eta} \tag{9.13}$$

converge (in the sense of Pringsheim) for each $\eta \neq \mathbf{0}, \mathbf{1}$. *Moreover, for any* $\mathbf{0} \leq \eta \leq \mathbf{1}$, *we have*

$$\sum_{\mathbf{k}=\mathbf{1}}^{\infty} a_{\mathbf{k}} b_{\mathbf{k}} = \sum_{\mathbf{k}=\mathbf{1}}^{\infty} \Delta^{\eta} a_{\mathbf{k}} B_{\mathbf{k}_{1-\eta}}^{\mathbf{k}_\eta}. \tag{9.14}$$

We stress that the convergence of $\sum a_{\mathbf{k}} b_{\mathbf{k}}$ is always regular, as it is a series with nonnegative terms. However, the terms of the series (9.13), involving differences of $\{a_{\mathbf{k}}\}$, may have arbitrary signs. For such series, the weaker Pringsheim convergence (along with the rest of the hypotheses) is enough in order to guarantee the convergence of $\sum a_{\mathbf{k}} b_{\mathbf{k}}$. Note also that Theorem 9.36 with $b_{\mathbf{k}} = 1$ yields Theorem 9.35.

Proof of Theorem 9.36 Note that (9.9) implies

$$B_{\mathbf{k}_{1-\eta}}^{\mathbf{n}_\eta} \lesssim \sum_{\mathbf{m}_\eta = \lfloor \mathbf{n}_\eta/2 \rfloor}^{\mathbf{n}_\eta} b_{\mathbf{m}_\eta, \mathbf{k}_{1-\eta}},$$

for all $\mathbf{0} \leq \eta \leq \mathbf{1}$, $\mathbf{k}_{1-\eta} \in \mathbb{N}^{1-\eta}$, and $\mathbf{n}_\eta \in \mathbb{N}^{\eta}$.

Let us show that the convergence of (9.10) implies that for every η with $1 \leq |\eta| < d$,

$$\sum_{\mathbf{k}_{1-\eta}=\mathbf{1}}^{\infty} a_{\mathbf{n}_\eta, \mathbf{k}_{1-\eta}} B_{\mathbf{k}_{1-\eta}}^{\mathbf{n}_\eta} \to 0 \qquad \text{as } |\mathbf{n}_\eta| \to \infty. \tag{9.15}$$

Indeed, if we define

$$c_{\mathbf{m}_\eta} = \sum_{\mathbf{k}_{1-\eta}=\mathbf{1}}^{\infty} b_{\mathbf{m}_\eta, \mathbf{k}_{1-\eta}} a_{\mathbf{m}_\eta, \mathbf{k}_{1-\eta}},$$

the convergence of $\sum_{\mathbf{m}_\eta=\mathbf{1}}^{\infty} c_{\mathbf{m}_\eta}$ (which is just the series in (9.10)) implies that

$$\sum_{\mathbf{k}_{1-\eta}=\mathbf{1}}^{\infty} \sum_{\mathbf{m}_\eta = \lfloor \mathbf{n}_\eta/2 \rfloor}^{\mathbf{n}_\eta} b_{\mathbf{m}_\eta, \mathbf{k}_{1-\eta}} a_{\mathbf{m}_\eta, \mathbf{k}_{1-\eta}} = \sum_{\mathbf{m}_\eta = \lfloor \mathbf{n}_\eta/2 \rfloor}^{\mathbf{n}_\eta} c_{\mathbf{m}_\eta} \to 0, \tag{9.16}$$

as $|\mathbf{n}_\eta| \to \infty$. Then, using $\{a_{\mathbf{k}}\} \in WMS$ and the assumption on $\{b_{\mathbf{k}}\}$, we find that

$$\begin{aligned} \sum_{\mathbf{m}_\eta = \lfloor \mathbf{n}_\eta/2 \rfloor}^{\mathbf{n}_\eta} c_{\mathbf{m}_\eta} &\gtrsim \sum_{\mathbf{k}_{1-\eta}=\mathbf{1}}^{\infty} a_{\mathbf{n}_\eta, \mathbf{k}_{1-\eta}} \sum_{\mathbf{m}_\eta = \lfloor \mathbf{n}_\eta/2 \rfloor}^{\mathbf{n}_\eta} b_{\mathbf{m}_\eta, \mathbf{k}_{1-\eta}} \\ &\gtrsim \sum_{\mathbf{k}_{1-\eta}=\mathbf{1}}^{\infty} a_{\mathbf{n}_\eta, \mathbf{k}_{1-\eta}} B_{\mathbf{k}_{1-\eta}}^{\mathbf{n}_\eta}, \end{aligned} \tag{9.17}$$

and our assertion follows. We note that, in particular, (9.16) implies (9.11) for the corresponding η. Combining (9.15) and (9.16) proves (9.12) for $\zeta = \mathbf{0}$.

We now prove (9.12) for $\zeta \neq \mathbf{0}$. To this end, we fix η with $1 \leq |\eta| < d$. We will distinguish between two cases, $|\eta| < d-1$ and $|\eta| = d-1$. In the first case we proceed by induction on $|\zeta|$. Let i be such that $\eta_i = 0$ and put $\zeta = \mathbf{e}_i$. By Abel's transformation, we have

$$\begin{aligned}
&\sum_{\mathbf{k}_{1-\eta-\mathbf{e}_i}=\mathbf{1}}^{\mathbf{n}_{1-\eta-\mathbf{e}_i}} \sum_{k_i=1}^{n_i} a_{\mathbf{n}_\eta,\mathbf{k}_{1-\eta}} B_{\mathbf{k}_{1-\eta}}^{\mathbf{n}_\eta} \\
&\quad = \sum_{\mathbf{k}_{1-\eta-\mathbf{e}_i}=\mathbf{1}}^{\mathbf{n}_{1-\eta-\mathbf{e}_i}} a_{\mathbf{n}_{\eta+\mathbf{e}_i},\mathbf{k}_{1-\eta-\mathbf{e}_i}} B_{\mathbf{k}_{1-\eta-\mathbf{e}_i}}^{\mathbf{n}_{\eta+\mathbf{e}_i}} \\
&\quad + \sum_{\mathbf{k}_{1-\eta-\mathbf{e}_i}=\mathbf{1}}^{\mathbf{n}_{1-\eta-\mathbf{e}_i}} \sum_{k_i=1}^{n_i-1} \Delta^{\mathbf{e}_i} a_{\mathbf{n}_\eta,\mathbf{k}_{1-\eta}} B_{\mathbf{k}_{1-\eta-\mathbf{e}_i}}^{\mathbf{n}_\eta,\mathbf{k}_{\mathbf{e}_i}}.
\end{aligned} \tag{9.18}$$

Now, letting $\mathbf{n}_{1-\eta} \to \infty$, we note that the left-hand side and the first series on the right-hand side of (9.18) vanish as $|\mathbf{n}_\eta| \to \infty$, by (9.16) and (9.17). Here the assumption $|\eta| < d-1$ is needed in order to apply (9.17) with $\eta + \mathbf{e}_i$ in place of η; note that in this case $|\eta + \mathbf{e}_i| < d$. This establishes the first step of the inductive argument for the case $|\eta| < d-1$.

Now assume that (9.12) holds for every choice of η with $|\mathbf{1} - \eta| < d$ and $|\eta| < d-1$, and every choice of ζ with $\zeta \leq \mathbf{1} - \eta$ and $1 \leq |\zeta| < |\mathbf{1} - \eta|$. We then prove that (9.12) holds for $\zeta + \mathbf{e}_i$ in place of ζ, where i is such that $\zeta_i = \eta_i = 0$. By Abel's transformation,

$$\begin{aligned}
&\sum_{\mathbf{k}_{1-\eta-\mathbf{e}_i}=\mathbf{1}}^{\mathbf{n}_{1-\eta-\mathbf{e}_i}} \sum_{k_i=1}^{n_i} \Delta^\zeta a_{\mathbf{n}_\eta,\mathbf{k}_{1-\eta}} B_{\mathbf{k}_{1-\eta-\zeta}}^{\mathbf{n}_\eta,\mathbf{k}_\zeta} \\
&\quad = \sum_{\mathbf{k}_{1-\eta-\mathbf{e}_i}=\mathbf{1}}^{\mathbf{n}_{1-\eta-\mathbf{e}_i}} \Delta^\zeta a_{\mathbf{n}_{\eta+\mathbf{e}_i},\mathbf{k}_{1-\eta-\mathbf{e}_i}} B_{\mathbf{k}_{1-\eta-\zeta-\mathbf{e}_i}}^{\mathbf{n}_{\eta+\mathbf{e}_i},\mathbf{k}_\zeta} \\
&\quad + \sum_{\mathbf{k}_{1-\eta-\mathbf{e}_i}=\mathbf{1}}^{\mathbf{n}_{1-\eta-\mathbf{e}_i}} \sum_{k_i=1}^{n_i-1} \Delta^{\zeta+\mathbf{e}_i} a_{\mathbf{n}_\eta,\mathbf{k}_{1-\eta}} B_{\mathbf{k}_{1-\eta-\zeta-\mathbf{e}_i}}^{\mathbf{n}_\eta,\mathbf{k}_{\zeta+\mathbf{e}_i}}.
\end{aligned} \tag{9.19}$$

By induction hypothesis, the series on the left-hand side of (9.19) and the first series on the right-hand side of (9.19) vanish as $|\mathbf{n}_\eta| \to \infty$ and as $|\mathbf{n}_{\eta+\mathbf{e}_i}| \to \infty$, respectively. Thus, the last series in (9.19) also vanishes as $|\mathbf{n}_\eta| \to \infty$, as desired.

The case $|\eta| = d-1$ is simpler, as we only need to prove (9.12) for $\zeta = \mathbf{e}_i$, where i is the only index such that $\eta_i = 0$. Abel's transformation gives

$$\sum_{k_i=1}^{n_i} a_{\mathbf{n}_{1-\mathbf{e}_i},\mathbf{k}_{\mathbf{e}_i}} B_{\mathbf{k}_{\mathbf{e}_i}}^{\mathbf{n}_{1-\mathbf{e}_i}} = a_\mathbf{n} B^\mathbf{n} + \sum_{k_i=1}^{n_i-1} \Delta^{\mathbf{e}_i} a_{\mathbf{n}_{1-\mathbf{e}_i},\mathbf{k}_{\mathbf{e}_i}} B^{\mathbf{n}_{1-\mathbf{e}_i},\mathbf{k}_{\mathbf{e}_i}}.$$

The left-hand side vanishes as $|\mathbf{n}_{\mathbf{1}-\mathbf{e}_i}| \to \infty$, by (9.15). On the other hand, the first term on the right-hand side vanishes as $|\mathbf{n}| \to \infty$, by (9.11), which we have proved above. Therefore, the last series on the right-hand side also vanishes as $|\mathbf{n}_{\mathbf{1}-\mathbf{e}_i}| \to \infty$.

Finally, it remains to prove that (9.14) holds for any $\mathbf{0} \leq \eta \leq \mathbf{1}$. To this end, we use a similar inductive argument on $|\eta|$, beginning with the case $|\eta| = 1$. First, for any $1 \leq i \leq d$, Abel's transformation yields

$$\sum_{\mathbf{k}=\mathbf{1}}^{\mathbf{n}} a_{\mathbf{k}} b_{\mathbf{k}} = \sum_{\mathbf{k}_{\mathbf{1}-\mathbf{e}_i}=\mathbf{1}}^{\mathbf{n}_{\mathbf{1}-\mathbf{e}_i}} a_{\mathbf{n}_{\mathbf{e}_i},\mathbf{k}_{\mathbf{1}-\mathbf{e}_i}} B_{\mathbf{k}_{\mathbf{1}-\mathbf{e}_i}}^{\mathbf{n}_{\mathbf{e}_i}} + \sum_{\mathbf{k}_{\mathbf{1}-\mathbf{e}_i}=\mathbf{1}}^{\mathbf{n}_{\mathbf{1}-\mathbf{e}_i}} \sum_{k_i=1}^{n_i-1} \Delta^{\mathbf{e}_i} a_{\mathbf{k}} B_{\mathbf{k}_{\mathbf{1}-\mathbf{e}_i}}^{\mathbf{k}_{\mathbf{e}_i}}.$$

Letting $\mathbf{n} \to \boldsymbol{\infty}$, we obtain, by (9.12),

$$\sum_{\mathbf{k}=\mathbf{1}}^{\boldsymbol{\infty}} a_{\mathbf{k}} b_{\mathbf{k}} = \sum_{\mathbf{k}=\mathbf{1}}^{\boldsymbol{\infty}} \Delta^{\mathbf{e}_i} a_{\mathbf{k}} B_{\mathbf{k}_{\mathbf{1}-\mathbf{e}_i}}^{\mathbf{k}_{\mathbf{e}_i}},$$

which establishes the first inductive step. Now assume that (9.14) holds for some η with $1 \leq |\eta| < d$. Then we show that (9.14) holds for $\eta + \mathbf{e}_i$ in place of η, where i is such that $\eta_i = 0$. Indeed, by Abel's transformation,

$$\begin{aligned}\sum_{\mathbf{k}=\mathbf{1}}^{\mathbf{n}} \Delta^{\eta} a_{\mathbf{k}} B_{\mathbf{k}_{\mathbf{1}-\eta}}^{\mathbf{k}_{\eta}} &= \sum_{\mathbf{k}_{\mathbf{1}-\mathbf{e}_i}=\mathbf{1}}^{\mathbf{n}_{\mathbf{1}-\mathbf{e}_i}} \Delta^{\eta} a_{\mathbf{n}_{\mathbf{e}_i},\mathbf{k}_{\mathbf{1}-\mathbf{e}_i}} B_{\mathbf{k}_{\mathbf{1}-\eta-\mathbf{e}_i}}^{\mathbf{k}_{\eta},\mathbf{n}_{\mathbf{e}_i}} \\ &\quad + \sum_{\mathbf{k}_{\mathbf{1}-\mathbf{e}_i}=\mathbf{1}}^{\mathbf{n}_{\mathbf{1}-\mathbf{e}_i}} \sum_{k_i=1}^{n_i-1} \Delta^{\eta+\mathbf{e}_i} a_{\mathbf{k}} B_{\mathbf{k}_{\mathbf{1}-\eta-\mathbf{e}_i}}^{\mathbf{k}_{\eta+\mathbf{e}_i}}.\end{aligned}$$

Letting $\mathbf{n} \to \boldsymbol{\infty}$ and using (9.12), we get

$$\sum_{\mathbf{k}=\mathbf{1}}^{\boldsymbol{\infty}} \Delta^{\eta} a_{\mathbf{k}} B_{\mathbf{k}_{\mathbf{1}-\eta}}^{\mathbf{k}_{\eta}} = \sum_{\mathbf{k}=\mathbf{1}}^{\boldsymbol{\infty}} \Delta^{\eta+\mathbf{e}_i} a_{\mathbf{k}} B_{\mathbf{k}_{\mathbf{1}-\eta-\mathbf{e}_i}}^{\mathbf{k}_{\eta+\mathbf{e}_i}},$$

which completes the inductive argument and therefore the proof of the "only if" part.

To prove the "if" part, it suffices to do the above process backwards. Note that in this case we do not even need the *WMS* property. □

We also have an extended Abel–Olivier test for decreasing sequences $\{a_{\mathbf{k}}\}$, with hypotheses different from those of Theorem 9.36. Weak monotonicity is equivalent to the corresponding condition in each variable separately (cf. Remark 9.3). Thus, monotonicity of $\{a_{\mathbf{k}}\}$ in each variable provides a stronger condition than $\{a_{\mathbf{k}}\} \in$ *WMS*, which in turn allows us to weaken the hypotheses on $\{b_{\mathbf{k}}\}$ (in particular, the condition (9.9) can be omitted) and arrive at the same conclusion as in Theorem 9.36. In contrast, we observe that (9.9) cannot be omitted if $\{a_{\mathbf{k}}\} \in$ *WMS*, as we have shown in one dimension through the examples $a_n = \sqrt{n}$ and $b_n = n^{-2}$.

Theorem 9.37 *Let $\{a_{\mathbf{k}}\}$ be a nonnegative d-tuple sequence decreasing in each variable such that $a_{\mathbf{k}} \to 0$ as $|\mathbf{k}| \to \infty$. Let $\{b_{\mathbf{k}}\}$ be a nonnegative d-tuple sequence. Then the conclusion of Theorem 9.36 holds.*

Proof Note that the hypotheses of Theorem 9.36 are used only for proving (9.15). Thus, it suffices to establish such a property under our new hypotheses. In particular, we will prove that for every $1 \leq j \leq d$ such that $\eta_j = 1$ one has

$$\sum_{\mathbf{k}_{1-\eta}=\mathbf{1}}^{\infty} a_{\mathbf{n}_\eta,\mathbf{k}_{1-\eta}} B_{\mathbf{k}_{1-\eta}}^{\mathbf{n}_\eta} \to 0, \qquad \text{as } n_j \to \infty,$$

for any $\mathbf{n}_{\eta-\mathbf{e}_j}$, which is equivalent to (9.15). Defining $c_{\mathbf{m}_\eta}$ as in the proof of Theorem 9.36, since the series

$$\sum_{\mathbf{m}_\eta=\mathbf{1}}^{\infty} c_{\mathbf{m}_\eta}$$

converges (in the regular sense, as it is a series of nonnegative terms), for any $\varepsilon > 0$ we can find $M \in \mathbb{N}$ such that for every $\mathbf{n}_\eta \geq \boldsymbol{\ell}_\eta$ with $|\boldsymbol{\ell}_\eta| \geq M$

$$\begin{aligned} \varepsilon > \sum_{\mathbf{m}_\eta=\boldsymbol{\ell}_\eta}^{\mathbf{n}_\eta} c_{\mathbf{m}_\eta} &= \sum_{\mathbf{k}_{1-\eta}=\mathbf{1}}^{\infty} \sum_{\mathbf{m}_\eta=\boldsymbol{\ell}_\eta}^{\mathbf{n}_\eta} a_{\mathbf{m}_\eta,\mathbf{k}_{1-\eta}} b_{\mathbf{m}_\eta,\mathbf{k}_{1-\eta}} \\ &\geq \sum_{\mathbf{k}_{1-\eta}=\mathbf{1}}^{\infty} a_{\mathbf{n}_\eta,\mathbf{k}_{1-\eta}} \sum_{\mathbf{m}_\eta=\boldsymbol{\ell}_\eta}^{\mathbf{n}_\eta} b_{\mathbf{m}_\eta,\mathbf{k}_{1-\eta}}. \end{aligned} \tag{9.20}$$

The last inequality follows from the fact that $\{a_\mathbf{k}\}$ is decreasing in each variable, and thus $a_{\mathbf{m}_\eta,\mathbf{k}_{1-\eta}} \geq a_{\mathbf{n}_\eta,\mathbf{k}_{1-\eta}}$ for every $\boldsymbol{\ell}_\eta \leq \mathbf{m}_\eta \leq \mathbf{n}_\eta$. Applying (7.1) to the $|\eta|$-dimensional sum $\sum_{\mathbf{m}_\eta=\boldsymbol{\ell}_\eta}^{\mathbf{n}_\eta} b_{\mathbf{m}_\eta,\mathbf{k}_{1-\eta}}$, we get

$$\varepsilon > \sum_{\mathbf{k}_{1-\eta}=\mathbf{1}}^{\infty} a_{\mathbf{n}_\eta,\mathbf{k}_{1-\eta}} \left(B_{\mathbf{k}_{1-\eta}}^{\mathbf{n}_\eta} - \sum_{\substack{0\leq\theta\leq\eta \\ \theta\neq\eta}} \sum_{\mathbf{m}_\theta=\boldsymbol{\ell}_\theta}^{\mathbf{n}_\theta} \sum_{\mathbf{m}_{\eta-\theta}=\mathbf{1}}^{(\boldsymbol{\ell}-\mathbf{1})_{\eta-\theta}} b_{\mathbf{m}_\eta,\mathbf{k}_{1-\eta}} \right)$$

(recall that the sum $\sum_{\mathbf{m}_{\eta-\theta}=\mathbf{1}}^{(\boldsymbol{\ell}-\mathbf{1})_{\eta-\theta}} b_{\mathbf{m}_\eta,\mathbf{k}_{1-\eta}}$ is taken to be equal to zero if there is $1 \leq i \leq d$ such that $(\eta-\theta)_i = 1$ and $\boldsymbol{\ell}_i = 1$). We note that the last inequality is valid for arbitrary $\mathbf{n}_\eta$ and $\boldsymbol{\ell}_\eta$, as long as $\mathbf{n}_\eta \geq \boldsymbol{\ell}_\eta$ and $|\boldsymbol{\ell}_\eta| \geq M$. We can rewrite such an inequality as

$$\sum_{\mathbf{k}_{1-\eta}=\mathbf{1}}^{\infty} a_{\mathbf{n}_\eta,\mathbf{k}_{1-\eta}} B_{\mathbf{k}_{1-\eta}}^{\mathbf{n}_\eta} < \varepsilon + \sum_{\substack{0\leq\theta\leq\eta \\ \theta\neq\eta}} \sum_{\mathbf{m}_\theta=\boldsymbol{\ell}_\theta}^{\mathbf{n}_\theta} \sum_{\mathbf{m}_{\eta-\theta}=\mathbf{1}}^{(\boldsymbol{\ell}-\mathbf{1})_{\eta-\theta}} \sum_{\mathbf{k}_{1-\eta}=\mathbf{1}}^{\infty} a_{\mathbf{n}_\eta,\mathbf{k}_{1-\eta}} b_{\mathbf{m}_\eta,\mathbf{k}_{1-\eta}}.$$

We now proceed to obtaining an upper bound for the right-hand side proportional to ε. To do this, we distinguish two subcases for the sum in θ: those where $\theta_j = 0$ and those where $\theta_j = 1$. Moreover, we fix $\ell_j = M$. Since $a_\mathbf{k}$ is decreasing in each variable, we have, for $\mathbf{n}_\eta \geq \boldsymbol{\ell}_\eta$,

$$\sum_{\substack{0\le\theta\le\eta\\ \theta\ne\eta,\,\theta_j=1}} \sum_{\mathbf{m}_\theta=\boldsymbol{\ell}_\theta}^{\mathbf{n}_\theta} \sum_{\mathbf{m}_{\eta-\theta}=\mathbf{1}}^{(\boldsymbol{\ell}-\mathbf{1})_{\eta-\theta}} \sum_{\mathbf{k}_{\mathbf{1}-\eta}=\mathbf{1}}^{\infty} a_{\mathbf{n}_\eta,\mathbf{k}_{\mathbf{1}-\eta}} b_{\mathbf{m}_\eta,\mathbf{k}_{\mathbf{1}-\eta}}$$

$$\le \sum_{\substack{0\le\theta\le\eta\\ \theta\ne\eta,\,\theta_j=1}} \sum_{\mathbf{m}_\theta=\boldsymbol{\ell}_\theta}^{\mathbf{n}_\theta} \sum_{\mathbf{m}_{\eta-\theta}=\mathbf{1}}^{(\boldsymbol{\ell}-\mathbf{1})_{\eta-\theta}} \sum_{\mathbf{k}_{\mathbf{1}-\eta}=\mathbf{1}}^{\infty} a_{\mathbf{m}_\eta,\mathbf{k}_{\mathbf{1}-\eta}} b_{\mathbf{m}_\eta,\mathbf{k}_{\mathbf{1}-\eta}}$$

$$\le \sum_{\substack{0\le\theta\le\eta\\ \theta\ne\eta,\,\theta_j=1}} \sum_{m_j=M}^{n_j} \sum_{\mathbf{k}_{\mathbf{1}-\mathbf{e}_j}=\mathbf{1}}^{\infty} a_{\mathbf{m}_{\mathbf{e}_j},\mathbf{k}_{\mathbf{1}-\mathbf{e}_j}} b_{\mathbf{m}_{\mathbf{e}_j},\mathbf{k}_{\mathbf{1}-\mathbf{e}_j}} < 2^{|\eta|-1}\varepsilon,$$

by the first inequality in (9.20). Let $\theta_j = 0$. Since $a_{\mathbf{k}} \to 0$ as $|\mathbf{k}| \to \infty$ and is decreasing in each variable, there exists $N \in \mathbb{N}$ such that if $n_j > N$ then

$$a_{\mathbf{n}_\eta,\mathbf{k}_{\mathbf{1}-\eta}} \le \varepsilon a_{(\boldsymbol{\ell}-\mathbf{1})_{\mathbf{e}_j},\mathbf{n}_{\eta-\mathbf{e}_j},\mathbf{k}_{\mathbf{1}-\eta}}.$$

Hence, if $\mathbf{m}_{\eta-\mathbf{e}_j} \le \mathbf{n}_{\eta-\mathbf{e}_j}$ and $m_j \le \ell_j - 1$,

$$a_{\mathbf{n}_\eta,\mathbf{k}_{\mathbf{1}-\eta}} \le \varepsilon a_{(\boldsymbol{\ell}-\mathbf{1})_{\mathbf{e}_j},\mathbf{n}_{\eta-\mathbf{e}_j},\mathbf{k}_{\mathbf{1}-\eta}} \le \varepsilon a_{\mathbf{m}_\eta,\mathbf{k}_{\mathbf{1}-\eta}}.$$

Thus,

$$\sum_{\substack{0\le\theta\le\eta\\ \theta\ne\eta,\,\theta_j=0}} \sum_{\mathbf{m}_\theta=\boldsymbol{\ell}_\theta}^{\mathbf{n}_\theta} \sum_{\mathbf{m}_{\eta-\theta}=\mathbf{1}}^{(\boldsymbol{\ell}-\mathbf{1})_{\eta-\theta}} \sum_{\mathbf{k}_{\mathbf{1}-\eta}=\mathbf{1}}^{\infty} a_{\mathbf{n}_\eta,\mathbf{k}_{\mathbf{1}-\eta}} b_{\mathbf{m}_\eta,\mathbf{k}_{\mathbf{1}-\eta}}$$

$$\le \varepsilon \sum_{\substack{0\le\theta\le\eta\\ \theta\ne\eta,\,\theta_j=0}} \sum_{\mathbf{m}_\theta=\boldsymbol{\ell}_\theta}^{\mathbf{n}_\theta} \sum_{\mathbf{m}_{\eta-\theta}=\mathbf{1}}^{(\boldsymbol{\ell}-\mathbf{1})_{\eta-\theta}} \sum_{\mathbf{k}_{\mathbf{1}-\eta}=\mathbf{1}}^{\infty} a_{\mathbf{m}_\eta,\mathbf{k}_{\mathbf{1}-\eta}} b_{\mathbf{m}_\eta,\mathbf{k}_{\mathbf{1}-\eta}} \le \varepsilon \sum_{\substack{0\le\theta\le\eta\\ \theta\ne\eta,\,\theta_j=0}} \sum_{\mathbf{k}=\mathbf{1}}^{\infty} a_{\mathbf{k}} b_{\mathbf{k}}$$

$$= C2^{|\eta|-1}\varepsilon,$$

where $C = \sum a_{\mathbf{k}} b_{\mathbf{k}} < \infty$, since the series converges. In conclusion, if $n_j \ge \max\{M, N\}$, then

$$\sum_{\mathbf{k}_{\mathbf{1}-\eta}=\mathbf{1}}^{\infty} a_{\mathbf{n}_\eta,\mathbf{k}_{\mathbf{1}-\eta}} B_{\mathbf{k}_{\mathbf{1}-\eta}}^{\mathbf{n}_\eta} < (C2^{|\eta|-1} + 2^{|\eta|-1} + 1)\varepsilon,$$

as desired. □

In Abel–Oliver's test, if one wishes to derive faster decay of the sequence $\{a_{\mathbf{k}}\}$ (cf. condition (9.11)), it is necessary to impose stronger monotonicity assumptions on such a sequence. The optimality of condition (9.11) in Abel–Olivier's test is shown in detail in Subsection 9.2.2 below. In particular, we will show in Remark 9.45 that for any $d \ge 2$ we can find a monotone d-tuple sequence $\{a_{\mathbf{n}}\}$ such that the series $\sum_{\mathbf{k}=\mathbf{1}}^{\infty} a_{\mathbf{k}}$ converges while $\|\mathbf{n}\|^d a_{\mathbf{n}} \not\to 0$.

We are interested in finding appropriate assumptions on $\{a_{\mathbf{k}}\}$ so that in condition (9.11) the weight $n_1 \cdots n_d$ can be replaced by the (stronger) radial weight $\|\mathbf{n}\|^d$.

To this end, it is more convenient to deal with functions rather than sequences. Before introducing the corresponding class we recall the definition of a weak monotone function (see Definition 9.10). A nonnegative function f defined on $\mathbb{R}^d_+$ is *WM* if for any $\mathbf{x} \in \mathbb{R}^d_+$,

$$f(\mathbf{t}) \le Cf(\mathbf{x}), \quad \text{with} \quad \mathbf{x} \le \mathbf{t} \le 2\mathbf{x},$$

where $C > 0$ is a constant independent of $\mathbf{x}$. We now introduce a related class.

Definition 9.38 Let $f \colon \mathbb{R}^d_+ \to \mathbb{R}_+$ $(d > 1)$. We say that f is *radially weak monotone*, written $f \in WM_{\mathrm{Rad}}$, if for any $\mathbf{x} \in \mathbb{R}^d_+$,

$$f(\mathbf{t}) \le Cf(\mathbf{x}) \quad \text{if } \|\mathbf{x}\| \le \|\mathbf{t}\| \le 2\|\mathbf{x}\|, \tag{9.21}$$

where $C > 0$ is a constant independent of $\mathbf{x}$.

Remark 9.39 $WM_{\mathrm{Rad}} \subsetneq WM$. The inclusion is clear. Its properness may be shown by means of a simple counterexample as the one given in two dimensions in Remark 6.33.

Proposition 9.40 *We have $f \in WM_{\mathrm{Rad}}$ if and only if there exists a radial function $g(\mathbf{x}) = g_0(\|\mathbf{x}\|)$, $g_0 \in WM$, such that, for some $0 < C_1 < C_2$,*

$$C_1 f(\mathbf{x}) \le g(\mathbf{x}) \le C_2 f(\mathbf{x}) \tag{9.22}$$

for all $\mathbf{x}$.

Proof First of all, by definition, $g_0 \in WM$ if and only if $g \in WM_{\mathrm{Rad}}$. Assume $f \in WM_{\mathrm{Rad}}$. From (9.21) it follows that if $\|\mathbf{x}\| = \|\mathbf{y}\|$ then

$$f(\mathbf{x}) \le Cf(\mathbf{y}) \le C^2 f(\mathbf{x}).$$

Thus, if we put

$$g(\mathbf{x}) := g_0(\|\mathbf{x}\|) := f\left(\frac{\|\mathbf{x}\|}{\sqrt{d}}, \ldots, \frac{\|\mathbf{x}\|}{\sqrt{d}}\right),$$

then $f \in WM_{\mathrm{Rad}}$ implies $g_0 \in WM$, and therefore $g \in WM_{\mathrm{Rad}}$.

Assume f is such that (9.22) holds with $g_0 \in WM$ (again $g \in WM_{\mathrm{Rad}}$, with constant, say, $K > 0$). For any $\mathbf{x} \in \mathbb{R}^d_+$ and any $\|\mathbf{x}\| \le \|\mathbf{t}\| \le 2\|\mathbf{x}\|$, we have

$$f(\mathbf{t}) \le \frac{1}{C_1} g(\mathbf{t}) \le \frac{K}{C_1} g(\mathbf{x}) \le \frac{KC_2}{C_1} f(\mathbf{x}),$$

as desired. □

We are now in a position to obtain a radial version of Abel–Olivier's test.

Proposition 9.41 *Let $f \in WM_{\mathrm{Rad}}$ and assume that the integral*

$$\int_{(\mathbf{0},\infty)} f(\mathbf{x})\, d\mathbf{x} \tag{9.23}$$

converges. Then $\|\mathbf{x}\|^d f(\mathbf{x}) \to 0$ as $|\mathbf{x}| \to \infty$ (or as $\|\mathbf{x}\| \to \infty$, which is equivalent).

Proof By Proposition 9.40, we know that there exists a radial function $g \in WM_{\mathrm{Rad}}$ such that

$$\int_{(\mathbf{0},\boldsymbol{\infty})} g(\mathbf{x})\, d\mathbf{x}$$

converges or diverges simultaneously with (9.23). If we put $g_0(\|\mathbf{x}\|) = g(\mathbf{x})$, then $g_0(r) \in WM$, as well as $r^{d-1}g_0(r) \in WM$, as functions of one variable. Moreover, the latter integral converges or diverges simultaneously with

$$\int_0^\infty r^{d-1} g_0(r)\, dr,$$

and we deduce that $r^d g_0(r) \to 0$ as $r \to \infty$. This is due to the one-dimensional Abel–Olivier test for *WM* functions (see Theorem 2.99, which is about series but is similarly valid for functions). Therefore, we have

$$\|\mathbf{x}\|^d f(\mathbf{x}) \to 0 \quad \text{as } \|\mathbf{x}\| \to \infty,$$

or, equivalently, as $|\mathbf{x}| \to \infty$. □

To conclude this section, we state and prove a multidimensional analog of the one- and two-dimensional Theorems 2.110 and 6.36.

Theorem 9.42 *Let $\{a_{\mathbf{k}}\}$ be a d-tuple WMS and assume the series $\sum_{\mathbf{k}=\mathbf{1}}^{\infty} a_{\mathbf{k}}$ diverges. Then the series*

$$\sum_{\mathbf{k}=\mathbf{1}}^{\infty} \min\left\{a_{\mathbf{k}}, \frac{1}{k_1 \cdots k_d}\right\} \tag{9.24}$$

also diverges.

Proof For a given sequence $\{a_{\mathbf{k}}\}$, there are two possibilities. First, if there exists $n \in \mathbb{N}$ such that for every $\mathbf{k} \in \mathbb{N}^d$ with $|\mathbf{k}| \geq n$ there holds

$$a_{\mathbf{k}} \leq \frac{1}{k_1 \cdots k_d}.$$

In this case the series (9.24) diverges, by the divergence of $\sum_{\mathbf{k}=\mathbf{1}}^{\infty} a_{\mathbf{k}}$.

Secondly, if for every $n \in \mathbb{N}$ we can find $\mathbf{k} \in \mathbb{N}^d$ with $|\mathbf{k}| \geq n$ and such that

$$a_{\mathbf{k}} \geq \frac{1}{k_1 \cdots k_d},$$

then

$$k_1 \cdots k_d \min\left\{a_{\mathbf{k}}, \frac{1}{k_1 \cdots k_d}\right\} \not\to 0 \qquad \text{as } |\mathbf{k}| \to \infty.$$

Now the divergence of (9.24) follows by Abel–Olivier's test, since the d-tuple sequence $\min\{a_{\mathbf{k}}, 1/(k_1 \cdots k_d)\}$ is *WMS*, by (v) of Proposition 9.5. □

Remark 9.43 The conclusion of Theorem 9.42 cannot be derived from the classical Abel–Olivier test for decreasing sequences as done here for *WMS*, since for decreasing $\{a_{\mathbf{k}}\}$ the sequence

$$\min\left\{a_{\mathbf{k}}, \frac{1}{k_1\cdots k_d}\right\}$$

need not be decreasing.

9.2.2 Optimality of Abel–Olivier's test

In this subsection we discuss the optimality of condition (9.11) in Theorem 9.36 (even for decreasing d-tuple sequences $\{a_{\mathbf{k}}\}$, rather than *WMS*). We consider the condition

$$(u_{n_j})a_{\mathbf{n}}B^{\mathbf{n}} \to 0 \quad \text{as} \quad \mathbf{n}\to\infty, \quad \text{or as} \quad |\mathbf{n}|\to\infty. \tag{9.25}$$

Here and from now on, we restrict ourselves to the case $b_{\mathbf{k}} = 1$ so that $B^{\mathbf{n}} = n_1\cdots n_d$.

Proposition 9.44 *Let $\{u_{\mathbf{n}}\}$ be a positive increasing d-tuple sequence and assume that there exist $\mathbf{0}\le\eta\le\mathbf{1}$, $\eta\ne\mathbf{0}$, and $\mathbf{k}^0_{\mathbf{1}-\eta}\in\mathbb{N}^{\mathbf{1}-\eta}$ such that*

$$\limsup_{\mathbf{n}_\eta\to\infty} u_{\mathbf{n}_\eta,\mathbf{k}^0_{\mathbf{1}-\eta}} = \infty. \tag{9.26}$$

Then there exists a positive convergent d-tuple series $\sum_{\mathbf{n}} a_{\mathbf{n}}$ with $\{a_{\mathbf{n}}\}$ decreasing such that for any fixed $\mathbf{k}_{\mathbf{1}-\eta}\in\mathbb{N}^{\mathbf{1}-\eta}$, $\mathbf{k}_{\mathbf{1}-\eta}\ge\mathbf{k}^0_{\mathbf{1}-\eta}$,

$$u_{\mathbf{n}_\eta,\mathbf{k}_{\mathbf{1}-\eta}}\left(\prod_{j:\eta_j=1} n_j\right)a_{\mathbf{n}_\eta,\mathbf{k}_{\mathbf{1}-\eta}} \not\to 0 \quad as\ \mathbf{n}_\eta\to\infty. \tag{9.27}$$

In particular, if

$$\limsup_{\mathbf{n}\to\infty} u_{\mathbf{n}} = \infty,$$

then there exists a positive convergent d-tuple series $\sum_{\mathbf{n}} a_{\mathbf{n}}$ with $\{a_{\mathbf{n}}\}$ decreasing and such that

$$u_{\mathbf{n}}n_1\cdots n_d a_{\mathbf{n}} \not\to 0 \quad as\ \mathbf{n}\to\infty. \tag{9.28}$$

Proof Without loss of generality we assume that $\limsup_{\mathbf{n}_\eta\to\infty} u_{\mathbf{n}_\eta,\mathbf{1}_{\mathbf{1}-\eta}} = \infty$ (in fact, it would suffice to consider $\{u_{\mathbf{n}_\eta}\}$ as an $|\eta|$-dimensional sequence, since the rest of indices in (9.26) are fixed). Then, since $\{u_{\mathbf{n}}\}$ is increasing, we can find a strictly increasing sequence of indices $\{\mathbf{m}_r\}_{r=1}^\infty$, $\mathbf{m}_r = \left(m_r^{(1)}, m_r^{(2)}, \ldots, m_r^{(d)}\right)$, such that $\mathbf{m}_1 = \mathbf{1}$,

$$\limsup_{r\to\infty} u_{(\mathbf{m}_{r+1})_\eta,\mathbf{1}_{\mathbf{1}-\eta}} = \infty, \tag{9.29}$$

and

$$\sum_{r=1}^{\infty} \frac{1}{\sqrt{u_{(\mathbf{m}_{r+1})_\eta,\mathbf{1}_{1-\eta}}}} = A < \infty. \tag{9.30}$$

For $r \in \mathbb{N}$ we define the set

$$U_r = \left\{\mathbf{n} \in \mathbb{N}^d : 1 \le n_\ell \le m_r^{(\ell)}, \text{ for } \ell \text{ such that } \eta_\ell = 1\right\}.$$

Let $a_{\mathbf{1}} = 1$ and for $\mathbf{n} \in U_{r+1} \backslash U_r$, $r \in \mathbb{N}$,

$$a_{\mathbf{n}} = \left(\prod_{\ell:\eta_\ell=1} \frac{1}{m_{r+1}^{(\ell)}}\right)\left(\prod_{\ell:\eta_\ell=0} \frac{1}{n_\ell^2}\right)\frac{1}{\sqrt{u_{(\mathbf{m}_{r+1})_\eta,\mathbf{n}_{1-\eta}}}}.$$

Then, by the definition of the sets U_r and the fact that $\{u_{\mathbf{n}}\}$ is increasing, and condition (9.30), we get

$$\begin{aligned}
\sum_{\mathbf{n}=\mathbf{1}}^{\infty} a_{\mathbf{n}} &= a_{\mathbf{1}} + \sum_{r=1}^{\infty} \sum_{\mathbf{n}\in U_{r+1}\backslash U_r} a_{\mathbf{n}} = a_{\mathbf{1}} + \sum_{\mathbf{n}_{1-\eta}=\mathbf{1}}^{\infty} \sum_{r=1}^{\infty} \left(\sum_{\mathbf{n}_\eta=\mathbf{1}}^{(\mathbf{m}_{r+1})_\eta} a_{\mathbf{n}} - \sum_{\mathbf{n}_\eta=\mathbf{1}}^{(\mathbf{m}_r)_\eta} a_{\mathbf{n}} \right) \\
&\le a_{\mathbf{1}} + \sum_{\mathbf{n}_{1-\eta}=\mathbf{1}}^{\infty} \left(\prod_{\ell:\eta_\ell=0} \frac{1}{n_\ell^2}\right) \sum_{r=1}^{\infty} \left(\prod_{\ell:\eta_\ell=1} \frac{m_{r+1}^{(\ell)}}{m_{r+1}^{(\ell)}}\right) \frac{1}{\sqrt{u_{(\mathbf{m}_{r+1})_\eta,\mathbf{n}_{1-\eta}}}} \\
&\le a_{\mathbf{1}} + \sum_{\mathbf{n}_{1-\eta}=\mathbf{1}}^{\infty} \left(\prod_{\ell:\eta_\ell=0} \frac{1}{n_\ell^2}\right) \sum_{r=1}^{\infty} \frac{1}{\sqrt{u_{(\mathbf{m}_{r+1})_\eta,\mathbf{1}_{1-\eta}}}} \\
&= a_{\mathbf{1}} + A \sum_{\mathbf{n}_{1-\eta}=\mathbf{1}}^{\infty} \left(\prod_{\ell:\eta_\ell=0} \frac{1}{n_\ell^2}\right) < \infty.
\end{aligned}$$

On the other hand, for fixed $\mathbf{k}_{1-\eta} \in \mathbb{N}^{\mathbf{1}-\eta}$ and $\mathbf{n}_\eta = (\mathbf{m}_{r+1})_\eta$, $r \in \mathbb{N}$, we have $\mathbf{n}_\eta, \mathbf{k}_{1-\eta} \in U_{r+1} \backslash U_r$, and hence

$$\begin{aligned}
&u_{\mathbf{n}_\eta,\mathbf{k}_{1-\eta}} \left(\prod_{\ell:\eta_\ell=1} n_\ell\right) a_{\mathbf{n}_\eta,\mathbf{k}_{1-\eta}} \\
&= u_{(\mathbf{m}_{r+1})_\eta,\mathbf{k}_{1-\eta}} \left(\prod_{\ell:\eta_\ell=1} m_{r+1}^{(\ell)}\right)\left(\prod_{\ell:\eta_\ell=1} \frac{1}{m_{r+1}^{(\ell)}}\right)\left(\prod_{\ell:\eta_\ell=0} \frac{1}{k_\ell^2}\right) \frac{1}{\sqrt{u_{(\mathbf{m}_{r+1})_\eta,\mathbf{k}_{1-\eta}}}} \\
&\ge \left(\prod_{\ell:\eta_\ell=0} \frac{1}{k_\ell^2}\right)\sqrt{u_{(\mathbf{m}_{r+1})_\eta,\mathbf{k}_{1-\eta}}} \ge \left(\prod_{\ell:\eta_\ell=0} \frac{1}{k_\ell^2}\right)\sqrt{u_{(\mathbf{m}_{r+1})_\eta,\mathbf{1}_{1-\eta}}},
\end{aligned}$$

and therefore

$$\begin{aligned}
&\limsup_{\mathbf{n}_\eta\to\infty} u_{\mathbf{n}_\eta,\mathbf{k}_{1-\eta}} \left(\prod_{\ell:\eta_\ell=1} n_\ell\right) a_{\mathbf{n}_\eta,\mathbf{k}_{1-\eta}} \\
&\ge \left(\prod_{\ell:\eta_\ell=0} \frac{1}{k_\ell^2}\right) \limsup_{r\to\infty} \sqrt{u_{(\mathbf{m}_{r+1})_\eta,\mathbf{1}_{1-\eta}}} = \infty,
\end{aligned}$$

by (9.29). Since $\{u_{\mathbf{n}}\}$ is increasing, (9.27) follows, which concludes the proof. Note that in particular (9.27) and (9.28) imply that

$$\limsup_{|\mathbf{n}|\to\infty} u_{\mathbf{n}} n_1 \cdots n_d a_{\mathbf{n}} > 0,$$

i.e., the decay condition (9.8) in Abel–Olivier's test is sharp. □

Remark 9.45 It easily follows from Proposition 9.44 that there is a decreasing sequence $\{a_{\mathbf{k}}\}$ such that $\sum a_{\mathbf{k}}$ converges, but $\|\mathbf{n}\|^d a_{\mathbf{n}} \not\to 0$ as $|\mathbf{n}| \to \infty$. Indeed, letting

$$u_{\mathbf{n}} = \frac{\|\mathbf{n}\|^d}{n_1 \cdots n_d},$$

one has $\limsup_{|\mathbf{n}|\to\infty} u_{\mathbf{n}} = \infty$, and so Proposition 9.44 applies. Note also that

$$n_1 \cdots n_d \le \|\mathbf{n}\|^d.$$

Hence, Abel–Olivier's test for WM_{Rad} (Proposition 9.41) provides a faster decay than that in the usual Abel–Olivier test, namely $n_1 \cdots n_d a_{\mathbf{n}} \to 0$ as $|\mathbf{n}| \to \infty$.

We now turn our attention toward sharpness of the Abel–Olivier test for WM_{Rad}.

Proposition 9.46 *Let $\varphi : \mathbb{R}^d_+ \to (0,\infty)$ be a function satisfying*

$$\limsup_{|\mathbf{x}|\to\infty} \varphi(\mathbf{x}) = \infty.$$

Then there exists a function $F \in WM_{Rad}$ such that the integral $\int_{(\mathbf{0},\boldsymbol{\infty})} F(\mathbf{x})\, d\mathbf{x}$ converges, but

$$\varphi(\mathbf{x})\|\mathbf{x}\|^d F(\mathbf{x}) \not\to 0 \quad \textit{as } |\mathbf{x}| \to \infty.$$

Proof Consider a sequence $\{\mathbf{x}_n\}_{n=1}^{\infty} \subset \mathbb{R}^d_+$ such that $\|\mathbf{x}_n\| = n$ for every n and

$$\limsup_{n\to\infty} \varphi(\mathbf{x}_n) = \infty.$$

Defining $\psi(n) = \varphi(\mathbf{x}_n)$, it follows from Proposition 2.109 that there exists a decreasing sequence $\{f(n)\}$ such that $\sum f(n)$ converges and

$$\psi(n) n f(n) \not\to 0 \quad \text{as } n \to \infty.$$

Set, for $r \in (0,\infty)$,

$$f(r) = f(n), \qquad n-1 < r \le n, \quad n \in \mathbb{N},$$

and note that $\int_0^\infty f(r)\,dr < \infty$ by the Maclaurin–Cauchy integral test. Further, let $F(\mathbf{x}) = \|\mathbf{x}\|^{1-d} f(\|\mathbf{x}\|)$. Then, $F \in WM_{\text{Rad}}$ and

$$\int_{(\mathbf{0},\boldsymbol{\infty})} F(\mathbf{x})\, d\mathbf{x} \asymp \int_0^\infty r^{d-1} \frac{f(r)}{r^{d-1}}\, dr = \int_0^\infty f(r)\, dr < \infty.$$

Moreover, we have

$$\limsup_{n\to\infty} \varphi(\mathbf{x}_n)\|\mathbf{x}_n\|^d F(\mathbf{x}_n) = \limsup_{n\to\infty} \psi(n) n f(n) \neq 0,$$

as desired. □

Given a d-tuple sequence $\{a_{\mathbf{k}}\} \in WMS$, if we additionally impose weak monotonicity assumptions on the sequence with condensed indices $\{2^{|\mathbf{k}|}a_{2^{\mathbf{k}}}\}$, we may obtain an extra logarithmic factor in condition (9.11). More precisely, assuming that $\{2^{|\mathbf{k}|}a_{2^{\mathbf{k}}}\} \in WMS_j$ for some $1 \le j \le d$ (cf. Definition 9.2), we will show that

$$(\log n_j) n_1 \cdots n_d a_{\mathbf{n}} \to 0 \quad \text{as} \quad |\mathbf{n}| \to \infty.$$

We note that we assume $\{2^{|\mathbf{k}|}a_{2^{\mathbf{k}}}\} \in WMS_j$ rather than WMS. The idea is to show how these new assumptions may allow certain choices of u_{n_k} so that (9.25) holds (with $B^{\mathbf{n}} = n_1 \cdots n_d$), but the extension to the case $\{2^{|\mathbf{k}|}a_{2^{\mathbf{k}}}\} \in WMS$, or to the case $\{2^{|\mathbf{k}|}a_{2^{\mathbf{k}}}\} \in WMS_j$ for certain values of $1 \le j \le d$, is immediate.

Lemma 9.47 *Let $\{a_{\mathbf{k}}\}$ be a nonnegative d-tuple sequence and assume that $\{a_{\mathbf{k}}\} \in WMS_j$ for some $1 \le j \le d$. If the d-tuple series $\sum_{\mathbf{k}=\mathbf{1}}^{\infty} a_{\mathbf{k}}$ converges, then*

$$n_j a_{\mathbf{n}} \to 0 \quad \text{as} \quad |\mathbf{n}| \to \infty.$$

Proof Using the WMS_j condition we have, for any $\mathbf{n} \in \mathbb{N}^d$,

$$n_j a_{\mathbf{n}} \le C \sum_{k_j=n_j}^{2n_j} a_{k_{\mathbf{e}_j}, \mathbf{n}_{1-\mathbf{e}_j}} \to 0, \qquad \text{as } |\mathbf{n}| \to \infty,$$

where the fact that the last sum vanishes as $|\mathbf{n}| \to \infty$ follows from the convergence of the d-tuple series $\sum_{\mathbf{k}=\mathbf{1}}^{\infty} a_{\mathbf{k}}$. □

Proposition 9.48 *If the d-tuple series $\sum_{\mathbf{k}=\mathbf{1}}^{\infty} a_{\mathbf{k}}$, with $\{a_{\mathbf{k}}\} \in WMS$, is convergent, and for some $1 \le j \le d$ the d-tuple sequence $\{2^{|\mathbf{k}|}a_{2^{\mathbf{k}}}\}$ is WMS_j, then*

$$(\log n_j) n_1 \cdots n_d a_{\mathbf{n}} \to 0 \quad \text{as} \quad |\mathbf{n}| \to \infty.$$

Proof By Theorem 9.55 below (the multidimensional Cauchy condensation test for WMS) the series $\sum_{\mathbf{k}=\mathbf{0}}^{\infty} 2^{|\mathbf{k}|}a_{2^{\mathbf{k}}}$ converges, so by Lemma 9.47 we have $k_j 2^{|\mathbf{k}|}a_{2^{\mathbf{k}}} \to 0$ as $|\mathbf{k}| \to \infty$, and hence

$$\log 2^{k_j} 2^{|\mathbf{k}|} a_{2^{\mathbf{k}}} \to 0 \quad \text{as} \quad |\mathbf{k}| \to \infty.$$

Given $\mathbf{n} \in \mathbb{N}^d$, let $\mathbf{r} \in \mathbb{N}^d$ be such that $2^{r_i-1} \le n_i < 2^{r_i}$ $(i = 1, \ldots, d)$. We observe that

$$\begin{aligned}(\log n_j) n_1 \cdots n_d a_{\mathbf{n}} &\le C(\log 2^{r_j}) 2^{r_1} \cdots 2^{r_d} a_{2^{\mathbf{r}-\mathbf{1}}} \\ &= C 2^d \left(\frac{r_j}{r_j - 1}\right)(\log 2^{r_j-1}) 2^{|\mathbf{r}-\mathbf{1}|} a_{2^{\mathbf{r}-\mathbf{1}}} \to 0 \quad \text{as } |\mathbf{r}| \to \infty.\end{aligned}$$

Therefore, $(\log n_j) n_1 \cdots n_d a_{\mathbf{n}} \to 0$ as $|\mathbf{n}| \to \infty$. □

9.2.3 Applications to Orlicz-type classes

Let us present some interesting consequences of Abel–Olivier's test (in fact, they are consequences of condition (9.8) alone, rather than of the extended Abel–Olivier test itself).

The following results are useful in functional analysis, in particular in the theory of Orlicz and Zygmund spaces (see Notes-89 in Chapter 2). These were presented in one and two dimensions (see Theorems 2.115 and 2.116 for the one-dimensional case, and Theorems 6.43 and 6.44 for the two-dimensional case). We now state and prove their multidimensional analogs.

Theorem 9.49 *Let $\{a_{\mathbf{n}}\}$ be a nonnegative d-tuple sequence and let $\varphi\colon [0,\infty) \to [0,\infty)$ be nondecreasing such that*

$$\varphi(x^2) \le C\varphi(x), \qquad x > 1. \tag{9.31}$$

(i) *If the series*

$$\sum_{\mathbf{n}=\mathbf{1}}^{\infty} \frac{a_{\mathbf{n}}}{(n_1 \cdots n_d)^2}\varphi(a_{\mathbf{n}}) \tag{9.32}$$

converges, then the series

$$\sum_{\mathbf{n}=\mathbf{1}}^{\infty} \frac{a_{\mathbf{n}}}{(n_1 \cdots n_d)^2}\varphi(n_1 \cdots n_d) \tag{9.33}$$

does as well.

(ii) *If there exists $C > 0$ such that*

$$a_{\mathbf{n}} \le \frac{C}{n_1 \cdots n_d} \sum_{\mathbf{k}=\lfloor \mathbf{n}/2 \rfloor}^{\mathbf{n}} a_{\mathbf{k}}$$

for all $\mathbf{n} \in \mathbb{N}^d$, then the convergence of (9.33) *implies that of* (9.32).

In particular, if $\{a_{\mathbf{n}}\} \in WMS$, then the series (9.32) *and* (9.33) *converge or diverge simultaneously.*

Proof Assume that the series $\sum_{\mathbf{n}=\mathbf{1}}^{\infty} \frac{a_{\mathbf{n}}}{(n_1 \cdots n_d)^2}\varphi(a_{\mathbf{n}})$ converges and define

$$E = \{\mathbf{n} \in \mathbb{N}^d : a_{\mathbf{n}} \ge \sqrt{n_1 \cdots n_d}\}.$$

We then have

$$\begin{aligned}\sum_{\mathbf{n}=\mathbf{1}}^{\infty} \frac{a_{\mathbf{n}}}{(n_1\cdots n_d)^2}\varphi(n_1\cdots n_d) &\le \sum_{\mathbf{n}\in E} \frac{a_{\mathbf{n}}}{(n_1\cdots n_d)^2}\varphi(a_{\mathbf{n}}^2) \\ &\quad + \sum_{\mathbf{n}\in\mathbb{N}^d\setminus E} \frac{\sqrt{n_1\cdots n_d}}{(n_1\cdots n_d)^2}\varphi(n_1\cdots n_d) \\ &\le C\sum_{\mathbf{n}\in E} \frac{a_{\mathbf{n}}}{(n_1\cdots n_d)^2}\varphi(a_{\mathbf{n}}) + \sum_{\mathbf{n}\in\mathbb{N}^d\setminus E} \frac{\varphi(n_1\cdots n_d)}{(n_1\cdots n_d)^{3/2}},\end{aligned}$$

where the last inequality follows from (9.31). At this point, in order to prove the result, it suffices to show that the series $\sum_{\mathbf{n}\in\mathbb{N}^d\setminus E} \frac{\varphi(n_1\cdots n_d)}{(n_1\cdots n_d)^{3/2}}$ converges. But this follows just from the fact that condition (9.31) implies the estimate

$$\varphi(x) \lesssim \log^{\gamma} x, \qquad x \ge 2,$$

for some $\gamma > 0$ (see the proof of Theorem 2.115 for details).

Conversely, suppose that the series $\sum_{\mathbf{n}=\mathbf{1}}^{\infty} \frac{a_{\mathbf{n}}}{(n_1\cdots n_d)^2}\varphi(n_1\cdots n_d)$ converges. Since φ is nondecreasing, the series $\sum_{\mathbf{n}=\mathbf{1}}^{\infty} \frac{a_{\mathbf{n}}}{(n_1\cdots n_d)^2}$ converges as well, and therefore

$$\frac{a_{\mathbf{n}}}{n_1\cdots n_d} \le C \sum_{\mathbf{k}=\lfloor \mathbf{n}/2\rfloor}^{\mathbf{n}} \frac{a_{\mathbf{k}}}{(k_1\cdots k_d)^2} \to 0 \qquad \text{as } |\mathbf{n}| \to \infty.$$

Note that if $\{a_{\mathbf{n}}\} \in WMS$, such a decay of $\{a_{\mathbf{n}}\}$ is a direct consequence of Abel–Olivier's test (Corollary 9.35). Therefore, there exists $m \in \mathbb{N}$ such that if $|\mathbf{n}| \ge m$ the estimate $a_{\mathbf{n}} \le n_1\cdots n_d$ holds. Again, by the fact that φ is nondecreasing, we conclude

$$\sum_{|\mathbf{n}|\ge m} \frac{a_{\mathbf{n}}}{(n_1\cdots n_d)^2}\varphi(a_{\mathbf{n}}) \le \sum_{|\mathbf{n}|\ge m} \frac{a_{\mathbf{n}}}{(n_1\cdots n_d)^2}\varphi(n_1\cdots n_d),$$

which implies the convergence of $\sum_{\mathbf{n}=\mathbf{1}}^{\infty} \frac{a_{\mathbf{n}}}{(n_1\cdots n_d)^2}\varphi(a_{\mathbf{n}})$. □

As already mentioned in the one-dimensional part, one of the functions φ satisfying the hypotheses of Theorem 9.49 is, for instance,

$$\varphi(x) = \log^{\beta}(1+x), \qquad \beta \ge 0.$$

The counterpart of Theorem 9.49 for functions reads as follows.

Theorem 9.50 *Let $f\colon (\mathbf{0},\mathbf{1}] \to [0,\infty)$ and let $\varphi\colon [0,\infty) \to [0,\infty)$ be nondecreasing such that*

$$\varphi(x^2) \le C\varphi(x), \qquad x > 1.$$

(i) *If the integral*

$$\int_{(\mathbf{0},\mathbf{1})} f(\mathbf{x})\varphi(f(\mathbf{x}))\, d\mathbf{x} \tag{9.34}$$

converges, then the integral

$$\int_{(\mathbf{0},\mathbf{1})} f(\mathbf{x})\varphi\Big(\frac{1}{x_1\cdots x_d}\Big)\,d\mathbf{x} \tag{9.35}$$

does as well.

(ii) *If there exists $C>0$ such that the estimate*

$$f(\mathbf{x}) \le \frac{C}{x_1\cdots x_d}\int_{(\mathbf{x}/2,\mathbf{x})} f(\mathbf{t})\,d\mathbf{t}$$

holds for all $\mathbf{x}\in(\mathbf{0},\mathbf{1}]$, then the convergence of (9.35) *implies that of* (9.34).

In particular, if $f\in WM$, both integrals (9.34) *and* (9.35) *converge or diverge simultaneously.*

Proof Rather than giving the full proof (which would essentially repeat that of Theorem 9.49), we outline the main ideas: in order to prove the first part, one should define the set

$$E=\Big\{\mathbf{x}\in(\mathbf{0},\mathbf{1}] : f(\mathbf{x})\ge \frac{1}{\sqrt{x_1\cdots x_d}}\Big\}$$

and then estimate from above the corresponding integrals over E and $(\mathbf{0},\mathbf{1}]\backslash E$ like in the proof of Theorem 9.49. The procedure for the second part is basically a repetition of the estimates used in the proof of Theorem 9.49. □

Remark 9.51 In the one-dimensional prototypes of Theorems 9.49 and 9.50 (Theorems 2.115 and 2.116, respectively), the assertion holds for sequences (or functions) from the class $WMS\cup\overline{GMS}$ (or $WM\cup\overline{GM}$). The reason is that in one dimension there holds

$$a_n \le \frac{C}{n}\sum_{k=\lfloor n/2\rfloor}^{n} a_k, \qquad \{a_k\}\in WMS\cup\overline{GMS},$$

$$f(x)\le \frac{C}{x}\int_{x/2}^{x} f(t)\,dt, \qquad f\in WM\cup\overline{GM}.$$

These estimates are rather straightforward for $\{a_k\}\in WMS$ ($f\in WM$) and for general monotone sequences and functions; the corresponding estimates are given in (1.17) and (1.26), respectively. In several dimensions, similar estimates hold in the weak monotone case, namely

$$a_{\mathbf{n}} \le \frac{C}{n_1\cdots n_d}\sum_{\mathbf{k}=\lfloor \mathbf{n}/2\rfloor}^{\mathbf{n}} a_{\mathbf{k}}, \qquad \{a_{\mathbf{k}}\}\in WMS,$$

$$f(\mathbf{x})\le \frac{C}{x_1\cdots x_d}\int_{(\mathbf{x}/2,\mathbf{x})} f(\mathbf{t})\,d\mathbf{t}, \qquad f\in WM.$$

However, considering general monotonicity ($\{a_{\mathbf{k}}\}\in GMS$ and $f\in GM$, which are subclasses of $\overline{GMS}$ and $\overline{GM}$, respectively; cf. Remark 9.26), these inequalities are not valid in general. For a two-dimensional counterexample, see Remark 6.45.

9.2.4 Variations of Dvoretzky's test

Here we present an extended Dvoretzky type test. More precisely, we extend Theorems 2.112 and 6.46 to any dimension and only then derive a generalization of the original Dvoretzky test.

To this end, we first define the following concept. It can be understood as a classification of a divergent multidimensional series.

Definition 9.52 Let $\sum_{\mathbf{k}} a_{\mathbf{k}}$ be a nonnegative divergent d-tuple series. The *dimension of divergence* χ of such a series is defined as

$$\chi = \min\left\{1 \le s \le d : \text{there exists } \eta \text{ with } |\eta| = s \text{ and } \mathbf{n_{1-\eta}} \in \mathbb{N}^{\mathbf{1}-\eta} \right.$$
$$\left. \text{such that } \sum_{\mathbf{k}_\eta=\mathbf{1}}^{\infty} a_{\mathbf{k}_\eta,\mathbf{n_{1-\eta}}} = \infty \right\}.$$

Note that if a divergent series $\sum_{\mathbf{k}} a_{\mathbf{k}}$ has dimension of divergence $1 < \chi \le d$, then for any $1 \le \kappa < \chi$, any η with $|\eta| = \kappa$, and any $\mathbf{n_{1-\eta}} \in \mathbb{N}^{\mathbf{1}-\eta}$, the series

$$\sum_{\mathbf{k}_\eta=\mathbf{1}}^{\infty} a_{\mathbf{k}_\eta,\mathbf{n_{1-\eta}}}$$

converges.

For instance, in three dimensions, the series

$$\sum_{n_1=1}^{\infty}\sum_{n_2=1}^{\infty}\sum_{n_3=1}^{\infty} \frac{1}{(n_1+n_2+n_3)^3} \tag{9.36}$$

has dimension of divergence 3. Indeed, fixing one of the indices (for instance, n_3), we have

$$\sum_{n_1=1}^{\infty}\sum_{n_2=1}^{\infty} \frac{1}{(n_1+n_2+n_3)^3} < \infty,$$

but the series (9.36) diverges (see Example 9.71 (i) below for details). Similarly, it is easy to see that the series

$$\sum_{n_1=1}^{\infty}\sum_{n_2=1}^{\infty}\sum_{n_3=1}^{\infty} \frac{1}{(n_1+n_2+n_3)^2}$$

has dimension of divergence 2.

More generally, in arbitrary dimension, an example of a d-tuple series with dimension of divergence $1 \le \chi \le d$ is the series with general term

$$a_{\mathbf{k}} = \frac{1}{|\mathbf{k}|^{\chi}}.$$

The case $\chi = 1$ is evident (any one-dimensional series corresponds to a harmonic series), so we assume $\chi \geq 2$. By Example 9.71 (i) below, one has that for any $1 \leq \ell < \chi$, any $\mathbf{0} \leq \eta \leq \mathbf{1}$ with $|\eta| = \ell$, and any fixed $\mathbf{n}_{\mathbf{1}-\eta} \in \mathbb{N}^{\mathbf{1}-\eta}$, the series

$$\sum_{\mathbf{k}_\eta=\mathbf{1}}^{\infty} a_{\mathbf{k}_\eta,\mathbf{n}_{\mathbf{1}-\eta}} = \sum_{\mathbf{k}_\eta=\mathbf{1}}^{\infty} \frac{1}{(|\mathbf{k}_\eta| + |\mathbf{n}_{\mathbf{1}-\eta}|)^\chi}$$

converges. On the other hand, if $\mathbf{0} \leq \zeta \leq \mathbf{1}$ is such that $|\zeta| = \chi$, for every $\mathbf{n}_{\mathbf{1}-\zeta} \in \mathbb{N}^{\mathbf{1}-\zeta}$ one has

$$\sum_{\mathbf{k}_\zeta=\mathbf{1}}^{\infty} a_{\mathbf{k}_\zeta,\mathbf{n}_{\mathbf{1}-\zeta}} = \infty.$$

Theorem 9.53 *Let $\{a_\mathbf{k}\}, \{b_\mathbf{k}\} \in WMS$ and assume that the series $\sum_\mathbf{k} a_\mathbf{k}$ and $\sum_\mathbf{k} b_\mathbf{k}$ converge and diverge, respectively, where $\sum_\mathbf{k} b_\mathbf{k}$ has dimension of divergence χ, $1 \leq \chi \leq d$. Then there exist η with $|\eta| = \chi$ and $\mathbf{n}_{\mathbf{1}-\eta} \in \mathbb{N}^{\mathbf{1}-\eta}$ such that for all increasing sequences of natural numbers $\{u_k^1\}, \dots, \{u_k^\chi\}$ satisfying $u_{k+1}^i - u_k^i = O(u_k^i - u_{k-1}^i)$, $i = 1, \dots, \chi$, there exist infinitely many χ-tuples $\mathbf{k}$ such that, for every $\mathbf{u_k} \leq \mathbf{m} \leq \mathbf{u_{k+1}}$,*

$$a_{\mathbf{m}_\eta,\mathbf{n}_{\mathbf{1}-\eta}} < b_{\mathbf{m}_\eta,\mathbf{n}_{\mathbf{1}-\eta}}.$$

Proof We can assume without loss of generality that $\chi = d$. Indeed, given η with $|\eta| = \chi$ and $\mathbf{n}_{\mathbf{1}-\eta} \in \mathbb{N}^{\mathbf{1}-\eta}$ such that $\sum_{\mathbf{k}_\eta=\mathbf{1}}^{\infty} b_{\mathbf{k}_\eta,\mathbf{n}_{\mathbf{1}-\eta}} = \infty$ (which exist since the dimension of divergence of the series $\sum_\mathbf{k} b_\mathbf{k}$ is χ), the χ-tuple sequence $c_{\mathbf{k}_\eta} = b_{\mathbf{k}_\eta,\mathbf{n}_{\mathbf{1}-\eta}}$ has maximal dimension of divergence.

We first put

$$\Sigma_\mathbf{k} = \big\{\mathbf{j} = (j_1, \dots, j_d) \in \mathbb{N}^d : \mathbf{u_k} \leq \mathbf{j} \leq \mathbf{u_{k+1}}\big\}.$$

Define also

$$\alpha_\mathbf{k} = \sum_{\mathbf{j}\in\Sigma_\mathbf{k}} a_\mathbf{j} \qquad \text{and} \qquad \beta_\mathbf{k} = \sum_{\mathbf{j}\in\Sigma_\mathbf{k}} b_\mathbf{j}.$$

Recall that the condition $u_{k_i+1}^i - u_k^i = O(u_{k_i}^i - u_{k_i-1}^i)$ implies $u_{k_i+1}^i = O(u_{k_i}^i)$. Now, if $a_\mathbf{n} \geq b_\mathbf{n}$ for some $\mathbf{n} \in \Sigma_\mathbf{k}$, $\mathbf{k} > \mathbf{1}$, the *WMS* condition together with $u_{k_i+1}^i = O(u_{k_i}^i)$ implies that $a_\mathbf{n} \leq C_1 a_{\boldsymbol{\ell}}$ for all $\boldsymbol{\ell} \in \Sigma_{\mathbf{k}-\mathbf{1}}$, and also $b_\mathbf{m} \leq C_2 b_\mathbf{n}$ for all $\mathbf{m} \in \Sigma_{\mathbf{k}+\mathbf{1}}$. By this, $b_\mathbf{m} \leq C_1 C_2 a_{\boldsymbol{\ell}}$, and therefore the corresponding inequality between averages

$$\frac{\alpha_{\mathbf{k}-\mathbf{1}}}{\prod_{i=1}^d (u_{k_i-1}^i - u_{k_i-2}^i + 1)} \geq C_1 C_2 \frac{\beta_{\mathbf{k}+\mathbf{1}}}{\prod_{i=1}^d (u_{k_i}^i - u_{k_i-1}^i + 1)} \tag{9.37}$$

holds as well.

Now, (9.37) implies, by the condition $u^i_{k_i+1} - u^i_{k_i} = O(u^i_{k_i} - u^i_{k_i-1})$, that

$$\alpha_{\mathbf{k-1}} \geq C_1 C_2 \beta_{\mathbf{k+1}} \frac{\prod_{i=1}^{d} (u^i_{k_i-1} - u^i_{k_i-2} + 1)}{\prod_{i=1}^{d} (u^i_{k_i} - u^i_{k_i-1} + 1)} \geq C_3 \beta_{\mathbf{k+1}}. \tag{9.38}$$

Note that the series

$$\sum_{\substack{\mathbf{0}\leq\eta\leq\mathbf{1}\\ \eta\neq\mathbf{0},\mathbf{1}}} \sum_{\mathbf{k}_\eta=\mathbf{1}}^{\infty} \sum_{\substack{\mathbf{1}\leq\mathbf{k}_{\mathbf{1}-\eta}\leq\mathbf{3}\\ \mathbf{k}_{\mathbf{1}-\eta}\neq\mathbf{3}}} \beta_{\mathbf{k}}$$

converges. Indeed, it is a sum of at most $(d-1)$-tuple series, and, by assumption, $\sum_{\mathbf{k}} b_{\mathbf{k}}$ has dimension of divergence d. Therefore, the series $\sum_{\mathbf{k}=\mathbf{3}}^{\infty} \beta_{\mathbf{k}}$ diverges. Finally, since the series $\sum_{\mathbf{k}=\mathbf{1}}^{\infty} \alpha_{\mathbf{k}}$ converges, there must be infinitely many d-tuples $\mathbf{k}$ for which (9.38) fails to hold, and hence, there are infinitely many d-tuples $\mathbf{k}$ for which $a_{\mathbf{j}} < b_{\mathbf{j}}$ for all $\mathbf{j} \in \Sigma_{\mathbf{k}}$. □

As shown in Remark 6.48 for the case $d = 2$, the dimension of divergence of a multiple series plays a fundamental role in Theorem 9.53. A similar example may be constructed in arbitrary dimension.

The multidimensional version of the original Dvoretzky theorem can now be easily derived.

Corollary 9.54 *If $\{a_{\mathbf{k}}\}, \{b_{\mathbf{k}}\} \in WMS$, and the series $\sum_{\mathbf{k}} a_{\mathbf{k}}$ and $\sum_{\mathbf{k}} b_{\mathbf{k}}$ converge and diverge, respectively, where $\sum_{\mathbf{k}} b_{\mathbf{k}}$ has dimension of divergence χ, $1 \leq \chi \leq d$. Then, for every fixed χ-tuple $\mathbf{M} = (M_1, \ldots, M_\chi) > \mathbf{1}$, there exist η with $|\eta| = \chi$ and $\mathbf{n}_{\mathbf{1}-\eta} \in \mathbb{N}^{\mathbf{1}-\eta}$ for which we may find infinitely many χ-tuples $\mathbf{k}$ such that, for all $\mathbf{m}$ with $\mathbf{M}^{\mathbf{k}} \leq \mathbf{m} \leq \mathbf{M}^{\mathbf{k}+\mathbf{1}}$, we have*

$$a_{\mathbf{m}_\eta, \mathbf{n}_{\mathbf{1}-\eta}} < b_{\mathbf{m}_\eta, \mathbf{n}_{\mathbf{1}-\eta}}.$$

The proof of this statement simply follows from Theorem 9.53 with $\mathbf{u}_{\mathbf{k}} = \mathbf{M}^{\mathbf{k}}$ (compare with Theorem 2.112 and Remark 2.113).

9.2.5 Cauchy's condensation test and its generalizations

After presenting generalizations of the standard Cauchy condensation test and its versions with the assumption of weak monotonicity for double series (see Subsection 6.2.5), our goal is to prove similar statements for higher dimensions.

One of the simplest and most important convergence tests of series, the Cauchy condensation test (which was stated in Theorem 8.18), reads as follows:

$$\sum_{\mathbf{k}=\mathbf{1}}^{\infty} a_{\mathbf{k}} \quad \text{converges if and only if} \quad \sum_{\mathbf{k}=\mathbf{0}}^{\infty} 2^{|\mathbf{k}|} a_{\mathbf{2^k}} \quad \text{does,}$$

provided $\{a_{\mathbf{k}}\}$ is a d-tuple nonnegative decreasing sequence. We now proceed to obtain a generalization in two different ways, namely by considering the *WMS* condition in place of monotonicity and by considering condensing sequences different from $\mathbf{2^k}$.

A natural replacement for the condensation indices $\mathbf{2^k}$ is given by sequences that simulate certain properties of lacunary sequences. In dimension one, this is done in Theorem 2.91, where it is shown that if $\{a_k\} \in WMS$, $\{u_k\} \in \Lambda$ (cf. (2.55)), and $u_{k+1} = O(u_k)$, then the series

$$\sum_{k=1}^{\infty} a_k, \quad \sum_{k=1}^{\infty} u_k |a_{u_k} - a_{u_{k+1}}|, \quad \text{and} \quad \sum_{k=1}^{\infty} u_k a_{u_k}$$

are equiconvergent. More than that, the equivalence

$$\sum_{k=1}^{\infty} u_k a_{u_k} \asymp \sum_{k=u_1}^{\infty} a_k \tag{9.39}$$

is valid.

Let us prove a multidimensional analog of this result.

Theorem 9.55 *Let $\{a_{\mathbf{k}}\}$ be a d-tuple WMS, $\{u_k^1\}, \ldots, \{u_k^d\} \in \Lambda$ and assume that $u_{k+1}^{\ell} = O(u_k^{\ell})$ for $\ell = 1, \ldots, d$. Then the series*

$$\sum_{\mathbf{j}=\mathbf{u_1}}^{\infty} a_{\mathbf{j}},$$

$$\sum_{\mathbf{k}_\eta=\mathbf{1}}^{\infty} \sum_{\mathbf{j}_{\mathbf{1}-\eta}=(\mathbf{u_1})_{\mathbf{1}-\eta}}^{\infty} \left(\prod_{i:\eta_i=1} u_{k_i}^i \right) a_{(\mathbf{u_k})_\eta, \mathbf{j}_{\mathbf{1}-\eta}}, \quad \textit{for any } \eta \neq \mathbf{0}, \tag{9.40}$$

and

$$\sum_{\mathbf{k}_\eta=\mathbf{1}}^{\infty} \sum_{\mathbf{j}_{\mathbf{1}-\eta}=(\mathbf{u_1})_{\mathbf{1}-\eta}}^{\infty} \left(\prod_{i:\eta_i=1} u_{k_i}^i \right) |\Delta^{\zeta} a_{(\mathbf{u_k})_\eta, \mathbf{j}_{\mathbf{1}-\eta}}|, \quad \textit{for any } \zeta, \eta \neq \mathbf{0},\ \zeta \leq \eta, \tag{9.41}$$

converge or diverge simultaneously.

Proof Given $\eta \neq \mathbf{0}$, the equiconvergence of $\sum_{\mathbf{j}=\mathbf{u_1}}^{\infty} a_{\mathbf{j}}$ and (9.40) follows easily by applying the equivalence (9.39) for the indices i such that $\eta_i = 1$.

Since for any $1 \leq i \leq d$ we have $u_{k+1}^i = O(u_k^i)$, the *WMS* condition yields

$$|\Delta^{\zeta} a_{(\mathbf{u_k})_\eta, \mathbf{j}_{\mathbf{1}-\eta}}| \lesssim a_{(\mathbf{u_k})_\eta, \mathbf{j}_{\mathbf{1}-\eta}}, \quad \zeta, \eta \neq \mathbf{0},\ \zeta \leq \eta. \tag{9.42}$$

Indeed,

$$|\Delta^{\mathbf{e}_i} a_{(\mathbf{u_k})_\eta, \mathbf{j}_{\mathbf{1}-\eta}}| \leq a_{(\mathbf{u_k})_\eta, \mathbf{j}_{\mathbf{1}-\eta}} + a_{(\mathbf{u_{k+e_i}})_\eta, \mathbf{j}_{\mathbf{1}-\eta}} \lesssim a_{(\mathbf{u_k})_\eta, \mathbf{j}_{\mathbf{1}-\eta}}$$

for any $\zeta \leq \eta$ and any $1 \leq i \leq d$ such that $\zeta_i = 1$. Hence, (9.42) follows by applying this estimate iteratively in the indices i such that $\zeta_i = 1$. Therefore, for any such choice of ζ and η, we have

$$\sum_{\mathbf{k}_\eta=\mathbf{1}}^{\infty} \sum_{\mathbf{j}_{\mathbf{1}-\eta}=(\mathbf{u}_1)_{\mathbf{1}-\eta}}^{\infty} \left(\prod_{i:\eta_i=1} u^i_{k_i}\right) |\Delta^\zeta a_{(\mathbf{u}_\mathbf{k})_\eta, \mathbf{j}_{\mathbf{1}-\eta}}|$$
$$\lesssim \sum_{\mathbf{k}_\eta=\mathbf{1}}^{\infty} \sum_{\mathbf{j}_{\mathbf{1}-\eta}=(\mathbf{u}_1)_{\mathbf{1}-\eta}}^{\infty} \left(\prod_{i:\eta_i=1} u^i_{k_i}\right) a_{(\mathbf{u}_\mathbf{k})_\eta, \mathbf{j}_{\mathbf{1}-\eta}}.$$

Thus, the convergence of (9.40) for a fixed η implies the convergence of (9.41) for any $\zeta \le \eta$ with $\zeta \ne \mathbf{0}$.

For the reverse estimate, recall that $\Delta^{\mathbf{0}} a_{(\mathbf{u}_\mathbf{k})_\eta, \mathbf{j}_{\mathbf{1}-\eta}} = a_{(\mathbf{u}_\mathbf{k})_\eta, \mathbf{j}_{\mathbf{1}-\eta}}$. Let $\eta \ne \mathbf{0}$ and $\zeta \le \eta$ satisfy $0 \le |\zeta| < |\eta|$. If we prove that, for any ℓ such that $\eta_\ell = 1$ and $\zeta_\ell = 0$, we have

$$\sum_{\mathbf{k}_\eta=\mathbf{1}}^{\infty} \sum_{\mathbf{j}_{\mathbf{1}-\eta}=(\mathbf{u}_1)_{\mathbf{1}-\eta}}^{\infty} \left(\prod_{i:\eta_i=1} u^i_{k_i}\right) |\Delta^\zeta a_{(\mathbf{u}_\mathbf{k})_\eta, \mathbf{j}_{\mathbf{1}-\eta}}|$$
$$\lesssim \sum_{\mathbf{k}_\eta=\mathbf{1}}^{\infty} \sum_{\mathbf{j}_{\mathbf{1}-\eta}=(\mathbf{u}_1)_{\mathbf{1}-\eta}}^{\infty} \left(\prod_{i:\eta_i=1} u^i_{k_i}\right) |\Delta^{\zeta+\mathbf{e}_\ell} a_{(\mathbf{u}_\mathbf{k})_\eta, \mathbf{j}_{\mathbf{1}-\eta}}|,$$

then we are done. To prove this fact, we just apply Abel's transformation to the ℓth one-dimensional partial sums and obtain, for any $N_\ell \in \mathbb{N}$,

$$\begin{aligned}
\sum_{k_\ell=1}^{N_\ell} u^\ell_{k_\ell} |\Delta^\zeta a_{(\mathbf{u}_\mathbf{k})_\eta, \mathbf{j}_{\mathbf{1}-\eta}}| &= \sum_{k_\ell=1}^{N_\ell-1} \left(\sum_{s=1}^{k_\ell} u^\ell_s\right) \Delta^{\mathbf{e}_\ell} |\Delta^\zeta a_{(\mathbf{u}_\mathbf{k})_\eta, \mathbf{j}_{\mathbf{1}-\eta}}| \\
&\quad + \left(\sum_{s=1}^{N_\ell} u^\ell_s\right) \sum_{k_\ell=N_\ell}^{\infty} \Delta^{\mathbf{e}_\ell} |\Delta^\zeta a_{(\mathbf{u}_\mathbf{k})_\eta, \mathbf{j}_{\mathbf{1}-\eta}}| \\
&\lesssim \sum_{k_\ell=1}^{N_\ell-1} u^\ell_{k_\ell} |\Delta^{\zeta+\mathbf{e}_\ell} a_{(\mathbf{u}_\mathbf{k})_\eta, \mathbf{j}_{\mathbf{1}-\eta}}| \\
&\quad + u^\ell_{N_\ell} \sum_{k_\ell=N_\ell}^{\infty} |\Delta^{\zeta+\mathbf{e}_\ell} a_{(\mathbf{u}_\mathbf{k})_\eta, \mathbf{j}_{\mathbf{1}-\eta}}| \\
&\lesssim \sum_{k_\ell=1}^{\infty} u^\ell_{k_\ell} |\Delta^{\zeta+\mathbf{e}_\ell} a_{(\mathbf{u}_\mathbf{k})_\eta, \mathbf{j}_{\mathbf{1}-\eta}}|,
\end{aligned}$$

which yields the desired estimate. □

In particular, the multidimensional version of the classical Cauchy condensation test for *WMS* follows from Theorem 9.55 with $u^j_k = 2^k$ for $j = 1, \dots, d$ and $\eta = \mathbf{1}$.

Corollary 9.56 *Let $\{a_\mathbf{k}\}$ be a d-tuple WMS. Then the assertion of Theorem 9.55 holds true if one takes $u^i_k = 2^k$, $i = 1, \dots, d$. In particular,*

$$\sum_{\mathbf{k}=\mathbf{1}}^{\infty} a_\mathbf{k}, \quad \sum_{\mathbf{k}=\mathbf{0}}^{\infty} 2^{|\mathbf{k}|} a_{2^\mathbf{k}}, \quad \text{and} \quad \sum_{\mathbf{k}=\mathbf{0}}^{\infty} 2^{|\mathbf{k}|} |\Delta^{\mathbf{1}} a_{2^\mathbf{k}}|$$

converge or diverge simultaneously.

9.2.6 $\ell^p - \ell^q$ inequalities and Stechkin's lemma

The following result is an improvement of the convergence test given by the classical $\ell^q \subset \ell^p$ inequality:

$$\left(\sum_{\mathbf{k}=\mathbf{1}}^{\infty} a_{\mathbf{k}}^p\right)^{\frac{1}{p}} \le \left(\sum_{\mathbf{k}=\mathbf{1}}^{\infty} a_{\mathbf{k}}^q\right)^{\frac{1}{q}}, \qquad 0 < q < p \le \infty. \tag{9.43}$$

In particular, it implies that if the series $\sum a_{\mathbf{k}}^q$ converges, then so does the series $\sum a_{\mathbf{k}}^p$. This last statement can be improved for $\{a_{\mathbf{k}}\} \in WMS$ as follows.

Theorem 9.57 *Let $\{a_{\mathbf{k}}\}$ be a d-tuple WMS and $0 < q < p < \infty$. Then the inequality*

$$\left(\sum_{\mathbf{k}=\mathbf{1}}^{\infty} a_{\mathbf{k}}^p\right)^{\frac{1}{p}} \le C^2 3^{\frac{d}{q}} \left(\sum_{\mathbf{k}=\mathbf{1}}^{\infty} a_{\mathbf{k}}^q (k_1 \cdots k_d)^{\frac{q}{p}-1}\right)^{\frac{1}{q}}$$

holds, where C is the WMS constant of $\{a_{\mathbf{k}}\}$. In particular, the convergence of the series

$$\sum_{\mathbf{k}=\mathbf{1}}^{\infty} a_{\mathbf{k}}^q (k_1 \cdots k_d)^{\frac{q}{p}-1}$$

implies that of

$$\sum_{\mathbf{k}=\mathbf{1}}^{\infty} a_{\mathbf{k}}^p.$$

For $p = \infty$ and $0 < q < \infty$, we have

$$\sup_{\mathbf{k}} a_{\mathbf{k}} \le 2^{\frac{d}{q}} C \left(\sum_{\mathbf{k}=\mathbf{1}}^{\infty} \frac{a_{\mathbf{k}}^q}{k_1 \cdots k_d}\right)^{\frac{1}{q}}.$$

Proof of Theorem 9.57 The *WMS* condition implies the estimate

$$\sum_{\mathbf{k}=\mathbf{1}}^{\infty} a_{\mathbf{k}}^p = \sum_{\mathbf{k}=\mathbf{0}}^{\infty} \sum_{\mathbf{n}=\mathbf{2^k}}^{\mathbf{2^{k+1}-1}} a_{\mathbf{n}}^p \le C^p \sum_{\mathbf{k}=\mathbf{0}}^{\infty} 2^{|\mathbf{k}|} a_{\mathbf{2^k}}^p.$$

Applying the inequality (9.43) (with 1 in place of p and $\frac{q}{p}$ in place of q) and the decomposition (7.1) (with $\boldsymbol{\ell} = \mathbf{0}$, $\mathbf{m} = \mathbf{1}$, and $\mathbf{n} = \infty$), we obtain

$$\left(\sum_{\mathbf{k}=\mathbf{0}}^{\infty} 2^{|\mathbf{k}|} a_{\mathbf{2^k}}^p\right)^{\frac{1}{p}} \le \left(\sum_{\mathbf{k}=\mathbf{0}}^{\infty} 2^{|\mathbf{k}|\frac{q}{p}} a_{\mathbf{2^k}}^q\right)^{\frac{1}{q}} = \left(\sum_{0 \le \theta \le 1} \sum_{\mathbf{k}_\theta = \mathbf{0}_\theta}^{\mathbf{0}_\theta} \sum_{\mathbf{k}_{1-\theta}=\mathbf{1}}^{\infty} 2^{|\mathbf{k}|\frac{q}{p}} a_{\mathbf{2^k}}^q\right)^{\frac{1}{q}}$$

$$= \left(\sum_{0 \le \theta \le 1} \sum_{\mathbf{k}_{1-\theta}=\mathbf{1}}^{\infty} 2^{|\mathbf{k}|\frac{q}{p}} a_{\mathbf{1}_\theta, (\mathbf{2^k})_{1-\theta}}^q\right)^{\frac{1}{q}}$$

$$= \left(\sum_{0 \le \theta \le 1} 2^{|\theta|} \sum_{\mathbf{k}_{1-\theta}=\mathbf{1}}^{\infty} 2^{|\mathbf{k}-\theta|} a_{\mathbf{1}_\theta, (\mathbf{2^k})_{1-\theta}}^q 2^{|\mathbf{k}|(\frac{q}{p}-1)}\right)^{\frac{1}{q}}$$

$$\le C\Big(\sum_{\mathbf{0}\le\theta\le\mathbf{1}} 2^{|\theta|}\sum_{\mathbf{k}=\mathbf{1}}^{\infty} a_{\mathbf{k}}^q (k_1\cdots k_d)^{\frac{q}{p}-1}\Big)^{\frac{1}{q}}$$

$$= C\Big(\sum_{\mathbf{0}\le\theta\le\mathbf{1}} 2^{|\theta|}\Big)^{\frac{1}{q}}\Big(\sum_{\mathbf{k}=\mathbf{1}}^{\infty} a_{\mathbf{k}}^q (k_1\cdots k_d)^{\frac{q}{p}-1}\Big)^{\frac{1}{q}}.$$

We note that

$$\sum_{\mathbf{0}\le\theta\le\mathbf{1}} 2^{|\theta|} = \sum_{k=0}^{d}\binom{d}{k}2^k = (2+1)^d = 3^d,$$

where the second equality follows from the binomial relation

$$(a+b)^n = \sum_{k=0}^{n}\binom{n}{k}a^k b^{n-k}, \qquad n\in\mathbb{N}, \qquad a,b\in\mathbb{C}.$$

Finally, combining all the above estimates together yields the desired inequality.

For the case $p=\infty$ and $0<q<\infty$, we observe that for any $\mathbf{0}\le\eta\le\mathbf{1}$ and any $\mathbf{n}_\eta\in\mathbb{N}^\eta$, the *WMS* condition implies

$$\frac{1}{2^{|\eta|}}\Big(\prod_{j:\eta_j=1} n_j\Big)a^q_{\mathbf{n}_\eta,\mathbf{1}_{1-\eta}} \le C^q \sum_{\mathbf{k}_\eta=\lfloor\mathbf{n}_\eta/2\rfloor}^{\mathbf{n}_\eta} a^q_{\mathbf{k}_\eta,\mathbf{1}_{1-\eta}}.$$

Hence,

$$a_{\mathbf{n}_\eta,\mathbf{1}_{1-\eta}} \le \Big(2^{|\eta|}C^q \sum_{\mathbf{k}_\eta=\lfloor\mathbf{n}_\eta/2\rfloor}^{\mathbf{n}_\eta} a^q_{\mathbf{k}_\eta,\mathbf{1}_{1-\eta}}\Big(\prod_{j:\eta_j=1} k_j\Big)^{-1}\Big)^{\frac{1}{q}}$$

$$\le 2^{\frac{d}{q}}C\Big(\sum_{\mathbf{k}=\mathbf{1}}^{\infty}\frac{a^q_{\mathbf{k}}}{k_1\cdots k_d}\Big)^{\frac{1}{q}}.$$

Since this estimate is valid for arbitrary η and $\mathbf{n}_\eta$ and the last term does not depend on such choices, the proof is complete. □

The counterpart of Theorem 9.57 for functions also holds.

Theorem 9.58 *Let $f\in WM$ and $0<q<p<\infty$. Then the inequality*

$$\Big(\int_{(\mathbf{0},\infty)} f(\mathbf{x})^p\,d\mathbf{x}\Big)^{\frac{1}{p}} \le 2^{\frac{d}{q}}C^2\Big(\int_{(\mathbf{0},\infty)} f(\mathbf{x})^q (x_1\cdots x_d)^{\frac{q}{p}-1}\,d\mathbf{x}\Big)^{\frac{1}{q}}$$

holds, where C is the WM constant of f. In particular, the convergence of the integral

$$\int_{(\mathbf{0},\infty)} f(\mathbf{x})^q (x_1\cdots x_d)^{\frac{q}{p}-1}\,d\mathbf{x}$$

implies that of

$$\int_{(\mathbf{0},\infty)} f(\mathbf{x})^p\,d\mathbf{x}.$$

For $p = \infty$ and $0 < q < \infty$, we have

$$\sup_{\mathbf{x}\in(\mathbf{0},\infty)} f(\mathbf{x}) \le C2^{\frac{d}{q}}\left(\int_{(\mathbf{0},\infty)} \frac{f(\mathbf{x})^q}{x_1\cdots x_d}\,d\mathbf{x}\right)^{\frac{1}{q}}.$$

In particular, the convergence of the last integral implies the boundedness of f.

Proof The *WM* condition implies that

$$\int_{(\mathbf{0},\infty)} f(\mathbf{x})^p\,d\mathbf{x} \le C^p \sum_{\mathbf{k}=-\infty}^{\infty} 2^{|\mathbf{k}|} f(\mathbf{2}^{\mathbf{k}})^p.$$

Applying the inequality (9.43) (with 1 in place of p and $\frac{q}{p}$ in place of q) and the *WM* inequality, we get

$$\begin{aligned}\left(\int_{(\mathbf{0},\infty)} f(\mathbf{x})^p\,d\mathbf{x}\right)^{\frac{1}{p}} &\le C\left(\sum_{\mathbf{k}=-\infty}^{\infty} 2^{|\mathbf{k}|\frac{q}{p}} f(\mathbf{2}^{\mathbf{k}})^q\right)^{\frac{1}{q}}\\ &= C\left(\sum_{\mathbf{k}=-\infty}^{\infty} 2^{-|\mathbf{k}|+d} 2^{|\mathbf{k}|\frac{q}{p}} \int_{[\mathbf{2}^{\mathbf{k}-1},\mathbf{2}^{\mathbf{k}}]} f(\mathbf{2}^{\mathbf{k}})^q\,d\mathbf{x}\right)^{\frac{1}{q}}\\ &\le 2^{\frac{d}{q}} C^2\left(\sum_{\mathbf{k}=-\infty}^{\infty} \int_{[\mathbf{2}^{\mathbf{k}-1},\mathbf{2}^{\mathbf{k}}]} f(\mathbf{x})^q (x_1\cdots x_d)^{\frac{q}{p}-1}\,d\mathbf{x}\right)^{\frac{1}{q}}\\ &= 2^{\frac{d}{q}} C^2\left(\int_{(\mathbf{0},\infty)} f(\mathbf{x})^q (x_1\cdots x_d)^{\frac{q}{p}-1}\,d\mathbf{x}\right)^{\frac{1}{q}},\end{aligned}$$

as desired.

Let us now turn to the case $p = \infty$ and $0 < q < \infty$. For any $\mathbf{t} \in (\mathbf{0},\infty)$, the *WM* condition implies that

$$\frac{t_1\cdots t_d}{2^d} f(\mathbf{t})^q \le C^q \int_{[\mathbf{t}/2,\mathbf{t}]} f(\mathbf{x})^q\,d\mathbf{x}.$$

Therefore,

$$f(\mathbf{t}) \le 2^{\frac{d}{q}} C\left(\int_{[\mathbf{t}/2,\mathbf{t}]} \frac{f(\mathbf{x})^q}{x_1\cdots x_d}\,d\mathbf{x}\right)^{\frac{1}{q}} \le 2^{\frac{d}{q}} C\left(\int_{(\mathbf{0},\infty)} \frac{f(\mathbf{x})^q}{x_1\cdots x_d}\,d\mathbf{x}\right)^{\frac{1}{q}}.$$

Since this estimate is valid for arbitrary $\mathbf{t} \in (\mathbf{0},\infty)$ and the last term does not depend on $\mathbf{t}$, the claim follows. □

We also have a version of Theorem 9.58 for WM_{Rad} functions.

Theorem 9.59 *Let $f \in WM_{\mathrm{Rad}}$ and $0 < q < p < \infty$. Then there exists a constant $C > 0$ (depending on the constant in condition* (9.21)*) such that*

$$\left(\int_{(\mathbf{0},\infty)} f(\mathbf{x})^p\,d\mathbf{x}\right)^{\frac{1}{p}} \le C\left(\int_{(\mathbf{0},\infty)} f(\mathbf{x})^q \|\mathbf{x}\|^{d(\frac{q}{p}-1)}\,d\mathbf{x}\right)^{\frac{1}{q}}.$$

In particular, the convergence of the integral

$$\int_{(\mathbf{0},\infty)} f(\mathbf{x})^q \|\mathbf{x}\|^{d(\frac{q}{p}-1)}\, d\mathbf{x}$$

implies that of

$$\int_{(\mathbf{0},\infty)} f(\mathbf{x})^p\, d\mathbf{x}.$$

For $p = \infty$ *and* $0 < q < \infty$, *we have*

$$\sup_{\mathbf{x}\in(\mathbf{0},\infty)} f(\mathbf{x}) \le C\left(\int_{(\mathbf{0},\infty)} \frac{f(\mathbf{x})^q}{\|\mathbf{x}\|^d}\, d\mathbf{x} \right)^{\frac{1}{q}},$$

for some constant C (depending on the constant in condition (9.21)*). In particular, the convergence of the last integral implies the boundedness of* f.

Proof By Proposition 9.40, there exists a function of one variable $g_0 \in WM$ such that $f(\mathbf{x}) \asymp g_0(\|\mathbf{x}\|)$ for all $\mathbf{x} \in \mathbb{R}_+^d$. Passing to spherical coordinates, we get

$$\int_{(\mathbf{0},\infty)} f(\mathbf{x})^p\, d\mathbf{x} \asymp \int_0^\infty r^{d-1} g_0(r)^p\, dr,$$

and the conclusion follows by applying the one-dimensional Theorem 2.118 to the last integral and then passing back to Euclidean coordinates.

For the case $p = \infty$ and $0 < q < \infty$, the conclusion similarly follows from the fact that

$$\sup_{\mathbf{x}\in(\mathbf{0},\infty)} f(\mathbf{x}) \asymp \sup_{r\in(0,\infty)} g_0(r)$$

and the one-dimensional Theorem 2.118. Passing back to Euclidean coordinates completes the proof. □

Finally, we present the multidimensional analog of Stechkin's lemma for $QMS(\alpha)$ (cf. Definition 9.8). The one- and two-dimensional statements are given by Theorems 2.119 and 6.54, respectively. Stechkin's lemma has important applications in approximation theory. For the corresponding references, see Notes-91 in Chapter 2.

Theorem 9.60 *Let* $\alpha \ge 0$ *and* $\{a_{\mathbf{k}}\} \in QMS(\alpha)$. *For* $0 < q < p < \infty$ *and any* $\mathbf{n} \le \mathbf{N} \in \mathbb{N}^d$, *we have*

$$\left(\sum_{\mathbf{k}=\mathbf{n}}^{\mathbf{N}} (k_1\cdots k_d)^{\alpha(p-q)} a_{\mathbf{k}}^p\right)^{\frac{1}{p}}$$
$$\le C(n_1\cdots n_d)^{-\alpha(p-q)+\frac{1}{p}-\frac{1}{q}}\left(\sum_{\mathbf{k}=\mathbf{1}}^{\mathbf{N}} (k_1\cdots k_d)^{\alpha(p-q)} a_{\mathbf{k}}^q\right)^{\frac{1}{q}},$$

where

$$C = \sup_n \left(n^{-\alpha p-1}\sum_{k=1}^{n} k^{\alpha p}\right)^{-\frac{d}{q}}$$

(in particular, if $\alpha = 0$, then $C = 1$). For the case $p = \infty$ and $0 < q < \infty$, we have

$$\max_{\mathbf{n}\le\mathbf{k}\le\mathbf{N}} (k_1\cdots k_d)^{-\alpha} a_{\mathbf{k}} \le (n_1\cdots n_d)^{-\frac{1}{q}} \left(\sum_{\mathbf{k}=1}^{\mathbf{N}} (k_1\cdots k_d)^{-\alpha q} a_{\mathbf{k}}^q\right)^{\frac{1}{q}}.$$

Proof The proof simply consists in rewriting that of Theorem 2.119 in arbitrary dimension (thus, we omit some calculations; details can be found in the proof of that statement). Since $\{(k_1\cdots k_d)^{-\alpha q} a_{\mathbf{k}}^q\}$ is decreasing, we have

$$\left(\sum_{\mathbf{k}=1}^{\mathbf{N}} (k_1\cdots k_d)^{\alpha(p-q)} a_{\mathbf{k}}^q\right)^{\frac{1}{q}} \ge \frac{a_{\mathbf{n}}}{(n_1\cdots n_d)^{\alpha}} \left(\sum_{\mathbf{k}=1}^{\mathbf{n}} (k_1\cdots k_d)^{\alpha p}\right)^{\frac{1}{q}}$$
$$= (n_1\cdots n_d)^{\alpha(\frac{p}{q}-1)+\frac{1}{q}} a_{\mathbf{n}} \left((n_1\cdots n_d)^{-\alpha p-1} \sum_{\mathbf{k}=1}^{\mathbf{n}} (k_1\cdots k_d)^{\alpha p}\right)^{\frac{1}{q}}$$
$$\ge C_0 (n_1\cdots n_d)^{\alpha(\frac{p}{q}-1)+\frac{1}{q}} a_{\mathbf{n}},$$

where

$$C_0 = \inf_{\mathbf{n}} \left((n_1\cdots n_d)^{-\alpha p-1} \sum_{\mathbf{k}=1}^{\mathbf{n}} (k_1\cdots k_d)^{\alpha p}\right)^{\frac{1}{q}} = \inf_{n} \left(n^{-\alpha p-1} \sum_{k=1}^{n} k^{\alpha p}\right)^{\frac{d}{q}}.$$

Putting

$$C = C_0^{-1} = \sup_{n} \left(n^{-\alpha p-1} \sum_{k=1}^{n} k^{\alpha p}\right)^{-\frac{d}{q}}$$

we can rewrite the above estimate as

$$a_{\mathbf{n}} \le C (n_1\cdots n_d)^{-\alpha(\frac{p}{q}-1)-\frac{1}{q}} \left(\sum_{\mathbf{k}=1}^{\mathbf{N}} (k_1\cdots k_d)^{\alpha(p-q)} a_{\mathbf{k}}^q\right)^{\frac{1}{q}}. \tag{9.44}$$

On the other hand, the fact that $\{(k_1\cdots k_d)^{-\alpha(p-q)} a_{\mathbf{k}}^{p-q}\}$ is decreasing yields

$$\sum_{\mathbf{k}=\mathbf{n}}^{\mathbf{N}} a_{\mathbf{k}}^p \le \frac{a_{\mathbf{n}}^{p-q}}{(n_1\cdots n_d)^{\alpha(p-q)}} \sum_{\mathbf{k}=1}^{\mathbf{N}} (k_1\cdots k_d)^{\alpha(p-q)} a_{\mathbf{k}}^q.$$

Applying the estimate for $a_{\mathbf{n}}$ given by (9.44) in the last inequality yields the desired conclusion.

For the case $p = \infty$ and $0 < q < \infty$, it is enough to show the corresponding inequality for decreasing sequences $\{a_{\mathbf{k}}\}$. The monotonicity of $\{a_{\mathbf{k}}\}$ implies

$$\left(\sum_{\mathbf{k}=1}^{\mathbf{N}} a_{\mathbf{k}}^q\right)^{\frac{1}{q}} \ge \left(\sum_{\mathbf{k}=1}^{\mathbf{n}} a_{\mathbf{k}}^q\right)^{\frac{1}{q}} \ge (n_1\cdots n_d)^{\frac{1}{q}} a_{\mathbf{n}} = (n_1\cdots n_d)^{\frac{1}{q}} \max_{\mathbf{n}\le\mathbf{k}\le\mathbf{N}} a_{\mathbf{k}},$$

which completes the proof. □

The classical Stechkin lemma in arbitrary dimension corresponds to Theorem 9.60 with $\alpha = 0$.

Corollary 9.61 *Let $\{a_{\mathbf{k}}\}$ be a nonnegative decreasing d-tuple sequence. For $0 < q < p < \infty$ and any $\mathbf{n} \le \mathbf{N} \in \mathbb{N}^d$, we have*

$$\left(\sum_{\mathbf{k}=\mathbf{n}}^{\mathbf{N}} a_{\mathbf{k}}^p\right)^{\frac{1}{p}} \le (n_1 \cdots n_d)^{\frac{1}{p}-\frac{1}{q}} \left(\sum_{\mathbf{k}=\mathbf{1}}^{\mathbf{N}} a_{\mathbf{k}}^q\right)^{\frac{1}{q}}.$$

For the case $p = \infty$ and $0 < q < \infty$, we have

$$\max_{\mathbf{n} \le \mathbf{k} \le \mathbf{N}} a_{\mathbf{k}} \le (n_1 \cdots n_d)^{-\frac{1}{q}} \left(\sum_{\mathbf{k}=\mathbf{1}}^{\mathbf{N}} a_{\mathbf{k}}^q\right)^{\frac{1}{q}}.$$

We present yet another generalization of Stechkin's lemma (different from that in Theorem 9.60) for sequences satisfying a rather general estimate. Its one- and two-dimensional analogs are Theorems 2.121 and 6.56, respectively.

Theorem 9.62 *Let $0 < q < p < \infty$ and let $\{a_{\mathbf{k}}\}$ be a nonnegative d-tuple sequence. Assume that for any $\mathbf{n} \in \mathbb{N}^d$ the estimate*

$$a_{\mathbf{n}} \le A\left(\frac{1}{n_1 \cdots n_d} \sum_{\mathbf{k}=\mathbf{1}}^{\mathbf{n}} a_{\mathbf{k}}^{\alpha}\right)^{\frac{1}{\alpha}} + B\left(\sum_{\mathbf{k} \in \mathbb{N}^d \setminus [\mathbf{1},\mathbf{n}]} \frac{a_{\mathbf{k}}^{\beta}}{k_1 \cdots k_d}\right)^{\frac{1}{\beta}} \tag{9.45}$$

holds with some $0 < \alpha, \beta \le q$, and constants $A, B \ge 0$. Then there exists $C > 0$ depending on A, B, p, q, and β, such that for every $\mathbf{n} \in \mathbb{N}^d$

$$\left(\sum_{\mathbf{k}=\mathbf{n}}^{\infty} a_{\mathbf{k}}^p\right)^{\frac{1}{p}} \le C(n_1 \cdots n_d)^{\frac{1}{p}-\frac{1}{q}} \left(\sum_{\mathbf{k}=\mathbf{1}}^{\infty} a_{\mathbf{k}}^q\right)^{\frac{1}{q}}.$$

In particular, if (9.45) *holds with $B = 0$, we have*

$$\left(\sum_{\mathbf{k}=\mathbf{n}}^{\infty} a_{\mathbf{k}}^p\right)^{\frac{1}{p}} \le A(n_1 \cdots n_d)^{\frac{1}{p}-\frac{1}{q}} \left(\sum_{k=1}^{\infty} a_{\mathbf{k}}^q\right)^{\frac{1}{q}}.$$

For the case $p = \infty$ and $0 < q < \infty$,

$$\max_{\mathbf{k} \ge \mathbf{n}} a_{\mathbf{k}} \le C(n_1 \cdots n_d)^{-\frac{1}{q}} \left(\sum_{\mathbf{k}=\mathbf{1}}^{\infty} a_{\mathbf{k}}^q\right)^{\frac{1}{q}}.$$

Remark 9.63 We observe that d-tuple sequences decreasing in each index satisfy the estimate (9.45) with $A = 1$, $B = 0$, and any $\alpha > 0$. In this case, we recover the conclusion from Corollary 9.61 (with $\mathbf{N} = \boldsymbol{\infty}$).

Similarly, for $\{a_{\mathbf{k}}\} \in WMS$, (9.45) holds with $B = 0$, arbitrary α, and $A = 2^{\frac{d}{\alpha}} C$, where C is the WMS constant of $\{a_{\mathbf{k}}\}$. Indeed, since $\{a_{\mathbf{k}}^{\alpha}\} \in WMS$ with constant C^{α}, it follows that

$$a_{\mathbf{n}}^{\alpha} \le \frac{2^d C^{\alpha}}{(n_1 \cdots n_d)} \sum_{\mathbf{k}=\lfloor \mathbf{n}/2 \rfloor}^{\mathbf{n}} a_{\mathbf{k}}^{\alpha}$$

(this estimate was already used in the proof of Theorem 9.57).

Proof of Theorem 9.62 We will omit some of the details of the proof and only outline the main steps. Such details can be found in the proof of the one-dimensional statement (Theorem 2.121) and are analogous in the multidimensional case.

Using Hölder's inequality (2.81) (or rather its multivariate version), we first estimate the two sums on the right-hand side of (9.45). On the one hand,

$$\left(\frac{1}{n_1\cdots n_d}\sum_{\mathbf{k}=\mathbf{1}}^{\mathbf{n}} a_{\mathbf{k}}^{\alpha}\right)^{\frac{1}{\alpha}} \le \left(\frac{1}{n_1\cdots n_d}\sum_{\mathbf{k}=\mathbf{1}}^{\mathbf{n}} a_{\mathbf{k}}^{q}\right)^{\frac{1}{q}},$$

which is also known as the generalized mean inequality. On the other hand,

$$\left(\sum_{\mathbf{k}\in\mathbb{N}^d\setminus[\mathbf{1},\mathbf{n}]} \frac{a_{\mathbf{k}}^{\beta}}{k_1\cdots k_d}\right)^{\frac{1}{\beta}} \le C_0\left(\frac{1}{n_1\cdots n_d}\sum_{\mathbf{k}\in\mathbb{N}^d\setminus[\mathbf{1},\mathbf{n}]} a_{\mathbf{k}}^{q}\right)^{\frac{1}{q}},$$

where the constant C_0 depends on q and β. Combining these two estimates with (9.45), we obtain

$$\begin{aligned} a_{\mathbf{n}} &\le A\left(\frac{1}{n_1\cdots n_d}\sum_{\mathbf{k}=\mathbf{1}}^{\mathbf{n}} a_{\mathbf{k}}^{q}\right)^{\frac{1}{q}} + C_0 B\left(\frac{1}{n_1\cdots n_d}\sum_{\mathbf{k}\in\mathbb{N}^d\setminus[\mathbf{1},\mathbf{n}]} a_{\mathbf{k}}^{q}\right)^{\frac{1}{q}} \\ &\le C\left(\frac{1}{n_1\cdots n_d}\sum_{\mathbf{k}=\mathbf{1}}^{\infty} a_{\mathbf{k}}^{q}\right)^{\frac{1}{q}}, \end{aligned}$$

for some constant $C > 0$ depending on all the previous ones (note that the case $p = \infty$ follows directly from the last estimate). Thus, for $p < \infty$,

$$\begin{aligned} \left(\sum_{\mathbf{k}=\mathbf{n}}^{\infty} a_k^{p}\right)^{\frac{1}{p}} &= \left(\sum_{\mathbf{k}=\mathbf{n}}^{\infty} a_{\mathbf{k}}^{q} a_{\mathbf{k}}^{p-q}\right)^{\frac{1}{p}} \le C\left(\sum_{\mathbf{k}=\mathbf{n}}^{\infty} a_{\mathbf{k}}^{q}\left(\frac{1}{k_1\cdots k_d}\sum_{\boldsymbol{\ell}=\mathbf{1}}^{\infty} a_{\boldsymbol{\ell}}^{q}\right)^{\frac{p-q}{q}}\right)^{\frac{1}{p}} \\ &\le C(n_1\cdots n_d)^{\frac{1}{p}-\frac{1}{q}}\left(\sum_{\mathbf{k}=\mathbf{n}}^{\infty} a_{\mathbf{k}}^{q}\right)^{\frac{1}{p}}\left(\sum_{\boldsymbol{\ell}=\mathbf{1}}^{\infty} a_{\boldsymbol{\ell}}^{q}\right)^{\frac{1}{q}-\frac{1}{p}} \\ &\le C(n_1\cdots n_d)^{\frac{1}{p}-\frac{1}{q}}\left(\sum_{\mathbf{k}=\mathbf{1}}^{\infty} a_{\mathbf{k}}^{q}\right)^{\frac{1}{q}}, \end{aligned}$$

as desired. □

9.2.7 Maclaurin–Cauchy integral test

In this subsection we extend the Maclaurin–Cauchy integral test to the instance of weak monotone functions and sequences in any dimension. For two-dimensional variants, see Subsection 6.2.5.

Theorem 9.64 *Let f be a WM function.*

(i) *If there exists $\mathbf{x} \in \mathbb{R}^d_+$ such that the integral*

$$\int_{[\mathbf{x},\infty)} f(\mathbf{t})\, d\mathbf{t} \tag{9.46}$$

converges, then for any $\mathbf{m} > \mathbf{x}$, $\mathbf{m} \in \mathbb{N}^d$, the series $\sum_{\mathbf{k}=\mathbf{m}}^{\infty} f(\mathbf{k})$, as well as for each $\eta \neq \mathbf{0}, \mathbf{1}$, the series

$$\sum_{\mathbf{k}_\eta=\mathbf{m}_\eta}^{\infty} f(\mathbf{k}_\eta, \mathbf{n}_{\mathbf{1}-\eta}), \tag{9.47}$$

with fixed $\mathbf{n}_{\mathbf{1}-\eta} \ge \mathbf{m}_{\mathbf{1}-\eta}$, converge.

(ii) *For any $\mathbf{m} \in (\mathbb{N} \cup \{0\})^d$, if the series $\sum_{\mathbf{k}=\mathbf{m}}^{\infty} f(\mathbf{k})$ converges, then the integral*

$$\int_{(\mathbf{m},\infty)} f(\mathbf{t})\, d\mathbf{t}, \tag{9.48}$$

as well as for each $\eta \neq \mathbf{0}, \mathbf{1}$, the integral

$$\int_{(\mathbf{m},\infty)^\eta} f(\mathbf{t}_\eta, \mathbf{n}_{\mathbf{1}-\eta})\, d\mathbf{t}_\eta, \tag{9.49}$$

with fixed $\mathbf{n}_{\mathbf{1}-\eta} \ge \mathbf{m}_{\mathbf{1}-\eta}$, converge.

Proof (i) Given $\mathbf{x}$ and $\mathbf{m}$, define the vector $\mathbf{y}$ by

$$y_j = \begin{cases} x_j, & \text{if } m_j - x_j \le 1/2, \\ m_j - 1/2, & \text{otherwise.} \end{cases}$$

Since $\mathbf{x} < \mathbf{m}$, we have $\mathbf{x} \le \mathbf{y} < \mathbf{m}$. If we define $z = \min_j\{m_j - y_j\}$, then $0 < z \le 1/2$. Also, the convergence of the integral (9.46) implies the convergence of

$$\int_{(\mathbf{y},\infty)} f(\mathbf{t})\, d\mathbf{t}. \tag{9.50}$$

For any $\mathbf{k} \in \mathbb{N}^d$, we define an auxiliary vector $\widetilde{\mathbf{k}}$ as follows:

$$\widetilde{k}_j = \begin{cases} k_j, & \text{if } k_j \ge m_j, \\ y_j, & \text{otherwise.} \end{cases}$$

Then, the convergence of (9.50) is equivalent to the boundedness of the d-tuple sequence

$$\sum_{\mathbf{k}=\mathbf{m}-\mathbf{1}}^{\mathbf{n}} \int_{(\widetilde{\mathbf{k}},\widetilde{\mathbf{k}+\mathbf{1}})} f(\mathbf{t})\, d\mathbf{t}$$

uniformly in $\mathbf{n}$. In virtue of

$$\frac{z^d}{C} f(\mathbf{k}+\mathbf{1}) \le \int_{(\widetilde{\mathbf{k}},\widetilde{\mathbf{k}+\mathbf{1}})} f(\mathbf{t})\, d\mathbf{t},$$

the convergence of the integral (9.50) implies the convergence of the series

$$\sum_{\mathbf{k}=\mathbf{m}}^{\infty} f(\mathbf{k}).$$

Since f is a nonnegative function, the convergence of the latter implies the convergence of (9.47).

(ii) If the series $\sum_{\mathbf{k}=\mathbf{m}}^{\infty} f(\mathbf{k})$ converges, the inequality

$$\int_{(\mathbf{k},\mathbf{k}+\mathbf{1})} f(\mathbf{t})\, d\mathbf{t} \leq C f(\mathbf{k})$$

implies the convergence of (9.48). Since f is a nonnegative function, the convergence of $\sum_{\mathbf{k}=\mathbf{m}}^{\infty} f(\mathbf{k})$ implies, for $\eta \neq \mathbf{0}, \mathbf{1}$, the convergence of each of the series (9.47) for any fixed $\mathbf{n}_{\mathbf{1}-\eta} \geq \mathbf{m}_{\mathbf{1}-\eta}$.

Applying part (b) of the theorem for a dimension lower than d, we get the convergence of the integrals (9.49), which completes the proof. □

We present the following immediate corollary of Theorem 9.64.

Corollary 9.65 *Let f be a WM function and $\mathbf{m} \in (\mathbb{N} \cup \{0\})^d$. Then the series $\sum_{\mathbf{k}=\mathbf{m}}^{\infty} f(\mathbf{k})$ converges if and only if the integral $\int_{(\mathbf{m},\infty)} f(\mathbf{t})\, d\mathbf{t}$ does, and for any $\eta \neq \mathbf{0}, \mathbf{1}$, each of the integrals*

$$\int_{(\mathbf{m},\infty)^{\eta}} f(\mathbf{t}_\eta, \mathbf{m}_{\mathbf{1}-\eta})\, d\mathbf{t}_\eta \tag{9.51}$$

converges.

Proof If the series $\sum_{\mathbf{k}=\mathbf{m}}^{\infty} f(\mathbf{k})$ converges, then the convergence of the integral

$$\int_{(\mathbf{m},\infty)} f(\mathbf{t})\, d\mathbf{t}$$

and of the integrals (9.51) for any $\eta \neq \mathbf{0}, \mathbf{1}$ follows from Theorem 9.64 (ii).

Suppose now that the integrals $\int_{(\mathbf{m},\infty)} f(\mathbf{t})\, d\mathbf{t}$ and (9.51) for $\eta \neq \mathbf{0}, \mathbf{1}$ are convergent. By Theorem 9.64 (i), the series $\sum_{\mathbf{k}=\mathbf{m}+\mathbf{1}}^{\infty} f(\mathbf{k})$ converges. Moreover, applying the same argument in lower dimensions, we derive that for each $\eta \neq \mathbf{0}, \mathbf{1}$ the convergence of the integral

$$\int_{(\mathbf{m},\infty)^{\eta}} f(\mathbf{t}_\eta, \mathbf{n}_{\mathbf{1}-\eta})\, d\mathbf{t}_\eta$$

implies the convergence of the series

$$\sum_{\mathbf{k}_\eta=(\mathbf{m}+\mathbf{1})_{\mathbf{1}-\eta}}^{\infty} f(\mathbf{k}_\eta, \mathbf{m}_{\mathbf{1}-\eta}).$$

By (7.1),

$$\sum_{\mathbf{k}=\mathbf{m}}^{\infty} f(\mathbf{k}) = \sum_{\mathbf{0}\leq\eta\leq\mathbf{1}} \sum_{\mathbf{k}_\eta=(\mathbf{m}+\mathbf{1})_\eta}^{\infty} f(\mathbf{k}_\eta, \mathbf{m}_{\mathbf{1}-\eta}).$$

Note that in case $\eta = \mathbf{0}$, the inner sum is $\sum_{\mathbf{k}_\eta=(\mathbf{m}+\mathbf{1})_\eta}^{\infty} f(\mathbf{k}_\eta, \mathbf{m}_{\mathbf{1}-\eta}) = f(\mathbf{m})$. Now, the conclusion follows, since all the series on the right-hand side converge. □

Remark 9.66 It is interesting to notice that in Theorem 9.64 (i) the condition $\mathbf{m} > \mathbf{x}$ is sharp, in the sense that the conclusion is no longer true if $\mathbf{x} = \mathbf{m}$. Indeed, suppose that $x_1 = m_1 = 1$ and $x_j \le m_j$ for all $2 \le j \le d$. If we define the function

$$f(\mathbf{y}) = \begin{cases} e^{-|\mathbf{y}|}, & \text{if } \mathbf{y} \in (\mathbf{1}, \boldsymbol{\infty}), \\ \prod_{j:\eta_j=0} \frac{1}{y_j}, & \text{if } \mathbf{y}_\eta \in (\mathbf{0}, \mathbf{1}]^\eta \text{ and } \mathbf{y}_{\mathbf{1}-\eta} \in (\mathbf{1}, \boldsymbol{\infty})^{\mathbf{1}-\eta},\ \eta \neq \mathbf{0}, \mathbf{1}, \\ 1, & \text{if } \mathbf{y} \in (\mathbf{0}, \mathbf{1}], \end{cases}$$

then it is clearly a *WM* function (in fact, monotone in each variable). Moreover, in this case the integral (9.64) (with $\mathbf{m} = \mathbf{1}$) converges. However, for any fixed $m_3, \ldots, m_d \in \mathbb{N}$, we have

$$\sum_{k=m_2}^{\infty} f(1, k, m_3, \ldots, m_d) = \sum_{k=m_2}^{\infty} \frac{1}{k} = \infty,$$

and therefore the series (9.47) diverges.

Remark 9.67 The assumption of Theorem 9.64 (i) is sufficient to ensure the convergence of (9.47), but not necessary. Indeed, if we consider the function f in Remark 9.66, then it is clear that the integral

$$\int_{(\mathbf{x},\boldsymbol{\infty})} f(\mathbf{t})\, d\mathbf{t}$$

diverges for any $\mathbf{x} < \mathbf{1}$. However, in this case, for any $\mathbf{0} \le \eta \le \mathbf{1}$ with $\eta \neq \mathbf{0}, \mathbf{1}$, the series (9.47) converges whenever $\mathbf{m} \ge \mathbf{1}$.

We can also prove an equiconvergence theorem for functions from the class WM_{Rad} (see Definition 9.38).

Theorem 9.68 *Let $f \in WM_{\mathrm{Rad}}$. Then the series*

$$\sum_{\mathbf{k}=\mathbf{1}}^{\infty} f(\mathbf{k})$$

and the integral

$$\int_{(\mathbf{1},\boldsymbol{\infty})} f(\mathbf{x})\, d\mathbf{x}$$

converge or diverge simultaneously.

Proof By Proposition 9.40, there exists a radial function g equivalent to f at each point. Now,

$$\int_{(\mathbf{1},\boldsymbol{\infty})} f(\mathbf{x})\, d\mathbf{x} \asymp \int_1^{\infty} r^{d-1} g_0(r)\, dr = \sum_{n=0}^{\infty} \int_{2^n}^{2^{n+1}} r^{d-1} g_0(r)\, dr \lesssim \sum_{n=0}^{\infty} 2^{dn} g_0(2^n),$$

where $g(\mathbf{x}) = g_0(\|\mathbf{x}\|)$ and g_0 is a *WM* function of one variable. Let us denote, for $n \geq 0$,

$$R_n := \{\mathbf{x} \in \mathbb{R}_+^d : 2^n \leq |\mathbf{x}| < 2^{n+1}\}.$$

Note that for every $n \in \mathbb{N}$, one has $|\mathbb{N}^d \cap R_n| \asymp 2^{dn}$. Using this and the fact that $g_0 \in WM$, we get

$$\sum_{n=1}^{\infty} 2^{dn} g_0(2^n) \lesssim \sum_{n=1}^{\infty} \sum_{\mathbf{k} \in \mathbb{N}^d \cap R_n} g_0(\|\mathbf{k}\|) = \sum_{n=1}^{\infty} \sum_{\mathbf{k} \in \mathbb{N}^d \cap R_n} g(\mathbf{k}) \asymp \sum_{\mathbf{k}=\mathbf{1}}^{\infty} f(\mathbf{k}).$$

For the converse, we similarly have

$$\sum_{\mathbf{k}=\mathbf{1}}^{\infty} f(\mathbf{k}) \asymp \sum_{n=0}^{\infty} \sum_{\mathbf{k} \in \mathbb{N}^d \cap R_n} g(\mathbf{k}) \lesssim \sum_{n=0}^{\infty} 2^{dn} g_0(2^n).$$

Finally, Proposition 9.40 yields

$$\sum_{n=0}^{\infty} 2^{dn} g_0(2^n) \lesssim g_0(1) + \sum_{n=1}^{\infty} \int_{2^{n-1}}^{2^n} r^{d-1} g_0(r)\, dr \asymp \int_{(\mathbf{1},\infty)} f(\mathbf{x})\, d\mathbf{x}. \qquad \square$$

Remark 9.69 Contrary to the nonradial case (Theorem 9.64), in Theorem 9.68 we do not assume the convergence of lower-dimensional integrals in order to derive the convergence of the corresponding series. This is because in the radial case the lower-dimensional integrals are dominated by the full-dimensional integral. For details, we refer to the discussion given for the two-dimensional case in Remark 6.60 (the extension of the argument to arbitrary dimension is immediate).

Before inspecting examples around Theorem 9.64, we present the following corollary. It relates the integral test with the Cauchy condensation test.

Corollary 9.70 *Let f be a WM function. The series*

$$\sum_{\mathbf{k}=\mathbf{0}}^{\infty} 2^{|\mathbf{k}|} f(\mathbf{2^k})$$

converges if and only if, for all $\eta \neq \mathbf{0}$, the integrals

$$\int_{(\mathbf{1},\infty)^\eta} f(\mathbf{t}_\eta, \mathbf{1}_{\mathbf{1}-\eta})\, d\mathbf{t}_\eta \tag{9.52}$$

converge.

Proof By Corollary 9.56, the series

$$\sum_{\mathbf{k}=\mathbf{0}}^{\infty} 2^{|\mathbf{k}|} f(\mathbf{2^k})$$

converges or diverges simultaneously with

$$\sum_{\mathbf{k}=1}^{\infty} f(\mathbf{k}),$$

and the latter series converges if and only if the integrals (9.52) converge for all $\eta \neq \mathbf{0}$, by Corollary 9.65. □

Example 9.71 (i) The series

$$\sum_{\mathbf{k}=1}^{\infty} |\mathbf{k}|^{-\gamma} \tag{9.53}$$

converges if and only if $\gamma > d$. Indeed, by Theorem 9.68, this series converges if and only if the integral

$$\int_{(\mathbf{1},\infty)} \frac{1}{|\mathbf{x}|^{\gamma}}\, d\mathbf{x} \asymp \int_1^{\infty} r^{d-1-\gamma}\, dr$$

converges, and this is the case if and only if $\gamma > d$.

(ii) Consider the integral

$$\int_{(\mathbf{2},\infty)} \frac{1}{x_1 \cdots x_d \log^{\gamma}(x_1 \cdots x_d)} d\mathbf{x}, \quad \gamma > 0. \tag{9.54}$$

It converges if and only if the series (9.53) does or, equivalently, if and only if $\gamma > d$.

(iii) The series

$$\sum_{n_1=1}^{\infty} \cdots \sum_{n_d=1}^{\infty} \frac{(n_1^{\alpha_1} + \cdots + n_d^{\alpha_d})^{\gamma}}{n_1^{\beta_1} \cdots n_d^{\beta_d}}, \quad \alpha_j, \beta_j, \gamma > 0, \quad j = 1, \ldots, d,$$

converges if and only if

$$\alpha_j \gamma - \beta_j < -1 \quad \text{for all} \quad j = 1, \ldots, d. \tag{9.55}$$

Indeed, since

$$\sum_{n_k=1}^{\infty} \frac{(n_k^{\alpha_k} + d - 1)^{\gamma}}{n_k^{\beta_k}}$$

converges if and only if $\alpha_k \gamma - \beta_k < -1$, the condition (9.55) is necessary for the considered series to converge. On the other hand, from the inequality $x_1 + \cdots + x_d \leq d x_1 \cdots x_d$ (valid whenever $x_j \geq 1$ for all j), we deduce that (9.55) is also a sufficient condition to guarantee the convergence. In the same way one can prove that

$$\sum_{n_1=1}^{\infty} \cdots \sum_{n_d=1}^{\infty} \frac{(n_1^{\alpha_1} + \cdots + n_d^{\alpha_d})^{\gamma} \log^{\delta_1} n_1 \cdots \log^{\delta_d} n_d}{n_1^{\beta_1} \cdots n_d^{\beta_d}}, \quad \alpha_j, \beta_j, \gamma, \delta_j > 0,$$

converges if and only if (9.55) holds, for any choice of $\delta_1, \ldots, \delta_d$.

(iv) We can also characterize the convergence of the series

$$\sum_{n_1=1}^{\infty} \cdots \sum_{n_d=1}^{\infty} \frac{n_1^{\alpha_1} \cdots n_d^{\alpha_d}}{(n_1^{\beta_1} + \cdots + n_d^{\beta_d})^{\gamma}}, \tag{9.56}$$

$$\alpha_j > -1,\ \beta_j > 0, \gamma > 0, \quad \text{for } j = 1, \ldots, d.$$

This series generalizes (9.53), which corresponds to the case $\alpha_j = 0$, $\beta_j = 1$ for all j. A necessary and sufficient condition for the convergence of (9.56) is

$$\sum_{i=1}^{d} \frac{\alpha_i + 1}{\beta_i} < \gamma. \tag{9.57}$$

In order to prove this assertion, we need the following estimate.

Lemma 9.72 *Let $b, \rho > 0$, $\alpha > -1$, and $\gamma > 0$. If $\alpha - \gamma < -1$, then*

$$\int_{\rho}^{\infty} \frac{x^{\alpha}}{(x+b)^{\gamma}}\, dx \asymp (\max\{b, \rho\})^{1+\alpha-\gamma}.$$

Proof We distinguish between two cases. First, if $b \le \rho$,

$$\int_{\rho}^{\infty} \frac{x^{\alpha}}{(x+b)^{\gamma}}\, dx \asymp \int_{\rho}^{\infty} x^{\alpha-\gamma}\, dx \asymp \rho^{1+\alpha-\gamma}.$$

On the other hand, for $b > \rho$,

$$\begin{aligned} b^{1-\alpha-\gamma} &\asymp \int_{b}^{\infty} \frac{x^{\alpha}}{(x+b)^{\gamma}}\, dx \le \int_{\rho}^{\infty} \frac{x^{\alpha}}{(x+b)^{\gamma}}\, dx \\ &= \int_{\rho}^{b} \frac{x^{\alpha}}{(x+b)^{\gamma}}\, dx + \int_{b}^{\infty} \frac{x^{\alpha}}{(x+b)^{\gamma}}\, dx \asymp \int_{\rho}^{b} \frac{x^{\alpha}}{(x+b)^{\gamma}}\, dx + b^{1+\alpha-\gamma}. \end{aligned}$$

Finally, since

$$\int_{\rho}^{b} \frac{x^{\alpha}}{(x+b)^{\gamma}}\, dx \le b^{-\gamma} \int_{0}^{b} x^{\alpha}\, dx \asymp b^{1+\alpha-\gamma},$$

the result follows. □

If we define

$$f(\mathbf{x}) = \frac{x_1^{\alpha_1} \cdots x_d^{\alpha_d}}{(x_1^{\beta_1} + \cdots + x_d^{\beta_d})^{\gamma}},$$

then $f \in WM$, and therefore we can apply Theorem 9.64. Thus, the convergence of (9.56) is equivalent to the convergence of the integral

$$\int_{(\mathbf{1},\infty)} f(\mathbf{x})\, d\mathbf{x},$$

where, as in (ii), the convergence of the lower-dimensional integrals follows from the convergence of the d-dimensional integral. Substituting $y_j = x_j^{\beta_j}$, we can rewrite the latter integral (up to a constant multiple) as

$$\int_1^\infty y_d^{\frac{\alpha_d+1}{\beta_d}-1}\int_1^\infty y_{d-1}^{\frac{\alpha_{d-1}+1}{\beta_{d-1}}-1}\cdots\int_1^\infty y_1^{\frac{\alpha_1+1}{\beta_1}-1}\frac{1}{(y_1+\cdots+y_d)^\gamma}\,d\mathbf{y}. \tag{9.58}$$

Let (9.57) hold. We are going to prove by induction that for any $1 \le \ell < d$

$$\int_1^\infty y_\ell^{\frac{\alpha_\ell+1}{\beta_\ell}-1}\cdots\int_1^\infty y_1^{\frac{\alpha_1+1}{\beta_1}-1}\frac{1}{(y_1+\cdots+y_d)^\gamma}\,dy_1\cdots dy_\ell$$
$$\asymp (y_{\ell+1}+\cdots+y_d)^{\sum_{i=1}^{\ell}\frac{\alpha_i+1}{\beta_i}-\gamma}. \tag{9.59}$$

For $\ell = 1$, because of (9.57), we have

$$\frac{\alpha_1+1}{\beta_1}-1-\gamma < -1.$$

Then, by Lemma 9.72 with $\rho = 1$, $b = y_2+\cdots+y_d$, and

$$\alpha = \frac{\alpha_1+1}{\beta_1}-1,$$

there holds

$$\int_1^\infty y_1^{\frac{\alpha_1+1}{\beta_1}-1}\frac{1}{(y_1+\cdots+y_d)^\gamma}\,dy_1 \asymp (y_2+\cdots+y_d)^{\frac{\alpha_1+1}{\beta_1}-\gamma}.$$

Assume now that (9.59) is valid for some $\ell-1 < d-1$. Then, the left-hand side of (9.59) is equivalent to

$$\int_1^\infty y_\ell^{\frac{\alpha_\ell+1}{\beta_\ell}-1}(y_\ell+\cdots+y_d)^{\sum_{i=1}^{\ell-1}\frac{\alpha_i+1}{\beta_i}-\gamma}\,dy_\ell,$$

by induction hypothesis. Also, by (9.57), we have

$$\sum_{i=1}^{\ell}\frac{\alpha_i+1}{\beta_i}-1-\gamma < -1.$$

Therefore, applying Lemma 9.72 with $b = y_{\ell+1}+\cdots+y_d$,

$$\alpha = \frac{\alpha_\ell+1}{\beta_\ell}-1,$$

and

$$\gamma' = \gamma - \sum_{i=1}^{\ell-1}\frac{\alpha_i+1}{\beta_i}$$

in place of γ, we can estimate the latter integral from above by

$$(y_{\ell+1}+\cdots+y_d)^{\sum_{i=1}^{\ell}\frac{\alpha_i+1}{\beta_i}-\gamma},$$

which completes the proof of (9.59). Finally, applying (9.59) with $\ell = d-1$, we majorize (9.58) by

$$\int_1^\infty y_d^{\frac{\alpha_d+1}{\beta_d}-1}\,y_d^{\sum_{i=1}^{d-1}\frac{\alpha_i+1}{\beta_i}-\gamma}\,dy_d.$$

The latter integral converges if and only if

$$\sum_{i=1}^{d} \frac{\alpha_i + 1}{\beta_i} - 1 - \gamma < -1,$$

which holds true in virtue of (9.57). In order to prove the necessity of (9.57), we proceed as follows. Assume (9.57) is not true and let

$$m = \min \left\{ n \in \{1, \dots, d\} : \sum_{i=1}^{n} \frac{\alpha_i + 1}{\beta_i} \geq \gamma \right\}.$$

We assume $m > 1$ (if $m = 1$, then the integral

$$\int_1^\infty \frac{x_1^{\alpha_1} \cdots x_d^{\alpha_d}}{(x_1^{\beta_1} + \cdots + x_d^{\beta_d})^\gamma} \, dx_1$$

is divergent for all fixed $x_2, \dots, x_d$, and so is the d-dimensional integral). By (9.59) we have

$$\int_1^\infty y_{m-1}^{\frac{\alpha_{m-1}+1}{\beta_{m-1}}-1} \cdots \int_1^\infty y_1^{\frac{\alpha_1+1}{\beta_1}-1} \frac{1}{(y_1 + \cdots + y_d)^\gamma} \, dy_1 \cdots dy_{m-1}$$
$$\asymp (y_m + \cdots + y_d)^{\sum_{i=1}^{m-1} \frac{\alpha_i+1}{\beta_i} - \gamma}.$$

Now,

$$\int_1^\infty y_m^{\frac{\alpha_m+1}{\beta_m}-1} \cdots \int_1^\infty y_1^{\frac{\alpha_1+1}{\beta_1}-1} \frac{1}{(y_1 + \cdots + y_d)^\gamma} \, dy_1 \cdots dy_m$$
$$\asymp \int_1^\infty y_m^{\frac{\alpha_m+1}{\beta_m}-1} (y_m + \cdots + y_d)^{\sum_{i=1}^{m-1} \frac{\alpha_i+1}{\beta_i} - \gamma} \, dy_m$$
$$\asymp \int_{y_{m+1}+\cdots+y_d}^\infty y_m^{\sum_{i=1}^{m} \frac{\alpha_i+1}{\beta_i} - \gamma - 1} \, dy_m = \infty,$$

since

$$\sum_{i=1}^{m} \frac{\alpha_i + 1}{\beta_i} - \gamma - 1 \geq -1.$$

This in turn implies the divergence of the d-dimensional integral.

(v) If $\alpha_1, \dots, \alpha_d \geq 0$, the series

$$\sum_{\mathbf{n}=\mathbf{1}}^{\infty} \frac{1}{|\mathbf{n}|^d \prod_{j=1}^{d} \log^{\alpha_j}(n_j + 1)}$$

converges if and only if $\max_j \alpha_j > 1$. To prove this claim, we first note that by the Maclaurin–Cauchy integral test this series converges if and only if the integral

$$\int_{(\mathbf{1},\infty)} \frac{1}{(x_1 + \cdots + x_d)^d \prod_{j=1}^{d} \log^{\alpha_j}(x_j + 1)} \, d\mathbf{x} \tag{9.60}$$

does (again, the convergence of the full dimensional integral yields convergence of the lower-dimensional ones). In order to prove convergence, let $1 \leq \ell \leq d$ be such that $\max_j \alpha_j = \alpha_\ell > 1$. We first apply the trivial estimate

$$\int_{(\mathbf{1},\infty)} \frac{1}{(x_1 + \cdots + x_d)^d \prod_{j=1}^d \log^{\alpha_j}(x_j + 1)} \, d\mathbf{x}$$
$$\lesssim \int_{(\mathbf{1},\infty)} \frac{1}{(x_1 + \cdots + x_d)^d \log^{\alpha_\ell}(x_\ell + 1)} \, d\mathbf{x}.$$

Applying Lemma 9.72 repeatedly, as in (iv), we find that

$$\int_{(\mathbf{1},\infty)} \frac{1}{(x_1 + \cdots + x_d)^d \log^{\alpha_\ell}(x_\ell + 1)} \, d\mathbf{x} \asymp \int_1^\infty \frac{1}{x_\ell \log^{\alpha_\ell}(x_\ell + 1)} \, dx_\ell,$$

which clearly converges, since $\alpha_\ell > 1$. In order to prove the divergence of integral (9.60) in the case $\max_j \alpha_j \leq 1$, we first note that

$$\int_{(\mathbf{1},\infty)} \frac{1}{(x_1 + \cdots + x_d)^d \prod_{j=1}^d \log^{\alpha_j}(x_j + 1)} \, d\mathbf{x}$$
$$\gtrsim \int_{(\mathbf{1},\infty)} \frac{1}{(x_1 + \cdots + x_d)^d \prod_{j=1}^d \log(x_j + 1)} \, d\mathbf{x}. \qquad (9.61)$$

In order to continue, we are in need of the following estimate.

Lemma 9.73 *Let $\gamma > 1$ and $b > e$. Then the estimate*

$$\int_1^\infty \frac{1}{(x + b)^\gamma \log(x + 1)} \, dx \asymp \frac{1}{b^{\gamma - 1}}$$

holds.

Proof The upper estimate is trivial, since

$$\int_1^\infty \frac{1}{(x + b)^\gamma \log(x + 1)} \, dx \leq \frac{1}{\log 2} \int_b^\infty \frac{1}{x^\gamma} \, dx \asymp \frac{1}{b^{\gamma - 1}}.$$

For the lower estimate, we have

$$\int_1^\infty \frac{1}{(x + b)^\gamma \log(x + 1)} \, dx > \int_b^\infty \frac{1}{(x + b)^\gamma \log(x + 1)} \, dx$$
$$\asymp \int_{b+1}^\infty \frac{1}{x^\gamma \log x} \, dx.$$

Integrating the last integral by parts, we obtain

$$\int_{b+1}^\infty \frac{1}{x^\gamma \log x} \, dx = \frac{\log\log(b + 1)}{(b + 1)^{\gamma - 1}} + (\gamma - 1) \int_{b+1}^\infty \frac{\log\log x}{x^\gamma} \, dx \gtrsim \frac{1}{b^{\gamma - 1}},$$

where in the last inequality we use that $b > e$, and hence $\log\log(b+1) > 0$. □

In order to show that the integral (9.61) diverges, it suffices to show that such an integral diverges when its domain of integration is $(\mathbf{3},\boldsymbol{\infty})$, rather than $(\mathbf{1},\boldsymbol{\infty})$ (this is just a technical modification that will ensure that the hypotheses of Lemma 9.73 are satisfied). We can write the corresponding integral as

$$\int_3^\infty \cdots \int_3^\infty \left(\int_3^\infty \frac{1}{|\mathbf{x}|^d \log(x_1+1)}\, dx_1 \right) \frac{1}{\prod_{j=2}^d \log(x_j+1)}\, dx_2 \cdots dx_d$$

and then apply Lemma 9.73 to estimate the inner integral to obtain

$$\int_3^\infty \frac{1}{(x_1+\cdots+x_d)^d \log(x_1+1)}\, dx_1 \gtrsim \frac{1}{(x_2+\cdots+x_d)^{d-1}}.$$

Applying Lemma 9.73 successively, one arrives at the lower estimate of (9.61) (up to multiplicative constants)

$$\int_3^\infty \int_3^\infty \frac{1}{(x_{d-1}+x_d)^2 \log(x_{d-1}+1)\log(x_d+1)}\, dx_{d-1}\, dx_d$$
$$\gtrsim \int_3^\infty \frac{1}{x_d \log(x_d+1)}\, dx_d.$$

This last integral is clearly divergent, which completes the proof of the claim.

9.2.8 Equiconvergence results involving general monotonicity

In some problems one should consider a class of monotone-type sequences different from *WMS*. Sometimes it is more convenient to consider generalizations of monotonicity related to the bounded variation concept, which is precisely the case of *GMS*.

In the multivariate case, we would like to obtain analogues of the Cauchy condensation test (Theorem 9.55) for power-type sequences $\{u_k^i\}$, rather than $\{u_k^i\} \in \Lambda$. This will be achieved in Theorem 9.80 below. The equiconvergence property we are going to consider rests on the condition of general monotonicity.

Theorem 9.74 *Let $\{a_{\mathbf{k}}\} \in GMS$ be vanishing as $|\mathbf{k}| \to \infty$ and $\mathbf{m} \in \mathbb{N}^d$. Then the series*

$$\sum_{\mathbf{k}=\mathbf{m}}^{\infty} |a_{\mathbf{k}}| \tag{9.62}$$

and, for any fixed $\eta \neq \mathbf{0}$, the series

$$\sum_{\mathbf{k}=\mathbf{m}}^{\infty} \left(\prod_{j:\eta_j=1} k_j \right) |\Delta^\eta a_{\mathbf{k}}| \tag{9.63}$$

converge or diverge simultaneously. What is more, we have, for any $\eta \neq \mathbf{0}$,

$$\sum_{\mathbf{k}=\mathbf{m}}^{\infty} |a_{\mathbf{k}}| \asymp \sum_{\mathbf{k}=\mathbf{m}}^{\infty} \left(\prod_{j:\eta_j=1} k_j \right) |\Delta^\eta a_{\mathbf{k}}|.$$

Proof Since $\{a_{\mathbf{k}}\} \in GMS$ and $\mathbf{m} \in \mathbb{N}^d$ is fixed, we have

$$
\begin{aligned}
\sum_{\mathbf{k}=\mathbf{m}}^{\infty} |a_{\mathbf{k}}| &\gtrsim \sum_{\mathbf{k}=\mathbf{m}}^{\infty} \sum_{\mathbf{j}=\mathbf{k}}^{2\mathbf{k}-\mathbf{1}} |\Delta^{\mathbf{1}} a_{\mathbf{j}}| \\
&\asymp \sum_{\mathbf{k}=\mathbf{m}}^{\infty} \frac{1}{k_1 \cdots k_d} \sum_{\mathbf{j}=\mathbf{k}}^{2\mathbf{k}-\mathbf{1}} j_1 \cdots j_d |\Delta^{\mathbf{1}} a_{\mathbf{j}}| \\
&\gtrsim \sum_{\mathbf{k}=\mathbf{m}}^{\infty} k_1 \cdots k_d |\Delta^{\mathbf{1}} a_{\mathbf{k}}|, \qquad (9.64)
\end{aligned}
$$

where in the last estimate we have applied the inequality

$$
\sum_{k=m}^{\infty} \frac{1}{k} \sum_{j=k}^{2k-1} b_j \gtrsim \sum_{k=m}^{\infty} b_k, \qquad b_k \geq 0,
$$

in each index separately. We are now going to obtain the following estimate, valid for any $\eta \neq \mathbf{0}$:

$$
\sum_{\mathbf{k}=\mathbf{m}}^{\infty} k_1 \cdots k_d |\Delta^{\mathbf{1}} a_{\mathbf{k}}| \geq \sum_{\mathbf{k}=\mathbf{m}}^{\infty} \Bigg(\prod_{j:\eta_j=1} k_j \Bigg) |\Delta^{\eta} a_{\mathbf{k}}| \geq \sum_{\mathbf{k}=\mathbf{m}}^{\infty} |a_{\mathbf{k}}| \qquad (9.65)
$$

through an inductive argument without any assumption about general monotonicity. Recall that $\Delta^{\mathbf{0}} a_{\mathbf{k}} = a_{\mathbf{k}}$. First, let $1 \leq \ell \leq d$. Applying Abel's transformation, we get, for $N_\ell > m_\ell$,

$$
\begin{aligned}
\sum_{k_\ell=m_\ell}^{N_\ell} |a_{\mathbf{k}}| &\leq \sum_{k_\ell=m_\ell}^{N_\ell-1} k_\ell \Delta^{\mathbf{e}_\ell} |a_{\mathbf{k}}| + N_\ell |a_{\mathbf{k}_{\mathbf{1}-\mathbf{e}_\ell}, \mathbf{N}_{\mathbf{e}_\ell}}| \\
&\leq \sum_{k_\ell=m_\ell}^{N_\ell-1} k_\ell |\Delta^{\mathbf{e}_\ell} a_{\mathbf{k}}| + N_\ell \Bigg| \sum_{k_\ell=N_\ell}^{\infty} \Delta^{\mathbf{e}_\ell} a_{\mathbf{k}} \Bigg| \leq \sum_{k_\ell=m_\ell}^{\infty} k_\ell |\Delta^{\mathbf{e}_\ell} a_{\mathbf{k}}|.
\end{aligned}
$$

This estimate implies that

$$
\sum_{\mathbf{k}=\mathbf{m}}^{\infty} |a_{\mathbf{k}}| \leq \sum_{\mathbf{k}=\mathbf{m}}^{\infty} k_\ell |\Delta^{\mathbf{e}_\ell} a_{\mathbf{k}}|,
$$

which proves the base instance of the inductive argument. Let now η be such that $1 \leq |\eta| < d$ and ℓ be such that $\eta_\ell = 0$. As above, we note that for $\mathbf{N} > \mathbf{m}$

$$
\sum_{\mathbf{k}=\mathbf{m}}^{\mathbf{N}} \Bigg(\prod_{j:\eta_j=1} k_j \Bigg) |\Delta^{\eta} a_{\mathbf{k}}| = \sum_{\substack{m_j \leq k_j \leq N_j \\ j \neq \ell}} \Bigg(\prod_{j:\eta_j=1} k_j \Bigg) \sum_{k_\ell=m_\ell}^{N_\ell} |\Delta^{\eta} a_{\mathbf{k}}|,
$$

and estimate the latter single series. Applying Abel's transformation, we obtain

$$\begin{aligned}\sum_{k_\ell=m_\ell}^{N_\ell}|\Delta^\eta a_{\mathbf{k}}| &= \sum_{k_\ell=m_\ell}^{N_\ell-1}(k_\ell-m_\ell+1)\Delta^{\mathbf{e}_\ell}|\Delta^\eta a_{\mathbf{k}}| \\ &\quad + (N_\ell-m_\ell+1)|\Delta^\eta a_{\mathbf{k}_{\mathbf{1}-\mathbf{e}_\ell},\mathbf{N}_{\mathbf{e}_\ell}}| \\ &\le \sum_{k_\ell=m_\ell}^{N_\ell-1}k_\ell|\Delta^{\eta+\mathbf{e}_\ell}a_{\mathbf{k}}| + N_\ell\left|\sum_{k_\ell=N_\ell}^{\infty}\Delta^{\eta+\mathbf{e}_\ell}a_{\mathbf{k}}\right| \\ &\le \sum_{k_\ell=m_\ell}^{\infty}k_\ell|\Delta^{\eta+\mathbf{e}_\ell}a_{\mathbf{k}}|.\end{aligned}$$

Thus, denoting $\zeta = \eta + \mathbf{e}_\ell$, we have

$$\sum_{\mathbf{k}=\mathbf{m}}^{\infty}\Big(\prod_{j:\eta_j=1}k_j\Big)|\Delta^\eta a_{\mathbf{k}}| \le \sum_{\mathbf{k}=\mathbf{m}}^{\infty}\Big(\prod_{j:\zeta_j=1}k_j\Big)|\Delta^\zeta a_{\mathbf{k}}|.$$

In particular, we have proved that the series

$$\sum_{\mathbf{k}=\mathbf{m}}^{\infty}k_1\cdots k_d|\Delta^{\mathbf{1}}a_{\mathbf{k}}|$$

dominates the series (9.62) as well as each of the series (9.63) with $\eta \ne \mathbf{0},\mathbf{1}$. Therefore, estimate (9.65) holds, and the proof is now complete. □

Note that in Theorem 9.74 one cannot replace the $\{a_{\mathbf{k}}\} \in GMS$ assumption by $\{a_{\mathbf{k}}\} \in WMS$ (where in the latter case the sequence is assumed to be nonnegative). Indeed, there exists a multiple (positive) sequence $\{a_{\mathbf{k}}\} \in WMS$ but not GMS for which

$$\sum_{\mathbf{k}=\mathbf{1}}^{\infty}a_{\mathbf{k}} < \infty \qquad \text{and} \qquad \sum_{\mathbf{k}=\mathbf{1}}^{\infty}k_1\cdots k_d|\Delta^{\mathbf{1}}a_{\mathbf{k}}| = \infty.$$

Consider the example in Remark 9.17, where

$$a_{\mathbf{k}_{\mathbf{e}_1},\mathbf{1}_{\mathbf{1}-\mathbf{e}_1}} = \begin{cases}2\cdot 4^{-m}, & \text{if } 2^m \le k_1 < 2^{m+1} \text{ and } k_1 \text{ is even},\\ 4^{-m}, & \text{if } 2^m \le k_1 < 2^{m+1} \text{ and } k_1 \text{ is odd},\end{cases} \qquad m\in\mathbb{N},$$

and zero otherwise. Recall that $\{a_{\mathbf{k}}\} \in WMS\backslash GMS$. Now,

$$\sum_{\mathbf{k}=\mathbf{1}}^{\infty}a_{\mathbf{k}} = \sum_{m_1=0}^{\infty}\sum_{k_1=2^{m_1}}^{2^{m_1-1}-1}a_{\mathbf{k}_{\mathbf{e}_1},\mathbf{1}_{\mathbf{1}-\mathbf{e}_1}} \le 2\sum_{m_1=0}^{\infty}2^{-m_1} < \infty,$$

but by setting $\mathbf{m} = (m_1, 0, \ldots, 0)$, we have

$$\begin{aligned}\sum_{\mathbf{k}=\mathbf{1}}^{\infty} k_1 \cdots k_d |\Delta^{\mathbf{1}} a_{\mathbf{k}}| &\geq \sum_{k_1=1}^{\infty} k_1 |\Delta^{\mathbf{1}} a_{\mathbf{k}_{\mathbf{e}_1}, \mathbf{1}_{\mathbf{1}-\mathbf{e}_1}}| \\ &\geq \frac{1}{2} \sum_{m_1=1}^{\infty} 2^{m_1} \sum_{k_1=2^{m_1}}^{2^{m_1+1}-2} |\Delta^{\mathbf{1}} a_{\mathbf{k}_{\mathbf{e}_1}, \mathbf{1}_{\mathbf{1}-\mathbf{e}_1}}| \\ &\geq \frac{1}{2} \sum_{m_1=1}^{\infty} 2^{m_1} 2^{m_1-2} a_{(\mathbf{2^m})_{\mathbf{e}_1}, \mathbf{1}_{\mathbf{1}-\mathbf{e}_1}} \geq \frac{1}{8} \sum_{m_1=1}^{\infty} 1 = \infty.\end{aligned}$$

Let us prove a result for functions similar to Theorem 9.74. We first introduce the following notation: for $\mathbf{m}, \mathbf{n} \in \mathbb{R}^d_+$, we will write $\mathbf{n} \preceq \mathbf{m}$ if the following hold:

1. If $n_j > 0$, then $m_j > n_j$, $1 \leq j \leq d$;
2. If $n_j = 0$, then $m_j \geq 0$, $1 \leq j \leq d$.

Theorem 9.75 *Let f be a GM function. Then*

(i) *if for some $\mathbf{m} \in \mathbb{R}^d_+$ and $\eta \neq \mathbf{0}$ the integral*

$$\int_{[\mathbf{m},\infty)} \left(\prod_{j:\eta_j=1} s_j \right) |d^{\eta} f(\mathbf{s}_{\eta}, \mathbf{x}_{\mathbf{1}-\eta})| \, d\mathbf{x}_{\mathbf{1}-\eta} \tag{9.66}$$

converges, then the integral

$$\int_{[\mathbf{m},\infty)} |f(\mathbf{x})| \, d\mathbf{x}$$

converges, as well as the integrals

$$\int_{[\mathbf{m},\infty)} \left(\prod_{j:\zeta_j=1} s_j \right) |d^{\zeta} f(\mathbf{s}_{\zeta}, \mathbf{x}_{\mathbf{1}-\zeta})| d\mathbf{x}_{\mathbf{1}-\zeta}, \quad \zeta \leq \eta, \quad \zeta \neq \mathbf{0},$$

do.

(ii) *if for some $\mathbf{n} \in \mathbb{R}^d_+$ the integral*

$$\int_{[\mathbf{n},\infty)} |f(\mathbf{x})| \, d\mathbf{x}$$

converges, then the integrals (9.66) *converge for all $\eta \neq \mathbf{0}$, provided $\mathbf{n} \preceq \mathbf{m}$.*

(iii) *if $|f| \in WM$ (in particular if $f \in GM_E$ with $E = \{\mathbf{e}_1, \ldots, \mathbf{e}_d, \mathbf{1}\}$), then* (ii) *holds for $\mathbf{n} = \mathbf{m}$, provided that for all $\eta \neq \mathbf{0}, \mathbf{1}$ the integrals*

$$\int_{[\mathbf{2m},\infty)^{\eta}} |f(\mathbf{x}_{\eta}, \mathbf{m}_{\mathbf{1}-\eta})| \, d\mathbf{x}_{\eta}$$

are convergent.

Recall that for a set $E \subset \{\eta : \mathbf{0} \le \eta \le \mathbf{1},\ \eta \ne \mathbf{0}\}$, we say that $f \in GM_E$ if

$$\int_{[\mathbf{x},2\mathbf{x}]^\eta} |d^\eta f(\mathbf{s}_\eta, \mathbf{x}_{\mathbf{1}-\eta})| \le Cf(\mathbf{x}), \tag{9.67}$$

for all $\eta \in E$ and $\mathbf{x} \in \mathbb{R}^d_+$.

Proof of Theorem 9.75 The proof essentially goes along the same lines as that of Theorem 9.74. The first part can be done without any reference to the *GM* property: for any $1 \le j \le d$, there holds

$$\int_{[\mathbf{m},\infty)} |f(\mathbf{x})|\, d\mathbf{x} \le 2 \int_{[\mathbf{m},\infty)} s_j |d^{\mathbf{e}_j} f(\mathbf{s}_{\mathbf{e}_j}, \mathbf{x}_{1-\mathbf{e}_j})|\, d\mathbf{x}_{1-\mathbf{e}_j}.$$

Indeed, integrating by parts in the sense of Stieltjes, we have

$$\begin{aligned}\int_{m_j}^{N_j} |f(\mathbf{x})|\, d\mathbf{x}_{\mathbf{e}_j} &= \Big[s_j |f(\mathbf{s}_{\mathbf{e}_j}, \mathbf{x}_{\mathbf{1}-\mathbf{e}_j})| \Big]_{s_j=m_j}^{N_j} - \int_{s_j=m_j}^{N_j} s_j d^{\mathbf{e}_j} |f(\mathbf{s}_{\mathbf{e}_j}, \mathbf{x}_{1-\mathbf{e}_j})| \\ &\le N_j |f(\mathbf{N}_{\mathbf{e}_j}, \mathbf{x}_{\mathbf{1}-\mathbf{e}_j})| + \int_{s_j=m_j}^{\infty} s_j |d^{\mathbf{e}_j} f(\mathbf{s}_{\mathbf{e}_j}, \mathbf{x}_{1-\mathbf{e}_j})| \\ &\le 2 \int_{s_j=m_j}^{\infty} s_j |d^{\mathbf{e}_j} f(\mathbf{s}_{\mathbf{e}_j}, \mathbf{x}_{1-\mathbf{e}_j})|.\end{aligned}$$

Therefore,

$$\begin{aligned}\int_{[\mathbf{m},\mathbf{N})} |f(\mathbf{x})|\, d\mathbf{x} &= \int_{[\mathbf{m},\mathbf{N})^{\mathbf{1}-\mathbf{e}_j}} \left(\int_{m_j}^{N_j} |f(\mathbf{x})|\, d\mathbf{x}_{\mathbf{e}_j} \right) d\mathbf{x}_{\mathbf{1}-\mathbf{e}_j} \\ &\le 2 \int_{[\mathbf{m},\infty)} s_j |\, d^{\mathbf{e}_j} f(\mathbf{s}_{\mathbf{e}_j}, \mathbf{x}_{\mathbf{1}-\mathbf{e}_j})|\, d\mathbf{x}_{\mathbf{1}-\mathbf{e}_j}.\end{aligned}$$

Applying an inductive argument similar to that of Theorem 9.74 completes the proof of (i).

For (ii), we can assume without loss of generality that $\mathbf{n} \preceq \mathbf{m} \preceq \mathbf{2n}$. Indeed, taking larger values of m_j (for some j) makes the integrals in (9.66) smaller. It follows from the definition of *GM* that

$$\begin{aligned}\int_{[\mathbf{n},\infty)} |f(\mathbf{x})|\, d\mathbf{x} &\gtrsim \int_{[\mathbf{n},\infty)} \int_{[\mathbf{x},2\mathbf{x}]} |d^{\mathbf{1}} f(\mathbf{s})|\, d\mathbf{x} \\ &\asymp \int_{[\mathbf{n},\infty)} \frac{1}{x_1 \cdots x_d} \int_{[\mathbf{x},2\mathbf{x}]} s_1 \cdots s_d |d^{\mathbf{1}} f(\mathbf{s})|\, d\mathbf{x} \\ &= \int_{n_1}^{\infty} \frac{1}{x_1} \int_{x_1}^{2x_1} s_1 \ldots \int_{n_d}^{\infty} \frac{1}{x_d} \int_{x_d}^{2x_d} s_d |d^{\mathbf{1}} f(\mathbf{s})|\, d\mathbf{x}.\end{aligned} \tag{9.68}$$

Now, for any $1 \le j \le d$, we have two possibilities. On the one hand, if $n_j = 0$, then

$$\begin{aligned}\int_0^\infty \frac{1}{x_j} \int_{x_j}^{2x_j} s_j |d^{\mathbf{1}} f(\mathbf{s}_{\mathbf{e}_j}, \mathbf{x}_{\mathbf{1}-\mathbf{e}_j})|\, d\mathbf{x}_{\mathbf{e}_j} &= \int_0^\infty s_j |d^{\mathbf{1}} f(\mathbf{s}_{\mathbf{e}_j}, \mathbf{x}_{\mathbf{1}-\mathbf{e}_j})| \int_{s_j/2}^{s_j} \frac{1}{x_j}\, d\mathbf{x}_{\mathbf{e}_j} \\ &= \log 2 \int_0^\infty s_j |d^{\mathbf{1}} f(\mathbf{s}_{\mathbf{e}_j}, \mathbf{x}_{\mathbf{1}-\mathbf{e}_j})|.\end{aligned}$$

On the other hand, if $n_j > 0$,

$$\int_{n_j}^{\infty} \frac{1}{x_j} \int_{x_j}^{2x_j} s_j |d^{\mathbf{1}} f(\mathbf{s}_{\mathbf{e}_j}, \mathbf{x}_{\mathbf{1}-\mathbf{e}_j})| \, d\mathbf{x}_{\mathbf{e}_j}$$
$$= \int_{n_j}^{2n_j} s_j |d^{\mathbf{1}} f(\mathbf{s}_{\mathbf{e}_j}, \mathbf{x}_{\mathbf{1}-\mathbf{e}_j})| \int_{n_j}^{s_j} \frac{1}{x_j} \, d\mathbf{x}_{\mathbf{e}_j} + \int_{2n_j}^{\infty} s_j |d^{\mathbf{1}} f(\mathbf{s}_{\mathbf{e}_j}, \mathbf{x}_{\mathbf{1}-\mathbf{e}_j})| \int_{s_j/2}^{s_j} \frac{1}{x_j} \, d\mathbf{x}_{\mathbf{e}_j}$$
$$\gtrsim \int_{n_j}^{2n_j} s_j \log \frac{s_j}{n_j} |d^{\mathbf{1}} f(\mathbf{s}_{\mathbf{e}_j}, \mathbf{x}_{\mathbf{1}-\mathbf{e}_j})| + \int_{2n_j}^{\infty} s_j |d^{\mathbf{1}} f(\mathbf{s}_{\mathbf{e}_j}, \mathbf{x}_{\mathbf{1}-\mathbf{e}_j})|.$$

Since $n_j < m_j \le 2n_j$ is fixed, we can estimate the latter expression from below by

$$\log \frac{m_j}{n_j} \int_{m_j}^{2n_j} s_j |d^{\mathbf{1}} f(\mathbf{s}_{\mathbf{e}_j}, \mathbf{x}_{\mathbf{1}-\mathbf{e}_j})| + \int_{2n_j}^{\infty} s_j |d^{\mathbf{1}} f(\mathbf{s}_{\mathbf{e}_j}, \mathbf{x}_{\mathbf{1}-\mathbf{e}_j})|$$
$$\gtrsim \int_{m_j}^{\infty} s_j |d^{\mathbf{1}} f(\mathbf{s}_{\mathbf{e}_j}, \mathbf{x}_{\mathbf{1}-\mathbf{e}_j})|,$$

where the involved constant depends on n_j and m_j. Applying the previous argument to each variable in (9.68), we obtain, for any $\mathbf{n} \preceq \mathbf{m} \le 2\mathbf{n}$,

$$\int_{[\mathbf{n},\infty)} |f(\mathbf{x})| \, d\mathbf{x} \gtrsim \int_{[\mathbf{m},\infty)} s_1 \cdots s_d |d^{\mathbf{1}} f(\mathbf{s})|,$$

and thus the result follows by applying part (i) with $\eta = \mathbf{1}$.

For proving (iii), let $\mathbf{m} \in \mathbb{R}_d^+$. For $\mathbf{0} \le \eta \le \mathbf{1}$, defining

$$Q_\eta = [\mathbf{m}, 2\mathbf{m})^{\mathbf{1}-\eta} \times [2\mathbf{m}, \infty)^\eta,$$

we have

$$\int_{[\mathbf{m},\infty)} \left(\prod_{j=1}^{d} s_j \right) |d^{\mathbf{1}} f(\mathbf{s})| = \sum_{\mathbf{0} \le \eta \le \mathbf{1}} \int_{Q_\eta} \left(\prod_{j=1}^{d} s_j \right) |d^{\mathbf{1}} f(\mathbf{s})|.$$

Since

$$\int_{[\mathbf{m},\infty)} |f(\mathbf{x})| \, d\mathbf{x}$$

converges, the integral

$$\int_{Q_\mathbf{1}} \left(\prod_{j=1}^{d} s_j \right) |d^{\mathbf{1}} f(\mathbf{s})|$$

converges by part (ii), while

$$\int_{Q_\mathbf{0}} \left(\prod_{j=1}^{d} s_j \right) |d^{\mathbf{1}} f(\mathbf{s})|$$

converges because f is locally of Hardy bounded variation. For $\eta \neq \mathbf{0}, \mathbf{1}$, the *GM* condition yields

$$\int_{Q_\eta} \left(\prod_{j=1}^{d} s_j\right) |d^{\mathbf{1}} f(\mathbf{s})|$$

$$= \sum_{\substack{1 \le k_j < \infty \\ j:\eta_j = 1}} \int_{[\mathbf{m}, 2\mathbf{m})^{\mathbf{1}-\eta}} \int_{[2^{\mathbf{k}}\mathbf{m}, 2^{\mathbf{k}+\mathbf{1}}\mathbf{m})^{\eta}} \left(\prod_{j=1}^{d} s_j\right) |d^{\mathbf{1}} f(\mathbf{s})|$$

$$\lesssim \sum_{\mathbf{k}_\eta = \mathbf{1}}^{\infty} 2^{|\mathbf{k}_\eta|} |f(2^{\mathbf{k}_\eta}\mathbf{m}_\eta, \mathbf{m}_{\mathbf{1}-\eta})|, \tag{9.69}$$

where the product $2^{\mathbf{k}_\eta}\mathbf{m}_\eta$ is understood coordinate-wise (note that the involved constants depend on $\mathbf{m}$, which is fixed). Since the function $|f(\mathbf{x}_\eta, \mathbf{m}_{\mathbf{1}-\eta})|$ is a *WM* function of $|\eta|$ variables for fixed $\mathbf{m}_{\mathbf{1}-\eta}$, the series from (9.69) converges if and only if the integrals

$$\int_{[2\mathbf{m},\infty)^\zeta} |f(\mathbf{x}_\zeta, \mathbf{m}_{\mathbf{1}-\zeta})| \, \mathbf{dx}_\zeta, \quad \zeta \le \eta, \quad \zeta \neq \mathbf{0},$$

converge, in virtue of Corollary 9.70 (with $\mathbf{m}$ in place of $\mathbf{1}$). The proof is now complete. □

Remark 9.76 Part (iii) of Theorem 9.75 is not valid in general without the assumption that the partial integrals

$$\int_{[2\mathbf{m},\infty)^\eta} |f(\mathbf{x}_\eta, \mathbf{m}_{\mathbf{1}-\eta})| \, d\mathbf{x}_\eta$$

converge for $\eta \neq \mathbf{0}, \mathbf{1}$, as was shown in the two-dimensional case in Remark 6.64.

We note that Theorems 9.74 and 9.75 can be proved for a class larger than *GMS* and *GM*, respectively. If $\beta := \{\beta_{\mathbf{n}}\}$ is a d-tuple sequence of nonnegative numbers depending on $a := \{a_{\mathbf{n}}\}$, we say that the couple (a, β) determines a general monotone sequence a with majorant β (denoted $(a, \beta) \in \mathcal{GMS}$) if

$$\sum_{\mathbf{k}=\mathbf{n}}^{2\mathbf{n}-\mathbf{1}} |\Delta^{\mathbf{1}} a_{\mathbf{k}}| \le C\beta_{\mathbf{n}},$$

for all $\mathbf{n} \in \mathbb{N}^d$. Such a generalization is also applicable to functions, where the corresponding condition reads as

$$\int_{[\mathbf{x},2\mathbf{x}]} |d^{\mathbf{1}} f(\mathbf{t})| \le C\beta(\mathbf{x}),$$

for a nonnegative function of d variables β. In this case, we say that the couple (f, β) determines a general monotone function f with majorant β, written as $(f, \beta) \in \mathcal{GM}$.

We note that

$$GMS = \{a = \{a_{\mathbf{n}}\} : (a, |a|) \in \mathcal{GMS}\},$$

and the same holds for functions.

Proposition 9.77 *Let $(a, \beta) \in \mathcal{GMS}$ be such that*

$$\sum_{\mathbf{k}=\mathbf{m}}^{\infty} \beta_{\mathbf{k}} \le C \sum_{\mathbf{k}=\mathbf{m}}^{\infty} |a_{\mathbf{k}}|,$$

for some constant $C > 0$. Then the conclusion of Theorem 9.74 holds.

Proof The proof is a repetition of that of Proposition 9.74, with the difference that we replace (9.64) by

$$\sum_{\mathbf{k}=\mathbf{m}}^{\infty} k_1 \cdots k_d |\Delta^{\mathbf{1}} a_{\mathbf{k}}| \lesssim \sum_{\mathbf{k}=\mathbf{m}}^{\infty} \sum_{\mathbf{j}=\mathbf{k}}^{2\mathbf{k}-\mathbf{1}} |\Delta^{\mathbf{1}} a_{\mathbf{k}}| \lesssim \sum_{\mathbf{k}=\mathbf{m}}^{\infty} \beta_{\mathbf{k}} \lesssim \sum_{\mathbf{k}=\mathbf{m}}^{\infty} |a_{\mathbf{k}}|,$$

as desired. □

As an example, we consider the class $\mathcal{GMS}$ given by

$$\beta_{\mathbf{n}} = \frac{1}{n_1 \cdots n_d} \sum_{\mathbf{k}=\lfloor c^{-1}\mathbf{n} \rfloor}^{\lfloor c\mathbf{n} \rfloor} |a_{\mathbf{k}}|, \quad c > 1,$$

which corresponds to the class $\overline{GMS}$ (see Definition 9.25). It can easily be shown that in this case

$$\sum_{\mathbf{n}=\mathbf{1}}^{\infty} \beta_{\mathbf{n}} \le C \sum_{\mathbf{n}=\mathbf{1}}^{\infty} |a_{\mathbf{n}}|,$$

and therefore Corollary 9.77 holds for $\mathbf{m} = \mathbf{1}$.

Proposition 9.78 *Let $(f, \beta) \in \mathcal{GM}$ be such that*

$$\int_{[\mathbf{n},\infty)} \beta(\mathbf{x}) d\mathbf{x} \le C \int_{[\mathbf{n},\infty)} |f(\mathbf{x})| d\mathbf{x}$$

for some constant $C > 0$ and $\mathbf{n} \in \mathbb{R}_+^d$. Then the conclusion of Theorem 9.75 holds.

Proof The argument is as for Proposition 9.77, the only difference is that we have

$$\int_{[\mathbf{n},\infty)} \int_{[\mathbf{x},2\mathbf{x}]} |d^{\mathbf{1}} f(\mathbf{s})|\, d\mathbf{x} \lesssim \int_{[\mathbf{n},\infty)} \beta(\mathbf{x}) d\mathbf{x} \lesssim \int_{[\mathbf{n},\infty)} |f(\mathbf{x})| d\mathbf{x}$$

in place of (9.68) in part (ii). The rest of the proof is the same. □

Remark 9.79 We cannot expect part (iii) of Theorem 9.75 to be true by assuming the weaker

$$\int_{[\mathbf{x},2\mathbf{x}]^{\eta}} |d^{\eta} f(\mathbf{s}_{\eta}, \mathbf{x}_{\mathbf{1}-\eta})| \le C\beta_{\eta}(\mathbf{x}), \quad |\eta| = 1, \tag{9.70}$$

for general functions β_η, instead of (9.67). Indeed, the proof relies on the fact that whenever (9.67) holds, $|f| \in WM$. This is not true in general if (9.70) is satisfied. For instance, if $\eta = \mathbf{e}_j$ and

$$\beta_{\mathbf{e}_j}(\mathbf{x}) = \frac{1}{x_j} \int_{x_j/\lambda}^{\lambda x_j} f(\mathbf{x})\, dx_j,$$

then we have $f \in \overline{GM}_{\{\mathbf{e}_1,\dots,\mathbf{e}_d\}}$ (see Definition 9.32). This is equivalent to the fact that for any $1 \le j \le d$, if $\mathbf{x_{1-e_j}}$ is fixed, then $f(\mathbf{x}) \in \overline{GM}$ as a function of the variable x_j (cf. Definition 1.25). In this case we would need $\overline{GM} \subset WM$, but we know that in one dimension the classes $\overline{GM}$ and WM are incomparable (see Remark 1.35). However, condition (9.67) and the convergence of the partial integrals

$$\int_{[2\mathbf{m},\infty)^\eta} |f(\mathbf{x}_\eta, \mathbf{m}_{1-\eta})|\, d\mathbf{x}_\eta, \qquad \eta \ne \mathbf{0},$$

are used only for proving the convergence of the series

$$\sum_{\mathbf{k}_\eta=\mathbf{1}}^{\infty} 2^{|\mathbf{k}_\eta|} |f(2^{\mathbf{k}_\eta}\mathbf{m}_\eta, \mathbf{m}_{1-\eta})|.$$

Hence, we can replace these two assumptions by the convergence of the series

$$\sum_{\mathbf{k}_\eta=\mathbf{1}}^{\infty} 2^{|\mathbf{k}_\eta|} \beta(2^{\mathbf{k}_\eta}\mathbf{m}_\eta, \mathbf{m}_{1-\eta}), \qquad \eta \ne \mathbf{0},$$

and thus the conclusion of part (iii) holds for $(f, \beta) \in \mathcal{GM}$.

As an example, consider the class $(f, \beta) \in \mathcal{GM}$ with

$$\beta(\mathbf{x}) = \frac{1}{x_1 \cdots x_d} \int_{(c^{-1}\mathbf{x}, c\mathbf{x})} |f(\mathbf{t})|\, d\mathbf{t}, \qquad c > 1.$$

For β in question, we have

$$\begin{aligned} \int_{(\mathbf{0},\infty)} \beta(\mathbf{x})\, d\mathbf{x} &= \int_{(\mathbf{0},\infty)} \frac{1}{x_1 \cdots x_d} \int_{(c^{-1}\mathbf{x}, c\mathbf{x})} |f(\mathbf{t})|\, d\mathbf{t}\, d\mathbf{x} \\ &= \int_{(\mathbf{0},\infty)} |f(\mathbf{t})| \int_{(c^{-1}\mathbf{t}, c\mathbf{t})} \frac{1}{x_1 \cdots x_d}\, d\mathbf{x}\, d\mathbf{t} \asymp \int_{(\mathbf{0},\infty)} |f(\mathbf{t})|\, d\mathbf{t}, \end{aligned}$$

and Corollary 9.78 holds for $\mathbf{n} = \mathbf{0}$.

We conclude with a generalization of Theorem 9.74, where the multiplied differences $\left(\prod_{j:\eta_j=1} k_j\right)|\Delta^\eta a_{\mathbf{k}}|$ are replaced by more general ones (for being able to do so, we require the additional assumption $\{|a_{\mathbf{k}}|\} \in WMS$). Recall that $\overleftarrow{\Delta} u_k$ is defined in Section 2.4 as the "backward" difference $\overleftarrow{\Delta} u_k = u_{k+1} - u_k$.

Theorem 9.80 *Let $\{a_{\mathbf{k}}\}$ be a d-tuple GMS such that $\{|a_{\mathbf{k}}|\} \in$ WMS. Let $\{u_k^1\}, \dots, \{u_k^d\}$ be strictly increasing sequences such that*

$$\overleftarrow{\Delta} u_{k+1}^j = O(\overleftarrow{\Delta} u_k^j),$$

for all $j = 1, \dots, d$. Then, the series $\sum_{\mathbf{k}=\mathbf{u_1}}^{\infty} |a_{\mathbf{k}}|$ and

$$\sum_{\mathbf{k}=\mathbf{1}}^{\infty} \Big(\prod_{\ell:\eta_\ell=1} u_{k_\ell}^\ell \Big) \Big(\prod_{\ell:\eta_\ell=0} \overleftarrow{\Delta} u_{k_\ell}^\ell \Big) |\Delta^\eta a_{\mathbf{u_k}}|, \quad \textit{for any } \mathbf{0} \le \eta \le \mathbf{1}, \tag{9.71}$$

converge or diverge simultaneously.

Proof As in other results where successive differences are involved, we are going to prove the statement by means of an inductive argument. More precisely, we will estimate (9.71) for any $\mathbf{0} \le \eta \le \mathbf{1}$ and $\eta \ne \mathbf{1}$ from above by the corresponding series with higher-order differences (so that for any choice of $\mathbf{0} \le \eta \le \mathbf{1}$ with $\eta \ne \mathbf{1}$, (9.71) is bounded from above by the corresponding series with $\eta = \mathbf{1}$). Finally we will show that this largest series can be estimated from above by $\sum |a_{\mathbf{k}}|$.

Let $\eta \ne \mathbf{1}$. We fix the smallest $\ell \in \{1, \dots, d\}$ such that $\eta_\ell = 0$ and define $\zeta = \eta + \mathbf{e}_\ell$. Applying Abel's transformation to the ℓth single series of (9.71), putting $\Delta^{\mathbf{0}} a_{\mathbf{u_k}} = a_{\mathbf{u_k}}$, and using

$$|\Delta^\eta a_{(\mathbf{u_k})_{\mathbf{1}-\mathbf{e}_\ell}, \mathbf{M}_{\mathbf{e}_\ell}}| = \Big| \sum_{k_\ell=M_\ell}^{\infty} \Delta^{\eta+\mathbf{e}_\ell} a_{\mathbf{u_k}} \Big| = \Big| \sum_{k_\ell=M_\ell}^{\infty} \Delta^\zeta a_{\mathbf{u_k}} \Big|, \qquad \mathbf{M} \in \mathbb{N}^d,$$

we get

$$\sum_{k_\ell=1}^{M_\ell} \overleftarrow{\Delta} u_{k_\ell}^\ell |\Delta^\eta a_{\mathbf{u_k}}| \le \sum_{k_\ell=1}^{M_\ell-1} u_{k_\ell}^\ell \Delta^{\mathbf{e}_\ell} |\Delta^\eta a_{\mathbf{u_k}}| + u_{M_\ell}^\ell \Big| \sum_{k_\ell=M_\ell}^{\infty} \Delta^\zeta a_{\mathbf{u_k}} \Big|. \tag{9.72}$$

Since $\{u_k^j\}$ is increasing for any $1 \le j \le d$ and

$$\big|\Delta^{\mathbf{e}_\ell} |\Delta^\eta a_{\mathbf{u_k}}|\big| \le |\Delta^{\eta+\mathbf{e}_\ell} a_{\mathbf{u_k}}| = |\Delta^\zeta a_{\mathbf{u_k}}|,$$

applying this last estimate on the right side of (9.72), we find that it is bounded from above by

$$\sum_{k_\ell=1}^{M_\ell-1} u_{k_\ell}^\ell |\Delta^\zeta a_{\mathbf{u_k}}| + \sum_{k_\ell=M_\ell}^{\infty} u_{k_\ell}^\ell |\Delta^\zeta a_{\mathbf{u_k}}| = \sum_{k_\ell=1}^{\infty} u_{k_\ell}^\ell |\Delta^\zeta a_{\mathbf{u_k}}|.$$

Thus, applying this estimate iteratively shows that if (9.71) converges with a certain ζ, then it converges with any $\eta \le \zeta$ in place of ζ as well.

Now we will show that $\sum_{\mathbf{k}=\mathbf{u}_1}^{\infty} |a_{\mathbf{k}}|$ converges whenever (9.71) converges for some ζ. Indeed, we have already proved that in this case the series corresponding to $\eta = \mathbf{0}$,

$$\sum_{\mathbf{k}=\mathbf{1}}^{\infty} \left(\prod_{j=1}^{d} \overleftarrow{\Delta} u_{k_j}^j \right) |a_{\mathbf{u_k}}|, \tag{9.73}$$

converges (since $\mathbf{0} \leq \zeta$). For $\ell \in \{1, \ldots, d\}$, we estimate from below the ℓth single series in (9.73). Since $\overleftarrow{\Delta} u_{k_\ell+1}^\ell \lesssim \overleftarrow{\Delta} u_{k_\ell}^\ell$, we have

$$\sum_{k_\ell=1}^{M_\ell} \overleftarrow{\Delta} u_{k_\ell}^\ell |a_{\mathbf{u_k}}| \gtrsim \sum_{k_\ell=1}^{M_\ell} \overleftarrow{\Delta} u_{k_\ell+1}^\ell |a_{\mathbf{u_k}}| = \sum_{k_\ell=1}^{M_\ell} \sum_{t=u_{k_\ell}^\ell}^{u_{k_\ell+1}^\ell - 1} |a_{\mathbf{u_k}}|. \tag{9.74}$$

As we showed in the proof of Theorem 2.127, if a one-dimensional sequence $\{u_k\}$ satisfies $\overleftarrow{\Delta} u_{k+1} = O(\overleftarrow{\Delta} u_k)$, then $u_{k+1} = O(u_k)$. Since $\{|a_{\mathbf{k}}|\} \in WMS$, we have $|a_{\mathbf{m}}| \lesssim |a_{\mathbf{u_k}}|$ for every $\mathbf{u_k} \leq \mathbf{m} \leq \mathbf{u_{k+1}}$. Hence, we can estimate the right-hand side of (9.74) from below (up to a constant multiple) by

$$\sum_{k_\ell=1}^{M_\ell} \sum_{n_\ell=u_{k_\ell}^\ell}^{u_{k_\ell+1}^\ell - 1} |a_{(\mathbf{u_k})_{1-\mathbf{e}_\ell}, \mathbf{n}_{\mathbf{e}_\ell}}| = \sum_{n_\ell=u_1^\ell}^{u_{M_\ell+1}^\ell - 1} |a_{(\mathbf{u_k})_{1-\mathbf{e}_\ell}, \mathbf{n}_{\mathbf{e}_\ell}}|, \qquad \mathbf{M} \in \mathbb{N}^d.$$

Repeating this argument iteratively in each variable of the series (9.73) leads to

$$\sum_{\mathbf{k}=\mathbf{1}}^{\mathbf{M}} \left(\prod_{j=1}^{d} \overleftarrow{\Delta} u_{k_j}^j \right) |a_{\mathbf{u_k}}| \gtrsim \sum_{\mathbf{n}=\mathbf{u}_1}^{\mathbf{u}_{\mathbf{M}+1}-\mathbf{1}} |a_{\mathbf{n}}|.$$

Finally, we are going to show that $\sum_{\mathbf{n}=\mathbf{u}_1}^{\infty} |a_{\mathbf{n}}|$ dominates (9.71) with $\eta = \mathbf{1}$, up to a constant factor. Indeed, applying the estimate (2.84) iteratively to every single series, we derive

$$\sum_{\mathbf{k}=\mathbf{1}}^{\infty} \left(\prod_{j=1}^{d} u_{k_j}^j \right) |\Delta^{\mathbf{1}} a_{\mathbf{u_k}}| \leq \sum_{\mathbf{n}=\mathbf{u}_1}^{\infty} \left(\prod_{j=1}^{d} n_j \right) |\Delta^{\mathbf{1}} a_{\mathbf{n}}|.$$

As shown in the proof of Proposition 9.74, the latter series is dominated by $\sum_{\mathbf{n}=\mathbf{u}_1}^{\infty} |a_{\mathbf{n}}|$ (they actually equiconverge), which completes the proof. □

Remark 9.81 If we choose $u_k^\ell = k$ for all ℓ, the statement of Theorem 9.80 reduces to that of Proposition 9.74. What is more, Theorem 9.80 remains true for any $(a, \beta) \in \mathcal{GMS}$ with $\sum \beta_{\mathbf{k}} \lesssim \sum |a_{\mathbf{k}}|$. Indeed, in the proof we use the *GMS* condition only for invoking the statement of Proposition 9.74, which does hold for this class (a, β) (cf. Proposition 9.77).

Finally, we note that Theorem 9.80 does not hold if the condition

$$\overleftarrow{\Delta} u^j_{k+1} = O(\overleftarrow{\Delta} u^j_k)$$

is not satisfied. To prove this claim, one may easily extend the two-dimensional counterexample of the corresponding statement (Theorem 6.67).

9.3 Notes

Section 9.1 (Generalizations of monotonicity)

General monotone sequences

1. A notion of multidimensional *GMS* has been introduced in [91, 92]. Naturally, a variety of general monotone classes are feasible as in the one-dimensional case; see [177] for a detailed survey of the properties of general monotone sequences and functions in dimension one. In this section we consider the simplest definition of general monotone sequences in the multivariate case.

2. For example, in two dimensions, the class GMS_E (or rather generalizations of such a class, as, for instance, $\overline{GMS_E}$) with $E = \{(0,1),(1,0),(1,1)\}$, have been used in convergence problems of trigonometric series and integrals (see, e.g., [149] and the references therein). It is also worth mentioning that in many cases, depending on additional properties of the sequence $\{a_{\mathbf{k}}\}$, the class $GMS_{\{\mathbf{1}\}}$ already has enough regularity in a variety of problems (see, e.g., [71], or [92]).

3. In order to prove that the sequence in Example 9.15 (iii) is *GMS*, we use a "higher order" mean value theorem.

Theorem 9.82 *Let $h > 0$, $a \in \mathbb{R}$, and $n \in \mathbb{N}$. Assume that $f\colon [a, a+nh] \to \mathbb{R}$ is an n times continuously differentiable function on $(a, a+nh)$. Then there exists $c \in (a, a+nh)$ such that*

$$\Delta^n_h f(a) := \sum_{k=0}^{n} \binom{n}{k} (-1)^{n-k} f(a+kh) = h^n f^{(n)}(c),$$

where $f^{(n)}$ denotes the nth derivative.

The proof of Theorem 9.82 goes through Rolle's theorem, as in the classical mean value theorem.

Proof of Theorem 9.82 Let P be the unique polynomial of degree n that coincides with f at the $n+1$ points $a+kh$, $k = 0, 1, \dots, n$. By Rolle's theorem (Theorem 3.35), since $f - P$ vanishes on the $n+1$ points $a+kh$, $k = 0, 1, \dots, n$, we can find n points

$$a + jh < x^{(1)}_j < a + (j+1)h, \qquad j = 0, \dots, n-1,$$

so that $f' - P'$ vanishes on each $x^{(1)}_j$. By this, we can apply Rolle's theorem again and find $n-1$ points

$$x^{(1)}_j < x^{(2)}_j < x^{(1)}_{j+1}, \qquad j = 0, \dots, n-2,$$

such that $f'' - P''$ vanishes on each $x_j^{(2)}$. It is clear that by applying Rolle's theorem iteratively (up to n times in total) we find a point $x_0^{(n)} = c \in (a, a+nh)$ such that $f^{(n)}(c) - P^{(n)}(c) = 0$, or in other words, $f^{(n)}(c) = P^{(n)}(c)$ (note also that $P^{(n)}(x)$ is a constant since P is a polynomial of degree n). On the other hand, we have

$$\Delta_h^n\big(f(a) - P(a)\big) = \sum_{k=0}^{n} \binom{n}{k} (-1)^{n-k} \big(f(a+kh) - P(a+kh)\big) = 0$$

(i.e., $\Delta_h^n f(a) = \Delta_h^n P(a)$). Also, it is well known that for any polynomial Q of degree n, one has $\Delta_h^n Q(x) = h^n Q^{(n)}(x)$. This implies $\Delta_h^n f(a) = h^n P^{(n)}(a)$. Since $P^{(n)}$ is constant, we conclude from the above equality that

$$\Delta_h^n f(a) = h^n P^{(n)}(a) = h^n P^{(n)}(c) = h^n f^{(n)}(c). \qquad \square$$

Section 9.2 (Convergence tests for multiple series with monotone-type coefficients)

Variations of Dvoretzky's test

4. The proof of the multidimensional Dvoretzky's type test (Theorem 9.53) is just a simple adaptation of the original proof by Dvoretzky ([85, Theorem 1]) and, correspondingly, of [178, Theorem 4.2].

Maclaurin–Cauchy integral test

5. One of the very first attempts to generalize the Maclaurin–Cauchy integral test to several dimensions is due to Hurwitz [131] and dates back to 1894. The test, however, is not purely multidimensional, since the considered series is of radial type.

Theorem 9.83 *Let $f\colon \mathbb{R}_+ \to \mathbb{R}_+$ be a decreasing function in each variable. Then, the d-tuple series*

$$\sum_{\mathbf{k}=\mathbf{0}}^{\infty} f(\|\mathbf{k}\|^2)$$

and the integral

$$\int_0^{\infty} x^{\frac{d}{2}-1} f(x)\, dx$$

converge or diverge simultaneously.

6. Generalizations of the Maclaurin–Cauchy integral test for monotone double series and integrals can be found in [B-16, Ch. 4, §32] and [B-21, Ch. 2, §4].

7. In Proposition 7.57 of [180], it is claimed that if $f(x,y)$ is decreasing in each variable, then the double integral

$$\int_1^{\infty} \int_1^{\infty} f(x,y)\, dx\, dy$$

converges if and only if the series

$$\sum_{k=1}^{\infty} \sum_{\ell=1}^{\infty} f(k,\ell)$$

does. However, in view of Theorem 9.64 (or Theorem 6.57, in the two-dimensional case), such a statement is invalid. These pieces of subtlety explain certain difference in the formulations of the integral tests in [B-16, Ch. 4, §32] and [B-21, Ch. 2, §4].

8. The convergence of the series (9.53) in terms of γ in Example 9.71 (i) was already characterized by Jordan in his paper [139].

9. In relation to the series in Example 9.71 (iv), necessary conditions for its convergence were obtained in [24]. In some special cases, its convergence was characterized, namely when all the parameters α_j, β_j, and γ are negative [24], and in the case where $\beta_j = 1$ for all j, cf. [189].

Table of correspondences

Result	1 dim.	2 dim.	arbitrary dim.
Characterization of *GMS* (*HGMS*)	Prop. 1.9	Prop. 6.20	Prop. 9.23
Characterization of *GM*	Prop. 1.23	Rem. 6.28	Rem. 9.34
Main properties of *WMS*	Prop. 1.31	Prop. 6.4	Prop. 9.5
Correspondence between *WMS* and *QMS*	Th. 1.33	Th. 6.8	Th. 9.9
Comparison test	Th. 2.18	Th. 5.1	Th. 8.1
d'Alembert's/ratio test	Th. 2.24	Th. 5.2	Th. 8.2
*k*th root test	Th. 2.29	Th. 5.5	Th. 8.4
Generalized *k*th root test	Th. 2.30	Th. 5.6	Th. 8.5
Rearrangement of absolutely convergent series	Th. 2.48	Th. 4.51	Th. 7.45
Dedekind's test	Th. 2.62	Th. 5.8	Th. 8.12
Du Bois-Reymond's test	Th. 2.63	Th. 5.12	Th. 8.15
Dirichlet's test	Th. 2.64	Th. 5.9	Th. 8.13
Abel's test	Th. 2.65	Th. 5.13	Th. 8.16
Leibniz's test	Th. 2.66	Th. 5.11	Th. 8.14
Maclaurin–Cauchy integral test (monotone)	Th. 2.77	Th. 5.16	Th. 8.19
Maclaurin–Cauchy integral test (*WMS*)	Th. 2.78	Th. 6.57	Th. 9.64
Generalized Cauchy's condensation test	Th. 2.91	Th. 6.49	Th. 9.55
Extended Abel–Olivier's test	Th. 2.99	Th. 6.29	Cor. 9.35
Generalized Abel–Olivier's test	Th. 2.100	Th. 6.30	Th. 9.36

Result	**1 dim.**	**2 dim.**	**arbitrary dim.**
Weighted classical Abel–Olivier's test	Th. 2.102	Th. 6.31	Th. 9.37
Optimality of Abel–Olivier's test	Prop. 2.109	Prop. 6.38	Prop. 9.44
Wik's theorem	Th. 2.110	Th. 6.36	Th. 9.42
Dvoretzky's test	Rem. 2.113	Cor. 6.47	Cor. 9.54
Generalized Dvoretzky's test	Th. 2.112	Th. 6.46	Th. 9.53
Orlicz–Zygmund spaces-type test (for series)	Th. 2.115	Th. 6.43	Th. 9.49
Orlicz–Zygmund spaces-type test (for integrals)	Th. 2.116	Th. 6.44	Th. 9.50
Improved $\ell^q \subset \ell^p$ inequality (WMS)	Th. 2.117	Th. 6.51	Th. 9.57
Improved $\ell^q \subset \ell^p$ inequality (WM)	Th. 2.118	Th. 6.52	Th. 9.58
Improved $\ell^q \subset \ell^p$ inequality (WM_{Rad})		Th. 6.53	Th. 9.59
Stechkin's lemma (QMS)	Th. 2.119	Th. 6.54	Th. 9.60
Stechkin's lemma (monotone)	Cor. 2.120	Cor. 6.55	Cor. 9.61
Equiconvergence for GMS	Th. 2.123	Th. 6.62	Th. 9.74
Equiconvergence for GM	Th. 2.124	Th. 6.63	Th. 9.75
Equiconvergence for $\mathcal{GMS}$	Prop. 2.125	Prop. 6.65	Prop. 9.77
Equiconvergence for $\mathcal{GM}$	Prop. 2.126	Prop. 6.66	Prop. 9.78
Condensation test for GMS	Th. 2.127	Th. 6.67	Th. 9.80

List of symbols

One-dimensional part

Two-dimensional part

Multidimensional part

References: Monographs and textbooks

[B-1] A. Aasma, H. Dutta, and P. N. Natarajan, *An Introductory Course in Summability Theory*, John Wiley & Sons, Hoboken, NJ, 2017.

[B-2] N. H. Abel, *Abel on Analysis*, Kendrick Press, Heber City, UT, 2007.

[B-3] M. Abramowitz and I. Stegun, *Handbook of Mathematical Functions with Formulas, Graphs, and Mathematical Tables*, US Government Printing Office, Washington, DC, 1964.

[B-4] I. M. Alabdulmohsin, *Summability Calculus*, Springer, Cham, 2018.

[B-5] G. Alexits, *Convergence Problems of Orthogonal Series*, Pergamon Press, New York, 1961.

[B-6] T. M. Apostol, *Mathematical Analysis*, Second edition, Addison-, Reading, MA, 1974.

[B-7] S. Baron, *Introduction to the Theory of Summability of Series*, Valgus, Tallinn, 1977 (Russian).

[B-8] N. K. Bary, *A Treatise on Trigonometric Series*, Pergamon, Oxford, 1964.

[B-9] C. Bennett and R. Sharpley, *Interpolation of Operators*. Academic Press, Boston, MA, 1988.

[B-10] N. H. Bingham, C. M. Goldie, and J. L. Teugels, *Regular Variation*, Encycl. Math. Appl. **27**, 1989.

[B-11] D. D. Bonar and M. Khoury Jr., *Real Infinite Series*, Mathematical Association of America, Washington, DC, 2006.

[B-12] V. E. Bononcini and A. Forlani, *Serie, Serie di Taylor, Serie di Fourier*, Edizioni Calderini, Bologna, 1972.

[B-13] J. Boos, *Classical and Modern Methods in Summability*, Oxford University Press, Oxford, 2000.

[B-14] É. Borel, *Lectures on Divergent Series*, Los Alamos Scientific Laboratory, Los Alamos, NM, 1975.

[B-15] L. Bourchtein and A. Bourchtein, *Theory of Infinite Sequences and Series*, Birkhäuser, Cham, 2022.

[B-16] T. J. I. Bromwich, *An Introduction to the Theory of Infinite Series*, Macmillan, London, 1908.

[B-17] A. A. Bukhshtab, *Number Theory*, Moscow, Izdat. 'Prosveshchenie', 1966 (Russian).

[B-18] B. Candelpergher, *Ramanujan Summation of Divergent Series*, Springer, Cham, 2017.

[B-19] G. Casanova, *Les Séries Mathématiques*, Presses Universitaires de France, Paris, 1974.

[B-20] A.-L. Cauchy, *Cours d'analyse de l'École Royale Polytechnique. Première partie. Analyse algébrique*, Chez Debure frères, Paris, 1821.

[B-21] V. G. Chelidze, *Some Methods of Summation of Double Series and Double Integrals*, Tbilisi State University, Tbilisi, 1977 (Russian).

[B-22] Y. Chevallard, *Théorie des Séries, 1*, CEDIC, Paris, 1979.

[B-23] J. Combes, *Suites et Séries (Sequences and Series)*, Mathématiques Presses Universitaires de France, Paris, 1982.

[B-24] J. W. Dauben, *Georg Cantor. His Mathematics and Philosophy of the Infinite*, Princeton University Press, Princeton, NJ, 1990 (reprint of the orig. publ. by Harvard University Press, 1979).

[B-25] E. Delabaere, *Divergent Series, Summability and Resurgence III*, Springer, Cham, 2016.

[B-26] R. A. DeVore and G. G. Lorentz, *Constructive Approximation*, Springer-Verlag, Berlin, 1993.

[B-27] J. Dieudonné, *Foundations of Modern Analysis*, enlarged and corrected printing, Academic Press, New York, 1969.

[B-28] D. Dũng, V. Temlyakov, T. Ullrich, and S. Tikhonov (ed.), *Hyperbolic Cross Approximation*, Advanced Courses in Mathematics, Cham, 2018.

[B-29] W. L. Ferrar, *A Textbook of Convergence*, Oxford University Press, New York, 1980.

[B-30] G. Ferraro, *The Rise and Development of the Theory of Series up to the Early 1820s*, Springer, New York, 2008.

[B-31] G .M. Fichtengolz, *Infinite Series: Rudiments*, Gordon and Breach, New York, 1970.

[B-32] G. M. Fichtengolz, *Infinite Series: Ramifications*, Gordon and Breach, New York, 1970.

[B-33] M. Field, *Essential Real Analysis*, Springer, Cham, 2017.

[B-34] O. Furdui, *Limits, Series, and Fractional Part Integrals*, Springer, New York, 2013.

[B-35] J. Galambos, *Representations of Real Numbers by Infinite Series*, Springer-Verlag, Berlin, 1976.

[B-36] P. M. Gauthier, *Lectures on Several Complex Variables*, Birkhäuser/Springer, Cham, 2014.

[B-37] S. R. Ghorpade, B. V. Limaye, *A Course in Multivariable Calculus and Analysis*, Springer, London, 2010.

[B-38] É. Goursat, *Cours d'Analyse Mathématique,* Tome I, Gauthier-Villars, Paris, 1933.

[B-39] E. Grant, *A Source Book in Medieval Science*, Harvard University Press, Cambridge, MA, 1974.

[B-40] E. Grigorieva, *Methods of Solving Sequence and Series Problems*, Birkhäuser/Springer, Cham, 2016.

[B-41] G. H. Hardy, *Divergent Series*, Clarendon Press, Oxford, 1949.

[B-42] G. H. Hardy, *A Course of Pure Mathematics*, Cambridge University Press, Cambridge, 1993.

[B-43] G. H. Hardy, J. E. Littlewood, and G. Pólya, *Inequalities*, Cambridge University Press, Cambridge, 1952.

[B-44] T.-X. He, *Methods for the Summation of Series*, CRC Press, Boca Raton, FL, 2022.

[B-45] I. I. Hirschman, Jr., *Infinite Series*, Rinehart and Winston, New York, 1962.

[B-46] E. W. Hobson, *The Theory of Functions of a Real Variable and the Theory of Fourier Series*, Vol. II, Second edition, Cambridge University Press, Cambridge, 1926.

[B-47] R. Honsberger (ed.), *Mathematical Plums*, The Mathematical Association of America, Washington, DC, 1979.

[B-48] C. Houzel, *The Work of Niels Henrik Abel*, Springer-Verlag, Berlin, 2004.

[B-49] J. M. Hyslop, *Infinite Series*, Oliver and Boyd, Edinburgh; Interscience, New York, 1942.

[B-50] A. E. Ingham, *The Distribution of Prime Numbers*, Cambridge University Press, Cambridge, 1990.

[B-51] M. Kac, *Statistical Independence in Probability, Analysis and Number Theory*, Mathematical Association of America, Washington, DC, 1959.

[B-52] M. I. Kadets and V. M. Kadets, *Rearrangements of Series in Banach Spaces*, American Mathematical Society, Providence, RI, 1991.

[B-53] M. I. Kadets and V. M. Kadets, *Series in Banach Spaces*, Birkhäuser, Basel, 1997.

[B-54] K. Knopp, *Theory and Application of Infinite Series*, Blackie & Son, London, 1928.

[B-55] H.-H. Körle, *Infinite Series in a History of Analysis*, De Gruyter, Berlin, 2015.

[B-56] T. W. Körner, *Fourier Analysis*, Cambridge University Press, Cambridge, 1988.

[B-57] A. Levy, *Basic Set Theory*, Dover, Mineola, NY, 2002. Reprint of the 1979 original.

[B-58] N. I. Lobachevskii, *Algebra or Calculus of Finite Quantities*, Kazan, 1834 (Russian).

[B-59] M. Loday-Richaud, *Divergent Series, Summability and Resurgence II*, Springer, Cham, 2016.

[B-60] P. Loya, *Amazing and Aesthetic Aspects of Analysis*, Springer, New York, 2017.

[B-61] B. M. Makarov, M. G Goluzina, A. A. Lodkin, and A. N. Podkorytov, *Selected Problems in Real Analysis*, American Mathematical Society, Providence, RI, 1992. Translated from the Russian by H. H. McFaden.

[B-62] H. Meschkowski, *Unendliche Reihen (Infinite Series)*, Second edition, Bibliographisches Institut, Mannheim, 1982.

[B-63] P. Mengoli, *Novae Quadraturae Arithmeticae, seu de Additione Fractionum*, Jacopo Monti, Bologna, 1650.

[B-64] D. S. Mitrinović and D. D. Adamović, *Nizovi i Redovi (Sequences and Series)*, Naučna Knjiga, Belgrade, 1971.

[B-65] C. Mitschi and D. Sauzin, *Divergent Series, Summability and Resurgence I*, Springer, Cham, 2016.

[B-66] P. N. Natarajan, *Classical Summability Theory*, Springer, Singapore, 2017.

[B-67] G. Pólya and G. Szegö, *Problems and Theorems in Analysis I*, Springer-Verlag, Berlin, 1998.

[B-68] A. Pringsheim, *Vorlesungen über Zahlen- und Funktionenlehre. 1. Band, 2. Abteilung: Unendliche Reihen mit reellen Gliedern*, B.G. Teubner, Leipzig, 1916.

[B-69] F. Riesz and B. Sz.-Nagy, *Functional Analysis*, Blackie & Son, London, 1956.

[B-70] R. Rolland, *Théorie des Séries, 2*, CEDIC, Paris, 1979.

[B-71] V. I. Romanovsky, *Introduction to Analysis*, Tashkent, 1939 (Russian).

[B-72] R. Roy, *Series and Products in the Development of Mathematics*, Vol. 1 and 2, Cambridge University Press, Cambridge, 2021.

[B-73] W. Rudin, *Principles of Mathematical Analysis*, Third edition, McGraw-Hill, Düsseldorf, 1976.

[B-74] W. Rudin, *Functional Analysis*, Second edition, McGraw-Hill, New York, 1991.

[B-75] G. S. Salekhov, *The Computation of Series*, Gosudarstv. Izdat. Tehn.-Teor. Lit., Moscow, 1955 (Russian).

[B-76] H.-J. Schell, *Unendliche Reihen (Infinite Series)*, BSB B. G. Teubner Verlagsgesellschaft, Leipzig, 1982.

[B-77] L. L. Smail, *Elements of the Theory of Infinite Processes*, McGraw-Hill, New York, 1923.

[B-78] H. H. Sohrab, *Basic Real Analysis*, Birkhäuser. Boston, MA, 2003.

[B-79] H. Triebel, *Theory of Function Spaces*, Birkhäuser, Basel, 2010. Reprint of the 1983 original.

[B-80] R. M. Trigub and E. S. Belinsky, *Fourier Analysis and Approximation of Functions*, Kluwer, Dordrecht, 2004.

[B-81] C. I. Vălean, *More (Almost) Impossible Integrals, Sums, and Series – A New Collection of Fiendish Problems and Surprising Solutions*, Springer, Cham, 2023.

[B-82] R. K. Vallin, *The Elements of Cantor Sets: With Applications*, John Wiley & Sons, NJ, 2013.

[B-83] N. N. Vorob'ev, *Theory of Series*, Nauka, Moscow, 1986 (Russian).

[B-84] A. I. Yanushauskas, *Double Series*, Nauka, Sibirsk. Otdel., Novosibirsk, 1980 (Russian).

[B-85] L. V. Zhizhiashvili, *Some Problems of Multidimensional Harmonic Analysis*, Second edition, Tbilisi State University, Tbilisi, 1996 (Russian).

[B-86] A. Zygmund, *Trigonometric Series*, Cambridge University Press, Cambridge, 1959.

References: Articles

[1] N. H. Abel, *Untersuchungen über die reihe:* $1 + \frac{m}{1}x + \frac{m\cdot(m-1)}{1\cdot 2}x^2 \cdots$, J. Reine Angew. Math. **1** (1826), 311–339.

[2] N. H. Abel, *Note sur le mémoire de M. L. Olivier No. 4, du second tome de ce Journal, ajant pour titre "remarques sur les series infinies et leur convergence"*, J. Reine Angew. Math. **3** (1828), 79–82.

[3] V. M. Abramov, *Extension of the Bertrand–De Morgan test and its application*, Am. Math. Mon. **127** (5) (2020), 444–448.

[4] V. M. Abramov, *Necessary and sufficient conditions for the convergence of positive series*, J. Class. Anal. **19** (2) (2021), 117–125.

[5] V. M. Abramov, *Evaluating the sum of convergent positive series*, Publ. Inst. Math., Nouv. Sér. **111** (2022), 41–53.

[6] V. Abramov, M. Cadena, and E. Omey, *A new test for convergence of positive series*, Publ. Inst. Math., Nouv. Sér. **109** (2021), 61–76.

[7] Y. S. Abu-Mostafa, *A differentiation test for absolute convergence*, Math. Mag. **57** (1984), 228–231.

[8] C. R. Adams and J. A. Clarkson, *Properties of functions $f(x,y)$ of bounded variation*, Trans. Amer. Math. Soc. **36** (1934), 711–730.

[9] R. P. Agnew, *Ratio tests for convergence of series*, Pac. J. Math. **1** (1951), 1–3.

[10] R. P. Agnew, *Permutations preserving convergence of series*, Proc. Am. Math. Soc. **6** (1955), 563–564.

[11] S. A. Ali, *The mth ratio test: new convergence tests for series*, Am. Math. Mon. **115** (2008), 514–524.

[12] J.-P. Allouche and C. Morin, *Kempner-like harmonic series*, Am. Math. Mon. **131** (2024), 775–783.

[13] A. Ambroladze, *On spherical convergence of numerical and functional series*, Vestn. Mosk. Univ., **1** (1) (1994), 30–35 (Russian). English transl. in Moscow Univ. Math. Bull. **49** (1994), 25–28.

[14] L. Amerio, *Sulla convergenza delle serie doppie*, Atti Accad. Italia, Rend., Cl. Sci. Fis. Mat. Natur., **VII** (2) (1941), 684–698.

[15] F. G. Arutyunyan, *Representation of functions by multiple series*, Dokl. Arm. Akad. Nauk **64** (2) (1977), 72–76 (Russian).

[16] F. G. Arutyunyan, *Representation of measurable functions of several variables by multiple trigonometric series*, Mat. Sb. **126** (2) (1985), 267–285 (Russian). English transl. in Math. USSR-Sb **54** (1986), 259–277.

[17] E. A. Asadullin, *Determination of the region of convergence of multiple power series*, Izv. Vyssh. Uchebn. Zaved. Matematika **11** (4) (1959), 21–26 (Russian).

[18] E. A. Asadullin, *Tests for the convergence of multiple series with positive terms*, Dokl. Akad. Nauk SSSR **136** (1961), 529–531 (Russian). English transl. in Soviet Math. Dokl. **2** (1961), 86–88.

[19] J. M. Ash, *Neither a worst convergent series nor a best divergent series exists*, College Math. J. **28** (1997), 296–297.

[20] J. M. Ash, *The limit comparison test needs positivity*, Math. Magazine **85** (2012), 374–375.

[21] J. M. Ash, A. Berele, and S. Catoui, *Plausible and genuine extensions of L'Hospital's rule*, Math. Mag. **85** (2012), 52–60.

[22] D. Azevedo and T. P. Andrade, *Summability characterizations of positive sequences*, Commun. Math. **30** (2022), 81–90.

[23] K. I. Babenko, *Approximation by trigonometric polynomials in a certain class of periodic functions of several variables*, Dokl. Akad. Nauk SSSR **132** (1960), 982–985 (Russian). English transl. in Soviet Math. Dokl. **1** (1960), 672–675.

[24] W. N. Bailey, *On the convergence of a certain multiple series*, Math. Proc. Cambridge Philos. Soc. **25** (4) (1929), 410–411.

[25] R. Baillie, *Sums of reciprocals of integers missing a given digit*, Am. Math. Mon. **86** (5) (1979), 372–374.

[26] M. Bakhbukh, *Convergence of double series*, Mat. Zametki **13** (1973), 341–350 (Russian). English transl. in Math. Notes **13** (1973), 208–214.

[27] M. Banakiewicz, B. Hanson, P. Pierce, and F. Prus-Wiśniowski, *A Riemann-type theorem for segmentally alternating series*, Bull. Iran. Math. Soc. **44** (5) (2018), 1303–1314.

[28] E. J. Barbeau and P. J. Leah, *Euler's 1760 paper on divergent series*, Hist. Math. **3** (1976), 141–160.

[29] N. Batir, *Remarks on Vălean's master theorem of series*, J. Class. Anal. **13** (2018), 79–82.

[30] J. Bell and V. Blåsjö, *Pietro Mengoli's 1650 proof that the harmonic series diverges*, Math. Mag. **91** (5) (2018), 341–347.

[31] M. Bello Hernández, *Problem 438*, Gac. R. Soc. Mat. Esp. **25** (2022), 112.

[32] M. Bello Hernández, *Solution to problem 438*, Gac. R. Soc. Mat. Esp. **26** (2023), 125–126.

[33] A. S. Belov, M. I. Dyachenko, and S. Yu. Tikhonov, *Functions with general monotone Fourier coefficients*, Russ. Math. Surv. **76** (6) (2021), 951–1017.

[34] C. Bennett and K. Rudnick, *On Lorentz-Zygmund spaces*, Diss. Math. **175** (1980), 67p.

[35] S. N. Bernstein, *On majorants of finite or quasi-finite growth*, Doklady Akad. Nauk SSSR (N.S.) **65** (1949), 117–120 (Russian).

[36] R. Bil and H. Laue, *Investigations about the Euler-Mascheroni constant and harmonic numbers*, Analysis (Berlin) **36** (4) (2016), 223–230.

[37] R. P. Boas Jr., *Partial sums of infinite series, and how they grow*, Am. Math. Mon. **84** (1977), 237–258.

[38] R. P. Boas Jr. and J. W. Wrench Jr., *Partial sums of the harmonic series*, Am. Math. Mon. **78** (1971), 864–870.

[39] P. du Bois-Reymond, *Eine neue Theorie der Convergenz und Divergenz von Reihen mit positiven Gliedern*, J. Reine Angew. Math. **76** (1873), 61–91.

[40] R. Bojanic and E. Seneta, *A unified theory of regularly varying sequences*, Math. Z. **134** (1973), 91–106.

[41] B. Booton, *General monotone functions and their Fourier coefficients*, J. Math. Anal. Appl. **426** (2) (2015), 805–823.

[42] B. Booton, *Rearrangements of general monotone functions and of their Fourier transforms*, Math. Inequal. Appl. **21** (3) (2018), 871–883.

[43] D. Borwein and J. M. Borwein, *A note on alternating series in several dimensions*, Am. Math. Mon. **93** (1986), 531–539.

[44] D. Borwein and A. Meir, *Divergence criteria for positive series*, Am. Math. Mon. **79** (1972), 1104–1106.

[45] L. Brandolini, L. Colzani, S. Robins, and G. Travaglini, *An Euler-Maclaurin formula for polygonal sums*, Trans. Am. Math. Soc. **375** (2022), 151–172.

[46] J. Brasseur, *On restrictions of Besov functions*, Nonlin. Anal. **170** (2018), 197–225.

[47] H. E. Bray, *Elementary properties of the Stieltjes integral*, Ann. Math. (2) **20** (3) (1919), 177–186.

[48] B. Brighi and N. Chevallier, $\sum 1/n^{2+\cos n}$, Revue de la filière Mathématique **119** (2008–2009), 3–8.

[49] R. W. Brink, *A new integral test for the convergence and divergence of infinite series*, Trans. Amer. Math. Soc. **19** (2) (1918), 186–204.

[50] R. W. Brink, *A simplified integral test for the convergence of infinite series*, Am. Math. Mon. **38** (4) (1931), 205–208.

[51] T. J. I. Bromwich, *The relation between the convergence of series and of integrals*, Proc. London Math. Soc. (2) **6** (1908), 327–338.

[52] N. Bugaev (Bougaïev), *Note relative à la théorie de la convergence des séries*, Mat. Sb., **14** (2) (1889), 279–282. Transl. from Russian: Mat. Sbornik, **14** (2) (1889), 279–282.

[53] V. P. Burlachenko, *Analogs of Leibnitz's theorem for sign-alternating double series*, Ukraïn. Mat. Zh. **19** (2) (1967), 3–10.

[54] J. S. Byrnes, A. Giroux, and O. Shisha, *Riemann sums and improper integrals of step functions related to the prime number theorem*, J. Approx. Theory **40** (2) (1984), 180–192.

[55] P. Calabrese, *A note on alternating series*, Am. Math. Mon. **69** (3) (1962), 215–217.

[56] G. T. Cargo, *Some extensions of the integral test*, Am. Math. Mon. **73** (5) (1966), 521–525.

[57] S. A. Carrillo, *Sum of alternating-like series as definite integrals*, Int. J. Math. Educ. Sci. Technol. **53** (10) (2022), 2856–2866.

[58] H. Chand, *On some generalizations of Cauchy's condensation and integral tests*, Am. Math. Mon. **46** (1939), 338–341.

[59] B. Christianson, *Condensing a slowly convergent series*, Math. Mag. **68** (4) (1995), 298–301.

[60] P. Chunaev, *Hölder and Minkowski type inequalities with alternating signs*, J. Math. Inequal. **9** (2015), 61–71.

[61] J. A. Clarkson and C. R. Adams, *On definitions of bounded variation for functions of two variables*, Trans. Amer. Math. Soc. **35** (1934), 824–854.

[62] A. Cohen and R. DeVore, *Approximation of high-dimensional parametric PDEs*, Acta Numerica **24** (2015), 1–159.

[63] A. Cohen, R. DeVore, and C. Schwab, *Analytic regularity and polynomial approximation of parametric and stochastic elliptic PDEs*, Anal. Appl. **9** (1) (2011), 11–47.

[64] L. Comtet, *Problems and solutions: solutions of advanced problems: 5346*, Am. Math. Mon. **74** (2) (1967), p. 209.

[65] Y. Coudène, *A strange inequality concerning alternating series*, Am. Math. Mon. **125** (6), 554–557.

[66] D. Cruz-Uribe, *The relation between the root and ratio tests*, Math. Magazine **70** (3) (1997), 214–215.

[67] D. Cruz-Uribe, *News and letters*, Math. Magazine **70** (4) (1997), 310–311.

[68] J. d'Alembert, *Réflexions sur les suites divergentes ou convergentes*, Opuscules Mathématiques ou Mémoires sur différens sujets de géométrie, de méchanique, d'optique, d'astronomie **5** (1768), 171–183.

[69] L. De Carli, D. Gorbachev, and S. Tikhonov, *Pitt and Boas inequalities for Fourier and Hankel transforms*, J. Math. Anal. Appl. **408** (2) (2013), 762–774.

[70] C. J. de la Vallée-Poussin, *Recherches analytiques sur la theorie des nombres premiers I–III*, Ann. Soc. Sci. Bruxelles **20** (2) (1987), 183–256, 281–362, 363–397.

[71] A. Debernardi, *Uniform convergence of double sine transforms of general monotone functions*, Anal. Math. **43** (2) (2017), 193–217.

[72] A. Debernardi, *Uniform convergence of sine integrals with general monotone functions*, Math. Nachr. **290** (17–18) (2017), 2815–2825.

[73] A. Debernardi, *Hankel transforms of general monotone functions*, Topics in Classical and Modern Analysis: In Memory of Yingkang Hu, Applied and Numerical Harmonic Analysis Series, Birkhäuser/Springer, Basel, 87–104, 2019.

[74] A. Debernardi, *The Boas problem on Hankel transforms*, J. Fourier Anal. Appl. **25** (6) (2019), 3310–3341.

[75] A. Debernardi, *Weighted norm inequalities for generalized Fourier-type transforms and applications*, Publ. Mat. **64** (2020), 3–42.

[76] D. W. DeTemple and S.-H. Wang, *Half integer approximations for the partial sums of the harmonic series*, J. Math. Anal. Appl. **160** (1991), 149–156.

[77] R. A. DeVore and V. N. Temlyakov, *Some remarks on greedy algorithms*, Adv. Comput. Math. **5** (2–3) (1996), 173–187.

[78] U. Dini, *Memoria sulle serie a termini positivi*, Ann. Univ. Tosc. **9** (Part II) (1867), 41–80.

[79] L. Dirichlet, *Démonstration d'un théorème d'Abel*, J. Math. Pures Appl. (Ser. 2) **7** (1862), 253–255.

[80] O. Domínguez and S. Tikhonov, *New estimates for the maximal functions and applications*. Trans. Am. Math. Soc. **375** (6) (2022), 3969–4018.

[81] O. Domínguez and S. Tikhonov, *Function spaces of logarithmic smoothness: embeddings and characterizations*, Mem. Am. Math. Soc. **1939** (2023), vii, 166p.

[82] O. Dovgoshey, O. Martio, V. Ryazanov, and M. Vuorinen, *The Cantor function*, Expo. Math. **24** (2006), 1–37.

[83] A. L. Durán and R. Estrada, *An extension of the integral test*, Proc. Am. Math. Soc. **127** (1999), 1745–1751.

[84] K. Duzinkiewicz and B. Szal, *On the uniform convergence of double sine series*, Colloq. Math. **151** (2018), 71–95.

[85] A. Dvoretzky, *On monotone series*, Amer. J. Math. **70** (1948), 167–173.

[86] M. I. Dyachenko, *Some problems in the theory of multiple trigonometric series*, Uspekhi Mat. Nauk **47** (5) (1992), 97–162 (Russian). English transl. in Russ. Math. Surv. **47** (5) (1992), 103–171.

[87] M. I. Dyachenko, *U-convergence almost everywhere of double Fourier series*, Mat. Sb. **186** (1995), 47–64 (Russian). English transl. in Sb. Math. **186** (1995), 47–64.

[88] M. I. Dyachenko, *u-convergence of multiple Fourier series*, Izvestiya RAN: Ser. Mat. **59** (2) (1995), 129–142 (Russian). English transl. in Izvestiya: Math. **59** (1995), 353–366.

[89] M. Dyachenko, E. Liflyand, and S. Tikhonov, *Uniform convergence and integrability of Fourier integrals*, J. Math. Anal. Appl. **372** (2010), 328–338.

[90] M. Dyachenko, A. Mukanov, and S. Tikhonov, *Uniform convergence of trigonometric series with general monotone coefficients*, Canad. J. Math. **71** (6) (2019), 1445–1463.

[91] M. Dyachenko and S. Tikhonov, *Convergence of trigonometric series with general monotone coefficients*, C. R. Math. Acad. Sci. Paris **345**, 123–126 (2007).

[92] M. Dyachenko and S. Tikhonov, *A Hardy-Littlewood theorem for multiple series*, J. Math. Anal. Appl. **339** (2008), 503–510.

[93] M. Dyachenko and S. Tikhonov, *General monotone sequences and convergence of trigonometric series*, Topics in Classical Analysis and Applications in Honor of Daniel Waterman. World Scientific, Hackensack, NJ, 88–101 (2008).

[94] M. Dyachenko and S. Tikhonov, *Integrability and continuity of functions represented by trigonometric series: coefficients criteria*, Studia Math. **193** (3) (2009), 285–306.

[95] M. Dyachenko and S. Tikhonov, *Smoothness and asymptotic properties of functions with general monotone Fourier coefficients*, J. Fourier Anal. Appl. **24** (4) (2018), 1072–1097.

[96] O. Echi, A. Khalfallah, and D. Kroumi, *Sharp estimate of the remainder of some alternating series*, Math. Inequal. Appl. **26** (2023), 83–91.

[97] U. Elias, *Rearrangement of a conditionally convergent series*, Am. Math. Mon. **110** (1) (2003), p. 57.

[98] P. Erdős, I. Joó, and L. A. Székely, *Some remarks on infinite series*, Stud. Sci. Math. Hung. **22** (1–4) (1987), 395–400.

[99] V. Ermakov, *Théorie de la convergence des séries infinies et des intégrates définies*, Mat. Sb. **6** (1872), 39–76.

[100] R. Estrada and M. Pavlović, *L'Hôpital's monotone rule, Gromov's theorem, and operations that preserve the monotonicity of quotients*, Publ. Inst. Math., Nouv. Sér. **101** (2017), 11–24.

[101] L. Euler, *De Transformatione serierum*, Institutiones calculi differentialis cum eius usu in analysi finitorum ac doctrina serierum, 1755.

[102] B. Fahri, *A curious result related to Kempner's series*, Am. Math. Mon. **115** (10) (2008), 933–938.

[103] A. Faisant, G. Grekos, and L. Mišík, *Some generalizations of Olivier's theorem*, Math. Bohem. **141** (2016), 483–494.

[104] L. Feng, V. Totik, and S. P. Zhou, *Trigonometric series with a generalized monotonicity condition*, Acta Math. Sin. (Engl. Ser.) **30** (8) (2014), 1289–1296.

[105] R. J. Ferdinands, *Selective sums of an infinite series*, Math. Mag. **88** (3) (2015), 179–185.

[106] R. Filipów, A. Kwela, and J. Tryba, *The ideal test for the divergence of a series*, Rev. R. Acad. Cienc. Exactas Fís. Nat., Ser. A Mat., RACSAM **117** (3) (2023), Paper No. 98, 25p.

[107] F. Garibay, P. Greenberg, L. Resendis, and J. J. Rivaud, *The geometry of sum-preserving permutations*, Pac. J. Math. **135** (2) (1988), 313–322.

[108] M. Goar, *Olivier and Abel on series convergence: an episode from early 19th century analysis*, Math. Mag. **72** (5) (1999), 347–355.

[109] D. Gorbachev, E. Liflyand, and S. Tikhonov, *Weighted Fourier inequalities: Boas' conjecture in* $\mathbb{R}^n$, J. Anal. Math. **114** (2011), 99–120.

[110] R. A. Gordon, *Comments on "subsums of the harmonic series"*, Am. Math. Mon. **126** (3) (2019), 275–279.

[111] J. Hadamard, *Sur la distribution des zéros de la fonction* $\zeta(s)$ *et ses conséquences arithmétiques*, Bull. Soc. Math. France **24** (1896), 199–220.

[112] J. Hafner and P. Mancosu, *The varieties of mathematical explanation*, Visualization, Explanation and Reasoning Styles in Mathematics. Based on the Meeting on Mathematics as Rational Activity, Roskilde, Denmark, November 1–3, 2001, Springer, Dordrecht, 215–250 (2005).

[113] H. J. Hamilton, *On transformations of double series*, Bull. Am. Math. Soc. **42** (1936), 275–283.

[114] H. J. Hamilton, *Transformations of multiple sequences*, Duke Math. J. **2** (1936), 29–60.

[115] R. W. Hamming, *Convergent monotone series*, Am. Math. Mon. **52** (1945), 70–72.

[116] R. W. Hamming, *Monotone series*, Am. J. Math. **68** (1946), 133–136.

[117] C. N. B. Hammond, *Ratios, roots, and means: a quartet of convergence tests*, Elem. Math. **75** (3) (2020), 89–102.

[118] C. N. B. Hammond, *The case for Raabe's test*, Math. Mag. **93** (2020), 36–46.

[119] C. N. B. Hammond and E. Omey, *A second look at the second ratio test*, Am. Math. Mon. **128** (7) (2021), 579–596.

[120] G. H. Hardy, *A general theorem concerning absolutely convergent series*, Proc. Lond. Math. Soc. (2) **1** (1904), 285–290.

[121] G. H. Hardy, *On the convergence of certain multiple series*, Proc. London Math. Soc. **s2-1** (1904), 124–128.

[122] G. H. Hardy, *Note in addition to a former paper on conditionally convergent multiple series*, Proc. London Math. Soc. (2) **2** (1905), 190–191.

[123] G. H. Hardy, *The multiplication of conditionally convergent series*, Proc. London Math. Soc. **s2-6** (1908), 410–423.

[124] G. H. Hardy, *Theorems connected with Maclaurin's test for the convergence of series*, Proc. London Math. Soc. (2) **9** (1911), 126–144.

[125] G. H. Hardy, *Note on a theorem of Cesàro*, Messeng. Math. **41** (2) (1912), 17–22.

[126] G. H. Hardy, *On the convergence of certain multiple series*, Proc. Camb. Phil. Soc. **19** (1917), 86–95.

[127] G. H. Hardy and J. E. Littlewood, *Some problems of "Partitio numerorum". III: on the expression of a number as a sum of primes*, Acta Math. **44** (1923), 1–70.

[128] T. H. Hildebrandt, *Remarks on the Abel-Dini theorem*, Am. Math. Mon. **49** (1942), 441–445.

[129] N. S. Hoang, *A limit comparison test for general series*, Am. Math. Mon. **122** (9) (2015), 893–896.

[130] M.-C. Hu and J.-K. Wang, *On rearrangements of series*, Bull. Inst. Math., Acad. Sin. **7** (1979), 363–376.

[131] A. Hurwitz, *Ueber Riemann's Convergenzkriterium*, Math. Ann. **44** (1894), 83–88 (German).

[132] E. Huynh, *A second Raabe's test and other series tests*, Am. Math. Mon. **129** (9) (2022), 865–875.

[133] G. J. O. Jameson, *A note on series of positive and negative terms*, Math. Gaz. **102** (2018), 320–325.

[134] V. Jamet, *Sur les séries à termes positifs*, C. R. Acad. Sci., Paris **114** (1892), 57–60.

[135] B. Jasek, *Über Umordnung von Reihen*, Colloq. Math. **7** (1960), 257–259.

[136] B. Jasek, *Complex series and connected sets*, Diss. Math. **52** (1966), 47p.

[137] H. Jehle, *Eine bemerkung zum konvergenzkriterium von Weierstraß*, Math. Z. **52** (1950), 60–61 (German).

[138] R. Johnsonbaugh, *Summing an alternating series*, Am. Math. Mon. **86** (8) (1979), 637–648.

[139] C. Jordan, *Sur les conditions de convergence de certaines séries multiples*, Bull. Soc. Math. France **9** (1881), 113–115.

[140] M. I. Kadets, *On the absolute, perfect and unconditional convergence of double series*, Funktsional. Anal. i Prilozhen. **30** (2) (1996), 85–86 (Russian). English transl. in Funct. Anal. Appl. **30** (2) (1996), 140–142.

[141] M. I. Kadets, *On absolute, perfect, and unconditional convergences of double series in Banach spaces*, Ukr. Mat. Zh. **49** (1997), 1032–1041 (Russian). English transl. in Ukr. Math. J. **49** (1997), 1158–1168.

[142] A. J. Kempner, *A curious convergent series*, Am. Math. Mon. **21** (2) (1914), 48–50.

[143] P. Kesava Menon, *On a class of perfect sets*, Bull. Amer. Math. Soc. **54** (1948), 706–711.

[144] T. Kojima, *On the theory of double sequences*, Tôhoku Math. J. **21** (1922), 3–14.

[145] P. Kórus, *Remarks on the uniform and L^1-convergence of trigonometric series*, Acta Math. Hungar. **128** (4) (2010), 369–380.

[146] P. Kórus, *On the uniform convergence of double sine series with generalized monotone coefficients*, Period. Math. Hungar. **63** (2) (2011), 205–214.

[147] P. Kórus, *On the uniform convergence of special sine integrals*, Acta Math. Hungar. **133** (1–2) (2011), 82–91.

[148] P. Kórus, *Uniform convergence of double trigonometric series*, Math. Bohem. **138** (3) (2013), 225–243.

[149] P. Kórus, *Uniform convergence of double trigonometric integrals*, Colloq. Math. **154** (2018), 107–119.

[150] P. Kórus and F. Móricz, *On the uniform convergence of double sine series*, Studia Math. **193** (2009), 79–97.

[151] A. B. Kostin and V. B. Sherstyukov, *Asymptotic behavior of remainders of special number series*, J. Math. Sci., New York **251** (6) (2020), 814–838.

[152] P. Kostyrko, T. Šalát, and W. Wilczyński, *$\mathcal{I}$-convergence*, Real Anal. Exch. **26** (2) (2001), 669–685.

[153] S. G. Krantz and J. D. McNeal, *Creating more convergent series*, Am. Math. Mon. **111** (1) (2004), 32–38.

[154] J. M. Krause, *Über Fouriersche Reihen mit zwei veränderlichen Größen*, Ber. Sächs. Akad. Wiss. Leipzig **55** (1903), 164–197.

[155] D. Kressner and C. Tobler, *Low-rank tensor Krylov subspace methods for parametrized linear systems*, SIAM J. Matrix Anal. Appl. **32** (4) (2011), 1288–1316.

[156] A. Kronrod, *On permutation of terms of numerical series*, Mat. Sb., Nov. Ser. **18** (1946), 237–280 (Russian).

[157] E. E. Kummer, *Über die convergenz und divergenz der unendlichen reihen*, J. Reine Angew. Math. **13** (1835), 171–184.

[158] S. H. L. Kung, *A note on a form of ratio test*, Math. Mag. **41** (1968), 144–146.

[159] V. Kuperberg, *Sums of singular series with large sets and the tail of the distribution of primes*, Q. J. Math. **74** (4) (2023), 1457–1479.

[160] E. Laeng and V. Pata, *A convergence-divergence test for series of nonnegative terms*, Expo. Math. **29** (2011), 420–424.

[161] L. Leindler, *On cosine series with positive coefficients*, Acta Math. Acad. Sci. Hung. **22** (1972), 397–406.

[162] L. Leindler, *Inequalities of Hardy-Littlewood type*, Anal. Math. **2** (2) (1976), 117–123.

[163] L. Leindler, *On the uniform convergence and boundedness of a certain class of sine series*, Anal. Math. **27** (2001), 279–285.

[164] L. Leindler, *A new class of numerical sequences and its applications to sine and cosine series*, Anal. Math. **28** (4) (2002), 279–286.

[165] L. Leindler, *Embedding relations of classes of numerical sequences*, Acta Sci. Math. (Szeged) **68** (2002), 689–695.

[166] L. Leindler, *Embedding results regarding strong approximation of sine series*, Acta Sci. Math. (Szeged) **71** (2005), 91–103.

[167] L. Leindler, *On the relationships of seven numerical sequences*, Acta Math. Hungar. **114** (3) (2007), 227–234.

[168] L. Leindler, *On the uniform convergence of single and double sine series*, Acta Math. Hungar. **140** (3) (2013), 232–242.

[169] L. Leindler, *An addition to Tandori's summation theorem*, Acta Math. Hungar. **143** (2014), 201–207.

[170] F. Leja, *Sur les transformations linéaires des suites doubles et multiples*, Bulletin Acad. Polonaise (1930), 1–10.

[171] A. S. Leonov, *On the total variation for functions of several variables and a multidimensional analog of Helly's selection principle* (English. Russian original), Math. Notes **63** (1998), 61–71.

[172] F. W. Levi, *Rearrangement of convergent series*, Duke Math. J. **13** (1946), 579–585.

[173] E. Liflyand, U. Stadtmüller, and R. Trigub, *An interplay of multidimensional variations in Fourier analysis*, J. Fourier Anal. Appl. **17** (2) (2011), 226–239.

[174] E. Liflyand and U. Stadtmüller, *A multidimensional Euler-Maclaurin formula and an application*, Complex Analysis and Dynamical Systems V. Contemp. Math. **591** (2013), 183–194.

[175] E. Liflyand and S. Tikhonov, *Extended solution of Boas' conjecture on Fourier transforms*, C. R. Math. Acad. Sci. Paris **346** (21–22) (2008), 1137–1142.

[176] E. Liflyand and S. Tikhonov, *The Fourier transforms of general monotone functions*, Trends in Mathematics, Basel: Birkhäuser, 373–391 (2009).

[177] E. Liflyand and S. Tikhonov, *A concept of general monotonicity and applications*, Math. Nachr. **284** (8–9) (2011), 1083–1098.

[178] E. Liflyand, S. Tikhonov, and M. Zeltser, *Extending tests for convergence of number series*, J. Math. Anal. Appl. **377** (2011) 194–206.

[179] J. E. Littlewood, *Note on the convergence of series of positive terms*, Messeng. Math. **39** (1910), 191–192.

[180] B. V. Limaye and M. Zeltser, *On the Pringsheim convergence of double series*, Proc. Est. Acad. Sci. **58** (2009), 108–121.

[181] N. I. Lobachevskii, *Polnoe sobranie sochinenii.* Tom 5. Gosudarstv. Izdat. Tehn.-Teor. Lit., Moscow-Leningrad, 1951, pp. 252–255 (Russian).

[182] M. Longo and V. Valori, *The comparison test–not just for nonnegative series*, Math. Mag. **79** (3) (2006), 205–210.

[183] B. Lubeck and V. Ponomarenko, *Subsums of the harmonic series*, Am. Math. Mon. **125** (4) (2018), 351–355.

[184] A. A. Lunin, *Grouping the terms of numerical series*, Math. Notes **44** (5) (1988), 842–847.

[185] F. Mertens, *Ein Beitrag zur analytischen Zahlentheorie*, J. Reine Angew. Math. **78** (1874), 46–62.

[186] B. Meyer, *On the convergence of alternating double series*, Am. Math. Mon. **60** (1953), 402–404.

[187] J. Michelow, *A note on two convergence tests*, Am. Math. Mon. **67** (6) (1960), 581–583.

[188] H. Miller, *A ratio-type convergence test*, Am. Math. Mon. **77** (1970), 285–287.

[189] L. J. Mordell, *On the convergence of a certain multiple series*, Math. Proc. Cambridge Philos. Soc. **25** (4) (1929), 407–409.

[190] F. Móricz, *On the convergence in a restricted sense of multiple series*, Anal. Math. **5** (2) (1979), 135–147.

[191] F. Móricz, *Some remarks on the notion of regular convergence of multiple series*, Acta Math. Hung. **41** (1983), 161–168.

[192] F. Móricz, *On the uniform convergence of sine integrals*, J. Math. Anal. Appl. **354** (2009), 213–219.

[193] F. Móricz, *On the convergence of double integrals and a generalized version of Fubini's theorem on successive integration*, Acta Sci. Math. (Szeged) **78** (3–4) (2012), 469–487.

[194] F. Móricz, *On the regular convergence of multiple integrals of locally Lebesgue integrable functions over* $\overline{\mathbb{R}}^m_+$, C. R. Math., Acad. Sci. Paris **350** (9–10) (2012), 459–464.

[195] F. Móricz, *On the regular convergence of multiple series of numbers and multiple integrals of locally integrable functions over* $\overline{\mathbb{R}}^m_+$, Anal. Math. **39** (2) (2013), 135–152.

[196] R. K. Morley, *The remainder in computing by series*, Am. Math. Mon. **57** (8) (1950), 550–551.

[197] R. Mortini and R. Rupp, *Sums of infinite series involving the Riemann zeta function*, Arch. Math. **123** (2024), 163–172.

[198] Ch. H. Müntz, *Über den Approximationssatz von Weierstraß*, Mathematische Abhandlungen Hermann Amandus Schwarz, 1914, pp. 303–312.

[199] C. St. J. A. Nash-Williams and D. J. White, *An application of network flows to rearrangement of series*, J. Lond. Math. Soc., II. Ser. **59** (2) (1999), 637–646.

[200] C. St. J. A. Nash-Williams and D. J. White, *Rearrangement of vector series, I*, Math. Proc. Camb. Philos. Soc. **130** (2001), 89–109.

[201] C. St. J. A. Nash-Williams and D. J. White, *Rearrangement of vector series, II*, Math. Proc. Camb. Philos. Soc. **130** (2001), 111–134.

[202] R. B. Nelsen, *An improved remainder estimate for use with the integral test*, College Math. J. **34** (5) (2003), 397–399.

[203] D. J. Newman, *Simple analytic proof of the prime number theorem*, Am. Math. Mon. **87** (9) (1980), 693–696.

[204] D. J. Newman and H. S. Shapiro, *The Taylor coefficients of inner functions*, Michigan Math. J. **9** (1962), 249–255.

[205] T. A. Newton, *A note on a generalization of the Cauchy-Maclaurin integral test*, Am. Math. Mon. **61** (1954), 331–334.

[206] C. P. Niculescu and G. T. Prăjitură, *Some open problems concerning the convergence of positive series*, Ann. Acad. Rom. Sci., Math. Appl. **6** (2014), 92–107.

[207] J. R. Nurcombe, *A sequence of convergence tests*, Am. Math. Mon. **86** (1979), 679–681,

[208] J. R. Nurcombe, *Necessary conditions for the convergence of series of positive terms*, Math. Gaz. **98** (No. 543) (2014), 493–496.

[209] J. R. Nurcombe, *Converses of a "curious" theorem of Cesàro on conditionally convergent series*, Math. Gaz. **100** (2016), 113–118.

[210] E. D. Nursultanov, *On the coefficients of multiple Fourier series in L_p-spaces*, Izv. RAN. Ser. Mat. **64** (2000), 95–122 (Russian). English transl. in Izv. Math. **64** (2000), 93–120.

[211] E. D. Nursultanov, *Net spaces and inequalities of Hardy–Littlewood type*, Mat. Sb. **189** (3) (1998), 83–102 (Russian). English transl. in Sb. Math. **189** (3) (1998), 399–419.

[212] M. A. Nyblom, *Integer part of the partial sums for the divergent p-series in closed form*, Am. Math. Mon. **127** (6) (2020), 554–557.

[213] C. Oehring, *Asymptotics of singular numbers of smooth kernels via trigonometric transforms*, J. Math. Anal. Appl. **145** (1990), 573–605.

[214] M. L. Olivier, *Remarques sur les séries infinies et leur convergence*, J. Reine Angew. Math. **2** (1827), 31–44.

[215] L. L. Pennisi, *Elementary proof that e is irrational*, Am. Math. Mon. **60** (7) (1953), p. 474.

[216] I. E. Perlin, *Series with deleted terms*, Am. Math. Mon. **48** (1941), 93–97.

[217] P. A. B. Pleasants, *Rearrangements that preserve convergence*, J. Lond. Math. Soc., II. Ser. **15** (1977), 134–142.

[218] P. A. B. Pleasants, *Addendum: rearrangements that preserve convergence* J. Lond. Math. Soc., II. Ser. **18** (1978), 576.

[219] K. A. Post, *A combinatorial lemma involving a divergence criterion for series of positive terms*, Am. Math. Mon. **77** (1970), 1085–1087.

[220] A. Pringsheim, *Allgemeine Theorie der Divergenz und Convergenz von Reihen mit positiven Gliedern*, Math. Ann. **35** (1890), 297–394.

[221] A. Pringsheim, *Elementare Theorie der unendlichen Doppelreihen*, Sitsungsberichte Math.-Phys. Classe Akad. Wiss. München **27** (1897), 101–153.

[222] A. Pringsheim, *Zur Theorie der zweifach unendlichen Zahlenfolgen*, Math. Ann. **53** (1900), 289–321.

[223] F. Prus-Wiśniowski, *A refinement of Raabe's test*, Am. Math. Mon. **115** (3) (2009), 249–252.

[224] F. Prus-Wiśniowski, *Comparison of Raabe's and Schlömilch's tests*, Tatra Mt. Math. Publ. **42** (2009), 119–130.

[225] H. Rademacher, *Über die asymptotische Verteilung gewisser konvergenzerzeugender Faktoren*, Math. Z. **11** (1921), 276–288.

[226] C. T. Rajagopal, *On an integral test of R. W. Brink for the convergence of series*, Bull. Am. Math. Soc. **43** (1937), 405–412.

[227] C. T. Rajagopal, *Remarks on some generalizations of Cauchy's condensation and integral tests*, Am. Math. Mon. **48** (1941), 180–185.

[228] C. S. Rees, *On the integral modulus of continuity of Fourier series*, Ind. J. Math. **9** (1967), 489–497.

[229] G. M. Robinson, *Divergent double sequences and series*, Trans. Amer. Math. Soc. **28** (1926), 50–73.

[230] T. Šalát and V. Toma, *A classical Olivier's theorem and statistical convergence*, Ann. Math. Blaise Pascal **10** (2) (2003), 305–313.

[231] M. Saucedo, *Hardy inequalities for p-weakly monotone functions*, Eurasian Math. J. **14** (2) (2023), 94–106.

[232] P. Schaefer, *Sum-preserving rearrangements of infinite series*, Am. Math. Mon. **88** (1981), 33–40.

[233] T. Schmelzer and R. Baillie, *Summing a curious, slowly convergent series*, Am. Math. Mon. **115** (6) (2008), 525–540.

[234] M. Schramm, J. Troutman, and D. Waterman, *Segmentally alternating series*, Am. Math. Mon. **121** (8) (2014), 717–722.

[235] J. A. Sellers, *Beyond mere convergence*, Probl. Res. Iss. Math. Undergrad. Stud. **12** (2) (2002), 157–164.

[236] S. M. Shah, *A note on quasi-monotone series*, Math. Student **15** (1947), 19–24.

[237] I. M. Sheffer, *Convergence of multiply-infinite series*, Am. Math. Mon. **52** (1945), 365–376.

[238] V. Yu. Slyusarchuk, *Differential convergence criteria for operator improper integrals and series*, Ukr. Math. J. **73** (9) (2022), 1445–1461.

[239] H. K. Sørensen, *Exceptions and counterexamples: understanding Abel's comment on Cauchy's Theorem*, Historia Math. **32** (4) (2005), 453–480.

[240] A. Stadler, *A calculus exam misprint ten years later*, SIAM Probl. Sol. (2009), 1–4.

[241] O. E. Stanaitis, *Integral tests for infinite series*, Am. Math. Mon. **78** (1971), 164–170.

[242] S. B. Stechkin, *Trigonometric series with monotone type coefficients*, Approximation Theory. Asymptotical Expansions, suppl., MAIK Nauka/Interperiodica, Moscow, 214–224, 2001.

[243] H. Steinhaus, *Les probabilités dénombrables et leur rapport à la théorie de la mesure*, Fund. Math. **4** (1922), 286–310.

[244] G. Stoica and Y. Wardat, *A special form of slower divergent series*, Am. Math. Mon. **130** (2) (2023), p. 186.

[245] O. Stolz, *Über unendliche Doppelreihen*, Math. Ann. **24** (1884), 157–171.

[246] B. Szal, *A new class of numerical sequences and its applications to uniform convergence of sine series*, Math. Nachr. **284** (14–15) (2011), 1985–2002.

[247] B. Szal, *On L-convergence of trigonometric series*, J. Math. Anal. Appl. **373** (2) (2011), 449–463.

[248] B. Szal, *On the integral modulus of continuity of Fourier series*, Acta Math. Hungar. **131** (1–2) (2011), 138–159.

[249] O. Szász, *Über die Approximation stetiger Funktionen durch lineare Aggregate von Potenzen*, Math. Ann. **77** (1916), 482–496.

[250] O. Szász, *Quasi-monotone series*, Amer. J. Math. **70** (1948), 203–206 (1948).

[251] T. Tao, *The convergence of an alternating series of Erdős, assuming the Hardy-Littlewood prime tuples conjecture*, Commun. Am. Math. Soc. **4** (2024), 80–96.

[252] S. A. Telyakovskii, *On the Gauss test for convergence of series*, Math. Higher Ed. **11** (2013), 43–44 (Russian).

[253] V.N. Temlyakov, *Approximation of functions with bounded mixed derivatives*, Trudy MIAN **178** (1986), 3–113 (Russian). English transl. in Proc. Steklov Inst. Math. **178** (1989), 1–121.

[254] S. Tikhonov, *Embedding results in questions of strong approximation by Fourier series*, Acta Sci. Math. (Szeged) **72** (1–2) (2006), 117–128; published first as S. Tikhonov, *Embedding theorems of function classes, IV*, November 2005, CRM preprint.

[255] S. Tikhonov, *Trigonometric series with general monotone coefficients*, J. Math. Anal. Appl. **326** (2007), 721–735.

[256] S. Tikhonov, *Best approximation and moduli of smoothness: computation and equivalence theorems*, J. Approx. Theory **153** (2008), 19–39.

[257] S. Tikhonov, *On L_1-convergence of Fourier series*, J. Math. Anal. Appl. **347** (2) (2008), 416–427.

[258] S. Tikhonov and M. Zeltser, *Weak monotonicity concept and its applications*, Fourier Analysis. Trends in Mathematics: Pseudo-differential Operators, Time-Frequency Analysis and Partial Differential Equations, Birkhäuser, 2014, 357–374.

[259] J. Tong, *Kummer's test gives characterizations for convergence or divergence of all positive series*, Am. Math. Mon. **101** (5) (1994), 450–452.

[260] C. Torres Latorre, *Functions preserving general monotone sequences*, Anal. Math. **47** (2021), 211–227.

[261] L. Tóth and J. Bukor, *On the alternating series* $1 - \frac{1}{2} + \frac{1}{3} - \frac{1}{4} + \cdots$, J. Math. Anal. Appl. **282** (2003), 21–25.

[262] R. M. Trigub, *A generalization of the Euler-Maclaurin formula*, Mat. Zametki **61** (1997), 312–316 (Russian). English transl. in Math. Notes **61** (1997), 253–257.

[263] P. L. Ul'yanov, *The imbedding of certain function classes* H_p^ω Izv. Akad. Nauk SSSR Ser. Mat **32** (3) (1968), 649–686; Izv. Math. **2** (3) (1968), 601–637.

[264] P. L. Ul'yanov, *Imbedding theorems and relations between best approximations*, Math. USSR, Sb. **10** (1970), 103–126.

[265] C. I. Vălean, *A master theorem of series and an evaluation of a cubic harmonic series*, J. Class. Anal. **10** (2) (2017), 97–107.

[266] M. B. Villarino, *The error in an alternating series*, Am. Math. Mon. **125** (4) (2018), 360–364.

[267] S. S. Volosivets, *Cosine and sine series with generalized monotone coefficients in Lorentz spaces*, Anal. Math. **45** (4) (2017), 641–655.

[268] S. S. Volosivets, *Series in multiplicative systems in Lorentz spaces*, Mat. Zametki **102** (3) (2017), 339–354.

[269] S. Volosivets and V. Krivobok, *The generalized monotonicity class and integrability of power series*, Colloq. Math. **130** (2) (2013), 241–249.

[270] A. D. Wadhwa, *An interesting subseries of the harmonic series*, Am. Math. Mon. **82** (1975), 931–933.

[271] A. D. Wadhwa, *Some convergent subseries of the harmonic series*, Am. Math. Mon. **85** (1978), 661–663.

[272] M. Ward, *A generalized integral test for convergence of series*, Am. Math. Mon. **56** (1949), 170–172.

[273] I. Wik, *Criteria for absolute convegence of Fourier series of functions of bounded variation*, Trans. Amer. Math. Soc. **163** (1972), 1–24.

[274] A. Wilansky, *On the convergence of double series*, Bull. Am. Math. Soc. **53** (1947), 793–799.

[275] A. Wintner, *The sum formula of Euler-Maclaurin and the inversions of Fourier and Möbius*, Amer. J. Math. **69** (1947), 685–708.

[276] R. Wituła, *On the set of limit points of the partial sums of series rearranged by a given divergent permutation*, J. Math. Anal. Appl. **362** (2) (2010), 542–552.

[277] R. Wituła, *Permutations preserving the convergence or the sum of series – a survey*, Monograph on the occasion of 100th birthday anniversary of Zygmunt Zahorski, Wydawnictwo Politechniki Śląskiej, Gliwice, 169–190, 2015.

[278] R. Wituła and D. Słota, *The convergence classes of divergent permutations*, Demonstr. Math. **42** (4) (2009), 781–796.

[279] M. Yeşilkayagil and F. Başar, *On the paranormed space of bounded variation double sequences*, Bull. Malays. Math. Sci. Soc. (2) **43** (3) (2020), 2701–2712.

[280] R. M. Young, *The error in alternating series*, Math. Gaz. **69** (1985), 120–121.

[281] D. Yu, P. Zhou, and S. Zhou, *Mean bounded variation condition and applications in double trigonometric series*, Anal. Math. **38** (2) (2012), 83–104.

[282] M. Zeltser, *On equiconvergence of number series*, Math. Slovaca **63** (2013), 1333–1346.

[283] S. M. Zemyan, *On two conjectures concerning the partial sums of the harmonic series*, Proc. Amer. Math. Soc. **95** (1985), 83–86.

[284] L. V. Zhizhiashvili, *Some problems in the theory of simple and multiple trigonometric and orthogonal series*, Uspekhi Mat. Nauk **28** (1973), 65–119 (Russian). English transl. in Russian Math. Surveys **28** (1973), 65–127.

Index

For EU product safety concerns, contact us at Calle de José Abascal, 56–1°, 28003 Madrid, Spain or eugpsr@cambridge.org.

www.ingramcontent.com/pod-product-compliance
Lightning Source LLC
Chambersburg PA
CBHW070759120226
39428CB00051B/87
* 9 7 8 1 0 0 9 5 6 2 9 9 7 *